Ordinary Level Mathematics

SIXTH EDITION

with answers

By L. Harwood Clarke

MATHEMATICS ONE
MATHEMATICS TWO
MATHEMATICS THREE
MATHEMATICS FOUR
ADDITIONAL PURE MATHEMATICS
PURE MATHEMATICS AT ADVANCED LEVEL
A NOTE BOOK IN PURE MATHEMATICS
A NOTE BOOK IN APPLIED MATHEMATICS
HINTS FOR ORDINARY LEVEL MATHEMATICS
HINTS FOR ADVANCED LEVEL MATHEMATICS
MODERN MATHEMATICS AT ORDINARY LEVEL
EXERCISES AND WORKED EXAMPLES IN ORDINARY LEVEL
MATHEMATICS
TRIGONOMETRY AT ORDINARY LEVEL
FOUR-FIGURE MATHEMATICAL TABLES

With F. G. J. Norton
ADDITIONAL APPLIED MATHEMATICS
OBJECTIVE TESTS IN C.S.E. MATHEMATICS
OBJECTIVE TESTS IN ORDINARY LEVEL MATHEMATICS

By F. G. J. Norton
ADVANCED LEVEL APPLIED MATHEMATICS

All published by Heinemann Educational Books Ltd

Ordinary Level Mathematics

L. HARWOOD CLARKE

SIXTH EDITION
prepared by
F. G. J. NORTON
Head of the Mathematics Department,
Rugby School

with answers

HEINEMANN EDUCATIONAL BOOKS
LONDON

Heinemann Educational Books Ltd
22 Bedford Square, London WC1B 3HH

LONDON EDINBURGH MELBOURNE AUCKLAND
HONG KONG SINGAPORE KUALA LUMPUR NEW DELHI
IBADAN NAIROBI LUSAKA JOHANNESBURG
EXETER (NH) KINGSTON PORT OF SPAIN

ISBN 0 435 50220 4 (text)
ISBN 0 435 50221 2 (with answers)

First published 1958
Second Edition 1959
Reprinted with corrections 1960
Reprinted 1961 (twice)
Reprinted 1963 (with tables)
Third Edition 1964 (reset)
Reprinted (twice)
Fourth Edition 1969
Reprinted 1969
Fifth Edition 1971
Reprinted (three times)
Sixth Edition 1978
Reprinted 1979

Illustrations drawn by
Reginald Piggott
Set in 10/11 Monotype Times

Printed and bound in Great Britain by
Butler & Tanner Ltd, Frome and London

Foreword to the First Edition

This book is intended to be used in the last years of the course for Mathematics at Ordinary level in the General Certificate of Education. It assumes a knowledge of the fundamental processes of Arithmetic and Algebra and avoids long explanations which are best left to the teacher. Two parallel sets of exercises are provided on each topic and it is hoped that the teacher will find it helpful to do one set in class and to give the other set for homework.

BEDFORD, January 1958   L. H. C.

Foreword to the Sixth Edition

The numerous reprints of the original book testify to its quality; yet Mathematics syllabuses change, quite rightly. New topics are now examined at this level, other topics are studied no longer. Teachers have had the opportunity of using the experimental syllabuses for some years and recently most of the 'Modern' syllabuses have been revised and some of the earlier ideas pruned.

This edition covers the topics examined at Ordinary level by the various GCE Boards. Like the earlier editions, it is laid out for ease of reference, so that teachers and students can find quickly any topic they wish to study. As in the earlier editions, two exercises are provided in most cases, and miscellaneous exercises at the end of each chapter. These have been tested in school, and found suitable for the interests and ability of pupils at this stage.

The past twenty years have seen the rise and fall of the slide rule as it is replaced by the electronic calculator. But many students do not yet have access to calculators, nor are calculators allowed at present in all O-level examinations. A chapter on Aids to calculating uses slide rule or calculator, and another chapter explains the use of logarithms.

I should like to thank Mr Hamish MacGibbon, of Heinemann Educational Books, for his encouragement in revising this book; my colleague, Dr Philip Stephens, for reading many of the chapters and for commenting constructively on them; my family, for their readiness at all times to discuss mathematical problems; and finally, perhaps most of all, my pupils, for their persistent questioning has taught me much.

RUGBY, 1978   F. G. J. Norton

Contents

ARITHMETIC

vii

STATISTICS AND PROBABILITY

TRIGONOMETRY

CALCULUS

Notation, Formulae

ARITHMETIC

ARITHMETIC

Formulae

Areas and volumes

The area of a circle $= \pi r^2$

The curved surface area of a right circular cylinder $= 2\pi rh$

The volume of a right circular cylinder $= \pi r^2 h$

The volume of a sphere $= \frac{4}{3}\pi r^3$

The volume of material in a pipe $= \pi(R^2 - r^2)h$.

Percentage

$$\text{Percentage gain (or loss)} = \frac{\text{gain (or loss)}}{\text{cost price}} \times 100$$

$$\text{Percentage error} = \frac{\text{error}}{\text{true value}} \times 100$$

Simple interest

$$I = \frac{PRT}{100}$$

Compound interest

$$A = P\left(1 + \frac{r}{100}\right)^n$$

1 Numerals and Number Bases

Numerals

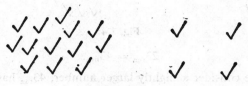

Fig. 1.1

How many √ are there in Fig. 1.1? As we count one, two, three, ... we have a different *number* as we include yet another √ in our total, and these numbers (the natural numbers, or counting numbers), are usually represented by *numerals*. We shall almost certainly have used the Arabic numerals 1, 2, 3, 4 ... if we wrote any symbols while counting, and a different symbol is required for each of the first nine numerals, we should soon run out of symbols!

Number bases

If we group our √ into bundles of say, 5, we can record how many bundles of five we have and how many are left over.

Fig. 1.2

Two bundles of five, and three extra, which we can write as 23, and this number has been written in base five.

We can now represent much larger numbers than nine using only the symbols 1, 2, 3, 4, 5, 6, 7, 8, 9, 0, by assigning a significance to the place in which each occurs (place value). Generally numbers are written in base ten, so that the √ would have been grouped

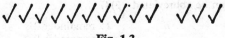

Fig. 1.3

and the numeral 13 signifies that we have one group of ten and three extra. If we have a number already written in base ten and we wish to change it into a number base five, we have to group in fives the number of objects it records. The number 23 in base ten (23_{ten}, whenever there is any doubt about which number base is being used) means that we have 2 groups of ten and three extra, that is, 4 groups of five and three extra,

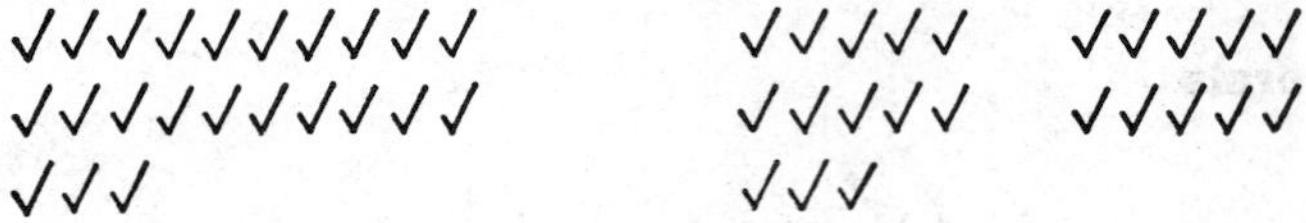

Fig. 1.4

so $$23_{\text{ten}} = 43_{\text{five}}.$$

But when we consider a slightly larger number, 43_{ten} has eight groups of five, i.e. 1 group of '5 fives' and 3 groups of five, with three extra so that

Fig. 1.5

$$43_{\text{ten}} = 133_{\text{five}}.$$

This regrouping is the reason for the successive division required in changing number bases.

Example 1. Express 47_{ten} in base three.

Successive division gives

$$3)47$$
$$15 \text{ remainder } 2$$
$$3)15$$
$$5 \text{ remainder } 0$$
$$3)5$$
$$1 \text{ remainder } 2$$
$$3)1$$
$$0 \text{ remainder } 1$$

This working can be abbreviated

$$3)47$$
$$15 \text{ remainder } 2$$
$$5 \text{ remainder } 0$$
$$1 \qquad 2$$
$$0 \qquad 1$$

so that $$47_{\text{ten}} = 1202_{\text{three}}.$$

We have used $47 = 27 + 18 + 2,$

i.e. $47 = 1(3)^3 + 2(3)^2 + 0(3) + 2.$

Example 2. Express 257_{ten} in base twelve.

Dividing as before,   12)257

$$21 \text{ remainder } 5$$
$$1 \text{ remainder } 9$$
$$0 \text{ remainder } 1$$
$$257_{\text{ten}} = 195_{\text{twelve}}.$$

Binary numbers (base two)

Because of their application to computers, numbers base two are widely used. Such numbers soon become very long,

e.g. $99_{\text{ten}} = 1\ 100\ 011_{\text{two}}$ and $999_{\text{ten}} = 1\ 111\ 100\ 111_{\text{two}}$

Octal numbers (base eight)

Octal numbers are only about one-third of the length of their binary equivalents, and binary numbers can easily be converted into octals, as the table shows:

Denary (base ten)
| 1 | 2 | 3 | 4 | 5 | 6 | 7 | 8 | 64 | 512 |

Octal (base eight)
| 1 | 2 | 3 | 4 | 5 | 6 | 7 | 10 | 100 | 1000 |

Binary (base two)
| 1 | 10 | 11 | 100 | 101 | 110 | 111 | 1000 | 1 000 000 | 1 000 000 000 |

Each group of three binary digits can be changed into the corresponding octal number (since $8 = 2^3$), so that

$$1\ 100\ 011_{\text{two}} = 143_{\text{eight}}$$

since
$$1\ 100\ 011_{\text{two}} = 1\ 000\ 000_{\text{two}} + 100\ 000_{\text{two}} + 011_{\text{two}}$$
$$= 100_{\text{eight}} + 40_{\text{eight}} + 3_{\text{eight}}$$
$$= 143_{\text{eight}}.$$

Similarly $1\ 101\ 100\ 011_{\text{two}} = 1543_{\text{eight}},$

since $101_{\text{two}} = 5_{\text{eight}}, 100_{\text{two}} = 4_{\text{eight}}$ and $011_{\text{two}} = 3_{\text{eight}}.$

Addition, subtraction in bases other than ten

The ordinary arithmetic operations, addition, subtraction, are carried out in exactly the same manner as in base ten, but we have to take extra care when 'carrying', since this requires considering the number base.

Example 3. Add the octal numbers 174 and 233.

There is no problem when adding the units digits, $4 + 3 = 7$, but when we add the 'eights' digits, $7 + 3 =$ ten, which we write as 12 in octals. Thus we write down a 2 and 'carry' 1.

$$
\begin{array}{r}
174 \\
233 \\
\hline
427
\end{array}
$$

Multiplication, division

The method for multiplication and for division in bases other than ten is the same as for base ten, but again care has to be taken in remembering and applying the new tables.

Example 4. Find the product of 214 and 23, both numbers being in base five.

$$
\begin{array}{r}
214 \\
23 \\
\hline
4330 \\
1202 \\
\hline
11032
\end{array}
$$

Notice that when multiplying by 2, $2 \times 4 = 13_{\text{five}}$, so we wrote down the 3 and carried 1. In the next line, $3 \times 4 = 22_{\text{five}}$, and similarly $3 \times 2 = 11_{\text{five}}$.

Binary multiplication and division are particularly easy, as we can multiply only by 0 or 1.

Example 5. Find the product of the binary numbers 1011 and 1101.

$$
\begin{array}{r}
1011 \\
1101 \\
\hline
1011000 \\
101100 \\
1011 \\
\hline
10001111
\end{array}
$$

Example 6. Divide the binary number 1 000 001 by 101.

$$
\begin{array}{r}
1101 \\
101\overline{)1000001} \\
101 \\
\hline
110 \\
101 \\
\hline
101 \\
101 \\
\hline
\end{array}
$$

Example 7. What number base has been used if 234 + 141 = 415?

Since the sum of 4 and 1 is 5 in this number base, the number base is greater than 5. But $3 + 4$ is not equal to 1, so that $3 + 4$ must equal 11 in the number base, which is therefore six. Notice that if we add more than two numbers together, we may 'carry' more than 1,

e.g. $14_{six} + 14_{six} + 15_{six} = 51_{six}$

Exercise 1.1a

1. Express the following denary (base ten) numbers in
   (i) base two   (ii) base three   (iii) base five   (iv) base eight   (v) base twelve:

   $$5, 7, 8, 9, 25, 27, 32, 60$$

2. Express the following binary numbers in base ten:

   $$11, 101, 1001, 1101, 10\ 000, 11\ 000$$

3. Express the following octal numbers in base ten:

   $$7, 11, 17, 100, 177, 1000$$

4. Express the following numbers base three in base ten:

   $$11, 101, 122, 222, 1000, 2000$$

5. Express the following binary numbers as octals:

   $$11, 101, 1111, 111\ 111, 1\ 101\ 101, 1\ 000\ 000\ 000$$

6. Express the following octal numbers in binary:

   $$11, 13, 17, 20, 100, 200$$

7. How can you tell at a glance that a binary number is divisible by eight?

8. The denary number 14 when written in another base ends with 0. What possible number-bases have been used?

9. If $265 + 443 = 718$, what number-base has been used?

10. In the following addition, the numbers are in a base other than ten, and one digit is missing, represented by *.

    $$\begin{array}{r} 653 \\ *44 \\ \hline 1217 \end{array}$$

    What base has been used? What is the missing digit?

11. All the numbers being binary, find the following:

(i) 111×101	(ii) 1101×111	(iii) 1111×1001
(iv) 10100×101	(v) 1001×1010	(vi) $1001 \div 11$
(vii) $11001 \div 101$	(viii) $100001 \div 11$	(ix) $1111101 \div 101$
(x) 11000×100	(xi) $11000 \div 100$	(xii) $11010000 \div 1000$

12. Find the following products and quotients:

(i) $132_{four} \times 2$	(ii) $132_{five} \times 2$
(iii) $132_{eight} \times 2$	(iv) $132_{four} \times 22_{four}$
(v) $132_{eight} \times 22_{eight}$	(vi) $132_{four} \div 2$
(vii) $132_{five} \div 2$	(viii) $132_{eight} \div 2$
(ix) $132_{four} \div 11_{four}$	(x) $3132_{eight} \div 6$

Exercise 1.1b

1. Express the following denary numbers in
   (i) base two   (ii) base three   (iii) base four   (iv) base eight   (v) base thirteen:

$$2, 11, 16, 18, 20, 32, 40, 100$$

2. Express the following binary numbers in base ten:

$$110, 1101, 1111, 10\ 000$$

3. Express the following octal numbers in base ten:

$$5, 12, 24, 144, 1728$$

4. Express the following octal numbers in binary:

$$5, 35, 105, 150, 177$$

5. What is the smallest binary number that must be added to 101 101 to make the sum divisible by sixteen?

6. Find the sum of the denary numbers $1 + 2 + 2^2 + 2^3 + 2^4 + 2^5$ as a binary number by first expressing each in binary. What number must be added to that sum to make it equal 2^6?

7. If $235 + 142 = 421$, what number base has been used?

8. If $15 \times 5 = 114$, what number base has been used?

9. If the numbers are in base three, divide 1021 by 2.

10. What base has been used in this subtraction, and what is the missing digit?

$$
\begin{array}{r}
5162 \\
2644 \\
\hline
2*15
\end{array}
$$

11. What is the largest denary number that can be expressed as a six-digit binary number?

12. How many binary digits has the number 37_{ten}?

Decimal fractions, decimals

Fractions with 10 (or powers of 10) in their denominator can easily be written as decimals, e.g.

$$\tfrac{1}{10} = 0.1, \quad \tfrac{3}{10} = 0.3, \quad \tfrac{37}{100} = 0.37,$$

and other fractions can be expressed as decimals by division, e.g.

$$\tfrac{3}{8} = 0.375, \quad \tfrac{3}{7} = 0.42857\dot{1}\ldots$$

Some of these decimal equivalents will terminate, e.g. $\tfrac{3}{8} = 0.375$, and some will give recurring decimals. We can see that the only ones which do terminate will be those whose denominators contain only powers of 2 or 5.

Binary fractions, bicimals

Similarly, this notation can be adapted for writing binary fractions, e.g. $\frac{1}{2}, \frac{1}{4}, \frac{1}{8}$ as bicimals. Clearly since

$$1 + 1 = 10 \text{ in binary,}$$
$$0.1 + 0.1 = 1,$$

so that 0.1 must represent $\frac{1}{2}$. Also $0.01 + 0.01 = 0.1$, so that 0.01 must represent $\frac{1}{4}$.

To write a fraction as a bicimal, if the denominator is a power of two, we have only to express both numerator and denominator as binary numbers, then division is easy.

$$\left(\frac{7}{8}\right)_{ten} = \left(\frac{111}{1000}\right)_{two} = 0.111_{two}$$

Fractions with other denominators require repeated division in binary, the same method as used when expressing decimal fractions as decimals. These bicimals will only terminate if the denominator is a power of 2.

Example 8. *Express $\frac{2}{3}$ as a binary fraction.*

Write 2 and 3 in binary,

$$\left(\frac{2}{3}\right)_{ten} = \left(\frac{10}{11}\right)_{two}$$

Divide

```
        0.10101
  11)10.000000000
      1.1
      0.100
        11
        100
         11
```

The sequence recurs, so that $\left(\frac{2}{3}\right)_{ten}$ as a bicimal is 0.101010 . . . i.e. $0.\dot{1}\dot{0}$.

Example 9. *Express the bicimal 0.1011 as a fraction base ten.*

Since $\qquad 0.1 = \frac{1}{2}, 0.001 = \frac{1}{8}$ and $0.0001 = \frac{1}{16}$,
$$0.1011 = \frac{1}{2} + \frac{1}{8} + \frac{1}{16},$$
$$= \frac{11}{16}.$$

Exercise 1.2a

1. Express in binary notation the base ten fractions
   (i) $\frac{1}{4}$ (ii) $\frac{3}{4}$ (iii) $\frac{1}{16}$ (iv) $\frac{5}{16}$
2. Express in binary notation the base ten decimals
   (i) 0.75   (ii) 0.125   (iii) 0.375   (iv) 0.3125

3. Express in binary the following base ten numbers

(i) $4\frac{1}{2}$ (ii) $8\frac{1}{8}$ (iii) 32.125   (iv) 17.0625

4. Find the sum of the bicimals 0.001 and 0.01.

5. Both numbers being in binary, multiply 0.01 by 101.

Exercise 1.2b

1. Express in binary notation the base ten fractions

(i) $\frac{3}{8}$ (ii) $\frac{7}{8}$ (iii) $\frac{1}{32}$ (iv) $\frac{31}{32}$

2. Express in binary notation the base ten decimals

(i) 0.25   (ii) 0.0625   (iii) 0.3   (iv) 0.$\dot{3}$

3. Express in binary notation the following written in base ten

(i) $\frac{1}{7}$ (ii) $\frac{1}{3}$ (iii) $\frac{5}{7}$ (iv) 0.$\dot{1}$

4. Both numbers being in binary, divide 10.101 by 1.1.

5. Find (a) the binary, (b) the denary equivalents of the base eight number 0.1.

Modular arithmetic

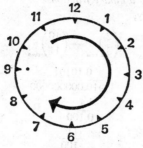

Fig. 1.6

'If you go to sleep at 11 o'clock, and wake up after 8 hours, what will the time be then?' 'Seven o'clock.' Notice that in this context we say

$$11 + 8 = 7.$$

'An aircraft leaves London at 11 a.m. to fly to Cape Town, a journey taking 23 hours. What is the scheduled arrival time?' '10 a.m. next day.' Again,

$$11 + 23 = 10.$$

In each case we have added the numbers first, then subtracted as many multiples of 12 as possible. This is arithmetic modulo 12.

Subtraction modulo 12 is carried out in the same way. 'If a cricket match finished at 6 p.m., after 7 hours play (including breaks), it started at 11 a.m. We do not say '− 1 p.m.'

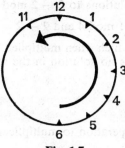

Fig. 1.7

We can display the results of modular arithmetic in a table, as below, which shows addition in arithmetic modulo 5

+	0	1	2	3	4
0	0	1	2	3	4
1	1	2	3	4	0
2	2	3	4	0	1
3	3	4	0	1	2
4	4	0	1	2	3

Multiplication is defined in the same way, e.g. multiplication modulo 5,

$$4 \times 4 = 16$$
$$= 1 \bmod 5,$$

after subtracting as many multiples of 5 as possible.

The tables below show multiplication mod 4 and mod 5, with 0 omitted, since $0 \times x = 0$ for all values of x, whatever modulo is used.

So that we have only a finite set of numbers, we do not consider fractions in modular arithmetic, and these tables show all the products that exist in each case.

Notice the difference in pattern, which we shall study in Chapter 21

×	1	2	3
1	1	2	3
2	2	0	2
3	3	2	1

×	1	2	3	4
1	1	2	3	4
2	2	4	1	③
3	3	1	4	2
4	4	3	2	1

We see from the table for multiplication mod 4 that

$$2 \times 1 = 2 \bmod 4 \text{ and } 2 \times 3 = 2 \bmod 4$$

so that there are two solutions to $2 \div 2 \bmod 4$,

$$2 \div 2 = 1 \bmod 4 \text{ and } 2 \div 2 = 3 \bmod 4$$

But there is no number which when multiplied by 2 mod 4 is equal to 3, so that $3 \div 2 \bmod 4$ has no solution in the set of whole numbers we are using.

Division

Division is the inverse operation to multiplication, so that to find

$$3 \div 2 \bmod 5$$

we must ask ourselves

'What number when multiplied by $2 = 3 \bmod 5$?'
Looking at the ringed entry in the table, we see

$$2 \times 4 = 3 \bmod 5$$
$$\therefore \ 3 \div 2 = 4 \bmod 5.$$

Sometimes there may be more than one quotient for division, and sometimes there may not be a quotient in the set of integers over which our modular arithmetic is defined.

Example 10. (i) *Find the sum of $3 + 4 + 1 + 2 + 3 \bmod 5$.*

$$3 + 4 + 1 + 2 + 3 = 13$$
$$= 3 \bmod 5$$

Example 10. (ii) *Find $4 - 7 \bmod 8$.*

$$4 - 7 = -3$$
$$= 5 \bmod 8$$

Example 10. (iii) *Find $4 \times 7 \bmod 8$.*

$$4 \times 7 = 28$$
$$= 4 \bmod 8, \text{ subtracting } 3 \times 8, \text{ i.e. } 24.$$

Any division problems other than the very easiest are best solved using multiplication.

Example 10. (iv) *Divide 5 by 7 in arithmetic modulo 9.*
Listing the 7 times table in modulo 9,

$7 \times 1 = 7$	$7 \times 2 = 14$	$7 \times 3 = 21$
	$= 5 \bmod 9,$	$= 3 \bmod 9,$
$7 \times 4 = 28$	$7 \times 5 = 35$	$7 \times 6 = 42$
$= 1 \bmod 9,$	$= 8 \bmod 9,$	$= 6 \bmod 9,$
$7 \times 7 = 49$	$7 \times 8 = 56$	$7 \times 9 = 63$
$= 4 \bmod 9,$	$= 2 \bmod 9,$	$= 0 \bmod 9.$

We see that 7×2 is the only product equal to 5 mod 9,

i.e.
$$7 \times 2 = 5 \text{ mod } 9$$
$$\therefore 5 \div 7 = 2 \text{ mod } 9.$$

If the quotient (7 in this example) is a factor of the modulo (here, 9), we can expect either more than one answer, or no answer at all, e.g.

$$3 \times 1 = 3 \text{ mod } 9, \qquad 3 \times 2 = 6 \text{ mod } 9, \qquad 3 \times 3 = 0 \text{ mod } 9$$
$$3 \times 4 = 3 \text{ mod } 9, \qquad 3 \times 5 = 6 \text{ mod } 9, \qquad 3 \times 6 = 0 \text{ mod } 9$$
$$3 \times 7 = 3 \text{ mod } 9, \qquad 3 \times 8 = 6 \text{ mod } 9, \qquad 3 \times 9 = 0 \text{ mod } 9$$
$$\therefore 3 \div 3 \text{ mod } 9 = 1 \text{ or } 4 \text{ or } 7,$$

whereas $2 \div 3$ mod 9 has no solution, in the set {0, 1, 2, 3, 4, 5, 6, 7, 8}.

Exercise 1.3a

1. Copy and complete the addition and multiplication tables for arithmetic modulo 6.

+	0	1	2	3	4	5
0						
1						
2						
3						
4						
5						

×	0	1	2	3	4	5
0						
1						
2						
3						
4						
5						

2. Copy and complete these parts of the addition and multiplication tables for arithmetic modulo 12

+	2	4	6	8	10
2					
4					
5					
7					
9					

×	1	3	5	7	9	11
2						
4						
5						
7						
9						

3. Find the missing entry (denoted by *) in each equation

(i) $3 + 4 = * \text{ mod } 5$ (ii) $3 + * = 1 \text{ mod } 5$

(iii) $3 - 4 = * \text{ mod } 5$ (iv) $2 - * = 1 \text{ mod } 5$

(v) $3 \times 3 = * \text{ mod } 5$ (vi) $3 \times * = 2 \text{ mod } 5$

4. List the squares mod 5 of the numbers 1, 2, 3, and 4.

5. Solve the following equations in the set of positive integers of each modular arithmetic.

(i) $3x + 4 = 7 \bmod 8$ (ii) $3x + 4 = 10 \bmod 9$
(iii) $4x - 3 = 6 \bmod 7$ (iv) $4x - 3 = 1 \bmod 8$
(v) $4x - 3 = 2 \bmod 8$ (vi) $x^2 = 1 \bmod 5$
(vii) $x^2 = 1 \bmod 6$ (viii) $x^2 - 2x + 2 = 0 \bmod 5$

Exercise 1.3b

1. Copy and complete the addition and multiplication tables for arithmetic modulo 4.

+	0	1	2	3
0				
1				
2				
3				

×	0	1	2	3
0				
1				
2				
3				

2. Compile the multiplication tables for arithmetic modulo 7 and modulo 8.

3. Compare the multiplication tables for arithmetic modulo 4, 5, 6 and 7. What are the differences in structure?

4. Verify that, in arithmetic modulo 5,
$$2^1 = 2, 2^2 = 4, 2^3 = 3, 2^4 = 1$$
Find the value of $2^1, 2^2, 2^3, 2^4, 2^5$ in arithmetic mod 6 and in arithmetic mod 8.

5. Solve the equation $3^x = 2$ in
    (i) mod 5   (ii) mod 6   (iii) mod 9.
Solve the equations $3^x = 1$ and $3^x = 3$ in each of those three arithmetics.

Exercise 1.4: Miscellaneous

1. Subtract 4821 from 5617 in base nine.

2. Express $\frac{2}{7}$ as a bicimal.

3. Is 31 in base four a prime number?

4. Is 6561 in base ten divisible by 9?

5. Is 3506 in base eight divisible by 7?

6. How can you tell that a number in base three is divisible by nine?

7. How can you tell that a number in base nine is divisible by ten?

8. Which is the larger, 312 in base four, or 257 in base eight?

9. If all the numbers are in base three, solve the equation
$$121 x + 11 = 1100$$

10. If all the numbers are in binary, solve for x the equation
$$\frac{x}{x + 101} = \frac{11}{1000}$$

11. If all the numbers are in binary, solve the equation
$$x^2 - 11x + 10 = 0$$

12. The number written 79 in an unknown base is not a prime. Find a possible value for this base.

13. Is $(3 \times 5) \times 4 = 3 \times (5 \times 4) \bmod 6$?

14. Is $3 \times (5 + 4) = 3 \times 5 + 3 \times 4 \bmod 6$?

15. Find n if $132_n = 70_{\text{eight}}$.

16. (i) What is the largest number that can be expressed in binary using only four digits?

(ii) A man has four biros of different colours. He takes at least one to work each day. For how many consecutive days can he take a different selection to work?

17. (i) What is the largest number that can be expressed in binary if only five digits may be used?

(ii) From a £1 note, a 50p piece, a 10p piece, a 5p piece and a penny, how many different sums of money can be made up?

18. Express $\left(\dfrac{1}{8.9}\right)_{\text{ten}}$ as a decimal, showing the first five digits after the decimal point. Express $\left(\dfrac{1}{7.8}\right)_{\text{nine}}$ $\left(\dfrac{1}{6.7}\right)_{\text{eight}}$ $\left(\dfrac{1}{5.6}\right)_{\text{seven}}$ in the same form.

Do the five digits look familiar?

19. Find the values of x less than 10 such that the equation
$$4y = 1 \bmod x$$
has (i) exactly one solution, (ii) no solutions, for integer values of y. Give the solutions where they exist.

20. Find the values of x less than 10 such that the equation
$$4z = 2 \bmod x$$
has (i) exactly one solution, (ii) no solutions, (iii) more than one solution, for integer values of z. Give the solutions where they exist.

2 Approximations

Estimates of calculations

It is most important that we always carry out a rough check of all our calculations, and to do this we shall want to take suitable approximations to the numbers in the calculation. Thus to check the calculation

$$346 \times 18 = 6228$$

we should say

$$300 \times 20 = 6000$$

and to check

$$\frac{6264}{27} = 232$$

we should say

$$\frac{6000}{30} = 200$$

In the multiplication we had a further check, for 18 is divisible by 9 and 6228 is divisible by 9, since the sum of the digits is divisible by 9.

Significant figures

Total government expenditure in 1975 was £37 874 000 000.* We may be inclined to think that the £4 000 000 was 'insignificant'! It is the first digit, and the powers of 10, that determine the size of a number. The first non-zero digit on the left of a number is called the first significant figure, and a number will be 'correct to 1 significant figure' if there is only one 'significant figure',

e.g.   3456 = 3000, correct to 1 significant figure.

           0.346 = 0.3, correct to 1 significant figure.

Generally, if the second digit is 5, the convention is that the number is corrected 'up', e.g. 350 = 400, correct to 1 significant figure (abbreviated to s.f.).

If more than 1 s.f. is required, we count all figures, including zeros, after the first non-zero digit.

Example 1.   *34 597 = 30 000, correct to 1 s.f.*
                    = 35 000, correct to 2 s.f.
                    = 34 600, correct to 3 s.f.
                    = 34 600, correct to 4 s.f.,
the first zero counting here as it follows the non-zero 6.

Example 2.   *0.003467 = 0.003, correct to 1 s.f.*
                      = 0.003 5, correct to 2 s.f.
                      = 0.003 47, correct to 3 s.f.

For significant figures, we start counting at the first non-zero digit.

* Source: Annual Abstract of Statistics, 1975.

Decimal places

To determine how many decimal places are given in a number, we start counting at the decimal marker (at present, a point (.), on the line, but a comma (,) is becoming more popular) e.g.,

$$0.012\ 34 = 0.01 \text{ correct to 2 decimal places}$$
$$= 0.012 \text{ correct to 3 decimal places}$$
$$= 0.012\ 3 \text{ correct to 4 decimal places.}$$

Notice that we count zeros after the decimal marker.

Example 3. $0.005\ 060\ 47 = 0.005$ correct to 3 d.p.
$$= 0.0051 \text{ correct to 4 d.p.}$$
$$= 0.005\ 06 \text{ correct to 5 d.p.}$$
$$= 0.005\ 060 \text{ correct to 6 d.p.}$$

Example 4. Write 45.46 (i) *correct to 1 s.f.*
              (ii) *correct to 1 d.p.*
$$45.46 = 50, \text{ correct to 1 s.f.}$$
$$45.46 = 45.5, \text{ correct to 1 d.p.}$$

Exercise 2.1a

1. Correct each number to 1 significant figure, and so estimate the value of the following:

   (i) 33×28 (ii) 47×61 (iii) 4.9×8.1
   (iv) 1.8×720 (v) 2.9×540 (vi) 4.1×0.082

   (vii) $\dfrac{330}{28}$ (viii) $\dfrac{610}{42}$ (ix) $\dfrac{8.1}{49}$

   (x) $\dfrac{4700}{32}$ (xi) $\dfrac{29}{0.074}$ (xii) $\dfrac{3685}{0.04}$

2. One of the following calculations in each set of three is wrong. Find by estimation which is wrong, each calculation being given to three significant figures.

   (i) $27 \times 53 = 1430$, $35 \times 74 = 239$, $25 \times 54 = 1350$.
   (ii) $2.6 \times 460 = 1200$, $496 \times 0.2 = 99.2$, $36 \times 75 = 2070$.

   (iii) $\dfrac{356}{48} = 7.42$, $\dfrac{685}{19} = 63.1$, $\dfrac{85.2}{0.43} = 198$.

   (iv) $\dfrac{0.407}{17} = 0.239$, $\dfrac{897}{0.31} = 2890$, $\dfrac{586}{0.26} = 153$.

3. Write each of the following correct to the required number of significant figures:

   (i) to 1 s.f.; 460; 6.5; 0.74; 0.0801; 0.0078
   (ii) to 2 s.f.; 467; 66.5; 0.747; 0.0801; 0.00775
   (iii) to 3 s.f.; 4766; 66.57; 0.7487; 0.08017; 0.007749

4. Correct 13. 4951 to

    (i) 1 s.f.   (ii) 2 s.f.   (iii) 3 s.f.   (iv) 4 s.f.   (v) 5 s.f.

5. Write each of the following correct to the required number of places of decimals:

    (i) to 1 d.p.; 16.74; 16.074; 16.047; 0.074; 0.047
    (ii) to 2 d.p.; 16.746; 16.074 5; 16.047; 0.070 4; 0.047 5
    (iii) to 3 d.p.; 16.704 6; 16.074 5; 16.0744; 16.004 7; 0.047 55

Exercise 2.1b

1. Correct each number to 1 significant figure, and so estimate the value of the following:

    (i) 38×58 (ii) 48×520 (iii) 0.59×89
    (iv) 5.7×680 (v) 788×0.006 (vi) 8580×0.002

    (vii) $\dfrac{470}{52}$ (viii) $\dfrac{685}{8.3}$ (ix) $\dfrac{8800}{0.074}$

    (x) $\dfrac{4600}{84}$ (xi) $\dfrac{8.006}{0.0057}$ (xii) $\dfrac{0.0073}{0.076}$

2. One of each pair of statements is correct, to 3 s.f. By using estimation find which is correct.

    (i) $59^2 = 4380$; $59^2 = 3480$
    (ii) $670^2 = 449\,000$; $670^2 = 44\,900$
    (iii) $3.9^3 = 59.3$; $3.9^3 = 69.3$
    (iv) $\sqrt{230} = 15.3$; $\sqrt{230} = 48.3$
    (v) $\sqrt{0.047} = 0.217$; $\sqrt{0.047} = 0.021\,7$

3. Write each of the following correct to the required number of significant figures:

    (i) to 1 s.f.; 567; 5.7; 0.057; 0.050 7; 5700
    (ii) to 2 s.f.; 567; 56.07; 0.567; 0.005 067; 0.005 007
    (iii) to 3 s.f.; 5667; 5607; 56 007; 0.056 07; 0.005 600 7

4. Correct 3.141 592 653 5 . . . (π) to

    (i) 3 s.f.   (ii) 4 s.f.   (iii) 5 s.f.
    (iv) 6 s.f.   (v) 7 s.f.

5. Write each of the following correct to the required number of decimal places:

    (i) to 1 d.p.; 0.707; 0.714; 0.714 9; 0.715 1; 0.715 9
    (ii) to 2 d.p.; 0.070 7; 0.007 07; 0.000 707; 0.700 7; 0.700 07
    (iii) to 3 d.p.; 0.007 07; 0.007 007; 0.700 07; 0.700 4; 0.700 7

Standard form

The speed of light in vacuo is about 300 000 000 m s^{-1}; the distance of the Earth from the Sun is about 149 500 000 km; the mass of an electron is 0.000 000 000 000 000 000 000 000 000 911 grammes. Numbers written like this are difficult both to write and to manipulate, and they are usually written in standard form. Many electronic calculators display their numbers in standard form.

$$3 \times 10^5, 5 \times 10^6, 7 \times 10^{-3}$$

are examples of numbers in standard form, i.e. the form $A \times 10^n$ where A is a number between 1 and 10 (including 1 but not 10) and n is an integer, positive or negative. Thus we write

$$3600 = 3.6 \times 10^3$$
$$360 = 3.6 \times 10^2$$
$$36 = 3.6 \times 10$$

and

$$0.36 = 3.6 \times 10^{-1}$$
$$0.036 = 3.6 \times 10^{-2}$$
$$0.003\,6 = 3.6 \times 10^{-3}$$
$$0.000\,36 = 3.6 \times 10^{-4} \text{ and so on.}$$

The speed of light in vacuo is about $3 \times 10^8 \text{ m s}^{-1}$; the Earth is about 1.495×10^8 km from the Sun; the mass of an electron is 9.11×10^{-28} grammes.

To express a number in standard form, put a decimal marker after the first non-zero digit, and then find the correct power of 10.

$$3600 = 3.6 \times 1000 = 3.6 \times 10^3$$
$$0.000\,036 = 3.6 \times 0.000\,01 = 3.6 \times 10^{-5}$$

remembering that
$$0.1 = 10^{-1}$$
$$0.01 = 10^{-2}$$
$$0.001 = 10^{-3}, \text{ etc,}$$

Many electronic calculators, at the time of writing, display the power of 10 separate from the other digits, 3.6×10^3 being

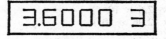

Fig. 2.1

Manipulation of numbers in standard form

The powers of 10 of course have to be treated as indices, so they are added when the numbers are multiplied, and subtracted when the numbers are divided. Thus

$$(3 \times 10^4) \times (2 \times 10^6) = 6 \times 10^{10}$$
$$(3 \times 10^4) \times (4 \times 10^6) = 12 \times 10^{10}$$
$$= 1.2 \times 10^{11}$$

and
$$(3 \times 10^4) \div (2 \times 10^6) = 1.5 \times 10^{-2}$$
$$(3 \times 10^4) \div (4 \times 10^6) = 0.75 \times 10^{-2}$$
$$= 7.5 \times 10^{-3}$$

Notice that in two of the examples above we did not at first obtain standard form. We had to write $12 = 1.2 \times 10$, so that $12 \times 10^{10} = (1.2 \times 10) \times 10^{10} = 1.2 \times 10^{11}$ and $0.75 = 7.5 \times 10^{-1}$ so that $0.75 \times 10^{-2} = (0.75 \times 10^{-1}) \times 10^{-2} = 7.5 \times 10^{-3}$. Care must be taken with negative indices.

Example 5. Carry out the following calculations, leaving the answers in standard form.

(i) $(4 \times 10^4) \times (6 \times 10^4) = 24 \times 10^8 = 2.4 \times 10^9$
(ii) $(4 \times 10^4) \times (6 \times 10^{-6}) = 24 \times 10^{-2} = 2.4 \times 10^{-1}$
(iii) $(4 \times 10^{-4}) \times (6 \times 10^{-6}) = 24 \times 10^{-10} = 2.4 \times 10^{-9}$
(iv) $(3 \times 10^6) \div (4 \times 10^4) = 0.75 \times 10^2 = 7.5 \times 10^1$
(v) $(3 \times 10^6) \div (4 \times 10^2) = 0.75 \times 10^4 = 7.5 \times 10^3$
(vi) $(3 \times 10^{-6}) \div (4 \times 10^{-4}) = 0.75 \times 10^{-2} = 7.5 \times 10^{-3}$

Exercise 2.2a

1. Write the following numbers in standard form:

(i) 7560	(ii) 756 000	(iii) 7 560 000 000
(iv) 75.6	(v) 0.075 6	(vi) 0.007 56
(vii) 0.000 000 756		(viii) 7.56

2. Find each of the following products, leaving your answer in standard form:

(i) $(2 \times 10^3) \times (3 \times 10^4)$
(ii) $(2 \times 10^3) \times (6 \times 10^4)$
(iii) $(2 \times 10^3) \times (6.5 \times 10^{-2})$
(iv) $(4 \times 10^3) \times (2.6 \times 10^2)$
(v) $(4 \times 10^3) \times (2.6 \times 10^{-4})$
(vi) $(4 \times 10^{-3}) \times (2.6 \times 10^4)$
(vii) $(4 \times 10^{-3}) \times (2.6 \times 10^{-4})$
(viii) $(4 \times 10^{-3}) \times (2.06 \times 10^3)$

3. Find each of the following quotients, leaving your answer in standard form:

(i) $\dfrac{3 \times 10^{10}}{2 \times 10^7}$ (ii) $\dfrac{3 \times 10^{10}}{4 \times 10^7}$

(iii) $\dfrac{4.8 \times 10^{10}}{6 \times 10^5}$ (iv) $\dfrac{4.8 \times 10^{10}}{6 \times 10^{-5}}$

(v) $(9 \times 10^{10}) \div (3 \times 10^{-3})$ (vi) $(9 \times 10^{10}) \div (1.8 \times 10^{-3})$
(vii) $(6.4 \times 10^3) \div (8 \times 10^{-4})$ (viii) $(6.4 \times 10^{-3}) \div (8 \times 10^{-4})$

4. The kinetic energy of a body mass m kilogrammes, velocity v metres per second is $\frac{1}{2}mv^2$ joules. Find in standard form the kinetic energy of bodies for which

(i) $m = 3 \times 10^4$, $\quad\quad$ $v = 2 \times 10^2$
(ii) $m = 3 \times 10^4$, $\quad\quad$ $v = 2 \times 10^{-2}$
(iii) $m = 3 \times 10^{-10}$, $\quad$ $v = 2 \times 10^2$
(iv) $m = 3 \times 10^{-20}$, $\quad$ $v = 2 \times 10^6$

Exercise 2.2b

1. Write each of the following numbers in standard form:

(i) 4380 $\quad\quad\quad\quad\quad$ (ii) 45 000 $\quad\quad\quad\quad$ (iii) 450 000 000
(iv) 43.8 $\quad\quad\quad\quad\quad$ (v) 0.004 5 $\quad\quad\quad\quad$ (vi) 0.000 040 05
(vii) 40 050 $\quad\quad\quad\quad$ (viii) 40 050 000 000

2. Find each of the following products, leaving your answer in standard form:

(i) $(2 \times 10^3) \times (6 \times 10^4)$ $\quad\quad\quad$ (ii) $(3 \times 10^3) \times (7 \times 10^2)$
(iii) $(2 \times 10^4) \times (5 \times 10^2)$ $\quad\quad\quad$ (iv) $(2 \times 10^4) \times (5 \times 10^{-2})$
(v) $(2 \times 10^{-4}) \times (5 \times 10^{-2})$ $\quad\quad$ (vi) $(4 \times 10^{-4}) \times (6 \times 10^2)$
(vii) $(2 \times 10^2) \times (5.4 \times 10^{-2})$ $\quad\quad$ (viii) $(2 \times 10^{-2}) \times (5.4 \times 10^{-4})$

3. Find each of the following quotients, leaving your answer in standard form.

(i) $\dfrac{5 \times 10^4}{2 \times 10^3}$ $\quad\quad\quad\quad\quad\quad$ (ii) $\dfrac{5 \times 10^4}{2 \times 10^{-2}}$

(iii) $\dfrac{2 \times 10^4}{5 \times 10^2}$ $\quad\quad\quad\quad\quad\quad$ (iv) $\dfrac{2 \times 10^4}{5 \times 10^{-2}}$

(v) $\dfrac{7 \times 10^{-4}}{2 \times 10^2}$ $\quad\quad\quad\quad\quad\quad$ (vi) $\dfrac{7 \times 10^4}{1.4 \times 10^2}$

(vii) $\dfrac{7 \times 10^4}{2.8 \times 10^{-2}}$ $\quad\quad\quad\quad\quad$ (viii) $\dfrac{7 \times 10^4}{3.5 \times 10^{-2}}$

4. In suitable units the heat generated per second by an electrical current in a wire resistance R ohms, voltage V volts is V^2/R. Find the heat generated per second when

(i) $R = 2.5 \times 10^{-1}$, $\quad\quad$ $V = 5 \times 10^2$
(ii) $R = 2.5 \times 10^2$, $\quad\quad\quad$ $V = 5 \times 10^2$
(iii) $R = 2.5 \times 10$, $\quad\quad\quad$ $V = 2 \times 10^{-2}$
(iv) $R = 2 \times 10^4$, $\quad\quad\quad\quad$ $V = 5 \times 10^{-2}$.

Degree of accuracy

A number given as 3720, correct to 3 s.f., must lie between the whole numbers 3715 and 3725. Since our convention is that a number ending 5 is 'rounded up', the number could have been 3715, but not 3725. Thus

$$x = 3720 \text{ to } 3 \text{ s.f.} \Leftrightarrow 3715 \leqslant x < 3725$$
and $\quad\quad\quad x = 3.7 \text{ to } 1 \text{ d.p.} \Leftrightarrow 3.65 \leqslant x < 3.75$

We shall say a number 'lies between e.g. 3.65 and 3.75' to include the possibility that it may be equal to the lower value. It is worth knowing that the rounding up of a 5 is only a convention, though a widespread one; other conventions are to 'throw to the even', e.g. $3.75 = 3.8$, since 8 is even, whereas $3.65 = 3.6$, as 6 is also even, and by contrast to 'throw to the odd', e.g. $3.75 = 3.7$, since 7 is odd and $3.65 = 3.7$. But these conventions are not very common, and should be ignored at present.

When we 'correct' a number, we lose some accuracy in any calculations using that number. The degree of accuracy lost will depend on the calculation we are making. Thus if the length x cm of a side of a square is '5 cm correct to the nearest cm', the exact length lies between 4.5 and 5.5, so the perimeter of the square is between 18 cm and 22 cm; the area of the square is between $(4.5)^2$ cm^2 and $(5.5)^2$ cm^2, i.e. between 20.25 cm^2 and 30.25 cm^2. The only way whereby we can determine the degree of accuracy in our results is to carry out the required calculations with the values at the extreme ends of our intervals.

Example 6. *The voltage V volts in an electrical circuit is equal to Ri, where R ohms is the resistance of the circuit and i amps the current in the circuit. If $R = 5$ and $i = 7$, each correct to 1 s.f., between what limits must V lie?*

Since $\qquad\qquad 4.5 \leqslant R < 5.5$ and $6.5 \leqslant i < 7.5$

$\qquad\qquad\qquad 4.5 \times 6.5 \leqslant V < 5.5 \times 7.5$

i.e. $\qquad\qquad\qquad 29.25 \leqslant V < 41.25$

Notice that we cannot obtain the value for V correct even to 1 s.f.

Example 7. *The length l of a rectangle is given as 5 cm and the area A as 20 cm^2, each correct to 1 s.f. Between what limits does the breadth b of the rectangle lie?*

Since $\qquad\qquad 4.5 \leqslant l < 5.5$ and $15 \leqslant A < 25$,

$$\frac{15}{5.5} < b < \frac{25}{4.5},$$

i.e. $\qquad\qquad 2.727\ldots < b < 5.55\ldots$

$\qquad\qquad 2.73 < b < 5.56$ correct to 3 s.f.

Notice that to find the smallest possible breadth we divide the smallest area by the greatest length, and vice-versa for the largest area. Since our convention is that l cannot have been equal to 5.5, nor can A be equal to 25, A does not attain the value of either end of the interval. In practice, this loss of equality is often hidden by the rounding of the numbers in the final division, as in this example.

Exercise 2.3

1. The side of a square is given as '6 cm, correct to 1 s.f.' Between what limits lie each of the following: (i) the perimeter   (ii) the area   (iii) the length of a diagonal?

2. The power P required to move a load against a resistance R with velocity v is given in suitable units by Rv. If $R = 500$ and $v = 30$, each correct to 1 s.f., between what limits does P lie?

3. The radius of a circle is given as '3 cm, correct to 1 s.f.' If the circumference is C cm and the area is A cm² between what limits do $\dfrac{C}{\pi}$ and $\dfrac{A}{\pi}$ lie? Hence find the limits, correct to 3 s.f., of C and A.

4. The length of an edge of a cube is given as '8 cm, correct to 1 s.f.' Between what limits, correct to 3 s.f., lie (i) the volume (ii) the surface area (iii) the length of a diagonal from one vertex to that opposite?

5. The base of a triangle is given as '8 cm' and the area of the triangle is given as '30 cm²', each correct to 1 s.f. Between what limits, correct to 2 s.f., does the height of the triangle lie?

6. Using a walking stick whose length is 0.9 m, correct to 1 s.f., a man measures the length of a lawn to be 36 m and the breadth to be 18 m. Between what limits lie the length, the breadth and the area of the lawn?

7. The mass in grammes of a body is equal to the product of the volume in cm³ and the density in g cm⁻³ of that body. If the volume is 50 cm³ and the density 8 grammes cm⁻³, each correct to 1 s.f., between what limits does the mass lie?

   If instead the mass is 400 grammes and the volume 50 cm each correct to 1 s.f., between what limits, correct to 2 s.f. does the density lie?

8. The length of a cylinder is 6 cm, and the radius is 4 cm, each correct to 1 s.f. Between what limits, correct to 2 s.f. lie the volume and the curved surface area?

9. The value of $\log_{10} 2$ is given in four-figure tables to be 0.3010. Between what values does $\log_{10} 2$ lie?

   Using $\log 4 = 2 \log 2$, between what limits does $\log_{10} 4$ lie? Between what limits does $\log_{10} 8$ lie?

   $\log_{10} 8$ is given in four-figure tables as 0.9031. Using this value, between what limits does $\log_{10} 2$ lie?

10. The average mass of each of the four oarsmen of a coxless four is 89 kg. Between what limits does the total mass of the four lie? If the total mass is 356 kg, correct to the nearest kg, between what limits does the average mass of the four oarsmen lie?

11. A firm employs 130 men, and the average weekly wage per man is £34, correct to the nearest £. What are the greatest and least possible weekly wage bills for that firm?

12. The death rate per year per 1000 of the population is calculated correct to 1 d.p. For a town with exactly 37 000 inhabitants, the death rate is given as 17.1 per thousand. Find the greatest and least possible number of deaths from which these figures could have been obtained.

   If the population of the town is '37 000, correct to the nearest 1000' what are now the greatest and least number of deaths consistent with a death rate of 17.1 per thousand per year?

3 Aids for Calculations

Abacus

Men have always, very wisely, made aids to help with calculations, and the earliest of these was the abacus, in which beads were threaded on a number of wires, usually fixed in a rectangular frame. This enabled addition and subtraction to be carried out rapidly and accurately.

Slide rule

The slide rule had a relatively short period of popularity, for it was rapidly replaced by the electronic calculator. It is still, however, permitted in some examinations in which the calculator is forbidden, and the exercises that follow are designed to give proficiency in the use of the slide rule; they are also suitable for electronic calculators. All slide rules are supplied with booklets of instructions, which should be kept and studied.

Multiplication

The markings on the A, B, C, and D scales of a slide rule correspond to the logarithms of the numbers, so that if we add the length corresponding to 2 to the length corresponding to 3, we obtain the length corresponding to 6. This is done by placing the 1 on the C scale over the 2 on the D scale, and reading under the 3 of the C scale. (Fig. 3.1).

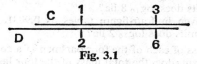

Fig. 3.1

Remember always to make a rough estimate first of the value of the product we wish to find.

If we need the product of 3 and 4, when we place the 1 on the C scale over the 3 on the D scale, the 4 of the C scale is off the right-hand

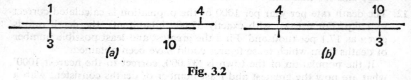

Fig. 3.2

end of the D scale (Fig. 3.2(a)) so we place the 10 of the C scale over the 3 on the D scale and now we read under the 4 of the C scale (Fig. 3.2(b)).

Reading intermediate values

The scale of the intermediate values usually changes along a slide rule, and care must be taken to ensure that the slide rule is read correctly.

Example 1. Find 1.9 × 2.8.

Rough estimate, 2 × 3 = 6.
Place the 1 on the C scale over the 1.9 on the D scale, and read under the 2.8 on the C scale. Notice that the intermediate markings shown in Fig. 3.3

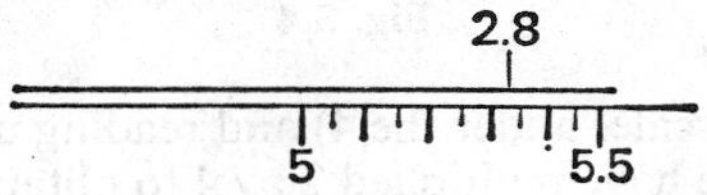

Fig. 3.3

(as on most slide rules) correspond to 5.3 and 5.35 and that the 2.7 on the C scale is nearer to 5.3. We write

$$1.9 \times 2.8 = 5.3, \text{ correct to 2 s.f.}$$

Exercise 3.1a

Make first a rough estimate, then find the value of each of the following as accurately as your slide rule or calculator permits
1. (i) 2.2 × 1.1   (ii) 2.2 × 3.4   (iii) 2.2 × 4.4   (iv) 2.2 × 5.4   (v) 2.2 × 7.4.
2. (i) 1.8 × 2.4   (ii) 1.8 × 3.4   (iii) 1.8 × 4.6   (iv) 1.8 × 5.4
   (v) 1.8 × 7.3.
3. (i) 1.75 × 2.4   (ii) 1.75 × 3.7   (iii) 1.75 × 6.1   (iv) 1.75 × 7.4
   (v) 1.75 × 8.8.
4. (i) 1.55 × 1.75   (ii) 1.55 × 3.46   (iii) 1.55 × 6.85   (iv) 1.55 × 7.85
   (v) 1.55 × 8.74.
5. (i) 1.04 × 2.44   (ii) 2.04 × 4.48   (iii) 3.15 × 4.56   (iv) 3.25 × 4.78
   (v) 4.55 × 4.65.

Exercise 3.1b

Make first a rough estimate, then find the value of each of the following as accurately as your slide rule or calculator allows.
1. (i) 24 × 1.2   (ii) 24 × 2.6   (iii) 24 × 5.2   (iv) 24 × 7.2   (v) 24 × 9.8.
2. (i) 28 × 2.4   (ii) 280 × 3.4   (iii) 0.28 × 4.6   (iv) 2800 × 0.54
   (v) 2.8 × 0.73.
3. (i) 1.25 × 2.4   (ii) 1.25 × 3.7   (iii) 1.25 × 6.1   (iv) 1.25 × 7.4
   (v) 1.25 × 8.8.
4. (i) 1.45 × 175   (ii) 14.5 × 346   (iii) 145 × 0.685   (iv) 1450 × 0.0785
   (v) 0.145 × 0.874.
5. (i) 10.6 × 28.4   (ii) 206 × 0.448   (iii) 31.5 × 45.8   (iv) 32.5 × 0.048
   (v) 0.455 × 98.4.

Division

8 divided by 4 is 2, since $4 \times 2 = 8$. So to find the result of dividing 8 by 4 using a slide rule, we ask ourselves 'What number when multiplied by 4 is equal to 8? Setting the slide rule as in Fig. 3.4, we know the pro-

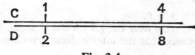

Fig. 3.4

duct is 8 (on the D scale, under the 4) and reading under the 1 on the C scale, we see that we have multiplied 2 by 4 to obtain 8.

Example 2. Divide 6.3 by 2.4.
   Rough estimate $6 \div 2 = 3$.

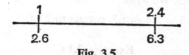

Fig. 3.5

What number, when multiplied by 2.4 equals 6.3?' Set the cursor of the slide rule over the 6.3 on the D scale, and put the 2.4 on the C scale over this 6.3. Read on the D scale the number under the 1 on the C scale. The number 2.6 has been multiplied by 2.4 to give 6.3,

i.e. $$\frac{6.3}{2.4} = 2.6 \text{ to 2 s.f.}$$

Example 3. Divide 18 by 5.6.
   Rough estimate, $20 \div 5 = 4$.
   When we set the 5.6 on the C scale over the 18 on the D scale, the 1 on the

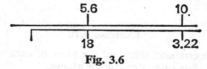

Fig. 3.6

C scale is off the left hand of the D scale. Read under the 10 on the C scale, to obtain 3.22.

$$\frac{5.6}{1.8} = 3.22, \text{ correct to 3 s.f.}$$

Exercise 3.2a

Make first a rough estimate, then find the value of each of the following as accurately as your slide rule or calculator allows.

1. (i) $7.2 \div 6$ (ii) $8.4 \div 8$ (iii) $9.3 \div 3$ (iv) $7.5 \div 5$ (v) $8.8 \div 4$
2. (i) $7.2 \div 6.4$ (ii) $8.4 \div 7.8$ (iii) $9.3 \div 2.9$ (iv) $7.5 \div 4.9$ (v) $8.8 \div 3.9$
3. (i) $7.2 \div 1.25$ (ii) $8.4 \div 1.25$ (iii) $9.3 \div 1.25$ (iv) $7.5 \div 1.25$
   (v) $8.8 \div 1.25$
4. (i) $7.25 \div 1.24$ (ii) $8.45 \div 1.44$ (iii) $9.35 \div 1.34$ (iv) $7.55 \div 1.74$
   (v) $8.85 \div 1.66$
5. (i) $72.5 \div 1.24$ (ii) $8.45 \div 144$ (iii) $0.935 \div 13.4$ (iv) $755 \div 174$
   (v) $8.85 \div 0.0166$

Exercise 3.2b

Make first a rough estimate and then find the value of each as accurately as your slide rule or calculator allows.

1. (i) $7.2 \div 9$ (ii) $4.8 \div 8$ (iii) $5.6 \div 7$ (iv) $7.8 \div 6$ (v) $4.5 \div 5$
2. (i) $7.2 \div 8.7$ (ii) $4.8 \div 8.2$ (iii) $5.6 \div 6.9$ (iv) $7.8 \div 5.7$ (v) $4.5 \div 5.2$
3. (i) $7.2 \div 1.45$ (ii) $4.8 \div 1.45$ (iii) $5.6 \div 1.45$ (iv) $7.8 \div 1.45$
   (v) $4.5 \div 1.45$
4. (i) $7.25 \div 1.44$ (ii) $4.84 \div 1.46$ (iii) $5.55 \div 1.42$ (iv) $7.85 \div 1.64$
   (v) $4.56 \div 1.34$
5. (i) $72.5 \div 1.44$ (ii) $484 \div 0.146$ (iii) $555 \div 142$ (iv) $0.785 \div 16.4$
   (v) $4560 \div 13.4$

Squares and square roots

The numbers marked on the A scale of slide rules are the squares of the corresponding numbers on the D scale, so that the square roots of the numbers on the A scale are the corresponding numbers on the D scale. The numbers on the A scale range from 1 to 100, and we have to express all numbers except the simplest in standard form, to find their squares.

Example 4. Find the squares of (i) 2, (ii) 2.3, (iii) 23.

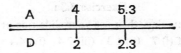

We see that 4 on the A scale is above 2 on the D scale: that 5.3 on the A scale is above 2.3 on the D scale: that $23 = 2.3 \times 10$,

so $2^2 = 4$, $2.3^2 = 5.3$, to 2 s.f.,

and $23^2 = (2.3 \times 10)^2 = 5.3 \times 10^2 = 530$, to 2 s.f.

Square roots

To find the square roots of a number between 1 and 100, on the A scale, we merely read the corresponding number on the D scale. Numbers other than these need to be written as multiples of EVEN powers of 10.

Example 5. Find the square root of (i) 16, (ii) 160, (iii) 0.016.

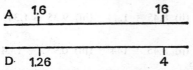

We see that 4 on the D scale is below 16 on the A scale: that $160 = 1.6 \times 10^2$ and 1.26 on the D scale is below 1.6 on the A scale: that $0.016 = 1.6 \times 10^{-2}$,

so $\sqrt{16} = 4$, $\sqrt{160} = \sqrt{1.6 \times 10^2} = 1.26 \times 10 = 12.6$,

and $\sqrt{0.016} = \sqrt{1.6 \times 10^{-2}} = 1.26 \times 10^{-1} = 0.126$

Reciprocals

Slide rules usually have a scale (C1) which gives the reciprocals of the numbers on the D scale, e.g., the reciprocal of 2 is $\frac{1}{2}$, i.e. 0.5, the reciprocal of 3 is $\frac{1}{3}$, i.e. 0.33, to 2 s.f. Care must be taken to read the intermediate markings from RIGHT to LEFT.

Example 6.Find the reciprocal of 7.4.
    Rough estimate, $\frac{1}{7} \approx 0.14$.

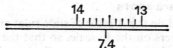

When the cursor is set over 7.4 on the D scale, it is between 1.4 and 1.3 on the C1 scale. Reading from right to left, the digits are 135 so that the reciprocal of 7.4 is 0.135.

Many calculators have square, square root and reciprocal keys and can be used in these exercises.

Exercise 3.3a

Make a rough estimate first, then use your slide rule or calculator.
1. Find the square of (i) 7   (ii) 1.3   (iii) 2.4   (iv) 4.4   (v) 6.4.
2. Find the square of (i) 50   (ii) 14   (iii) 260   (iv) 0.44   (v) 0.64.
3. Find the square root of (i) 5   (ii) 50   (iii) 1.3   (iv) 13   (v) 9.6.
4. Find the square root of (i) 500   (ii) 5000   (iii) 0.13   (iv) 0.013   (v) 9600.
5. Find the reciprocal of (i) 4   (ii) 2.4   (iii) 3.4   (iv) 4.4   (v) 8.4.

Exercise 3.3b

Make a rough estimate first, then use your slide rule or calculator.
1. Find the square of (i) 6   (ii) 1.5   (iii) 2.5   (iv) 7.3   (v) 8.8.
2. Find the square of (i) 80   (ii) 18   (iii) 270   (iv) 0.71   (v) 0.62.

3. Find the square root of (i) 8   (ii) 80   (iii) 2.1   (iv) 21   (v) 8.1.
4. Find the square root of (i) 800   (ii) 8000   (iii) 0.21   (iv) 0.021   (v) 0.81.
5. Find the reciprocal of (i) 0.22   (ii) 32   (iii) 4.3   (iv) 0.57   (v) 92.

Use of trigonometric scales

Most slide rules have scales marked for sines and tangents. Slide rules vary quite considerably in their marking of these scales, and the instruction booklets should be read for each slide rule. The exercises that follow give practice in the use of slide rule or calculator.

Example 7. Find 7.3 sin 28°.

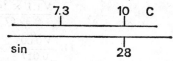

Rough estimate, sin 28° ≈ 0.5 so 8 × 0.5 = 4.
Set the cursor over 28 on the sine scale. Put the 10 of the C scale under the hair-line of the cursor. Read on the D scale under 7.3 of the C scale, 3.42,

so   7.3 sin 28° = 3.42 to 3 s.f.

Exercise 3.4a

Make a rough estimate first then evaluate:
1. (i) 8 sin 10°   (ii) 6 sin 20°   (iii) 4 sin 40°   (iv) 3.4 sin 60°   (v) 2.8 sin 80°.
2. (i) 8 cos 10°   (ii) 6 cos 20°   (iii) 4 cos 40°   (iv) 3.4 cos 60°   (v) 2.8 cos 80°.
3. (i) 8 sin 1°   (ii) 6 sin 2°   (iii) 4 sin 4°   (iv) 3.4 cos 6°   (v) 2.8 cos 8°.
4. (i) 8 tan 10°   (ii) 8 tan 20°   (iii) 4 tan 40°   (iv) 3.4 tan 60°   (v) 2.8 tan 80°.
5. (i) $\dfrac{8}{\sin 10°}$ (ii) $\dfrac{6}{\sin 20°}$ (iii) $\dfrac{4}{\cos 40°}$ (iv) $\dfrac{3.4}{\tan 60°}$ (v) $\dfrac{2.8}{\tan 80°}$.

Exercise 3.4b

1. (i) 7 sin 15°   (ii) 5 sin 25°   (iii) 3 sin 45°   (iv) 4.3 sin 65°   (v) 8.2 sin 85°.
2. (i) 7 cos 15°   (ii) 5 cos 25°   (iii) 3 cos 45°   (iv) 4.3 cos 65°   (v) 8.2 cos 85°.
3. (i) 7 sin 1.5°   (ii) 5 sin 2.5°   (iii) 3 sin 4.5°   (iv) 4.3 cos 6.5°
   (v) 8.2 cos 85°.
4. (i) 7 tan 15°   (ii) 5 tan 25°   (iii) 3 tan 45°   (iv) 4.3 tan 65°
   (v) 8.2 tan 85°.
5. (i) $\dfrac{7}{\sin 15°}$ (ii) $\dfrac{5}{\sin 25°}$ (iii) $\dfrac{3}{\sin 45°}$ (iv) $\dfrac{4.3}{\tan 65°}$ (v) $\dfrac{8.2}{\tan 85°}$.

Exercise 3.5: Miscellaneous

1. When $r = 6.6$, find (i) $2\pi r$ (ii) πr^2 (iii) $4\pi r^2$ (iv) $\tfrac{4}{3}\pi r^3$.
2. When $r = 0.23$, find (i) $2\pi r$ (ii) πr^2 (iii) $4\pi r^2$ (iv) $\tfrac{4}{3}\pi r^3$.

3. Find the value of $\dfrac{C}{2\pi}$ when $C =$ (i) 10   (ii) 120   (iii) 0.13   (iv) 75.

4. Find the value of $\sqrt{\dfrac{A}{\pi}}$ when $A =$ (i) 10   (ii) 120   (iii) 0.12   (iv) 750.

5. Find the value of $\sqrt{\dfrac{3V}{\pi h}}$ when (i) $V = 10$, $h = 17$ (ii) $V = 10$, $h = 1.7$

(iii) $V = 0.12$, $h = 45$ (iv) $V = 15$, $h = 0.064$.

Find the value of the expressions in Q.6 to 18.

6. $\dfrac{22.67 - 11.43}{22.67 \times 11.43}$

7. $(0.82)^3$

8. $(0.82)^4$

9. $\dfrac{\sqrt{1.1} \times 14.23}{39.67}$

10. $\dfrac{1 + \sqrt{0.072}}{1 - \sqrt{0.072}}$

11. $\dfrac{\sqrt{0.003}}{0.017}$

12. $\dfrac{8.072 \times \sqrt{0.74}}{0.084}$

13. $2\pi (5.4^2 - 3.4^2)$

14. $\sqrt[4]{(4)}$

15. $\sqrt[4]{(0.4)}$

16. $\dfrac{5.4 \sin 32°}{\sin 40°}$

17. $\dfrac{7.4 \sin 67°}{\sin 5°}$

18. $\frac{1}{2} \times 5.8 \times 7.9 \sin 23°$

19. Find A if $\sin A = \dfrac{5.4 \sin 32°}{8.6}$.

20. Find a if $a^2 = 5.4^2 + 5.6^2 - 2 \times 5.4 \times 5.6 \cos 40°$.

4 Logarithms

The theory of logarithms is included in the algebra section of this book and in this chapter no attempt is made to explain the reasons behind the operations. The rules for finding a logarithm are stated, and they are regarded merely as a useful device for shortening some calculations.

Very simple calculations should always be done without any calculating aids at all.

Numbers in standard form

First express the number in standard form, e.g., 3.6×10^3. The table of logarithms will give the logarithm of 3.6, i.e., 0.5563 in four-figure tables. This is called the *mantissa*. The power of 10 in the standard form is the *characteristic*, and

$$\log 3.6 \times 10^3 = 3.5563.$$

Negative characteristics

The logarithm of 3.6×10^{-2} is $-2 + 0.5563$. To show that only the characteristic is negative, we write the negative sign over the 2, and read it as 'bar 2', thus

$$\log (3.6 \times 10^{-2}) = \bar{2}.5563$$

Care must be taken especially when adding and subtracting numbers with negative characteristics. Remember that the bar means only that number is negative, and apply the ordinary rules for addition and subtraction of negative numbers.

Example 1.

1.4235	1.4235	$\bar{1}$.4235
+$\bar{2}$.6137	+$\bar{2}$.6137	+2.6137
0.0372	$\bar{2}$.0372	2.0372

$\bar{2}$.7831	2.7831	$\bar{2}$.7831
−$\bar{1}$.9256	−$\bar{1}$.9256	−1.9256
$\bar{2}$.8575	2.8575	$\bar{4}$.8575

Subtraction may be checked by the addition of the last two lines. The sum should equal the first number (e.g. $\bar{1}$.9256 + $\bar{2}$.8575 = $\bar{2}$.7837).

Antilogarithms

Given a logarithm, to find the number of which it is the logarithm, tables of antilogarithms are often used. Look up the decimal part, i.e.

the mantissa, in the antilog tables. This gives us the A of the standard form. The characteristic gives the n.

Example 2. (i) *To find the number whose logarithm is 2.3724.*

Look up 0.3724 in the antilog tables; we should find 2357, so that $A = 2.357$. The characteristic 2 gives n, so the number whose logarithm is 2.3724 is 2.357×10^2, which we may wish to write as 235.7.

(ii) *To find the number whose logarithm is $\bar{1}.0822$.*
Look up 0.0822 in the antilog tables. Notice that it begins .0822; we should find 1209. The negative characteristic $\bar{1}$ tells us the number required is 1.209×10^{-1}, which we may wish to write as 0.1209.

Multiplication

To multiply two numbers, add their logarithms. Their product is the antilog of this sum.

Example 3. Multiply 41.3 by 28.6.

$41.3 = 4.13 \times 10^1$ so log $41.3 = 1.6160$; similarly find log 28.6. Set out the working as below; the final arrow $\curvearrowright$ emphasises that we need to find the antilog of 3.0724. The tables will probably still be open at the table of logarithms, so take care not to use logarithms!

No	Log
41.3	1.6160
28.6	1.4564
1.181×10^3	3.0724

$$41.3 \times 28.6 = 1.181 \times 10^3$$
$$= 1180, \text{ to 3 s.f.}$$

It is a good plan to make a rough estimate of the result first as a check. Here, $40 \times 30 = 1200$. Remember also that logarithms will often only be accurate to 3 s.f., and so the result must be given as 1180 (to 3 s.f.).

Example 4. Evaluate $423 \times 0.000\ 72$.
(Approximation, $400 \times 0.000\ 7 = 0.28$.)

	No	Log
$423 = 4.23 \times 10^2$	423	2.6263
$0.000\ 72 = 7.2 \times 10^{-4}$	0.000 72	$\bar{4}.8573$
	3.045×10^{-1}	$\bar{1}.4836$

$$423 \times 0.000\ 72 = 3.045 \times 10^{-1}$$
$$= 0.305, \text{ correct to 3 s.f.}$$

Division

To divide one number by another, subtract their logarithms. The difference is the logarithm of the quotient.

Example 5. Evaluate 0.0741 ÷ 23.82.
(Approximation, 0.07 ÷ 20 = 0.0035.)

	No	Log
$0.0741 = 7.41 \times 10^{-2}$	0.0741	$\overline{2}.8698$
$23.82 = 2.382 \times 10^{1}$	23.82	1.3770
	3.111×10^{-2}	3.4928

$$\frac{0.0741}{23.82} = 3.111 \times 10^{-3}$$

$$= 0.003.11, \text{ to 3 s.f.}$$

With practice, it will not be necessary to write each number first in standard form; we shall be able to proceed straight to the logarithm.

Exercise 4.1a

Evaluate correct to 3 s.f.,

1. $(3.21 \times 10^2) \times (7.24 \times 10^{-2})$
2. $(6.27 \times 10^{-1}) \times (7.42 \times 10^{-1})$
3. $(3.41 \times 10) \times (8.1 \times 10^{-1})$
4. $(7.6 \times 10^{-3}) \times (3.42 \times 10^4)$
5. 5.28×84.21
6. $(3.21 \times 10^2) \div (7.24 \times 10^{-2})$
7. $(6.27 \times 10^{-1}) \div (7.24 \times 10^{-1})$
8. $(3.41 \times 10) \div (8.1 \times 10^{-1})$
9. $0.0076 \div 0.342$
10. $528 \div 8421$

Exercise 4.1b

Evaluate correct to 3 s.f.,

1. $(2.821 \times 10^2) \times (3.24 \times 10^{-2})$
2. $(7.2 \times 10^{-3}) \times (2.432 \times 10^4)$
3. $(7.16 \times 10^{-1}) \times (3.12 \times 10^{-1})$
4. $(7.18 \times 10^{-2}) \times (2.34 \times 10^{-2})$
5. 21.2×0.0667
6. $(0.0324 \times 10^{-2}) \div (2.821 \times 10^2)$
7. $(7.2 \times 10^{-3}) \div (2.431 \times 10^{-2})$
8. $(7.16 \times 10^{-1}) \div (3.12 \times 10^{-1})$
9. $0.0718 \div 0.0234$
10. $21.2 \div 0.0667$

Powers (exponents)

To square a number, double the logarithm. This gives the logarithm of the square.

To cube a number, treble the logarithm. This gives the logarithm of the cube.

To raise a number to any power n, multiply the logarithm by n. This gives the logarithm of the nth power.

Example 6. Evaluate $(0.621)^3$.
(Approximation, $(0.6)^3 = 0.216$.)

No	Log
0.621	$\bar{1}.7931$
2.395×10^{-1}	$\bar{1}.3793$ (i.e., $3 \times \bar{1}.7931$)

$$(0.621)^3 = 2.395 \times 10^{-1}$$
$$= 0.240 \text{ (to 3 s.f.).}$$

Example 7. Evaluate $(0.0275)^2$.
(Approximation, $(0.03)^2 = 0.0009$.)

No	Log
0.0275	$\bar{2}.4393$
7.561×10^{-4}	$\bar{4}.8786$ ($2 \times \bar{2}.4393$)

$$(0.0275)^2 = 7.561 \times 10^{-4}$$
$$= 0.000\,756 \text{ (to 3 s.f.).}$$

Roots

To find the square root of a number, divide the logarithm by 2. This gives the logarithm of the square root.

To find the cube root of a number, divide the logarithm by 3. This gives the logarithm of the cube root.

To find the nth root of a number, divide the logarithm by n. This gives the logarithm of the nth root.

When finding a root of a number less than 1 we have to perform a division in which the dividend has a negative characteristic. The method needs careful study. Increase the number under the bar to the next integer exactly divisible by the divisor. Balance by adding the appropriate whole number to the decimal part.

For example,

$$\frac{\bar{1}.7821}{3} = \frac{\bar{3} + 2.7821}{3} = \bar{1}.9274.$$

$$\frac{\bar{4}.6812}{7} = \frac{\bar{7} + 3.6812}{7} = \bar{1}.5259.$$

$$\frac{\bar{7}.2132}{3} = \frac{\bar{9} + 2.2132}{3} = \bar{3}.7377.$$

$$\frac{\bar{1}.6425}{5} = \frac{\bar{5} + 4.6425}{5} = \bar{1}.9285.$$

Example 8. Evaluate $\sqrt[5]{28.37}$.

No	Log
28.37	1.4529
1.953	0.2906 (1.4529 ÷ 5).

$$\therefore \sqrt[5]{28.37} = 1.95 \text{ (to 3 s.f.)}.$$

Example 9. Evaluate $\sqrt[3]{0.6274}$.

No	Log
0.6274	$\bar{1}.7976$
8.561 × 10⁻¹	$\bar{1}.9325$ ($\bar{1}.7976 \div 3$).

$$\therefore \sqrt[3]{0.6274} = 0.856 \text{ (to 3 s.f.)}.$$

Exercise 4.2a

Evaluate

1. $\bar{1}.3621 - \bar{3}.9815$.
2. $\bar{1}.3621 + \bar{3}.9815$.
3. $\bar{2}.7621 \times 3$.
4. $\bar{3}.7612 \div 4$.
5. $\bar{7}.1111 \div 3$.
6. $\sqrt{28.61}$.
7. $\sqrt[3]{0.0761}$.
8. $(0.721)^3$.
9. $\sqrt[3]{0.008}$.
10. $(0.2562)^3$.

Exercise 4.2b

Evaluate

1. $\bar{2}.2711 - \bar{3}.8826$.
2. $\bar{2}.2711 + \bar{3}.8826$.
3. $\bar{1}.7111 \times 3$.
4. $\bar{3}.6222 \div 2$.
5. $\bar{3}.8888 \div 4$.
6. $\sqrt[3]{32.2}$.
7. $\sqrt[4]{0.7111}$.
8. $(0.955)^4$.
9. $\sqrt[3]{0.0621}$.
10. $(1.111)^7$.

Worked examples

We often wish to evaluate more complicated expressions when we need to use two or more of the methods considered. Some examples are given showing how the work should be set out.

Example 10. Evaluate, to three significant figures,

$$2\pi \sqrt{\frac{98.1}{32.2}}, \text{ taking } \pi = 3.142.$$

(Approximation: $6\sqrt{3} \simeq 6(1.73) = 10.3$.)

No	Log
98.1	1.9917
32.2	1.5089
Quotient	0.4838
$\sqrt{\dfrac{98.1}{32.2}}$	0.2419 (dividing by 2).
2	0.3010
π	0.4972
10.96	1.0401

$$\therefore 2\pi \sqrt{\frac{98.1}{32.2}} = 10.96 = 11.0 \text{ (to 3 s.f.)}.$$

Example 11. *Evaluate, to three significant figures,*

$$\frac{\sqrt[3]{0.0072} \times (81.3)^2}{\sqrt{23\,140}}.$$

$$\left(\text{Approximation } \frac{0.2 \times 6400}{150} = \frac{128}{15} \simeq 8.5.\right)$$

No	Log
$\sqrt[3]{0.0072}$	$\bar{3}.8573 \div 3 = \bar{1}.2858$
$(81.3)^2$	$1.9101 \times 2 = 3.8202$
Product	3.1060
$\sqrt{23\,140}$	$4.3643 \div 2 = 2.1822$
8.39	0.9238

The value of the expression is 8.39 (to 3 s.f.).

Example 12. *Evaluate, to three significant figures,*

$$\frac{1 + \sqrt[3]{0.0075}}{1 - \sqrt[3]{0.0075}}$$

Remember that logarithms cannot be used for addition or subtraction, and in an example such as this, the work cannot be done in one operation.

First calculate $\sqrt[3]{0.0075}$.

No	Log
0.0075	$\bar{3}.8751$
0.1957	$\bar{1}.2917$ ($\bar{3}.8751 \div 3$)

$$\text{The expression} = \frac{1 + 0.1957}{1 - 0.1957} = \frac{1.1957}{0.8043}$$

Now evaluate this fraction by logarithms.

No	Log
1.1957	0.0778
0.8043	$\overline{1}.9055$
1.487	0.1723

Therefore the given expression is equal to 1.487 or 1.49 (to 3 s.f.).

Exercise 4.3: Miscellaneous

(Give your answers correct to 3 significant figures: take π to be 3.142.)

1. Find the value of $\dfrac{12.82 - 6.41}{12.82 \times 6.41}$.

2. Evaluate $\sqrt{\dfrac{s(s-b)}{(s-a)(s-c)}}$, where $a = 18.4$, $b = 12.6$, $c = 11.4$ and $2s = a + b + c$.

3. The distance of the horizon from a point h metres above the earth's surface is $\sqrt{\dfrac{2rh}{1000}}$ km, where r is the radius of the earth in km. If $r = 6370$, calculate the distance in km when $h = 186$.

4. The volume of a sphere is $\frac{4}{3}\pi r^3$. Find the volume of a sphere of radius 10.6 cm.

5. Find the value of $(0.0823)^{\frac{2}{3}}$.

6. Evaluate $\dfrac{0.81 + \sqrt{0.81}}{1.81}$.   **7.** Evaluate $\dfrac{(1.31)^5 + 1}{(1.31)^5 - 1}$.

8. The amount A to which a principal P amounts at $r\%$ compound interest for n years is given by the formula $A = P\left(1 + \dfrac{r}{100}\right)^n$. Find A if $P = 126$, $r = 4$ and $n = 5$.

9. Evaluate $\dfrac{187\{(1.3)^{12} - 1\}}{3426}$.   **10.** Evaluate $\sqrt{\dfrac{8.621 \times 27.34}{52.18 \times 0.0724}}$.

11. Evaluate $(0.647)^{\frac{2}{3}}$.

12. The area of a triangle is given by $\sqrt{s(s-a)(s-b)(s-c)}$, where $2s = a + b + c$. Find the area of a triangle given that $a = 14.2$, $b = 12.6$ and $c = 8.4$.

13. Evaluate $\dfrac{0.2463 \times (0.1721)^2}{\sqrt{0.7621}}$.

14. If $\frac{4}{3}\pi r^3 = 128.1$, calculate $4\pi r^2$.

15. Find x given that $12x = 13.41 - \frac{1}{9}\log 4.82$.

16. Find x given that $10x = 8.76 - \frac{1}{9}\log 0.6$.

17. Evaluate $\log\left(\dfrac{28.3}{13.4} + \dfrac{12.6}{\sqrt{27.1}}\right)$.   **18.** Evaluate $\left(\dfrac{2347}{641} - \dfrac{751}{864}\right)^3$.

19. Evaluate $\sqrt[3]{0.777}$.

5 Areas and Volumes

The triangle

The area (Δ) of a triangle is equal to half the product of its base and height. In the figure, $\Delta = \frac{1}{2}ah$. If one of the angles of the triangle is a right angle, the area is equal to half the product of the sides containing the right angle.

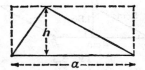

Fig. 5.1

Notice that any one side of the triangle may be taken as the base. There are thus three possible expressions for the area of the triangle. By equating two such expressions, the height of the triangle may be found as in the following example.

Example 1. A triangle ABC has AB = 3 cm, AC = 4 cm and the angle A a right angle. Find the length of the perpendicular from A to BC.

The area of the triangle $ABC = \frac{1}{2}(3)(4) = 6 \text{ cm}^2$.

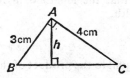

Fig. 5.2

By Pythagoras, $BC^2 = 3^2 + 4^2$. $\therefore BC = 5$.
The area of the triangle is

$$\frac{1}{2}h(BC) = \frac{5h}{2}$$

$$\therefore \frac{5h}{2} = 6,$$

or $\qquad h = 2.4$. $\therefore$ Height $= 2.4$ cm.

The area of a triangle in terms of the sides

Another useful formula for the area of a triangle is

$$\sqrt{s(s-a)(s-b)(s-c)},$$

where a, b, c are the lengths of the sides and $s = \frac{1}{2}(a + b + c)$.

The parallelogram

A parallelogram is divided by a diagonal into two congruent triangles. The area of the parallelogram is therefore twice the area of one of these triangles.

The area of a parallelogram = base × height, where the base may be taken as any one of the sides.

Example 2. The parallelogram ABCD is such that AB = 4 cm, AD = 3 cm and the length of the perpendicular from D to AB is 1.5 cm. Find the length of the perpendicular from D to BC.

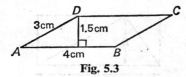

Fig. 5.3

If the length of the perpendicular required is h cm, the area of the parallelogram is equal to 4(1.5) or to $3h$ cm².

$$\therefore 3h = 6$$
$$h = 2. \quad \therefore \text{ Length of perpendicular is 2 cm.}$$

The rectangle

The rectangle is a particular case of the parallelogram and the area of a rectangle = length × breadth.

From this it follows that the length of a rectangle is equal to its area divided by its breadth.

The box or cuboid

The volume of a box is equal to the product of the three edges of the box.

Volume = length × breadth × height.

Given the volume of a box and the dimensions of the base, to find the height, divide the volume by the area of the base.

The cube

The cube is a box with all its edges equal and so the volume of a cube is equal to the cube of an edge.

The area of the walls of a room

The perimeter of a room is twice the sum of length and breadth, i.e. $p = 2(l + b)$.

The total area of the walls of a room is equal to the product of perimeter and height, i.e. $A = p \times h = 2h(l + b)$.

Solid with uniform cross section

If, when a solid is cut by a plane perpendicular to its length, the area of the cross section is always exactly the same, the solid is said to be a body of uniform cross section. The volume of such a body is equal to the area of the cross section × length.

$$\text{Volume} = (\text{area of cross section}) \times \text{length}.$$

The prism

Generally, a prism is any body of uniform cross section, but in common usage the term refers to a body having a uniform triangular section.

$$\text{Volume} = (\text{area of triangular base}) \times \text{length}.$$

The tetrahedron

The tetrahedron is the solid figure obtained by joining the vertices of a triangle to another point not in the same plane. The volume of a tetrahedron is equal to one third the product of the area of the triangle and the perpendicular distance of the fourth point from the plane of the triangle. A similar formula holds for any solid formed by joining the points of a plane figure to another point not in its plane.

$$\text{Volume} = \tfrac{1}{3}(\text{base area})(\text{height}).$$

Units

An area is measured in square units and a volume in cubic units. The relation between units of length may be changed into relations between units of area and units of volume in the following way.

$$
\begin{aligned}
1 \text{ m} &= 100 \text{ cm} \\
1 \text{ m}^2 &= 100^2 \text{ cm}^2 = 10\ 000 \text{ cm}^2 \\
1 \text{ m}^3 &= 100^3 \text{ cm}^3 = 1\ 000\ 000 \text{ cm}^3 \\
1 \text{ hectare} &= 10\ 000 \text{ m}^2 \\
1 \text{ km} &= 1000 \text{ m} \\
1 \text{ km}^2 &= 1000^2 \text{ m}^2 \\
&= 1\ 000\ 000 \text{ m}^2 \\
&= 100 \text{ hectares.}
\end{aligned}
$$

Region between two rectangles

Example. A rectangular carpet measures 16 m by 12 m and it has a stained border of width 2 m surrounding it. What is the area of the border?

Always treat an area such as this as the difference between two rectangles.

The sides of the longer rectangle will each be 4 m longer than the parallel sides of the smaller rectangle.

Therefore the area of stained region $= (20 \times 16) - (16 \times 12)$
$$= 16(20 - 12)$$
$$= 128 \text{ m}^2.$$

Space between two boxes

Example 3. A closed box is made of wood everywhere $\frac{1}{2}$ cm thick and its external dimensions are 6 cm by 5 cm by 4 cm. Find the volume of wood used in making the box.

Always treat such a volume as the difference between two volumes. The edge of the internal box is 1 cm shorter than the parallel edge of the larger box.

The external volume $= 6 \times 5 \times 4 \text{ cm}^3$
The internal volume $= 5 \times 4 \times 3 \text{ cm}^3$
The volume of wood $= 6 \times 5 \times 4 - 5 \times 4 \times 3 \text{ cm}^3$
$$= 60 \text{ cm.}^3$$

If the box had no lid, the internal height would be $5\frac{1}{2}$ cm (assuming that the height is 6 cm).
The internal volume is $5\frac{1}{2} \times 4 \times 3 \text{ cm}^3$.
The volume of wood $= 6 \times 5 \times 4 - 5\frac{1}{2} \times 4 \times 3$
$$= 54 \text{ cm}^3.$$

Density

The density of a body is its mass per unit volume. It is measured in kg/m^3 or g/cm^3.

$$\text{Density} = \frac{\text{Mass}}{\text{Volume}}.$$

Relative density

The relative density of a body is equal to the mass of the body divided by the mass of an equal volume of water. It is the number of times the body is heavier than water, volume for volume.

$$1 \text{ kg} = 1000 \text{ g.}$$
$$1 \text{ tonne} = 1000 \text{ kg.}$$
$$1 \text{ litre} = 1000 \text{ cm}^3.$$

Exercise 5.1a

1. Find the area of a triangle whose sides are 4.2 cm, 5.8 cm and 6.0 cm.
2. A triangle has sides 5 cm, 12 cm and 13 cm. Find the length of the perpendicular from the opposite vertex to the side of length 13 cm.
3. The area of a rectangle is 33.12 cm². Given that one side is of length 4.6 cm, find the length of the other side.
4. A picture which measures 48 cm by 40 cm is surrounded by a frame which is 1 cm wide. Find the area of the frame.

5. A box of volume 90 cm³ has a base of area 20 cm². Find the height of the box.

6. The external dimensions of an open box are 8 cm by 5 cm by 4 cm high. Find the volume of wood used in making the box, if the wood is $\frac{1}{2}$ cm thick.

7. A tetrahedron has a base 3 m by 4 m by 5 m and a height of 10 m. Find its volume.

8. A solid wooden box of sides 10 cm, 8 cm and 7 cm is of mass 504 g. Find the relative density of the wood.

9. A tree trunk of uniform cross section of area 0.5 m² is made of wood of relative density 0.8. Find the mass of a length of 1 m of trunk.

10. Find the number of cm² in 4.05 m².

Exercise 5.1b

1. Find the area of a triangle whose sides are 9 cm, 10 cm and 11 cm.

2. A parallelogram has sides of 12 cm and 8 cm. The distance between the 12 cm sides is 4 cm. Find the distance between the 8 cm sides.

3. A box with a square base and a height of 8 cm has a volume of 1352 cm³. Find the length of a side of the base.

4. The external dimensions of a closed box are 7 cm by 6 cm by 5 cm. If the wood used is $\frac{1}{2}$ cm thick, find the volume of wood in the box.

5. A square room has a square carpet symmetrically placed in it. This leaves an area uncovered of 9 m² and the area of the whole room is 25 m². Find the length of one side of the carpet.

6. A tetrahedron has a base of 5 m by 12 m by 13 m and a height of 10 m. Find its volume.

7. A right prism of length 10 cm has as its cross section an equilateral triangle of side 6 cm. Find its volume.

8. A room 5 m by 4 m by 3 m high is to be papered with paper 50 cm wide. What length of paper is required?

9. A swimming pool is 30 m long and 10 m wide. The water at the deep end is 3 m deep and at the shallow end is 1 m deep. Find the volume of water in the pool.

10. Find the mass of 1 m³ of water.

The circumference of a circle

The circumference of a circle is equal to $2\pi r$, where r is the radius.

$$C = 2\pi r \quad \text{or} \quad \pi d \ (d = \text{diameter}).$$

π is a constant which cannot be expressed accurately as a decimal or as a fraction. Its value correct to 4 significant figures is 3.142.

N.B. The value of π is not $\frac{22}{7}$. This is an approximation for π, but we only use this value in our work when we are particularly instructed to do so.

The area of a circle

The area of a circle is πr^2.

Example 4. The area of a circle of radius 4 cm is equal to
$$\pi(4)^2 = 16 \times 3.142 = 50.3 \ cm^2.$$

The area of an ellipse

The area of an ellipse $= \pi ab$, where a and b are the lengths of the two semi-axes of the ellipse.

Example 5. The area of an ellipse whose axes are 10 cm and 8 cm in length is
$$\pi(5)(4) = 20\pi \ cm^2 \approx 62.8 \ cm^2.$$

The surface area of a right circular cylinder

The surface area of a cylinder is equal to the product of the circumference of the base and the height.

$$S = 2\pi rh.$$

This gives the area of the curved surface only. If the cylinder has a base but no top, the total surface area is $2\pi rh + \pi r^2$; if the cylinder has both base and top, the total surface area is

$$2\pi rh + 2\pi r^2, \quad \text{or} \quad 2\pi r(h + r).$$

Example 6. Find the total surface area of a cylinder of height 6 cm and radius of base 4 cm.

The area of covered surface $= 2\pi rh = 2\pi(4)(6) = 48\pi \ cm^2$.
The area of each end $\quad = \pi(4)^2 = 16\pi \ cm^2$.
The total surface area $\quad = 48\pi + 2(16)\pi = 80\pi \ cm^2$
$\qquad\qquad\qquad\qquad\qquad = 251 \ cm^2$ (to 3 s.f.).

N.B. Use π as marked on the slide-rule, or take $\pi = 3.142$ when substitution can no longer be avoided.

The volume of a right circular cylinder

The volume of a right circular cylinder is equal to the product of the area of the base and the height.

$$\therefore \ V = \pi r^2 h.$$

This is a particular case of a solid of uniform cross section (see page 43).

Example 7. The volume of a cylinder whose height is 4 cm and whose base radius is 5 cm is equal to
$$\pi(5)^2(4) = 100\pi = 314 \ cm^3 \ (to \ 3 \ s.f.).$$

The volume of a circular cone

This is a particular case of a solid obtained by joining the points of a plane figure to another point not in the same plane (see page 43).

The volume of a cone is equal to one third the product of its base area and height.

$$V = \tfrac{1}{3}\pi r^2 h.$$

Example 8. The volume of a cone of height 6 cm and base radius 5 cm is $\tfrac{1}{3}\pi(5)^2(6) = 50\pi = 157\ cm^3$ *(to 3 s.f.).*

The surface area of a cone

The surface area of a cone (S) is given in terms of the radius of the base (r) and the slant height (l) by the formula

$$S = \pi r l.$$

This gives the area of the curved surface only.

If the cone has a base, the total surface area equals

$$\pi r l + \pi r^2 \quad \text{or} \quad \pi r(r + l).$$

Example 9. Find the area of the curved surface of a cone whose base radius is 3 cm and whose height is 4 cm.

First find the slant height.
From Fig. 5.4.

$$l^2 = h^2 + r^2.$$
$$\therefore l^2 = 4^2 + 3^2 = 25 \quad \text{and} \quad l = 5.$$
$$\therefore S = \pi r l = \pi(3)(5) = 15\pi = 47.1\ cm^2 \text{ (to 3 s.f.).}$$

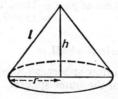

Fig. 5.4

The sphere

The formulae for the surface area and volume of a sphere are

$$S = 4\pi r^2;$$
$$V = \tfrac{4}{3}\pi r^3.$$

V is the volume, S the surface area and r the radius.

The area of a ring

Treat the area of a ring as the difference in area between two circles.

Fig. 5.5

If R is the radius of the larger circle and r the radius of the smaller, the area of the ring equals

$$\pi(R^2 - r^2), \quad \text{or} \quad \pi(R + r)(R - r).$$

The volume of material in a pipe

The hollow pipe has a uniform cross section. The cross section is a ring and if its outer and inner radii are R and r respectively, the area of the ring is $\pi(R^2 - r^2)$. The volume of the pipe is equal to the product of the area of the cross section and the length.

$$V = \pi(R^2 - r^2)l, \quad \text{or} \quad \pi(R + r)(R - r)l.$$

Example 10. A pipe, made of metal 1 cm thick, has an external radius of 11 cm. Find the volume of metal used in making 2.4 m of pipe.

$$\text{The external radius} = 11 \text{ cm.}$$
$$\text{The internal radius} = 10 \text{ cm.}$$
$$\text{The area of cross section} = \pi(11^2 - 10^2) = 21\pi \text{ cm}^2.$$
$$\text{The volume of 240 cm of pipe} = 21\pi(240) \text{ cm}^3$$
$$= 15\ 800 \text{ cm}^3 \text{ (to 3 s.f.).}$$

The volume of a spherical shell

To find the volume of a hollow sphere, subtract the volume of the inner sphere from the volume of the outer sphere.

$$\therefore V = \tfrac{4}{3}\pi R^3 - \tfrac{4}{3}\pi r^3 = \tfrac{4}{3}\pi(R^3 - r^3).$$

where R and r are the external and internal radii.

Discharge of water from a pipe

If water is flowing through a pipe at V m/s, then in every second the water contained in a length of V metres of pipe will be discharged. If the pipe is always full and the area of cross section of the pipe is A m², the volume discharged per second is AV m³.

Example 11. *Water flows through a circular pipe of internal radius of 10 cm at 5 m/s. If the pipe is always half full, find the number of m³ discharged in half an hour.*

The area of the circular cross section $= \pi(10)^2 = 100\pi$ cm².
The area of cross section of the water $= 50\pi$ cm²

$$= \frac{50\pi}{100 \times 100} \text{ m}^2.$$

The volume of water discharged per second $= \dfrac{250\pi}{100 \times 100}$ m³

$$= \frac{\pi}{40} \text{ m}^3.$$

The number of m³ discharged in half an hour

$$= \frac{(30)(60)\pi}{40}$$

$$= 45\pi$$

$$= 141 \text{ (to 3 s.f.).}$$

The areas of similar figures

The areas of similar figures are in the ratio of the squares on their corresponding linear dimensions. For example, if the radius of one circle is twice that of another, its area is four times as large; if a triangle is cut by a line parallel to its base so that each side of the smaller triangle is one third the corresponding size of the larger, its area is one ninth that of the larger; if the model of a sailing ship is made on a scale of 1 in 20, the deck area of the ship is 400 times as great as that of the model.

The volumes of similar solids

The volumes of similar solids are in the ratio of the cubes on their corresponding linear dimensions. For example, if the radius of a sphere is doubled, its volume is increased by a factor 8; if a cone is cut by a plane parallel to its base so that the height of the smaller cone is one third that of the larger, the volume of the smaller cone is one twenty-seventh the volume of the larger.

(*N.B.* If the top of a cone is cut off by a plane parallel to the base, the remainder is called **a frustum** of the cone.)

Exercise 5.2a

1. The circumference of a circle is 28 cm. Find its radius.
2. The area of a circle is 42 cm². Find its radius.
3. A cylinder of base radius 4 cm has a volume of 100 cm³. Find the height.

4. A solid cone of base radius 3 cm and height 4 cm is made of metal of density 7 g/cm³. Find its mass.

5. A pipe of thickness ½ cm has an external diameter of 12 cm. Find the volume of 2.4 m of pipe.

6. Find the total surface area of a cylinder of base radius 5 cm and length 7 cm.

7. The volume of a sphere is 827 cm³. Find its radius.

8. Find the total surface area of a cone of base radius 5 cm and height 12 cm.

9. Water flows through a circular pipe of radius 10 cm at 6 m/s. How many m³ does it discharge per minute?

10. A cone of height 6 cm and radius of base 4 cm has its top cut off by a plane parallel to its base and 4 cm from it. Find the volume of the remaining frustum.

11. The area of an ellipse is 132 cm². The length of its major axis is 14 cm. Find the length of its minor axis.

12. A right cone of height 6 cm has as its cross-section an ellipse of axes 10 cm and 7 cm. Find the volume of the cone.

Exercise 5.2b

1. A window in the form of a semicircle with diameter as base has a radius of 30 cm. Find its perimeter.

2. Find the area of a circular path 3 m wide surrounding a circular plot of radius 20 m.

3. Find the volume in cm³ of 1 km of circular cable of radius 0.1 cm.

4. Find the volume of rubber used in making a hollow ball of radius 2 cm if the thickness of the rubber is 0.2 cm.

5. Find the total surface area of a cone radius of base 6 cm and height 8 cm.

6. Find the radius of a sphere whose surface area is 108 cm².

7. Find the radius of a sphere whose volume is 426 cm³.

8. Water is flowing through a circular channel at 8 m/s. Find the number of m³ discharged per minute if the channel, always full, has a radius of 8 cm.

9. Find the volume of a frustum of a cone given that its height is 4 cm and that the radii of its ends are 3 cm and 6 cm.

10. Find the volume of a sphere whose surface area is 1000 cm².

11. The area of an ellipse is 308 cm². Find the length of its major axis given that it is twice as long as the minor axis.

12. A right cylinder has a volume of 66 cm³. If the cross section of the cylinder is an ellipse whose axes are 4 cm and 3 cm long, find the length of the cylinder.

Worked examples

Example 12. A test tube consists of a cylinder and a hemisphere of the same radius. 282 cm³ of water are required to fill the whole tube and 262 cm³ are required to fill it to a level which is 1 cm below the top of the tube. Find the radius of the tube and the length of the cylindrical part.

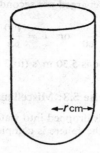

←*r* cm→

Fig. 5.6

If the radius of the tube is *r* cm, the volume of a cylinder of length 1 cm is πr^2 cm.

$$\therefore \pi r^2 = 282 - 262 = 20$$

$$r = \sqrt{\frac{20}{\pi}}$$

or by logs

$\simeq 2.52$, by slide rule or calculator.

No	Log
20	1.301
π	0.497
r^2	0.804
2.52←	0.402

The radius of the tube is 2.52 cm (to 3 s.f.).

Now find the volume of the hemispherical end. The volume of a hemisphere is $\frac{2}{3}\pi r^3$.

$$V = \frac{2}{3}\pi(2.52)^3 = 33.6 \text{ cm}^3.$$

The volume of the cylindrical part of the tube is $282 - 33.6$, i.e. 248.4 cm³, so that if the length of the cylinder is *l* cm,

$$\pi r^2 l = 248.4,$$

$$\therefore l = \frac{248.4}{\pi(2.52)^2} = 12.5, \text{ to 3 s.f.}$$

The length of the cylindrical part is therefore 12.5 cm.

Example 13. A tank holding 1 m³ of water is filled by a pipe in 10 minutes by a circular pipe diameter 2 cm. Find the speed of the water in the pipe.

The area of cross section of the pipe $= \pi$ cm²

$$= \frac{\pi}{100 \times 100} \text{m}^2.$$

If *v* m/s is the speed of water in the pipe, the volume discharged per second $= \dfrac{\pi v}{100 \times 100}$ m³.

But 1 m³ is discharged in 10 minutes.

$$\therefore \frac{1}{10 \times 60} \text{ m}^3 \text{ are discharged per second.}$$

$$\therefore \frac{\pi v}{100 \times 100} = \frac{1}{10 \times 60} \quad \text{or} \quad v = \frac{100}{6\pi} \simeq 5.30, \text{ by slide rule.}$$

The speed of water in the pipe is 5.30 m/s (to 3 s.f.).

Exercise 5.3: Miscellaneous

1. A sphere of radius 2 cm is dropped into water contained in a cylindrical vessel of radius 4 cm. If the sphere is completely immersed, find the rise in level of the water.

2. Four hundred metres of fencing are required to enclose a square field. What greater area can be enclosed by the same length of fencing if the enclosure is circular?

3. A swimming pool of length 30 m and width 12 m is 4 m deep at one end and 1 m deep at the shallow end. Find the number of m³ of water in the pool.

4. Printing paper is wrapped round a wooden core of diameter 12 cm. If the diameter of a roll is 36 cm and the length of paper 600 m, find the thickness of the paper.

5. A solid cone of height 4 cm and radius of base 2 cm is lowered into a cylindrical jar of radius 6 cm which contains water sufficient to submerge the cone completely. Find the rise in water level.

6. A cube has an edge of 4 cm. A triangle is formed by joining the middle point of one face of the cube to the ends of a diagonal of the opposite face. Calculate the area of the triangle.

7. Construct a triangle having its sides 3 cm, 5 cm and 6 cm long. Construct the altitudes from the vertices to the opposite sides, measure them and calculate the area of the triangle from each of these measurements. Find the average of these three results and compare it with the calculated area of the triangle.

8. Given that 1 in = 2.54 cm, calculate the number of km² in a square mile.

9. Calculate the mass of water which falls on a hectare of ground in a rainfall of 1 cm. Give your answer in tonnes.

10. The external dimensions of a closed wooden box are 30 cm by 20 cm by 18 cm. The thickness of the wood is 1 cm. If the relative density of the wood is 0.7, calculate the mass of the box.

11. A tank of rectangular cross section 2 m by 1 m and of height 1 m is filled by a pipe of cross section 10 cm². If the pipe delivers 1 m³ per minute, find (i) the rate of flow of water in the pipe, (ii) the time taken to fill the tank.

12. A tent has its base in the shape of a regular hexagon whose sides are 10 m. If the height of the tent is 12 m, find its volume.

13. A kite consists of an isosceles triangle with a semicircle on its base. If the

isosceles triangle has sides 13 cm, 13 cm and 10 cm, find the area of the cardboard forming the kite.

14. Taking the radius of the earth to be 6370 km, find the velocity of a place on the equator in km/h.

15. The dimensions of a hut are shown in Fig. 5.7. Calculate the volume of the hut and its total surface area, omitting the floor.

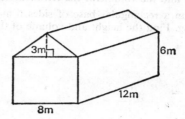

Fig. 5.7

16. The rainwater from a flat roof 15 m by 20 m drains into a tank 3 m deep on a base 4 m square. What depth of rainfall will fill the tank?

17. A hollow pipe is made of metal 0.5 cm thick. The volume of the metal in 30 m³ of pipe is 1.03 m³. Find the external diameter of the pipe.

18. A wire 2 m long is cut into two pieces which can be bent to make a square and the circle passing through the vertices of the square. Find the radius of the circle.

19. A wire encloses a circle of area 18 cm². The wire is now bent into the shape of a regular hexagon. Calculate the area of the hexagon.

20. A circular path has an external radius of 20 m and is 1 m wide. Find the total cost of gravelling the path at £2.50 per m² and fencing both sides of the path at £3.75 per m. Give your answer to the nearest £10.

21. A cylindrical tube when pushed vertically to the bottom of water contained in a cylindrical jar of radius 4 cm caused the water in the jar to rise from a height of 10.2 cm to a height of 10.32 cm. Assuming that some of the tube still projects from the water, calculate the diameter of the tube.

22. The distance round the Equator is 40 million metres. Calculate the radius of the earth in kilometres.

23. A wooden cube is of edge 6 cm. The corner A is cut off the cube by a plane which passes through the middle points of the edges through A. Find the volume of the wood removed.

24. 1000 lead shot each of diameter ¼ cm are melted down and recast into the form of a cube. Find the side of the cube.

25. A cylindrical vessel contains 250 g of a liquid whose density is 1.2 g/cm³. If the height of liquid in the vessel is 8 cm, find the base radius of the cylinder.

26. The cross section of a prism is an equilateral triangle of side 3 m. The length of the prism is 5 m. Find (i) its volume, (ii) the area of its surface.

27. Find, in tonnes, the mass of water required to fill a rectangular tank 3.2 m long, 2.4 m wide and 1 metre high.

28. An equilateral triangle has the same area as a square. Express the perimeter of the triangle as a fraction of the perimeter of the square.

29. *ABCD* is a tetrahedron. *ABC* is an equilateral triangle of side 4 m and *DA = DB = DC =* 6 m. *DN* is the perpendicular from *D* to the face *ABC*. Find *DN* and the volume of the tetrahedron.

30. A pyramid is on a rectangular base of sides 6 m and 4 m. Each slant edge is 8 m long. Find the height and volume of the pyramid.

6 Ratio and Percentage

RATIO

A common scale for a detailed map is 2 cm to 1 km. This means that 2 cm on the map represents an actual horizontal distance of 1 km. This scale may also be given as 1 in 50 000 or 1 : 50 000, which compares the distance on the map with the actual horizontal distance.

Two different quantities of the same kind may always be compared in this way. If one of the quantities is expressed as a fraction of the other quantity, this fraction is said to be the ratio of their sizes. Care must be taken to make sure that the two quantities are in the same units.

Example 1. (i) *If I am 45 years old and my son is 11 years old, the ratio of my age to his is 45:11; the ratio of his age to mine is 11:45.*

(ii) *The speed of a cyclist is 10 km/h and of a car is 35 km/h. The ratio of the speed of the cyclist to that of the car is 2:7.*

(iii) *The price of tea is increased from 240 pence per kg to 260 pence per kg. The ratio of the old price to the new price is 240:260 or 12:13.*

(iv) *The price of an article is increased from £1.25 to £2. The ratio of the old price to the new price is 125:200 or 5:8.*

Exercise 6.1a

Express each of the following ratios as simply as possible:

1. A length of 8 cm to a length of 3 m.
2. A speed of 12 km/h to a speed of 40 km/h.
3. A speed of 12 km/h to a speed of 10 m/s.
4. A cost of 4 pence per kg to a cost of 12 pence per kg.
5. A cost of 2 pence per g to a cost of £0.48 per g.
6. $62\frac{1}{2}$ pence: £5.
7. 4 cm²: 10 mm².
8. 100 m : 1 km.
9. 7 g : 1 kg.
10. 1 hectare : 100 m².

Exercise 6.1b

Express each of the following ratios as simply as possible:

1. A length of 9 cm to a length of 24 cm.
2. A speed of 15 km/h to a speed of 35 km/h.

3. A speed of 15 km/h to a speed of 40 m/s.

4. A cost of 80 pence per kg to a cost of 72 pence per kg.

5. A cost of 2 pence per g to a cost of £1.68 per kg.

6. £3.12½ : £4.37½.

7. 1 km : 82 m.

8. 20 m : 1 km.

9. 140 g : 1 kg.

10. 15 pence per g : £4 per kg.

Proportional parts

Three quantities A, B, C of the same kind may be expressed as ratios in the following way,

$$A : B : C = 3 : 4 : 5.$$

This means that the ratio of A to B is 3 to 4; that of A to C is 3 to 5 and that of B to C is 4 to 5.

The actual values of the quantities A, B, C must be $3k$, $4k$ and $5k$, where k is some number.

Example 2. A father leaves a legacy of £4500 to be divided between his three sons Arthur, Bernard and Charles in the ratios 3 : 5 : 7. What does each receive?

Method (i). If the money is divided so that Arthur receives 3 equal parts, Bernard 5 equal parts and Charles 7 equal parts, the total must be divided into $(3 + 5 + 7)$ equal parts, or 15 equal parts.

$$\tfrac{1}{15} \text{ of } £4500 = £300.$$
$$\text{Arthur receives } 3 \times £300 = £900.$$
$$\text{Bernard receives } 5 \times £300 = £1500.$$
$$\text{Charles receives } 7 \times £300 = £2100.$$

Method (ii). Suppose that Arthur receives £$3k$, Bernard £$5k$ and Charles £$7k$.

Then
$$3k + 5k + 7k = 4500.$$
$$\therefore 15k = 4500 \quad \text{or} \quad k = 300.$$

Arthur receives £900, Bernard £1500 and Charles £2100.

Exercise 6.2a

1. The sides of the triangle are in the ratios 4 : 7 : 8 and its perimeter is 38 cm. Find the sides.

2. Divide a line of length 60 cm into three parts whose ratios are 3 : 7 : 10.

3. Divide £5 into two parts so that one is two thirds of the other.

4. Three men provided capitals of £1000, £1500 and £2500 for a business on the understanding that the shares of the profits were proportional to the capital provided. If the profits were £350, what should each receive?

5. If A is half as old again as B and B is half as old again as C and the sum of their three ages is 114 years, find their ages.

6. A sum of money worth £13 consists of an equal number of coins worth 50 pence, 10 pence and 5 pence. How many coins of each kind are there?

7. Divide £140 between A, B and C so that A has twice as much as B and B has twice as much as C.

8. Archie does as much work in three hours as Bill does in four hours. Bill's son works half as fast as Bill. If the three working together are paid £8.50, how much should each receive?

9. If $p : q = 2 : 3$ and $q : r = 4 : 7$, find $p : q : r$.

10. If Alice deserves twice as many marks as Brenda and Brenda deserves half as many marks as Catharine, how many does each receive when their total marks are 125?

Exercise 6.2b

1. Divide 246 in the ratios $1\frac{1}{2} : 2 : 3\frac{1}{3}$.

2. A father leaves £1500 to be divided between his three sons in the ratios $\frac{1}{2} : \frac{1}{3} : \frac{1}{6}$. How much does each receive?

3. Three club servants agree to divide their Christmas fund in the ratios of their ages, which are 45 years, 48 years and 51 years. If the sum collected is £12, how much does each receive?

4. If the perimeter of a triangle is 36 cm and the ratios of the sides are $3 : 4 : 5$, what is the length of each side?

5. Leonard is half as old as Martin and Martin is half as old again as Norman. The sum of their ages is 91 years. How old is Leonard?

6. A sum of money worth £13.65 consists of an equal number of coins worth 50 pence, 10 pence and 5 pence. How many of each kind are there?

7. Divide £21 into two parts so that one is $\frac{3}{4}$ of the other.

8. Divide £15 between Jack, Jill, George and Georgina so that each boy has twice as much as each girl.

9. If Mark lends £400 for 9 months, Peter lends £300 for 8 months and the total interest paid is £25, how much should each receive?

10. If $x : y = 3\frac{1}{2} : 2\frac{1}{3}$ and $y : z = 1\frac{1}{4} : 2\frac{1}{2}$, find $x : y : z$.

PERCENTAGE

Suppose that, in a certain school, $\frac{2}{5}$ of the pupils are girls. The ratio of the number of girls in the school to the total number of pupils is $2 : 5$ or $40 : 100$. On the average, of every 100 pupils, 40 are girls and this fact may be more easily expressed by stating that 40 per cent (or 40%) of the pupils are girls. A percentage is simply a ratio in which the second number is arranged to be 100.

40% is equivalent to the ratio $40 : 100$ or $2 : 5$.
40% is equivalent to the fraction $\frac{40}{100}$ or $\frac{2}{5}$.
40% is equivalent to the decimal 0.4.

In the example we have considered, it is worth noting that 40% of the pupils in the school are girls and so the remaining 60% must be boys.

To express one quantity as a percentage of another quantity of the same kind, first express it as a fraction of the second quantity and then multiply by 100.

Some of the common percentages should be immediately associated with the fractions corresponding. For example,

$$75\% \text{ is } \tfrac{3}{4};$$
$$66\tfrac{2}{3}\% \text{ is } \tfrac{2}{3};$$
$$50\% \text{ is } \tfrac{1}{2};$$
$$33\tfrac{1}{3}\% \text{ is } \tfrac{1}{3};$$
$$25\% \text{ is } \tfrac{1}{4};$$
$$12\tfrac{1}{2}\% \text{ is } \tfrac{1}{8};$$
$$10\% \text{ is } \tfrac{1}{10};$$
$$5\% \text{ is } \tfrac{1}{20}.$$

Example 3. In a town of 4280 inhabitants, there were 56 births during 1964. Find the percentage birth rate.

The percentage rate is $\frac{56}{4280} \times 100 = \frac{560}{428} = 1.31\%$ (to 3 s.f.).

Exercise 6.3a

Express statements **1** to **6** in the form of a percentage.

1. 29 people out of 82 die before reaching the age of 45.
2. 5 eggs are bad in a box of 144.
3. Income tax is charged at £0.47½ in the pound.
4. The rates in a certain town are 74 pence in the pound.
5. In a school of 960 pupils, 42 are absent.
6. The interest on £125 is £5.50.
7. 30% of the inhabitants of a town are men and 32% are women. What is the percentage of children?
8. A man spends 12% of his income on rent and 52% on food. What percentage of his income is left?
9. Express $\tfrac{3}{8}$ as a percentage.
10. What is 15% of £28.50?

Exercise 6.3b

Express statements **1** to **6** in the form of a percentage.

1. In a town of 42 500 there are 504 births in a year.
2. A man has a weekly income of £34 and spends £15.50 on food.
3. A boy has 480 cigarette cards and gives 25 away.
4. A tax of 17½ pence in the pound.
5. A rate of 91½ pence in the pound.

6. In a room of area 216 m² there is a carpet of area 152 m².

7. At an election 42% vote Conservative, 37% vote Labour and the rest vote Liberal. What percentage vote Liberal?

8. Express 0.042 as a percentage.

9. What is $12\frac{1}{2}$% of £17?

10. What percentage is £3.75 of £42.50?

Percentage gain and loss

If a man buys a watch for £10 and sells it for £7.50 he has lost £2.50. This loss is incurred on a capital outlay of £10 and his percentage loss is $\dfrac{£2.50}{£10} \times 100 = 25\%$.

N.B. Percentage gain or loss is always expressed in terms of the cost price.

$$\text{Percentage gain} = \frac{\text{actual gain}}{\text{cost price}} \times 100.$$

$$\text{Percentage loss} = \frac{\text{actual loss}}{\text{cost price}} \times 100.$$

$$\text{Percentage error} = \frac{\text{actual error}}{\text{true value}} \times 100.$$

Percentage gain

A gain of 30% means that what was originally 100 is now 130.

The ratio of the new value to the old = 130 : 100.
The ratio of the new value to the gain = 130 : 30.
The ratio of the old value to the gain = 100 : 30.

Percentage loss

A loss of 30% means that what was originally 100 is now 70.

The ratio of the new value to the old = 70 : 100.
The ratio of the new value to the loss = 70 : 30.
The ratio of the old value to the loss = 100 : 30.

N.B. A gain of x% means that the new value is $\dfrac{100 + x}{100}$ of the old.

A loss of x% means that the new value is $\dfrac{100 - x}{100}$ of the old.

Examples are most easily worked by considering the ratio of the two quantities involved.

Example 4. A man sells a car for £420 at a gain of 5%. What did it cost him?

Cost price: selling price $= 100 : 105$.
Cost price $= \frac{100}{105}$ of selling price
$$= \frac{100}{105} \times £420 = £400.$$

Example 5. A man receives 10% discount for cash and pays £63.45. What discount does he receive?

The ratio of the discount to the original price $= 10 : 100$.
The ratio of the discount to the cash price is $10 : 90$.
Therefore the discount is $\frac{10}{90}$ of the cash price
$$= \frac{10}{90} \times £63.45$$
$$= £7.05.$$

Exercise 6.4a

1. A man bought a car for £800 and sold it for £640. Find his loss per cent.
2. I buy eggs at 4 pence each and sell for 5 pence. Find my gain per cent.
3. A shopkeeper lost 30% by selling an article for £1.40. What did he lose?
4. An article costing £5 is sold at a gain of $12\frac{1}{2}\%$. Find the selling price.
5. A dealer gains £4 when he sells an article to gain 8%. What is the selling price?
6. A line 8 cm long is measured as 8.04 cm. What is the percentage error?
7. A shopkeeper sells a carpet for £72 and gains 20%. What did it cost him?
8. A shopkeeper buys a desk for £50 and sells it for £56. What is his gain per cent?
9. A professional sells a golf bag for £22 and gains 10%. Find the cost to the professional.
10. I am 10% older than my wife. My wife is $x\%$ younger than I am. Find x.

Exercise 6.4b

1. I bought a car for £650 and sold it for £520. Find my loss per cent.
2. An increase in wages of 8% makes the weekly wage bill of a firm £4320. What is the amount of the increase?
3. Coal has increased in price from £26.40 per tonne to £36.30 per tonne. Find the percentage increase.
4. A shopkeeper gains £1.75 by selling an article for £6.25. What is his percentage gain?
5. A shopkeeper gains 8% by selling a table for £27. What did it cost him?
6. A sells to B at a gain of 20%; B sells to C at the price A paid. What does B lose as a percentage?
7. At a sale, goods are marked down by 5 pence in the pound. What was the original price of an article marked £2.85?
8. A television set is sold for £122.50 to gain $22\frac{1}{2}\%$. What is the actual gain?

9. The price of a share rose 25% yesterday and fell 25% today. What is the total rise or fall per cent?

10. Soap before the war was $2\frac{1}{2}d.$ per tablet. It is now 6 pence per tablet. Find the percentage increase in the price.

Worked Examples

Example 6. A shopkeeper marks his goods to gain 35%. He allows 10% discount for cash. Find his percentage profit when sold for cash.

Method (i)   Marked price : cost price = 135 : 100.
                            Cash price : marked price = 90 : 100.

Therefore the cash price = $\frac{90}{100}$ of the marked price

and   the marked price = $\frac{135}{100}$ of the cost price.

So   the cash price = $\frac{90}{100} \times \frac{135}{100}$ of the cost price

$$= \frac{121.5}{100} \text{ of the cost price.}$$

The cash price : cost price = $121\frac{1}{2}$: 100.

His gain per cent is $21\frac{1}{2}$.

Method (ii).   Goods costing £100 are marked at £135.
                 Discount at 10% ($\frac{1}{10}$ of £135) is   £13.50.
                 The cash price is   £121.50.

His gain per cent is $21\frac{1}{2}$.

Example 7. A man buys eggs at 55 pence a score. He finds that 10% of the eggs are unsaleable but sells the rest at 60 pence a dozen. Find his percentage profit.

Method (i). Choose any convenient number of eggs and work with this number. Suppose he buys 440 eggs.
   Cost of 440 eggs is

$$\frac{440}{20} + 55 \text{ pence} = £12.10$$

He sells   440 − 44 eggs = 396 or 33 dozen.

The selling price of the eggs is

$$33 \times 60 \text{ pence} = £19.80.$$

His gain is £7.70 on an outlay of £12.10.

$$\text{His percentage gain} = \frac{£7.70}{£12.10} \times 100$$

$$= \frac{770}{1210} \times 100$$

$$= \frac{700}{11} = 63.6 \text{ (to 3 s.f.)}.$$

(440 was chosen because, after 10% reduction, it is divisible by 12.)

Method (ii). Suppose he buys x eggs.

The cost price is $\dfrac{55x}{20}$ pence.

He sells $\dfrac{9x}{10}$ eggs.

The selling price is $\dfrac{9x}{10} \times \dfrac{60}{12}$ pence.

Therefore $\dfrac{\text{selling price}}{\text{cost price}} = \dfrac{9x \times 60 \times 20}{10 \times 12 \times 55x}$

$$= \dfrac{18}{11} = 1.636.$$

His gain per cent is 63.6 (to 3 s.f.).

Example 8. *A motorist reduces his annual distance travelled by $x\%$ when the price of petrol is increased by $y\%$. Find the increase per cent in his petrol bill.*

Suppose his original distance was A km and the original price of petrol B pence a litre. His original petrol bill was $\dfrac{AB}{M}$ pence annually, where M is the number of km travelled on 1 litre of petrol.

His distance is $\dfrac{100 - x}{100}A$ km and the new price of petrol $\dfrac{100 + y}{100}B$ pence a litre.

His new annual bill in pence is

$$\dfrac{(100 - x)(100 + y)}{100 \times 100} \dfrac{AB}{M}.$$

Therefore $\dfrac{\text{New charge}}{\text{Old charge}} = \dfrac{(100 - x)(100 + y)}{100 \times 100}$

$$= \dfrac{10\,000 - 100x + 100y - xy}{10\,000}$$

$$= 1 + \dfrac{y - x}{100} - \dfrac{xy}{10\,000}.$$

The new charge is $100\left(1 + \dfrac{y - x}{100} - \dfrac{xy}{10\,000}\right)\%$ of the old charge.

The percentage increase is $y - x - \dfrac{xy}{100}$.

Exercise 6.5: Miscellaneous

1. In making a television set, the costs for labour and materials are in the ratio of 3 : 2. The manufacturer sells for £150 to make a gain of 25% on his outlay. What is the cost of the materials for the set?

2. A legacy of £6000 is to be divided between Alfred, Arthur and Anne in the ratios 1 : 2 : 4, but first death duties of $12\frac{1}{2}$% must be paid. What does each receive?

3. Of the total runs made by his side during the season, Harry made 6% and Bill 11%. If Bill scored 120 runs more than Harry, how many runs did Bill make during the season?

4. A bookseller makes a profit of 20% by selling a certain book for 30 pence. When he has sold 90% of his stock, he finds he has to sell the rest at a sale price of 20 pence each. What percentage profit does he make on the transaction?

5. A dealer buys 30 wireless sets, all at the same price. He sells 20 of them at a profit of 16%, and has to sell the remaining 10 at a loss of 4%. What is the percentage profit on the deal?

6. The radius of a circle is increased from 10.0 to 10.4 cm. Find the percentage increase in its area.

7. The radius of a sphere is increased from 10.0 to 10.4 cm. Find the percentage increase in its volume.

8. For an examination, 1200 of the candidates were boys. If 50% of the boys and 40% of the girls were successful, find the number of girl candidates, given that 46% of the total number of candidates were successful.

9. By selling an article for £5.35, a shopkeeper gains 7%. What should the selling price be for a profit of 15%?

10. A motorist's cost for petrol and oil is in the ratio 15 : 1. An increase of 6% in the price of petrol and 4% in the price of oil increases his annual bill for running costs by £4.70. Find his annual bill for petrol before the increase.

11. Salmon sells his goods at the prices marked in his window. Barker allows 10% reduction from the marked price, which means that he sells for £99 goods that Salmon sells for £100. What price does Barker mark an article which Salmon sells for £5?

12. I sell 12 eggs at the price for which I buy 20 eggs. What is my profit per cent?

13. When the price of petrol is increased by 5%, a driver reduces his annual distance travelled by 5%. As a consequence, he finds that he saves 20 pence on his annual petrol bill. What was his annual petrol bill before the increase?

14. A greengrocer sells potatoes at 12 pence a kg, which should give him a profit of 20%. To turn the scales he finds he actually gives 1.1 kg when he sells 1 kg. Find his real percentage profit.

15. Brown sold his house to Smith at a profit of 10%. Smith sold it to Robinson at a gain of 5%. Robinson paid £1240 more for the house than Brown paid. What did Brown pay?

16. A circle is inscribed in a square. Express the area of the circle as a percentage of the area of the square.

17. A shopkeeper buys eggs at 30 pence a score. He finds $\frac{1}{4}$ of them are unsaleable but sells the rest at 48 pence a dozen. Find his profit per cent.

18. If a shopkeeper's weekly takings increase from £330 to £357.50 find the percentage increase. If he wishes to increase his profit by a further 5%, what must he aim to take?

19. A wholesaler sells goods to a retailer at a profit of $33\frac{1}{3}\%$. The retailer sells to the customer at a profit of 50%. Express the price the customer paid as a percentage of the cost to the wholesaler.

20. A wholesaler sells goods to a retailer at a profit of 20%. The retailer sells to the customer, who pays 80% more than the cost to the wholesaler. Find the retailer's percentage profit.

21. The cost of manufacturing a car is made up of three items: cost of materials, labour and overheads. In 1974, the cost of these items were in the ratio 4 : 3 : 2. In 1975, the cost of materials rose by 10%, the cost of labour increased by 8% but the overheads reduced by 5%. Find the increase per cent in the price of a car.

22. An alloy is formed by mixing metal A with metal B so that the ratio of their volumes is 6 : 5. The relative density of A is 8.4 and that of B is 9.6. Find the percentage mass of metal A in the alloy.

23. The lorries belonging to a company are valued at £14 500. Each year 12% of the value is written off for depreciation. Find the value of the lorries at the end of 2 years.

24. A dealer sells a car for £1540 at a profit of 40%. Find what he paid for the car and express his profit as a percentage of the selling price.

25. A salesman receives as commission 4% of the value of the goods he sells. Find the value of the goods sold when his commission is £62.

26. If $3\frac{1}{3}\%$ of a sum of money is £2.40, find $3\frac{3}{4}\%$ of the sum.

27. A shopkeeper marks his goods to gain 45% but allows 5% discount for cash. By selling a wireless set, he makes a profit of £18.87$\frac{1}{2}$ on a cash deal. Find what the shopkeeper paid for the set.

28. A man bought 12 hens at £1.50 each. During a year he obtained from them 2400 eggs which he sold at 40 pence a dozen. The cost of feeding them for the year was £30. At the end of the year he sold the 10 surviving hens at £1.25 pence each. Find his percentage profit.

29. The cost of printing a book was £5000 for 4000 copies. The publisher sold to a bookshop at a profit of 15% and the bookseller sold the book at £2.10. Find the bookseller's profit per cent.

30. A dealer buys 100 oranges for £4. He is unable to sell 4 of the oranges but sells the rest at 90 pence a dozen. Find his percentage profit and the price per dozen at which he should have sold them to have gained 8% more.

31. Death duties of 20% are paid on a legacy of £4500. The eldest son takes 50%, the second son 30%, and the youngest the remainder. What percentage of the original legacy does the youngest son receive?

32. Two blends of tea costing £1.40 and £1.60 per kg respectively are mixed in the proportion of 2 : 3 by mass. The mixture is sold at £2.40 per kg. Find the gain per cent.

33. If a man runs a kilometre in 3 minutes, express his speed as a percentage of the speed of a train running at 80 km/h. If man and train both decrease their speeds by 5%, express the speed of the man now as a percentage of the new speed of the train.

34. Two partners Smith and Brown invest £3000 and £1800 respectively in a business. It is agreed that Brown should take 30% of the profits for running the business and that the remaining profit should be divided between them in the ratio of the capital investments. Find what percentage of the profits Smith receives.

35. A train is scheduled to go a certain distance in a certain time. Owing to stoppages, the train driver estimates that he must cover the distance in 75% of the scheduled time. By what percentage must he increase his speed?

7 Simple and Compound Interest

Simple interest

When money is lent, the lender expects the borrower to pay for the use of the money. The amount of money lent is called the **Principal** and the charge made for lending it is called the **Interest**. Interest on money borrowed is paid at definite intervals (monthly, half-yearly or yearly) and the money is said to be lent at simple interest. Simple interest is obviously proportional to the amount lent and will be the same amount on a given principal during equal intervals of time.

Compound interest

If the interest, when it falls due, is not paid direct to the lender but is added to the amount borrowed, then the principal will increase with the time. When money is lent at simple interest, assuming no repayment, the principal is constant. When the principal is increased each year (or at other intervals) by the addition of the interest, the money is said to be lent at **Compound Interest** and since the principal increases, the annual interest will also increase. The total sum owed after any interval of time (i.e. the new principal) is called the **Amount**.

The following abbreviations are traditional and will be used throughout the chapter:

P stands for principal;
I stands for interest;
A stands for amount;
R stands for rate per cent per annum;
T stands for the time in years.

SIMPLE INTEREST

Calculation of simple interest

Interest is usually expressed as a percentage of the principal. The interest is obviously proportional to the principal and to the rate of interest per annum. It is also proportional to the time. If money is lent at 4% per annum for 3 years, the total interest paid during that time will be 12% of the principal. Similarly if money is lent at $R\%$ per annum for T years, the interest is $RT\%$ of the principal.

$$\therefore I = RT\% \text{ of } P$$
$$= \frac{RT}{100} \text{ of } P;$$

and
$$I = \frac{PRT}{100}.$$

This formula, by simple algebraic manipulation, may be used to express any one of I, P, R, T in terms of the other letters.

We get
$$P = \frac{100I}{RT};$$

$$R = \frac{100I}{PT};$$

and
$$T = \frac{100I}{PR}.$$

Example 1. *Find the simple interest on £325 in 5 years at 3% per annum.*

$$I = \frac{PRT}{100} = £\frac{325 \times 5 \times 3}{100} = £\frac{195}{4} = £48.75$$

Example 2. *Find at what rate per annum simple interest £525 will amount to £588 in 3 years.*

$$I = A - P = £588 - £525 = £63.$$

$$R = \frac{100I}{PT} = \frac{100 \times 63}{525 \times 3} = \frac{4 \times 63}{21 \times 3} = 4\%.$$

To find the principal, given the amount

Example 3. *A sum of money invested at 3% per annum simple interest amounts after 4 years to £280. Find the sum invested.*

The interest in 4 years is 12% of the principal.
The amount is therefore 112% of the principal.

$$\therefore 112\% \text{ of the principal} = £280$$

and   the principal $= \frac{100}{112}$ of £280 = £250.

Exercise 7.1a

1. Find the simple interest on £680 in 4 years at 5% per annum.
2. Find the simple interest on £121 in 5 years at 3% per annum.
3. If the simple interest on £280 in 4 years is £39.20, find the rate per cent per annum.
4. If the simple interest on £500 for 3 years is £37.50, find the rate per cent per annum.
5. If the simple interest on £220 at 4% per annum is £44, find the time.
6. If the simple interest on £720.50 at 4% per annum is £72.05, find the time.
7. If the simple interest on a sum of money invested at 3% per annum for $2\frac{1}{2}$ years is £123, find the principal.
8. Find the principal which amounts at simple interest to £218 in 2 years at $4\frac{1}{2}$% per annum.

9. Find the principal which amounts at simple interest to £530 in 3 years at 2% per annum.

10. Find the principal which amounts at simple interest to £585.80 in 4 years at 4% per annum.

Exercise 7.1b

1. Find the simple interest on £505 in 4 years at 4% per annum.

2. Find the simple interest on £220 in $2\frac{1}{2}$ years at 4% per annum.

3. If the simple interest on £192 in $2\frac{1}{4}$ years is £12, find the rate per cent per annum.

4. If the simple interest on £205 in 4 years is £28.70, find the rate per cent per annum.

5. If £210 amounts to £235.20 at 3% per annum, find the time.

6. If £168 amounts to £189 at 5% per annum, find the time.

7. If the simple interest on a sum of money invested at $2\frac{1}{2}$% per annum for 6 years is £127.20, find the sum.

8. Find the principal which amounts at simple interest to £840 in 3 years at 4% per annum.

9. Find the principal which amounts at simple interest to £103.70 in $2\frac{1}{2}$ years at $2\frac{1}{2}$% per annum.

10. Find the principal which amounts at simple interest to £597.50 in $3\frac{1}{4}$ years at 6% per annum.

COMPOUND INTEREST

Calculation of compound interest

In calculating the compound interest, the principal at the beginning of each year must be found. The work is best set out in tabular form. There is a useful device for finding 4% (for example) of a sum of money. Multiply the sum of money by 4, at the same time moving the decimal point two places to the left. This move of the decimal point is equivalent to division by 100.

Example 4. Find the compound interest on £450 in 3 years at 4% per annum.

Principal for first year	=	£450
Interest for first year	=	18
Principal for second year	=	468
Interest for second year	=	18.72
Principal for third year	=	486.72
Interest for third year	=	19.4688
Final amount	=	506.1888
Original amount	=	450
Interest	=	£56.1888

The interest to the nearest penny is £56.19.

Example 5. Find the compound interest on £285.38 in 4 years at 2½% per annum.

2½% is $\frac{1}{40}$ and so the amount at the beginning of any one year is divided by 40 to give the interest for that year.

To find the interest to the nearest penny, we must obtain the result correct to 2 decimal places of £1. To ensure this keep 4 decimal places in the working and then correct to 2 d.p.

Principal for the first year	=	£285.38
Interest for the first year	=	7.1345
Principal for the second year	=	292.5145
Interest for the second year	=	7.3129
Principal for the third year	=	299.8274
Interest for the third year	=	7.4957
Principal for the fourth year	=	307.3231
Interest for the fourth year	=	7.6831
Final amount	=	315.0062
Original amount	=	285.38
Interest	=	29.6262
	=	£29.63 (to 2 decimal places)

The interest to the nearest penny is £29.63.

Formula for compound interest

Supposing that the rate of interest is 4% per annum, the amount at the end of any particular year will be 104% of the amount at the beginning of that year. At the end of the first year, the amount is $1.04P$ where P is the principal; at the end of the second year, the amount is $(1.04)^2 P$ and so on. Similarly if the rate of interest is r% per annum, the multiplying factor is $\left(1 + \dfrac{r}{100}\right)$. The amount after n years is $P\left(1 + \dfrac{r}{100}\right)^n$.

$$\therefore A = P\left(1 + \frac{r}{100}\right)^n.$$

This formula may be used in place of the tabular method.

Exercise 7.2a

(Give answers to the nearest penny)

1. Find the compound interest on £200 in 3 years at 4% per annum.
2. Find the compound interest on £360 in 4 years at 3% per annum.
3. Find the compound interest on £186.61 in 2 years at 2½% per annum.
4. Find the compound interest on £370.50 in 3 years at 4% per annum.
5. Find the compound interest on £300 in 3 years at 3% per annum.

Exercise 7.2b

(Give answers to the nearest penny)

1. Find the compound interest on £200 in 4 years at 5% per annum.
2. Find the compound interest on £280 in 2 years at 3% per annum.
3. Find the compound interest on £312.78 in 3 years at $2\frac{1}{2}$% per annum.
4. Find the compound interest on £181.31 in 2 years at 6% per annum.
5. Find the compound interest on £408.38 in 3 years at 3% per annum.

Periodical borrowing or repayment

If the principal varies due to fresh borrowing or repayment at the end of a year, the amount at the beginning of each year must be calculated separately and the interest for this amount worked out. The method is illustrated in the following examples.

Example 6. A man invests £100 on January 1st of each year at 4% per annum. Find the total amount of his investment at the end of 3 years.

He invests	£100
Interest for first year	= 4
Value at end of first year	= 104
He invests another £100	= 100
Value at beginning of second year	= 204
Interest for second year	= 8.16
Value at end of second year	= 212.16
He now invests	100
Value at beginning of third year	= 312.16
Interest for third year	= 12.4864
Value after three years	= £324.6464

The amount of his investment, to the nearest penny, is £324.65.

Example 7. A man borrows £600 from a corporation at 3% per annum interest. He repays £200 at the end of each year. How much does he still owe after the third repayment?

He borrows	£600
Interest for the first year	= 18
Amount now owing is	618
He repays	200
Amount owing at the beginning of second year	= 418
Interest for the second year	= 12.54
Amount owing after this is	430.54
He repays	200
Amount owing at beginning of third year	= 230.54
Interest for the third year	= 6.9162
Amount now owing	237.4562
He repays	200
Amount owing after third repayment	= £37.4562

The amount still owing is £37.46, to the nearest penny.

Exercise 7.3: Miscellaneous

1. A town borrows £2 000 000 at 5% per annum and repays £500 000 at the end of each year. How much is still owing after the fourth repayment?

2. The value of a car depreciates by 12% each year. A man pays £800 for his car. What will be its value after 3 years? Give your answer to the nearest pound.

3. Find to the nearest pound the amount of £100 invested at 4% per annum compound interest after 20 years.

4. A man invests £1000 at 4% per annum. The interest is invested at the same rate. What is the value of his holding after 4 years?

5. A saving certificate costing 75 pence is worth £1.02½ at the end of 10 years. To what rate per cent per annum compound interest is this equivalent?

6. How long would it take a sum of money to double itself at 5% per annum compound interest?

7. A bank lent £275 to a client at 4% per annum and £220 to another client at 5½% per annum. What is the bank's average percentage return on its capital?

8. A man borrows £700 at 4% per annum. He repays £200 each year. Find during which year the debt will be cleared.

9. A father leaves a legacy of £5000 to his son. After death duties of 10% have been paid, the money is invested for 4 years at 3% compound interest. What is now the value of the son's holding to the nearest pound?

10. A man borrowed £100 at 6½% per annum and had to pay £2.14 interest when he repaid the loan. For how many days did he borrow the money?

11. Smith and Brown invest £165 and £192 respectively at the same rate of interest. When the value of Smith's holding is £184.80, what is the value of Brown's?

12. £100 invested at compound interest amounts to £130.50 after 5 years. What is its value after another 5 years?

13. In order to pay his son's school fees, a father invests £5000 at 4% per annum. Assuming that the money is invested at the beginning of the boy's school career and that £1500 is withdrawn at the end of each year for the school fees, find how much is left after 3 years.

14. A man buys a house for £25 000 and after allowing for annual repairs of £400 wishes to make a 5% per annum return on his money. What should be the annual rent of the house? The tenant pays rates which are ¼ of the rent. He spends £4000 on furniture and sublets to get a clear return of 6% on his capital outlay. At what annual charge does he sublet?

15. How long will it take a sum of money invested at 5% per annum simple interest to increase in value by 40%?

16. Find the simple interest on £340.62½ in 3 years at 4½% per annum.

17. Find the simple interest on £280 for 3½ years at 3½% per annum.

18. A sum of £312.50 is invested at 4% per annum simple interest. After how many years will this sum amount to £500?

19. The population of a town increases by 4% each year. If the population at one census is 42 500 what will it be at the next census, 4 years later?

20. Find the sum of money which yields £7.15 simple interest in 4 years at $2\frac{3}{4}$% per annum.

21. A man buys a house valued at £18 000. He pays 30% of its value immediately, £6000 eighteen months later and the remainder after a further eighteen months. Simple interest is charged at $4\frac{1}{2}$% per annum. Find how much he pays altogether.

22. Find the sum of money which will amount to £826 at $4\frac{1}{2}$% per annum simple interest in 4 years.

23. If £400 invested for 5 years yields a simple interest of £38, what will be the interest on £240 invested at the same rate for $7\frac{1}{2}$ years?

24. Machinery worth £5200 was bought 3 years ago. What is its present value if depreciation is allowed for at 15% per annum?

25. Machinery valued at £4800 is revalued 3 years later at £2800. To what annual percentage depreciation is this equivalent?

26. Find the sum of money which invested at 4% per annum compound interest amounts after 5 years to £120.

27. A moneylender charges $1\frac{1}{2}$ pence per week interest on every £1 borrowed. To what annual rate per cent is this equivalent and what would be the interest on £1000 in 4 years at this rate?

28. A sum of money at compound interest increases from £x to £y in t years. What will be the amount of the investment after a further period of $3t$ years?

29. A sum of £100 is invested at the rate of 4% per annum. Find the increase in interest during a year if the interest is calculated each month instead of at the end of the year.

30. A man earns £2400 and invests 15% of it each year at 5% per annum simple interest. What is his total interest in the first three years?

31. A man invests £100 each year at 3% per annum compound interest. What is the value of his investment just before he invests the fourth £100?

32. A sum of money invested at 4% per annum simple interest amounts after 5 years to £720. What would be the amount of the same sum invested at $3\frac{1}{2}$% per annum for 6 years?

33. Find the simple interest on £450 at 3% per annum for 2 years 8 months.

34. A sum of money invested at 4% per annum simple interest amounts after 5 years to £960. How long will it take the same sum to amount to £1080 at $3\frac{1}{2}$% per annum simple interest?

35. The simple interest on a sum of money invested for 4 years at $3\frac{1}{2}$% per annum exceeds the simple interest on the same sum invested at $4\frac{1}{2}$% per annum for 3 years by £1.34. What sum was invested?

8 Rates and Taxes, Stocks and Shares

RATES

Every flat, house and office in a town is given by the Inland Revenue Valuation Office what is called a 'rateable value'. This rateable value depends both upon the size of the building and on its position. The rates of a town, which are fixed each year to bring in the amount of money estimated to be necessary, are levied at so much in the pound of this rateable value. The money brought in by the rates is used for the administration of the town to cover all costs of such things as education, road sweeping, town parks, town sanitation, libraries and so on. A penny rate means a charge of 1 penny in the pound on the rateable value of the town. The income of a town is changed as necessary by altering the rate and not the rateable value, although this may be reviewed from time to time.

Example 1. *Find the annual rates at $74\frac{1}{2}$ pence in the £ on a house whose rateable value is £340.*

Annual rates are $340 \times 74\frac{1}{2}$ pence.

$$340 \times 74\frac{1}{2} = 25\ 330$$

The annual rates are therefore 25 330 pence or £253.30.

Example 2. *The rateable value of a town is £438 400. The council has to estimate for an increase of £15 600 in education costs. What increase in the rates (to the nearest half penny) is necessary?*

£438 400 is the equivalent of a rate of 100 pence.

£15 600 is the equivalent of a rate of $\dfrac{100 \times 15\ 600}{438\ 400}$ pence

$$= 3.56 \text{ pence.}$$

To the nearest half penny, the increase necessary is $3\frac{1}{2}$ pence.

Exercise 8.1a

1. Find the rates at 86 pence in the pound on a house of rateable value £216.
2. Find the rates of 92 pence in the pound on a house of rateable value £312.
3. The rateable value of a town is £2 840 000 and its expenditure is £2 120 000. What are the rates, to the nearest penny?
4. The expenditure of a town is £3 274 000 and its rates are 93 pence in the

pound. The cost of the library is £124 000. What rate is necessary for the library's upkeep?

5. The annual rates at 83 pence in the pound on a house are £215.80. What is the rateable value of the house?

Exercise 8.1b

1. Find the rates at 94 pence in the pound on a house of rateable value £376.
2. Find the rates at $89\frac{1}{2}$ pence in the pound on a house of rateable value £264.
3. The rateable value of a town is £4 260 000 and its annual expenditure is £3 275 000. What rate must the town declare?
4. To raise an income of £4 176 000 a town declares a rate of 87 pence in the pound. What is the rateable value of the town?
5. The annual rates at 73 pence in the pound on a house are £262.80. What is the rateable value?

INCOME TAX

The Central Government of each country has to tax its subjects to pay for the services they expect, the Armed forces to defend the country, the social services, Education, Health and other Welfare services for the benefit of the country, One of the chief ways of raising revenue is Income Tax.

Everybody with an income above a certain minimum level has to pay part of it to the Government in Income Tax; this part is usually quoted as a percentage, e.g. 33%, which means that of every £, 33p is taken as tax. The level of tax varies from year to year, and is usually fixed, in Britain, each April in the Budget. At present (1976) the rates are

> 33% on the first £4500 of taxable income,
> 38% on the next £500 of taxable income,
> 43% on the next £1000,
> 48% on the next £1000,

with higher rates going up to 83% of taxable income.

Taxable income

Everybody who pays tax has certain allowances he can set against his income, before whatever remains is subject to tax. A married man may have to support his wife and family; men and women are encouraged to save by being allowed to deduct first the contributions they make to their pensions, and many other allowances can be obtained. When these have been deducted, whatever remains is called **taxable income** and it is on this that Income Tax has to be paid.

Example 3. A man with an annual salary of £3000 has allowances of £1200. If Income Tax is 35%, how much tax does he pay each year?

His annual salary is   £3000
His allowances are   £1200
So his taxable income is £1800
He pays tax at 35% on this, so his annual Income Tax is

$$\frac{35}{100} \times 1200 = £420.$$

Example 4. The Chairman of a large nationalised industry has a salary of £20 000. He has allowances of £5000. Find the annual tax he pays if the rates are   33% on the first £4000 of taxable income,
                      43% on the next £4000,
                      53% on the next,
and   63% on any remaining.

His annual salary is   £20 000.

His allowances are   £5000.

So his taxable income is £15 000.

On the first £4000 he pays $\frac{33}{100} \times 4000$, i.e. £1320,

On the next £4000 he pays $\frac{43}{100} \times 4000$, i.e. £1720,

On the next £4000 he pays $\frac{53}{100} \times 4000$, i.e. £2120,

and on the remaining £3000 he pays $\frac{63}{100} \times 3000$, i.e. £1890.

Thus the total he pays in tax is £1320 + £1720 + £2120 + £1890, i.e. £7050.

Exercise 8.2a

In these exercises, use the rates of Income Tax given on page 74.
1. A man with an annual salary of £4200 has allowances of £1400. How much does he pay each year in Income Tax?
2. A man with an annual salary of £3700 has allowances of £900. How much does he pay each year in Income Tax? What percentage is that of his total salary?
3. A man whose allowances are £800 finds that he pays 25% of his total salary in Income Tax. Find his total (i.e. gross) salary, and his (net) salary after tax.
4. A woman whose salary is £8000 has allowances of £1820. Find the total she pays each year in Income Tax.
5. A certain doctor has a salary of £11 500. His allowances, which include the expenses of running the practice, are £4800. Find the total he pays each year in tax.

Exercise 8.2b

1. A man with an annual salary of £2800 has allowances of £820. How much does he pay each year in tax?
2. A man with an annual salary of £5000 pays each year £1254 Income Tax. What allowances does he receive?
3. A lorry driver finds that he pays £1320 one year in Income Tax. What was his taxable income that year?
4. A teacher whose salary is £7000 a year has allowances of £1700. Find the Income Tax paid each year.
5. Find the taxable income of a person who pays 35.5% of his taxable income in Income Tax.

Value-added tax, duty

Another tax to raise revenue is Value-Added Tax. This tax requires that when anything is sold, a certain tax is added to the bill and has to be paid by the purchaser; when any bill for services, e.g. solicitors' work, is sent, a certain tax is added which again has to be paid by the customer. Some essential goods, such as food, are 'zero rated', which means that no tax is paid when they are sold; some items, such as petrol, have an extra large Value-Added Tax (VAT).

Duty

Duty is levied when certain goods (principally wines, spirits, perfumes and tobacco) are imported into the country. This used to be one of the main sources of tax, though smuggling reduced the amount of duty that was actually paid.

Example 5. A bill for work done by a garage is £22.00. To this must be added VAT at 8%. Find the VAT and the total bill paid by the customer.

$$8\% \text{ of } £22 = £1.76, \quad \text{so} \quad \text{VAT is } £1.76.$$

The bill paid by the customer is £22.00 + £1.76, i.e. £23.76.

Example 6. The duty on a bottle of wine is 25p, and this is 40% of the cost of the wine in France. How much did the wine cost?

$$40\% \text{ of the cost is } 25p,$$

so $$100\% \text{ of the cost is } \frac{100}{40} \times 25p, \text{ i.e. } 62\tfrac{1}{2}p.$$

Exercise 8.3a

1. If VAT is levied at 8%, find the VAT to be added to a bill of £64.
2. A spade is priced at £6.48, including VAT. If VAT was added at 8%, find the original price of the spade, and the amount of VAT paid.

3. A girl notices that her hairdresser adds 32p VAT to her bill. If VAT was charged at 8%, what was the bill before VAT was added? How much did the girl pay altogether?

4. Duty at 25% of the original price is charged on a gold ring. How much duty was paid if the value, including duty, of the ring is £55?

5. Estate duty of 15% had to be paid on an estate worth £30 000. How much duty was paid, and how much remained of the estate?

Exercise 8.3b

1. If VAT is levied at 25%, find the VAT to be added to a bill of £80.

2. A radio is priced at £21.60, including VAT. If VAT was added at 8%, find the price of the radio before VAT was added, and the amount of VAT paid.

3. The VAT on a certain crate of beer is 28p. If VAT was levied at 8%, find the total cost of the crate of beer, including VAT.

4. The price of an imported car is £1980. If this includes duty at 10% of the original value, find the value of the car before duty, and the duty paid.

5. Find, correct to the nearest penny, the VAT to be added to a bill of £54.35, if VAT is added at 8%.

STOCKS AND SHARES

A private individual often finds it impossible to produce sufficient capital for his business and may make it a public company. He issues shares in the company and invites the public to subscribe for these shares. These shares are issued generally in units of 5 pence, 10 pence, 25 pence, 50 pence or £1. This unit is called the **nominal value** of the share. At the end of the financial year, the company declares a dividend in the form of a percentage of the nominal value and the shareholder is paid the dividend due on his holding.

The buying and selling of shares is controlled by the Stock Exchange. If the company does well and declares a good dividend, the price of the shares will go up; if the company does not flourish, the quoted price of the shares will go down. The fluctuation in the price of the shares does not change the nominal value of the shares which always stays the same. The stock exchange prices for shares are quoted in most daily newspapers.

Preference and ordinary shares

There are several different types of shares of which the most usual are the preference and the ordinary. A preference share is issued at a fixed percentage—for example, Gaumont-British $5\frac{1}{2}$% Preference—and the payment of this dividend is the first charge on the profits of the company. The dividend on an ordinary share is not fixed and is declared by the directors of the company according to the profits made. If the preference dividend takes all or nearly all the profits of the company, the ordinary shareholders may get nothing. On the other hand, in a bumper year the ordinary shareholders may get a much larger dividend than the preference shareholders.

The preference share is therefore a safer holding but not so likely to appreciate (or depreciate) in capital value as is the ordinary share.

Difference between stock and share

If a company issues 25 penny shares, for example, a subscriber has to buy an exact number of these shares. Government securities, on the other hand, are quoted in terms of £100 nominal value and it is possible to buy any amount of stock required. It is possible to buy, say, £77.60 stock of $3\frac{1}{2}\%$ War Loan.

Nominal value and market value

It is very important to appreciate the difference between nominal value and market value. The dividend of a company is always quoted on the nominal value, and the market value is the price which governs the buying and selling of the shares.

Example 7. David Whitehead 5 penny shares are quoted at $7\frac{1}{2}$ pence and declare a dividend of 15%. What does it cost me to buy 4000 shares and what dividend do I expect?

$$\text{The cost} = 4000 \times 7\frac{1}{2} \text{ pence} = £300.$$
$$\text{The dividend} = 15\% \text{ of the nominal value}$$
$$= \frac{15}{100} \text{ of } £200$$
$$= £30.$$

Example 8. War Loan $3\frac{1}{2}\%$ stock is quoted at $37\frac{1}{2}$. What will be the cost of £400 stock and what dividend is due on that stock?

$$\text{Cost of £400 stock} = \frac{400}{100} \times £37.50 = £150.$$
$$\text{Dividend} = \frac{400}{100} \times £3\frac{1}{2} = £14.$$

Exercise 8.4a

1. I buy 300 Courtauld shares at £1.80 and sell at £2.12$\frac{1}{2}$. What is my gain?
2. Standard Motor 25p shares stand at 37$\frac{1}{2}$p and declare a dividend of 20%. What dividend shall I receive on an investment of £300 cash?
3. Lloyds Bank £1 shares stand at £3.50 and declare a dividend of 15%. What sum must I invest to get a dividend of £15?
4. Gaumont-British 5$\frac{1}{2}\%$ Preference shares of £1 nominal value are quoted at £0.72$\frac{1}{2}$. What will be the cost of 200 shares and what dividend is payable on them?

5. $2\frac{1}{2}\%$ Consols are quoted at 65. How much cash must I invest in them to produce a dividend of £50?

6. Imperial Tobacco £1 shares stand at £3.02$\frac{1}{2}$ and declare an interim dividend of $8\frac{1}{4}\%$. What would it cost me to buy 176 shares and what would be my interim dividend?

7. Fairey Aviation £0.50 shares are quoted at £2.10 and pay a dividend of 25%. What dividend shall I get on an investment of £882?

8. A share of nominal value 10p is quoted at 14p and pays 10%. What would be my income on an investment of £280 cash?

9. 4% Victory Loan is quoted at 105. How much cash must I invest to produce a dividend of £40?

10. Cow and Gate 5p shares are quoted at 11p. How many shares can I buy for £550?

Exercise 8.4b

1. I bought enough $3\frac{1}{2}\%$ War Loan at 88 to give me an income of £63. I sell my holding at 93. What is my gain?

2. What income do I get from investing £250 in £0.25 copper shares standing at £0.62$\frac{1}{2}$ if they declare a dividend of 24%?

3. How much cash must I invest in a $6\frac{1}{2}\%$ Preference share, nominal value £1, standing at £0.92$\frac{1}{2}$ to produce an income of £39?

4. I buy £300 stock standing at 90. When I sell, I make a profit of £51. What do I sell at?

5. A share of nominal value 25p stands at 30p and declares a dividend of 8%. How much cash is needed for a dividend of £18?

6. Barclay's Bank shares are quoted at £2.42$\frac{1}{2}$. How many shares can I buy for £388?

7. Foster Clark's £0.50 shares are quoted at £0.42$\frac{1}{2}$. If I invest £170, how much shall I gain if I sell at £0.52$\frac{1}{2}$?

8. A $4\frac{1}{2}\%$ preference share, of nominal value £1, is quoted at £0.92$\frac{1}{2}$. What dividend do I receive from an investment of £370 cash?

9. If I sell £300 4% stock at 90 and reinvest the proceeds in $3\frac{1}{2}\%$ stock at 70, what is my gain in income?

10. I buy 400 shares of nominal value 25p at 32$\frac{1}{2}$p. If the dividend declared is 4%, what is the income derived?

The yield

When an investor buys shares, one of the most important factors is the *yield*, that is the percentage return he gets on the cash invested. If he buys for example $5\frac{1}{2}\%$ preference shares, his yield will be $5\frac{1}{2}\%$ only if the shares stand at par, i.e. at their nominal value. If the shares are higher than their nominal value, his yield will be less than $5\frac{1}{2}\%$; if the shares are lower than the nominal value, his yield will be greater.

Example 9. Find the yield on $3\frac{1}{2}\%$ War Loan standing at 40.

£40 cash buys £100 stock and so gives £$3\frac{1}{2}$ interest.

$$\therefore \quad £100 \text{ cash gives } \frac{£3\frac{1}{2} \times 100}{40} \text{ interest} = \frac{£35}{4} = £8\frac{3}{4}.$$

The yield is $8\frac{3}{4}\%$.

Exercise 8.5a

Find the yield on the following investments:
1. 4% stock at 90.
2. 25p shares quoted at $32\frac{1}{2}$p, when the dividend is 8%.
3. Shares of nominal value 30p quoted at 40p, when the dividend is 10%.
4. 6% stock at 110.
5. £0.50 shares quoted at £0.42$\frac{1}{2}$, when the dividend declared is $4\frac{1}{2}\%$.

Exercise 8.5b

Find the yield on the following investments:
1. $3\frac{1}{2}\%$ stock at 75.
2. Shares of nominal value 10p, quoted at $12\frac{1}{2}$p, when the dividend declared is 8%.
3. $5\frac{1}{2}\%$ Preference shares of nominal value £1 quoted at £1.10.
4. 5% stock at 105.
5. £1 shares quoted at £1.12$\frac{1}{2}$ when the dividend declared is 9%.

Worked example

A worked example follows showing how to deal with the finance of a company.

Example 10. A company has issued 200 000 ordinary shares of £0.10 each and 50 000 4% preference shares of £1 each. In one year, after the company has paid the preference dividend and a dividend of 6% on the ordinary shares, its profits are just sufficient for it to put £2000 to reserve. What are the profits of the company for that year?

Nominal value of preference shares	= £50 000.
4% dividend on this capital	= £2000.
Nominal value of ordinary shares	= 200 000 × £0.10
	= £20 000.
6% dividend on this capital	= £1200.
Total dividend	= £3200.
Reserve	= £2000.

$\therefore$ Total profits are £5200.

Exercise 8.6: Miscellaneous

1. A man can either buy a house at £17 500, which he has to borrow paying 8% per annum interest, and then rates at £225 per annum; or he can rent a flat at £30 per week, inclusive of rates. Which is the cheaper, and by how much each year?

2. A man lives in a house of rateable value £340. He pays a rent of £8.10 a week and rates at $72\frac{1}{2}$p in the £. What is the annual cost of house in rent and rates?

3. When the rateable value of a house was £160, the rates were charged at 99p in the £. One year the rateable value was increased to £360, but the rates were reduced to 64p in the £. How much more did the householder have to pay in rates that year?

4. A man invests £4800 in a Bank which pays 6% per annum interest. If he has to pay Income Tax at 45% on his investment income, how much tax does he pay each year? How much income does he receive after tax? What percentage is this of the money he has invested?

5. Find the net income of a woman from an investment of £5000 cash in $2\frac{1}{2}$% Consols bought at 24, after she has paid tax at 45%.

6. A man is allowed £820 of his income tax free. On the rest, he pays tax at 45%. If he pays £810 in tax, find his gross income.

7. The VAT on a bill is '£6.51, correct to the nearest penny'. If VAT was charged at 8%, what was the value of the original bill before VAT was added?

8. The VAT on a fur coat is '£420.84, correct to the nearest penny'. If the VAT was charged at 25%, between what limits was the value of the coat?

9. £1200 is invested partly at 5% per annum and partly at 6% per annum. The interest in one year is £64. How much is invested at 5%?

10. £1800 is invested partly at 5% per annum and partly at $3\frac{1}{2}$% per annum. The interest in one year is £75. How much is invested at 5%?

11. At what price would $3\frac{1}{2}$% War Loan give a yield of $4\frac{1}{2}$%?

12. A holder of 250 £1 shares in the Imperial Tobacco Company receives a dividend of $8\frac{1}{2}$% together with a bonus of $7\frac{1}{2}$p per share. Tax is deducted on the dividend and bonus at 45p in the pound. What does he receive?

13. A man buys 200 7% preference shares, nominal value £1, at £1.22$\frac{1}{2}$ each. What does he pay for them and what is his net annual income if he pays tax at $32\frac{1}{2}$%?

14. A man holds 200 shares in one company paying a dividend of $3\frac{1}{2}$%, the £ shares of the company standing at £0.80; he also holds 400 shares in another company paying a dividend of 6% whose £ shares stand at £1.25. He sells both holdings and reinvests all the money in 5% shares of nominal value £1 standing at £1.10. What is his decrease in annual income?

15. A man bought 200 shares at £1.75 a share. He received a dividend of £0.22$\frac{1}{2}$ per share and a bonus of one share for every 5 held by him. He sold all his shares at £1.60 each. Find his gain (including his dividend).

ALGEBRA

ALGEBRA FORMULAE

Factors

$$ab + ac = a(b + c)$$
$$a^2 - b^2 = (a - b)(a + b)$$
$$a^2 - 2ab + b^2 = (a - b)(a - b)$$
$$a^2 + 2ab + b^2 = (a + b)(a + b)$$
$$a^2 + b^2 \quad \text{has no factors.}$$
$$a^3 + b^3 = (a + b)(a^2 - ab + b^2)$$
$$a^3 - b^3 = (a - b)(a^2 + ab + b^2)$$

Indices

$$a^m \times a^n = a^{m+n}$$
$$a^m \div a^n = a^{m-n}$$
$$(a^m)^n = a^{mn}$$
$$a^0 = 1$$
$$a^{-n} = \frac{1}{a^n}$$
$$a^{p/q} = (\sqrt[q]{a})^p$$

Solutions of quadratic equations

The solutions of $ax^2 + bx + c = 0$ are

$$x = \frac{-b \pm \sqrt{b^2 - 4ac}}{2a}$$

Logarithms

$$\log x + \log y = \log xy$$
$$\log x - \log y = \log \frac{x}{y}$$
$$\log x^n = n \log x$$
$$\log_b x = \frac{\log_a x}{\log_a b}$$

Arithmetic progressions

nth term is $a + (n - 1)d$.

Sum of the first n terms is $\frac{1}{2}n\{2a + (n - 1)d\}$ or $\frac{1}{2}n(a + l)$.

Geometric progressions

nth term   is $a\,r^{n-1}$.

Sum of the first n terms is $a\,\dfrac{1-r^n}{1-r}$ or $a\,\dfrac{r^n-1}{r-1}$.

9 Factors

Common factor

If two terms have the same factor, this is called a **common** factor. It is often helpful to find the common factor.

Example 1.
(i) $\qquad ab - ac = a(b - c)$
(ii) $\qquad ax^2 - a^2x = ax(x - a)$
(iii) $ab^2c^3 - a^2bc^2 = abc^2(bc - a)$

This often enables us to simplify fractions. Notice that we have to divide the numerator and denominator by the common factor.

Example 2.
(i) $\dfrac{ab - ac}{2a + 3a^2} = \dfrac{a(b - c)}{a(2 + 3a)} = \dfrac{(b - c)}{(2 + 3a)}$

(ii) $\dfrac{2a^2 + 6ab}{9ab^2 - 8a^2} = \dfrac{2a(a + 3b)}{a(9b^2 - 8a)} = \dfrac{2(a + 3b)}{(9b^2 - 8a)}$

Exercise 9.1a

Factorise the following:
1. $4a + 6ab - 8ab^2$.
2. $5h^2 + 10gh - 20g^2h$.
3. $2c^2d - 4cd^2$.
4. $6a^4b^2 - 18a^2b^3$.
5. $8x^2y - 24xy^2$.
6. $6u^2v + 12uv^3$.

Simplify each of the following fractions:
7. $\dfrac{2a + 6b}{6a - 8b}$.
8. $\dfrac{4a - 8b}{6a + 18b}$.
9. $\dfrac{6a - 9b}{4a - 6b}$.
10. $\dfrac{ax - ab}{a^2 - ax}$.
11. $\dfrac{ax}{a^2 - ax}$.
12. $\dfrac{ax + ab}{a^2 + ab}$.

Exercise 9.1b

Factorise the following:
1. $ax + ay + az$.
2. $ax - 2ay + 3az$.
3. $a^2x - 4ax + 3ax^2$.
4. $ab^2c - abc^2 - b^2c^2$.
5. $4a^2x^2 + 8a^3x^3$.
6. $6abc - 9a^2c$.

Simplify each of the following fractions:
7. $\dfrac{6x + 18y}{4x - 12y}$.
8. $\dfrac{5x + 10y}{10x - 5y}$.
9. $\dfrac{8x}{4x - 12y}$.
10. $\dfrac{xy - xz}{x^2 - xz}$.
11. $\dfrac{xy}{x^2 - xy}$.
12. $\dfrac{x^2 + xy}{x^2 - xy}$.

Difference of two squares

The fundamental identity is $A^2 - B^2 = (A + B)(A - B)$.
N.B. $A^2 + B^2$ has no rational factors.

Example 3. Factorise $9a^2 - 16x^2$.

The expression is the difference between the squares of $3a$ and $4x$.
The factors are $(3a + 4x)(3a - 4x)$.

Example 4. Factorise $(a - b)^2 - c^2$.

The expression is the difference between the squares of $(a - b)$ and c.
The factors are $(a - b + c)(a - b - c)$.

Example 5. Factorise $c^2 - (x - y)^2$.

The expression is the difference between the squares of c and $(x - y)$.
The factors are $[c + (x - y)][c - (x - y)]$ or $(c + x - y)(c - x + y)$.

Example 6. Factorise $16(a - b)^2 - 25(x - y)^2$.

The expression is the difference between the squares of $4(a - b)$ and
$5(x - y)$.
The factors are $[4(a - b) + 5(x - y)][4(a - b) - 5(x - y)]$
$\quad$ or $(4a - 4b + 5x - 5y)(4a - 4b - 5x + 5y)$.

Example 7. Factorise $9 - 36x^2$.

There is an obvious factor, 9.
The factors are $9(1 - 4x^2)$ or $9(1 + 2x)(1 - 2x)$.

Exercise 9.2a

Factorise:

1. $x^2 - 16$.

2. $p^2 - 9q^2$.

3. $25x^2 - y^2$.

4. $25a^2 - 16b^2$.

5. $a^2 - 4(b - c)^2$.

6. $9a^2 - 4(b - c)^2$.

7. $25(a - b)^2 - c^2$.

8. $(a - b)^2 - (c - d)^2$.

9. $4(a - b)^2 - (c - d)^2$.

10. $1 - 9x^2$.

11. $4 - 25a^2$.

12. $1 - (a - b)^2$.

13. $1 - 9(a - b)^2$.

14. $16 - (a - b)^2$.

15. $25 - 16(a - b)^2$.

16. $36 - z^2$.

17. $108 - 3z^2$.

18. $z^4 - 1$.

19. $x^4 - 16$.

20. $4(a - b)^2 - 25(c - d)^2$.

Exercise 9.2b

Factorise:

1. $x^2 - 49$.

2. $a^2 - 64b^2$.

3. $25p^2 - 64q^2$.

4. $p^2 - 9q^4$.

5. $1 - 49x^2$.

6. $9(a - b)^2 - c^2$.

7. $(a - b)^2 - 9(c - d)^2.$

8. $49 - 4x^2.$

9. $p^2 - 49(r - s)^2.$

10. $9 - (a - b)^2.$

11. $1 - 16x^4.$

12. $7 - 63x^2.$

13. $16p^2 - 49q^2.$

14. $36 - 25z^2.$

15. $16 - 49(a - b)^2.$

16. $x^8 - 16.$

17. $25(a - b)^2 - 49(c - d)^2.$

18. $a^2 - 81b^4.$

19. $(2a - b)^2 - 9(3c - d)^2.$

20. $16(3a + 2b)^2 - 25(p + 2q)^2.$

Trinomials

A trinomial is a three-termed expression such as $x^2 + 9x + 20$. Other examples are $3x^2 - 2xy - y^2$, $x^4 + 2x^2y + y^2$ and $1 - 3x - 2x^2$. The method of factorisation is that of trial and error.

The product of $(x + 4)$ and $(x + 5)$ is $x^2 + 9x + 20$.

Conversely, the factors of $(x^2 + 9x + 20)$ are $(x + 4)$ and $(x + 5)$.

To factorise an expression, we must find quantities which, when multiplied together, give the original expression.

When factorising, first look for an obvious factor.

Example 8. $ax^2 + 9ax + 20a = a(x^2 + 9x + 20) = a(x + 4)(x + 5).$

When we have factorised an expression, make sure that none of our factors will factorise further.

Example 9. $9x^2 - 81 = 9(x^2 - 9) = 9(x + 3)(x - 3).$

(a) *When the coefficient of x^2 is unity*

Example 10. Factorise $x^2 + 9x + 20$.

Look for two numbers whose product is 20 and whose sum is 9. These are obviously 5 and 4. The factors are therefore $(x + 4)(x + 5)$.

Example 11. Factorise $x^2 - 2x - 15$.

Look for two numbers whose product is -15 and whose sum is -2. These numbers are -5 and 3. The factors are $(x - 5)(x + 3)$.

(b) *When the coefficient of x^2 is not unity*

Example 12. Factorise $6x^2 + 11x - 10$.

Try pairs of factors of 6 and pairs of factors of -10, as shown in the following three cases:

$$\begin{array}{ccc}
\begin{vmatrix} 6 & 5 \\ 1 & -2 \end{vmatrix} & \begin{vmatrix} 3 & 5 \\ 2 & -2 \end{vmatrix} & \begin{vmatrix} 3 & -2 \\ 2 & 5 \end{vmatrix} \\
-12 + 5 & -6 + 10 & 15 - 4 \\
= -7 & = +4 & = +11
\end{array}$$

Cross multiply these numbers and add the two products. Look for an arrangement in which the sum of the products is equal to the coefficient of x. i.e. $+ 11$. The third arrangement satisfies this condition and the factors are $(3x - 2)(2x + 5)$.

Similarly the factors of $6x^4 + 11x^2y - 10y^2$ are $(3x^2 - 2y)(2x^2 + 5y)$.

Exercise 9.3a

Factorise:

1. $x^2 + 5x + 6$.
2. $x^2 - 8x - 20$.
3. $x^2 - x - 6$.
4. $x^2 - 5x + 6$.
5. $x^2 - 8x + 15$.
6. $x^2 - 4x - 12$.
7. $x^2 + 11x + 18$.
8. $x^4 + 10x^2 + 24$.
9. $x^2 + 4xy + 3y^2$.
10. $x^4 - 4x^2y - 5y^2$.
11. $2x^2 - x - 3$.
12. $2x^2 - xy - 6y^2$.
13. $3x^2 - 7x - 6$.
14. $5x^2 + 17xy + 6y^2$.
15. $6x^2 - 19x + 10$.
16. $6 - x - x^2$.
17. $12 + x - 6x^2$.
18. $2x^2 + 11x + 15$.
19. $2x^2 - x - 15$.
20. $12x^2 + 7x - 10$.

Exercise 9.3b

Factorise:

1. $x^2 + 6x + 8$.
2. $t^2 - 7t - 18$.
3. $u^2 - 5u - 6$.
4. $v^2 + 7v + 10$.
5. $y^2 - 2y - 24$.
6. $x^2 - 10xy - 11y^2$.
7. $x^2 + 10xy + 16y^2$.
8. $x^4 + 5x^2 + 4$.
9. $x^2 + 6xy + 5y^2$.
10. $z^4 - 2z^2x - 8x^2$.
11. $2p^2 - 5p - 7$.
12. $6a^2 + 19ab + 10b^2$.
13. $6l^2 - 17lm + 12m^2$.
14. $7x^2 - 19x - 6$.
15. $12x^4 + 11x^2 + 2$.
16. $12 - x - x^2$.
17. $3 + x - 2x^2$.
18. $12y^2 + 11y + 2$.
19. $15 + x - 2x^2$.
20. $10 - 7x - 12x^2$.

Grouping

To factorise an expression containing four terms, group in two pairs so that each pair has a common factor.

Find the other factor by division.

N.B. This method can be used only when there is a common factor.

Example 13. Factorise $ax + ay - bx - by$.
$$ax + ay - bx - by = (ax + ay) - (bx + by) = a(x + y) - b(x + y)$$
$$= (x + y)(a - b).$$

The common factor, $(x + y)$, should be written first and the second factor, $(a - b)$, is found by division.

Example 14. Factorise $x^2 - y^2 - 6x + 6y$.
$$x^2 - y^2 - 6x + 6y = (x - y)(x + y) - 6(x - y)$$
$$= (x - y)(x + y - 6).$$

Example 15. Factorise $ax - ay + 6y - 6x$.

$$ax - ay + 6y - 6x = a(x - y) + 6(y - x).$$

Here the common factor is not quite obvious, but remember that

$$(y - x) = -(x - y).$$

So $(x - y)$ is a factor. The other factor by division is $(a - 6)$.

$$\therefore ax - ay + 6y - 6x = (x - y)(a - 6).$$

Exercise 9.4a

Factorise:

1. $h(x + y) + (m + n)(x + y)$.
2. $px + pq - 6x - 6q$.
3. $x + y - ax - ay$.
4. $cx - dx + dq - cq$.
5. $ab + xy - ay - bx$.
6. $a^3 - a^2 - a + 1$.
7. $ah + bh + ch + ap + bp + cp$.
8. $x(2a - b) + 2a - b$.
9. $x(2a - b) - 2a + b$.
10. $ax^2 - ay^2 + bx^2 - by^2$.
11. $ab + ac - (b + c)^2$.
12. $ab - 2ac - 3b + 6c$.
13. $x^2 - y^2 - 6x - 6y$.
14. $x^2 - y^2 - 6x + 6y$.
15. $x^2 - y^2 - x - y$.
16. $x^2 - y^2 - x + y$.
17. $x^2 - (y + 4)x + 4y$.
18. $x^2 - (y - 5)x - 5y$.
19. $(a + b)(c + d) + a + b$.
20. $(a + b)(c + d) - a - b$.

Exercise 9.4b

Factorise:

1. $ap - 2aq + bp - 2bq$.
2. $ap - 2aq + 2bq - bp$.
3. $(a + b)(x + y) - 2x - 2y$.
4. $(a + b)(x + y) - (x + y)^2$.
5. $a^3 + a^2 + a + 1$.
6. $3ab + 3abc + 2c + 2c^2$.
7. $p^2 - q^2 + 4(p - q)$.
8. $p^2 - q^2 - 5p + 5q$.
9. $h^2 - k^2 - 6h - 6k$.
10. $a^2 - 4b^2 - ac - 2bc$.
11. $a^2 - 9b^2 - ax + 3bx$.
12. $a^3 + a^2 + a + a^2b + ab + b$.
13. $x - 1 - (x - 1)^2$.
14. $a(h - k) - b(k - h)$.
15. $a(h^2 - k^2) - b(k - h)$.
16. $x(y - 1) - y + 1$.
17. $a^2 - ab + ca - cb$.
18. $a^2 - a(b + c) + bc$.
19. $x^2 - x(2b + c) + 2bc$.
20. $z^2 - z(2a - b) - 2ab$.

Sum and difference of two cubes

The fundamental identities are

$$A^3 + B^3 = (A + B)(A^2 - AB + B^2)$$
and
$$A^3 - B^3 = (A - B)(A^2 + AB + B^2).$$

Example 16. Factorise $x^3 + 8y^3$.

$$x^3 + 8y^3 = (x)^3 + (2y)^3 = (x + 2y)(x^2 - 2xy + 4y^2).$$

Example 17. Factorise $8z^6 - 27x^3y^3$.

$$8z^6 - 27x^3y^3 = (2z^2)^3 - (3xy)^3 = (2z^2 - 3xy)(4z^4 + 6xyz^2 + 9x^2y^2).$$

Exercise 9.5a

Factorise:

1. $8a^3 + b^3$.   2. $z^3 + 1$.   3. $z^6 + 1$.   4. $z^6 - 1$.
5. $a^3 - b^3c^3$.   6. $a^3 - 27b^3c^3$.   7. $64h^3 - k^3$.   8. $64h^3 + 27$.
9. $a^3 - 8$.   10. $8 - 27b^3$.

Exercise 9.5b

Factorise:

1. $27a^3 + b^3$.   2. $z^3 + 8$.   3. $z^6 + 8$.   4. $z^6 - 8$.
5. $h^3 - 8m^3n^3$.   6. $1 - 27b^3c^3$.   7. $64 - h^3$.   8. $64h^3 - 27$
9. $125a^3 - 1$.   10. $8 - 125x^3$.

Expressions containing five terms

To factorise an expression containing five terms (or a four-termed expression which will not factorise by grouping), **look for the trinomial**.

Example 18. Factorise $a^2 + 2ab + b^2 + a + b$.

$$(a^2 + 2ab + b^2) + a + b = (a + b)^2 + (a + b) = (a + b)(a + b + 1).$$

Example 19. Factorise $a^2 - ab - 3a + 2 + 2b$.

The trinomial is $a^2 - 3a + 2$.

$$\begin{aligned}
a^2 - ab - 3a + 2 + 2b &= (a^2 - 3a + 2) - b(a - 2) \\
&= (a - 1)(a - 2) - b(a - 2) \\
&= (a - 2)(a - 1 - b).
\end{aligned}$$

Example 20. Factorise $a^2 - b^2 - 4a + 4$.

The trinomial is $a^2 - 4a + 4$.

$$\begin{aligned}
a^2 - b^2 - 4a + 4 &= (a^2 - 4a + 4) - b^2 \\
&= (a - 2)^2 - b^2 \\
&= (a - 2 + b)(a - 2 - b).
\end{aligned}$$

Exercise 9.6a

Factorise:

1. $a^2 + 2a + 1 - ax - x$.   2. $x^2 + ax + 4x + 3a + 3$.
3. $x^2 - z^2 + 2xy + y^2$.   4. $4z^2 - 4x^2 - 4x - 1$.

5. $z^2 + 3z + 2 + az + 2a.$
6. $ax + a - x^2 - 4x - 3.$
7. $x^2 + 2px - y^2 + p^2.$
8. $4x^2 - 4xy + 5x + 1 - y.$
9. $1 - a^2 + 2ab - b^2.$
10. $x^2 + 7x + yx + 10 + 5y.$

Exercise 9.6b

Factorise:

1. $a^2 + 5a + 6 + ax + 2x.$
2. $x^2 + ax + 3x + a + 2.$
3. $x^2 - z^2 + 4x + 4.$
4. $z^2 - x^2 - 6x - 9.$
5. $p^2 + pq - 5p - 6q - 6.$
6. $ay + a - y^2 - 2y - 1.$
7. $x^2 + 6px - 9y^2 + 9p^2.$
8. $2x^2 + 2xy + 3x + 1 + y.$
9. $16 - a^2 - 2ab - b^2.$
10. $x^2 + 7x + xy + 12 + 4y.$

Exercise 9.7: Miscellaneous

1. Factorise $y^2 - y - 42$.
2. Evaluate $100^2 - 99^2$.
3. Evaluate $96^2 - 36$.
4. Factorise $2a^2 - 15ab + 18b^2$.
5. What number must be added to $(x^2 + 6x)$ to make the result a perfect square?
6. What number must be added to $(x^2 - 18x)$ to make the result a perfect square?
7. What number must be added to $(4x^2 - 12x)$ to make the result a perfect square?
8. What number must be added to $(9x^2 + 12x)$ to make the result a perfect square?
9. Factorise $ax + 3a - xy - 3y$.
10. Factorise $ax^2 + 2ax + a + x + 1$.
11. Factorise $1 - 9p^2 - 6pq - q^2$.
12. What is the square root of $(x^2 + 2x + 1)(x^2 - 6x + 9)$?
13. Factorise $(x + y)^2 - (xy + 1)^2$.
14. Factorise $(x^2 + x + 1)^2 - (y^2 + y + 1)^2$.
15. Factorise $xy - 8x + 5y - 40$.
16. Simplify $\dfrac{(x^2 - x - 6)(x^2 + 4x + 3)}{(x^2 - 9)(x + 1)}$.
17. Simplify $\dfrac{(x^2 - 16)(x + 2)}{(x^2 - 2x - 8)(x^2 + 5x + 4)}$.
18. Factorise $9a^2 - 6 + \dfrac{1}{a^2}$.
19. Factorise $(R - 2r)^2 - r^2$.
20. Factorise $ab^3 - 8a$.

21. Factorise $xy^3 + 64x$.

22. Factorise $1 - 125a^3b^3$.

23. Factorise $27 + 8a^3b^3c^3$.

24. Factorise $(a - b)^2 - (x - y - z)^2$.

25. Factorise $16c^2 - (a - b - d)^2$.

26. Factorise $(9a - b)^2 - 49(c - d)^2$.

27. Factorise $a^2 - ax + 9a - 7x + 14$.

28. Factorise $(x^2 + 6x + 7)^2 - (3x + 7)^2$.

29. Factorise $(2a^2 + 5a + 6)^2 - (a^2 + a + 3)^2$.

30. Divide $ab^3 - a$ by $b - 1$.

31. Divide $x^3 - 9x$ by $x^2 + 3x$.

32. Factorise $a(a - 1) - x(x - 1)$.

33. Factorise $(ax - by)^2 + (bx + ay)^2$.

34. Factorise $(lx + my)^2 - 3lx - 3my$.

35. Factorise $4\pi(R + r)^3 - 4\pi R^3$.

36. Factorise $x^3 + 5x^2 + 4x + 20$.

37. Factorise $\frac{1}{2}m(v + 2u)^2 - \frac{1}{2}m(v + u)^2$.

38. Find k if $(x + 2)$ is a factor of $x^3 + kx^2 - 4x - 8$.

39. Factorise completely $x^3 - 6x^2 + 11x - 6$, given that $(x - 3)$ is one of the factors.

40. Factorise completely $x^3 + 4x^2 - x - 4$, given that $(x + 1)$ is one of the factors.

Factor theorem

It was difficult to factorise the expressions containing terms in x^3, and it is of course harder to factorise expressions containing higher powers of x. Suppose though we know the factors, so that

$$x^3 + ax^2 + bx + c = (x - p)(x - q)(x - r).$$

Then since the R.H.S. of the equation is zero when $x = p$ or when $x = q$ or when $x = r$,

$$p^3 + ap^2 + bp + c = 0$$

similarly for q and r. So that if we can find a value of x say $x = k$, which makes the L.H.S. zero, then $(x - k)$ is a factor of the L.H.S. The other quadratic factor can be found by division, and then its linear factors, if any found.

Example 21. Find the factors of $x^3 - 6x^2 + 11x - 6$.

Put $x = 1$, $x^3 - 6x^2 + 11x - 6 = 1 - 6 + 11 - 6 = 0$.
Therefore $(x - 1)$ is a factor.
By long division:

$$x - 1) \overline{x^3 - 6x^2 + 11x - 6} (x^2 - 5x + 6$$

$$\underline{x^3 - x^2}$$

$$\quad\quad - 5x^2 + 11x$$

$$\quad\quad \underline{- 5x^2 + 5x}$$

$$\quad\quad\quad\quad 6x - 6$$

$$\quad\quad\quad\quad \underline{6x - 6}$$

$$\therefore \; x^3 - 6x^2 + 11x - 6 = (x - 1)(x^2 - 5x + 6)$$
$$= (x - 1)(x - 2)(x - 3).$$

Remainder Theorem

A useful extension of this is the Remainder Theorem. Suppose that when $(x - p)$ is divided into $x^3 + ax^2 + bx + c$, there is a remainder R. Then $x^3 + ax^2 + bx + c = (x - p)\,Q + R$, where Q is the quotient, in this case an expression containing x^2. Putting $x = p$ in both sides of the equation

$$p^3 + ap^2 + bp + c = R.$$

Notice that the degree of the remainder is less than that of the divisor $(x - p)$, for if it is of the same degree or higher, then it will be possible to divide further. Since $(x - p)$ is of degree one, then the remainder must be of degree 0, that is, a constant.

Example 22. When divided by $(x + 1)$, the expression $ax^2 + bx + c$ leaves remainder 1. When divided by $(x + 2)$, the same expression leaves remainder -1. Given that $(x - 1)$ is a factor of the expression, find a, b and c.

Put $x = -1$:	$a - b + c = 1$	(i)
Put $x = -2$:	$4a - 2b + c = -1$	(ii)
Put $x = 1$:	$a + b + c = 0$	(iii)
Subtract (ii) from (i):	$-3a + b = 2.$	
Subtract (iii) from (i):	$-2b = 1.$	$\therefore \; b = -\tfrac{1}{2}.$
Substitute:	$-3a - \tfrac{1}{2} = 2$	
	$\therefore \; 3a = -2\tfrac{1}{2} \;$ and $\; a = -\tfrac{5}{6}.$	
Substitute in (iii):	$-\tfrac{5}{6} - \tfrac{1}{2} + c = 0.$	$\therefore \; c = \tfrac{4}{3}.$

Example 23. Find the value of k if $(x + 1)$ is a factor of
$$x^3 + kx^2 + 3x - 2.$$

Put $x = -1$, $-1 + k - 3 - 2 = 0.$
$$k = 6.$$

Exercise 9.8a

1. Find the remainder when $x^3 - 6x + 1$ is divided by $x - 3$.
2. Find the remainder when $x^3 + x^2 + x + 1$ is divided by $x + 3$.
3. Find the factors of $x^3 - 4x^2 - x + 4$.
4. Find the values of a and b if $(x - 1)$ and $(x - 2)$ are both factors of $x^3 + ax^2 + bx - 6$.

5. Find the value of k if $(x + 1)$ is a factor of $x^3 + kx^2 + 6x + 4$.

6. The expression $x^2 + bx + c$ leaves remainder 1 when divided by $(x - 1)$. When the expression is divided by $(x + 1)$, the remainder is 3. Find b and c.

7. Given that $(x + 2y)$ is a factor of $x^3 - 7xy^2 - 6y^3$, factorise the expression completely.

8. Factorise $x^3 - 7x + 6$.

9. Given that $x = 1$ and $x = -1$ are solutions of the equation $x^4 - 3x^3 + x^2 + 3x - 2 = 0$, find the other solutions.

10. Find the value of k if $(x + 1)$ is a factor of $x^3 + 5x^2 + kx + 3$. Then find the other factors of the expression.

Exercise 9.8b

1. Find the remainder when $x^4 - 6x^2 + 1$ is divided by $x - 3$.

2. Find the remainder when $x^3 + 5x^2 + x + 1$ is divided by $x + 3$.

3. Find the factors of $x^3 + 4x^2 + x - 6$.

4. Find the value of k if $(x + 2)$ is a factor of $x^3 + kx^2 - 2x + 4$.

5. The expression $ax^2 + bx + 1$ has remainder 2 when divided by $(x - 1)$. When it is divided by $(x + 1)$, the remainder is 4. Find a and b.

6. Find the values of a and b if $(x + 1)$ and $(x + 2)$ are both factors of $x^3 + ax^2 + bx + 8$.

7. Given that $(x + 3y)$ is a factor of $x^3 + 6x^2y + 11xy^2 + 6y^3$, factorise the expression completely.

8. Factorise $x^3 + 6x^2 + 5x - 12$.

9. Given that $x = 2$ and $x = -2$ are solutions of the equation $x^4 + 3x^3 - 2x^2 - 12x - 8 = 0$, find the other solutions.

10. Find the value of k if $(x - 1)$ is a factor of $x^3 + 4x^2 + kx - 6$. Find also the other factors of the expression.

10 Formulae; Expressions

Constructing formulae

If I walk for 2 hours at 6 km/h, I shall walk 12 km; if I walk for t hours at v km/h, I shall walk vt km. If the distance I walk is denoted by s km, then

$$s = vt$$

The following examples show how other formulae can be obtained, by regarding algebra as a generalised form of arithmetic.

Example 1. A workman is paid £n for each day he works and is fined £q for each day he is absent. If he works for x days of a six-day week, find how much he earns.

He works for x days. For this he is paid $£nx$.
He is absent for $(6 - x)$ days. He is fined $£q(6 - x)$.
His earnings in £ are

$$nx - q(6 - x) = (nx + qx - 6q).$$

Example 2. An author receives in royalties 10% of the selling price on the first 4000 copies of his book and 12½% on any further copies sold. If the book sells at £s a copy, find how much the author receives if x copies of the book are sold, x being greater than 4000.

On the first 4000 copies, he receives 10% of £4000s, i.e. £400s.
On the remaining $(x - 4000)$ copies, he receives 12½% of £$(x - 4000)s$, i.e. £$\frac{1}{8}(x - 4000)s$.
The author therefore receives £400s + $\frac{1}{8}(x - 4000)s$ i.e. £$(\frac{1}{8}x - 100)s$ pence.

Example 3. The postage required for an inland letter of mass 250 g is 20p. For every additional 50 g, 5p is charged. What is the postage required for a letter of mass 250n grammes, where n is greater than 1?

For 250 g, the charge is 20p.
The additional mass is $250(n - 1)$ grammes.
The additional charge is $25(n - 1)$p.
The charge in pence for 250n grammes is therefore $20 + 25(n - 1) = 5(5n - 1)$.

Exercise 10.1a

1. If I buy p books at P pence each and q other books at Q pence each, how many pence change shall I have from a £5 note?

2. The cost of a hotel for an adult is £2x a day. A child is taken at half-price. Find the total bill for a week for a party of 2 adults and 3 children.

3. The radius of the base of a right circular cylinder is r cm and its length is h cm. Find the volume of the cylinder.

4. A circular flower bed of radius x metres is surrounded by a path y metres wide. Find the area of the path in m².

5. Find a formula for the total surface area of a right circular cylinder closed at both ends, given that its radius is r and its length h.

6. A man buys eggs at x pence per dozen. Find the cost in £ of y dozen eggs.

7. Find the area of the walls of a room x metres long, y metres wide and z metres high.

8. Local time is 4 min ahead of Greenwich time for each degree of longitude east of Greenwich. What is the time at a place of longitude (a) $x°$ E., (b) $y°$ W., when it is midday at Greenwich?

9. The average age of x boys in a class is m months. What is the average age in months after a boy of age y years joins the class?

10. A man buys a car for £P and sells it for £Q. What is his loss per cent?

11. A greengrocer buys p tonnes of potatoes at £P per tonne and a further q tonnes at £Q per tonne. He mixes the potatoes and sells them to make a profit of $z\%$. At what price per kg must he sell them?

12. A rectangle on a map has sides h and k cm. If the scale of the map is x cm to 1 km, find in km² the area represented by the rectangle.

13. A car is driven x km at u km/h and a further y km at v km/h. How long does the journey take and what is the average speed?

14. The boundary of an enclosure consists of two parallel sides each of length p metres and two semicircles, each of radius r metres. Find a formula for the area of the enclosure.

15. A car is driven x km a week and does g km per litre of petrol. Find the cost of petrol for the week in pounds, if petrol costs s pence a litre.

Exercise 10.1b

1. A rectangular sheet of metal, x cm by y cm has a square of side z cm cut from each corner. The sheet is then bent to form a tray of depth z cm. Find the volume of the tray.

2. A classroom is x metres long, y metres wide and z metres high. If it is designed to accommodate n boys, find how many m³ of air are allowed per boy.

3. If x cm³ of a substance whose density is g g/cm³ are mixed with y cm³ of another substance of density h g/cm³, write down a formula for the total mass of the mixture.

4. How many castings each of mass x kg y grammes can be made from c kg of metal? (Assume there is an exact number.)

5. A man travels a distance of $2x$ km. He walks for half the distance at v km/h and runs the rest of the way at $2v$ km/h. How long does he take and what is his average speed?

6. A room has dimensions p metres by q metres by r metres high. It has a door x metres by y metres and two windows each c metres by d metres. How many m² of paper are needed for the walls of the room?

7. A man is paid x pence per hour for normal work and double rate for overtime. If he does a 50-hour week which includes w hours of overtime, find an expression for his weekly earnings.

8. During one week, a man posted x letters with 10p stamps and y letters with 8p stamps. Find his postage bill for the week.

9. An exam is taken by x boys and y girls. The boys score an average mark of p, the girls an average mark of q. Find the average mark of all the candidates.

10. In a school are p pupils of average age x years. At the end of the year, q pupils of average age y years leave and r new pupils of average age z years join the school. What is the average age of the school now?

11. A man walks x km at V km/h and finishes his journey by car at $6V$ km/h. If the ratio of the distance walked to that travelled by car is 2: 7, find an expression for the total time taken for the journey.

12. A party of A adults and C children go to a holiday camp for a week's holiday. The cost of a return train ticket is £a for an adult and the cost per day at the holiday camp for an adult is £b. The charges for a child on the train and at the holiday camp are half those for an adult. Find the total cost of the week's holiday.

13. A try at rugby football scores four points and a goal six points. By how many points does a team which scores x tries and y goals beat a team which scores p tries and q goals?

14. A man can dig a trench y metres long in x days working z hours a day. How many hours a day must p men work to finish digging a trench q metres long in s days?

15. A rectangular sheet of metal x metres long and y metres wide has a circular hole cut from it. What is the radius of the hole if the area of the remaining metal is A m²?

Simplifying expressions

Some of the expressions we obtain will be fairly complicated and in several of the previous questions we saw that we could simplify them by finding common factors. Take care not to alter the value of an expression by omitting a common factor, or by omitting the denominator when simplifying fractions.

Example 4. *Simplify* $\dfrac{x}{3} - \dfrac{x}{4} + \dfrac{5x}{6}$.

To add fractions, express them with the same denominators. The L.C.M. of 3, 4 and 6 is 12, so we have

$$\frac{x}{3} - \frac{x}{4} + \frac{5x}{6} = \frac{4x - 3x + 10x}{12}$$

$$= \frac{11x}{12}.$$

Lowest Common Multiple

We see that we have first to find the L.C.M. of the denominators of the fractions. The L.C.M. contains the factors of every denominator, so that we first have to factorise the denominators. Notice that x is not a factor of $x + 1$, nor of $x + a$, whatever the value of a (unless, of course, a is zero).

Example 5. Find the L.C.M. of $x^2 + x$, $x^2 - 1$, $x^2 - x$.

$$x^2 + x = x(x + 1)$$
$$x^2 - 1 = (x - 1)(x + 1)$$
and
$$x^2 - x = x(x - 1),$$
so the L.C.M. is $x(x - 1)(x + 1)$.

Example 6.

$(a - b)(a - c)$
$(b - c)(b - a)$ L.C.M. $= (a - b)(a - c)(b - c)$.
$(c - a)(c - b)$

The sign of the L.C.M. is of no great importance. $(b - a)$ divides into $(a - b)$, the quotient being -1. So we do not need both these factors in the L.C.M. We could have chosen $(b - a)$ instead of $(a - b)$. In fact, the L.C.M. could equally well have been $(b - a) (c - a) (c - b)$ or $(a - b) (c - a) (b - c)$.

Exercise 10.2a

Find the L.C.M. of:

1. $(x^2 - 1)$, $(x^2 - 3x + 2)$.
2. $6(x - 1)$, $3(x + 1)$.
3. $x^2 - a^2$, $x^2 - ax - bx + ab$.
4. $(x - a)^2$, $(x^2 - a^2)$.
5. $x^2 + a^2$, $x + a$, $x - a$.
6. $y^2 - 2y$, $y^2 - 3y + 2$.
7. $z^2 - z$, $z^2 - 2z + 1$.
8. $z^2 - z$, $z^2 - 1$, $z^2 + z$.
9. $t^2 - 1$, $t - 1$, $3(t + 1)$.
10. $p^2 - 3p + 2$, $p^2 - 4p + 3$, $p^2 - 5p + 6$.

Exercise 10.2b

Find the L.C.M. of:

1. $x^2 - 4$, $x^2 - 3x + 2$.
2. $3(x + 1)$, $6(x + 1)^2$.
3. $x^2 - a^2$, $x^2 - ax - x + a$.
4. $x^2 - 2ax + a^2$, $x^2 - ax$.
5. $x^2 + 1$, $(x + 1)^2$, $x + 1$.
6. $y^2 - y$, $y^2 - 1$.
7. $y^3 - y$, $3(y + 1)$.
8. $z^3 - z$, $z + 1$, $z(z - 1)$.
9. $a^2 - b^2$, $3(a - b)$, $2(a + b)$.
10. $x^2 - 1$, $x^2 - 3x + 2$, $x^2 + x - 2$.

Simplification of fractions

To add or subtract fractions, find the L.C.M. of the denominators. Express each fraction as a fraction with this common denominator and simplify the numerator. Factorise the numerator and divide by common factors where possible. (See Example 9.)

Example 7. *Simplify* $\dfrac{1}{x} - \dfrac{2}{x+1} + \dfrac{1}{x+2}$.

$$\dfrac{1}{x} - \dfrac{2}{x+1} + \dfrac{1}{x+2} = \dfrac{(x+1)(x+2) - 2x(x+2) + x(x+1)}{x(x+1)(x+2)}$$

$$= \dfrac{x^2 + 3x + 2 - 2x^2 - 4x + x^2 + x}{x(x+1)(x+2)}$$

$$= \dfrac{2}{x(x+1)(x+2)}.$$

Example 8. *Simplify* $\dfrac{x}{x^2 - 3x + 2} - \dfrac{x+3}{x^2 - 1}$.

$$\dfrac{x}{x^2 - 3x + 2} - \dfrac{x+3}{x^2 - 1} = \dfrac{x}{(x-1)(x-2)} - \dfrac{x+3}{(x-1)(x+1)}$$

$$= \dfrac{x(x+1) - (x+3)(x-2)}{(x-1)(x-2)(x+1)}$$

$$= \dfrac{x^2 + x - (x^2 + x - 6)}{(x-1)(x-2)(x+1)}$$

$$= \dfrac{6}{(x-1)(x-2)(x+1)}.$$

Example 9. *Simplify* $\dfrac{1}{x^2 + 3x + 2} + \dfrac{1}{x^2 + 5x + 6}$.

$$\dfrac{1}{x^2 + 3x + 2} + \dfrac{1}{x^2 + 5x + 6} = \dfrac{1}{(x+1)(x+2)} + \dfrac{1}{(x+2)(x+3)}$$

$$= \dfrac{(x+3) + (x+1)}{(x+1)(x+2)(x+3)}$$

$$= \dfrac{2x+4}{(x+1)(x+2)(x+3)}$$

$$= \dfrac{2(x+2)}{(x+1)(x+2)(x+3)}$$

$$= \dfrac{2}{(x+1)(x+3)}, \quad \text{dividing numerator and denominator by } (x+2).$$

Exercise 10.3a

Simplify:

1. $\dfrac{2}{3x} + \dfrac{3}{4x}$.

2. $\dfrac{1}{x+1} + \dfrac{1}{x-1}$.

3. $\dfrac{2}{2-x} - \dfrac{4}{4-x}.$

4. $\dfrac{x-2}{4} - \dfrac{x-3}{6}.$

5. $1 + \dfrac{1}{x(x+2)}.$

6. $\dfrac{x}{4x+8} \times \dfrac{x^2+5x+6}{3x^2}.$

7. $\dfrac{x+3}{x^2-1} - \dfrac{3}{2(x-1)}.$

8. $\dfrac{1}{(a-b)(a-c)} + \dfrac{1}{(b-c)(b-a)}.$

9. $\dfrac{2}{x^2+x} - \dfrac{1}{x^2+3x+2}.$

10. $\dfrac{1}{x^2+x} + \dfrac{1}{x^2+3x+2} + \dfrac{1}{x^2+2x}.$

Exercise 10.3b

Simplify:

1. $\dfrac{3}{2x} - \dfrac{4}{3x}.$

2. $\dfrac{1}{4} + \dfrac{1}{x+2}.$

3. $\dfrac{3}{3-x} - \dfrac{2}{1-x}.$

4. $\dfrac{5(2x-1)}{6} - \dfrac{3(x+1)}{4}.$

5. $\dfrac{1}{x(x+3)} + \dfrac{1}{x^2+5x+6}.$

6. $\dfrac{2x}{3x+9} \times \dfrac{x^2+4x+3}{x^2}.$

7. $\dfrac{x+1}{x^2-4} - \dfrac{5}{x+2}.$

8. $\dfrac{1}{(a-b)(a-c)} + \dfrac{1}{(b-c)(b-a)} + \dfrac{1}{(c-a)(c-b)}.$

9. $\dfrac{3}{x^2-x} - \dfrac{4}{x^2-1}.$

10. $\dfrac{1}{x^2-x} + \dfrac{1}{x^2-3x+2} + \dfrac{1}{x^2-2x}.$

Changing the subject of a formula

If we walk 12 km at an average speed of 6 km/h, it takes 2 hours; if we walk 12 km in 2 hours, our average speed is 6 km/h. Similarly if we walk s km at v km/h, it takes s/v hours; if we walk s km in t hours, our average speed is s/t km/h. We see that we have three formulae,

$$s = vt; \quad t = s/v; \quad v = s/t$$

and we can use whichever of the three has the unknown quantity as its subject.

If we are given a formula, say that for finding the volume of a solid right circular cylinder,

$$V = \pi r^2 h$$

and we wish to change it so that h is the subject of the formula, we carry out the ordinary processes of algebra.

Since
$$V = \pi r^2 h$$

dividing both sides of the equation by πr^2,

$$h = \frac{V}{\pi r^2}$$

If we wish to make r the subject of the formula, first divide both sides by πh,

$$r^2 = \frac{V}{\pi h}$$

Next take the square root of both sides,

$$r = \sqrt{\frac{V}{\pi h}}$$

Strictly, of course, we take only the *positive* square root, because the negative square root is meaningless in this context, as a radius cannot be negative.

Example 10. Find x from the equation $\dfrac{x}{x - a} = \dfrac{x - b}{x - c}$.

Multiply through by $(x - a)(x - c)$, the L.C.M.:
$$x(x - c) = (x - a)(x - b).$$

Remove brackets:
$$x^2 - xc = x^2 - ax - bx + ab.$$

Arrange with terms containing x on L.H.S.:
$$ax + bx - cx = ab.$$
or
$$x(a + b - c) = ab.$$

Divide by $(a + b - c)$, the coefficient of x:

$$x = \frac{ab}{a + b - c}.$$

Exercise 10.4a

Find x in questions 1 to 10.

1. $ax + b = cx + d$.

2. $\dfrac{x}{a} - \dfrac{x}{b} = 1$.

3. $\dfrac{a}{x} + \dfrac{b}{x} = 1$.

4. $x(x - a) = (x - b)(x - c)$.

5. $\dfrac{1}{x} + \dfrac{1}{x - a} = \dfrac{2}{x - b}$.

6. $\dfrac{x + a}{x + b} = \dfrac{l}{m}$.

7. $a(x + a) = b(x + b)$.

8. $\dfrac{x}{x + a} = \dfrac{p}{q}$.

9. $\dfrac{x}{a} + \dfrac{x}{b} = \dfrac{1}{c}.$

10. $\sqrt{x^2 - a^2} = b.$

11. Make h the subject of the formula $A = 2\pi r(r + h)$.

12. Make r the subject of the formula $A = \pi\{(r + t)^2 - r^2\}.$

13. Make u the subject of the formula $\dfrac{1}{u} + \dfrac{1}{v} = \dfrac{1}{f}.$

14. Make l the subject of the formula $T = 2\pi\sqrt{\dfrac{l}{g}}.$

15. Make l the subject of the formula $A = \pi r l + \pi r^2.$

16. Make f the subject of the formula $s = ut + \frac{1}{2}ft^2.$

17. Make u the subject of the formula $v^2 = u^2 + 2fs.$

18. Make x the subject of the formula $(x - a)^2 + (y - b)^2 = r^2.$

19. Make s the subject of the formula $x = \sqrt{\dfrac{s - a}{s - b}}.$

20. Make d the subject of the formula $S = \dfrac{n}{2}\{2a + (n - 1)d\}.$

Exercise 10.4b

Find x in questions **1** to **10.**

1. $a(x + 1) = b(x + 2).$

2. $x - \dfrac{x}{a} = 1.$

3. $\dfrac{a}{x} + b = 1.$

4. $x^2 = (x - c)(x - d).$

5. $\dfrac{1}{x - a} + \dfrac{1}{x - b} = \dfrac{2}{x - c}.$

6. $\dfrac{ax + 1}{bx + 1} = \dfrac{c}{d}.$

7. $a(x - a) = b(x - b).$

8. $\dfrac{x + a}{x + b} = \dfrac{p}{q}.$

9. $\dfrac{x - a}{b} + \dfrac{x - b}{c} = 1.$

10. $x^2 + y^2 = b^2.$

11. Make a the subject of the formula $S = \dfrac{n}{2}\{2a + (n - 1)d\}.$

12. Make b the subject of the formula $ax^2 + bx + c = 0.$

13. Make R the subject of the formula $I = \dfrac{PRT}{100}.$

14. Make r the subject of the formula $A = P\left(1 + \dfrac{r}{100}\right).$

15. Make R the subject of the formula $H = \dfrac{v(P - R)}{550}.$

16. Make f the subject of the formula $v^2 = u^2 + 2fs.$

17. Make f the subject of the formula $\dfrac{1}{u} + \dfrac{1}{v} = \dfrac{1}{f}$.

18. Make x the subject of the formula $\dfrac{x^2}{a^2} + \dfrac{y^2}{b^2} = 1$.

19. Make u the subject of the formula $s = ut + \frac{1}{2}ft^2$.

20. Make s the subject of the formula $v^2 = u^2 + 2fs$.

Exercise 10.5: Miscellaneous

1. A cylindrical tank has radius x cm and depth y cm. Find an expression of the number of litres the tank will hold.

2. What must be added to $\dfrac{x}{y}$ to make $\dfrac{2y}{x}$?

3. Find, in pence, the average cost of an egg if x eggs are bought for n pence and a further y eggs for £n.

4. A blanket was a m long and b metres wide. When washed it lost $12\frac{1}{2}\%$ of its length and 10% of its breadth. Find its percentage loss of area.

5. A line x cm long is divided into two parts in the ratio $m : n$. Find the length in cm of each part.

6. The average speed of a car from London to Coventry, x km apart, is p km/h. Its average speed from Coventry to Birmingham, y km apart is q km/h. Find its average speed from London to Birmingham.

7. A wire in the shape of an equilateral triangle encloses a region area A cm². If the same wire is bent to form a circle, find the area of the circle in cm².

8. A man bought E eggs at x pence a dozen and M more eggs at y pence a hundred. What was the average price he paid for an egg?

9. A bookseller buys $(x + y)$ copies of a certain book, all at the same price. He sells x of them at a profit of $p\%$ but has to sell the remainder at a loss of $q\%$. Find his percentage profit on the whole transaction.

10. If $y = \dfrac{1 + 2x}{x - 2}$, express $2x + y$ (i) in terms of x only; (ii) in terms of y only.

11. Simplify $\dfrac{1}{x - 1} - \dfrac{1}{x + 1} - \dfrac{3}{2(x^2 - 1)}$.

12. Simplify $\dfrac{1}{2ax + x^2} - \dfrac{1}{2ax - x^2} - \dfrac{2}{x^2 - 4a^2}$.

13. Simplify $1 + \dfrac{x}{2} + \dfrac{2}{x}$.

14. Simplify $\dfrac{1}{a(a - b)(a - c)} + \dfrac{1}{b(b - a)(b - c)} + \dfrac{1}{c(c - a)(c - b)}$.

15. Find the L.C.M. of $a^4 - x^4$, $a^2 - x^2$, $ax - x^2$.

11 Linear Equations and Inequalities

Equations

If two expressions are equal, after each expression has been multiplied by the same quantity, they remain equal. Similarly, if the same quantity is added to or subtracted from each expression, the resulting expressions are equal. This often enables us to solve equations containing an unknown.

Linear equations

An equation which contains only terms in x and constants is called a linear equation. These can always be solved by this method.

Example 1. Solve the equation: $\dfrac{x-2}{3} - \dfrac{3x-4}{4} = 1.$

Multiply through by 12 (the L.C.M.),
$$4(x-2) - 3(3x-4) = 12.$$
Remove brackets, $\quad 4x - 8 - 9x + 12 = 12.$
Collect terms, $\quad\quad\quad\quad\quad 4x - 9x = 12 + 8 - 12$
or $\quad\quad\quad\quad\quad\quad\quad\quad -5x = 8.$
Divide by -5 (the coefficient of x),
$$x = -\tfrac{8}{5} = -1\tfrac{3}{5}.$$

Check. When $x = -1\tfrac{3}{5}$, the L.H.S. of the equation
$$= \frac{-1\tfrac{3}{5} - 2}{3} - \frac{-4\tfrac{4}{5} - 4}{4}$$
$$= \frac{-3\tfrac{3}{5}}{3} + \frac{8\tfrac{4}{5}}{4} = -1\tfrac{1}{5} + 2\tfrac{1}{5} = 1.$$

There now follows an example of an equation which contains terms in y^2. These terms vanish and the equation reduces to a linear equation.

Example 2. Solve the equation: $\dfrac{2}{y-1} + \dfrac{3}{y+1} = \dfrac{5}{y}.$

Multiply through by $y(y-1)(y+1)$, the L.C.M.:
$$2y(y+1) + 3y(y-1) = 5(y-1)(y+1).$$
Remove brackets: $\quad 2y^2 + 2y + 3y^2 - 3y = 5y^2 - 5.$
Subtract $5y^2$ from both sides: $\quad\quad\quad 2y - 3y = -5.$
$$\therefore -y = -5.$$
Divide by -1: $\quad\quad\quad\quad\quad\quad\quad\quad y = 5.$
Check. When $y = 5$,
$$\text{L.H.S.} = \tfrac{2}{4} + \tfrac{3}{6} = \tfrac{1}{2} + \tfrac{1}{2} = 1.$$
$$\text{R.H.S.} = \tfrac{5}{5} = 1.$$

Exercise 11.1a

Solve the following equations:

1. $\dfrac{5x}{3} = 2.$

2. $5x - 1 = 4.$

3. $3(x - 3) = 6.$

4. $3x - 2 = 2x - 1.$

5. $3(x - 2) = 2(x - 1).$

6. $\dfrac{x}{3} - \dfrac{x}{4} = 1.$

7. $x + \dfrac{x}{2} = 3.$

8. $\dfrac{x}{2} + \dfrac{x}{3} + \dfrac{x}{4} = 1.$

9. $\dfrac{x - 1}{2} - \dfrac{x - 2}{3} = 1.$

10. $\dfrac{2(x - 1)}{3} - \dfrac{3(x - 2)}{4} = 1.$

11. $\dfrac{3y}{7} - \dfrac{2y}{5} = \dfrac{4}{35}.$

12. $\dfrac{p + 1}{5} - \dfrac{3(p - 1)}{10} = 2.$

13. $\dfrac{2(p - 1)}{5} - \dfrac{3(p + 1)}{10} = p.$

14. $\dfrac{2(z - 1)}{3} - \dfrac{3(2z + 1)}{4} = \dfrac{z - 2}{5}.$

15. $\dfrac{t + 1}{t - 1} = \dfrac{3}{4}.$

16. $\dfrac{2}{t} = \dfrac{3}{t + 1}.$

17. $\dfrac{1}{z} + \dfrac{1}{z + 1} = \dfrac{2}{z - 1}.$

18. $\dfrac{1}{y - 1} + \dfrac{2}{y + 1} = \dfrac{3}{y}.$

19. $\dfrac{z - 1}{z + 1} = \dfrac{2z - 3}{2z + 3}.$

20. $\dfrac{2z + 1}{z - 1} = \dfrac{6z + 1}{3z - 2}.$

Exercise 11.1b

Solve the following equations:

1. $\dfrac{3x}{4} = 2.$

2. $3x - 4 = 2.$

3. $4(x - 3) = 5.$

4. $4x - 2 = 3x - 1.$

5. $2(x - 3) = 3(x - 1).$

6. $\dfrac{x}{4} - \dfrac{x}{5} = 2.$

7. $x + \dfrac{x}{3} = 4.$

8. $\dfrac{x}{3} + \dfrac{x}{4} + \dfrac{x}{5} = 1.$

9. $\dfrac{2x - 1}{3} - \dfrac{3x - 1}{4} = 1.$

10. $\dfrac{3(x - 2)}{2} - \dfrac{x - 3}{4} = 2.$

11. $\dfrac{2y}{5} - \dfrac{3y}{4} = \dfrac{3}{10}.$

12. $\dfrac{p + 1}{7} - \dfrac{3(p - 2)}{14} = 1.$

13. $\dfrac{p + 2}{6} - \dfrac{p - 3}{3} = p.$

14. $\tfrac{3}{4}(z - 1) - \tfrac{2}{3}(3z + 1) = \tfrac{1}{5}(z + 1).$

15. $\dfrac{3t+1}{3t-1} = \dfrac{2}{3}$.

16. $\dfrac{3}{t-1} = \dfrac{4}{t+1}$.

17. $\dfrac{1}{z} + \dfrac{2}{z+1} = \dfrac{3}{z-1}$.

18. $\dfrac{3}{y-1} + \dfrac{1}{y+1} = \dfrac{4}{y}$.

19. $\dfrac{2z-1}{2z+1} = \dfrac{3z-1}{3z+2}$.

20. $\dfrac{2z+3}{z-2} = \dfrac{6z}{3z+2}$.

Inequalities

The same method can be used to solve inequalities, but we have now to be very careful when we multiply both sides of the inequality to ensure that the multiplier (or divisor) is positive. We notice that

$$2 < 3$$

so that adding 5 to each, $7 < 8$; subtracting 5 from each, $-3 < -2$; multiplying each by 5, $10 < 15$, and dividing each by 5, $\frac{2}{5} < \frac{3}{5}$. But if we multiply each by -5, we obtain -10 and -15, and $-10 > -15$. More generally,

if $\qquad a < b,$

$\qquad\qquad a + c < b + c$, for all c,

$\qquad\qquad ac < bc \qquad$ for all positive c,

but $\qquad\qquad ac > bc \qquad$ if c is negative.

Example 3. *Find the range of values of x for which*

$$\tfrac{1}{2}(x-1) + \tfrac{1}{3}(x+2) < 0$$

Multiply both sides by 6: $\qquad 3(x-1) + 2(x+2) < 0$

Remove the brackets: $\qquad\qquad\qquad 5x + 1 < 0$

Subtract 1 from both sides: $\qquad\qquad\qquad\quad 5x < -1,$

Divide both sides by 5: $\qquad\qquad\qquad\qquad x < -\tfrac{1}{5}.$

Example 4. *Find the range of values of x for which*

$$\tfrac{1}{2}(1-x) < 3$$

Multiply both sides by 2: $\qquad\qquad 1 - x < 6.$

Add x to both sides and subtract 6 from both sides: $-5 < x.$

Notice that in this example we could have subtracted 1 from both sides of the inequality $1 - x < 6$ to give $-x < 5$

When we now multiply both sides by -1, we change the inequality, and

$$x > -5.$$

Example 5. *Solve the inequality $a(x+b) < c$ to find the range of values of x, in terms of the other letters.*

If a is positive, divide both sides by a: $x + b < \dfrac{c}{a}$.

Subtract b from both sides: $\qquad\qquad\qquad x < \dfrac{c}{a} - b.$

But if a is negative, when we divide both sides by a, $x + b > \dfrac{c}{a}$

and $x > \dfrac{c}{a} - b.$

Notice that if a is zero, the inequality will be satisfied by all values of x if c is positive, but by no values of x if c is negative.

Exercise 11.2a

Find the range of values of x which satisfy the following inequalities:

1. $\dfrac{5x}{3} < 2.$ 2. $5x - 1 > 4.$

3. $3(x - 2) < 6.$ 4. $3x - 2 > 2x - 1.$

5. $3(x - 2) < 2(x - 1).$ 6. $3(2 - x) > 2(1 - x).$

7. $4(2 - x) < 3(3 - 2x).$ 8. $\dfrac{x}{3} > \dfrac{x}{4} + 1.$

9. $x < \frac{1}{2}x + 3.$ 10. $\dfrac{x}{2} + \dfrac{x}{3} + \dfrac{x}{4} > 1.$

Exercise 11.2b

Find the range of values of x which satisfy the following inequalities:

1. $\dfrac{3x}{4} < 2.$ 2. $3x - 4 > 2.$

3. $4(x - 3) < 5.$ 4. $4x - 2 > 3x - 1.$

5. $2(x - 3) < 3(x - 1).$ 6. $2(3 - x) > 3(1 - x).$

7. $5(3 - 2x) < 3(4 - 3x).$ 8. $\dfrac{x}{4} - \dfrac{x}{5} > 2.$

9. $x < \dfrac{x}{3} + 4.$ 10. $\dfrac{x}{3} + \dfrac{x}{4} + \dfrac{x}{5} > 1.$

Problems leading to linear equations

Make sure that the letter used represents a *number* and not a quantity, so that we say 'let the height of the tree be x metres' or 'let the farmer have x cows' and *not* 'let the height of the tree be x' or 'let the cows be x'.

The method of solving a problem by algebra may be divided into the following steps.

(1) Choose a letter (or letters) to represent the number (or numbers) required.

(2) Translate each statement given in the question to a statement containing your letter (or letters).

(3) By linking up the parts of the question, form an equation (or equations).

The resulting equation may be a linear equation or a quadratic equation. If two unknowns are used, we should arrive at simultaneous equations.

Example 6. A man drives from Bedford to Cambridge, a distance of 48 km, in 45 minutes. Where the surface is good, he drives at 72 km/h; where it is bad, at 48 km/h. Find the number of km of good surface.

Suppose there are x km of good surface.
Then there are $(48 - x)$ km of bad surface.
Over the good surface he drives at 72 km/h.
Therefore he drives x km at 72 km/h.
The time taken is $x/72$ hours.
Over the bad surface, he drives at 48 km/h.
Therefore he drives $(48 - x)$ km at 48 km/h.
The time taken is $(48 - x)/48$ hours.
The total time taken is 45 minutes or $\frac{3}{4}$ hour.

$$\therefore \frac{x}{72} + \frac{48 - x}{48} = \frac{3}{4}.$$

Multiply by 144:
$$2x + 3(48 - x) = 108$$
$$2x + 144 - 3x = 108$$
$$-x = -36$$
$$\therefore x = 36$$

There are 36 km of good surface.
Check. Always check from the question and not from the equations.
He drives 36 km at 72 km/h. Time taken = 30 min.
„   „   12 km „ 48 km/h.   „   „   = 15 min.
Total time taken = 45 min.

Example 7. A certain sum of money consists of 30 coins some of which are 10 penny pieces and the rest of which are 5 penny pieces. If the total value of the coins is £2, find the number of 10 penny pieces.

Let the number of ten penny pieces be x.
The number of five penny pieces must be $(30 - x)$.
The value of x ten penny pieces is $10x$ pence.
The value of $(30 - x)$ five penny pieces is $5(30 - x)$ pence.
The total value is 200 pence.

N.B. Always made sure that both sides of the equation are expressed in the same units.

$$\therefore 10x + 5(30 - x) = 200$$
$$10x - 5x = 200 - 150$$
$$x = 10.$$

The number of ten penny pieces is 10.

Check.   The value of 10 ten penny pieces is £1.
The value of 20 five penny pieces is £1.
The total value is £2.

Exercise 11.3a

1. A boy is paid 50 pence for each day he works and is fined 25 pence for each day he fails to work. After 20 days, he is paid £7. For how many days has he worked?

2. A sum of money is made up of an equal number of ten penny pieces and five penny pieces. If the number of ten penny pieces were doubled and the number of five penny pieces halved, the sum would be increased by £1.80. Find the number of five penny pieces originally.

3. A man has to go 10 km to catch a bus. He walks part of the way at 7 km/h and runs the rest of the way at 12 km/h. If he takes 1 hour 15 minutes to complete his journey, find how far he walks.

4. A shopkeeper bought tea at £2.20 per kg and mixed it with twice as much tea at £2 kg. He sold the mixture at £2.40 per kg and gained £40 on the transaction. Find how many kg of each kind of tea he bought.

5. A fraction not in its lowest terms is equal to $\frac{3}{4}$. If the numerator of the fraction were doubled, it would be 34 greater than the denominator. Find the fraction.

6. A father is three times as old as his son. In 12 years' time, he will be twice as old. How old is the father now?

7. Write down the value of the number having x as its unit digit and y as its ten digit.
   A number is such that its ten digit is twice its unit digit. Prove that the number itself must be seven times the sum of its digits.

8. From London to Coventry is 144 km. A cyclist starts from London towards Coventry at a steady speed of 16 km/h. An hour later, a motorist starts from Coventry for London and travels at an average speed of 48 km/h. How far from London do they meet?

9. A concert is attended by 300 people. Some paid 60 pence each and the rest 50 pence each. The total receipts were £158. Find how many dearer tickets were sold.

10. A sum of £2800 is invested, partly at 5% and partly at 4%. If the total income is £128 per annum, find the amount invested at 5%.

Exercise 11.3b

1. A man averages 32 km/h between Bournemouth and London and 48 km/h between London and Cambridge. If the whole journey of 240 km takes 6 hours 40 min, find how far it is from London to Bournemouth.

2. A sum of money in 50 penny pieces and 10 penny pieces amounts to £60. If there are, in all, 440 coins, how many of them are 50 penny pieces?

3. A sum of £3000 is invested partly at $3\frac{1}{2}\%$ and partly at 4%. If the annual income from the two investments is £112, find how much is invested at 4%.

4. A number of two digits is such that the sum of its digits is 10. When the digits are reversed, the number is increased by 36. Find the number.

5. A man who leaves home at 08.30h arrives at the station, 5 km away, at 09.08h. He walked part of the way at 6 km/h and ran the rest of the way at 10 km/h. How far did he walk?

6. Tickets for a concert cost either 50 pence each or 70 pence each. If the number of cheaper tickets sold was twice the number of dearer tickets and the total receipts were £340, find the number of cheaper tickets sold.

7. In a factory, the skilled men are paid £10 a day and the apprentices £8 a day. If there are 400 people employed and the daily wage bill is £3800, find the number of skilled men employed.

8. A train leaves Edinburgh for London at 64 km/h and an express to Edinburgh leaves London an hour later at 96 km/h. If the distance between London and Edinburgh is 624 km, how far are the trains from London when they meet?

9. The same number is added to both numerator and denominator of the fraction $\frac{7}{17}$. If the fraction is then equal to $\frac{2}{3}$, find the number added.

10. Find two consecutive numbers such that the difference of their squares is 53.

Problems involving inequalities

Problems involving inequalities are handled in a similar way.

Example 8. A woman buys x buns at 5p each and (x + 6) cakes at 8p each. If she wishes to have some change from a £1 note, form an inequality in x, and solve it to find the range of values of x.

The buns cost $5x$ pence and the cakes $8(x + 6)$ pence, so she spends $5x + 8(x + 6)$ pence, i.e. $13x + 48$ pence. As this is to be less than 100 p,

$$13x + 48 < 100$$
$$13x < 52$$
$$x < 4$$

If the woman insists on having some change, $x < 4$, but if she is willing to spend the whole of the £1, then $13x + 48 \leqslant 100$, so that $x \leqslant 4$.

Exercise 11.4a

In each of the following, form an inequality in x and solve it to find the range of values of x.

1. A man buys x newspapers at 8p and $(x + 4)$ magazines at 20p. He spends less than £2.

2. A man cuts some wire from a roll length 2 m. He forms it into a rectangle, breadth x cm, length $(x + 10)$ cm.

3. A boy walks for 3 hours at x km/h, then runs for half an hour at $(x + 2)$ km/h. He travels less than 30 km.

4. From an irregular piece of cardboard area 40 cm² a man cuts a rectangle length x cm, breadth 5 cm. (If the shape of the cardboard is very irregular, then x may be very much less than the maximum value obtained here.)

5. A girl buys x 5p stamps and $3x$ 6p stamps. She has some change from £1.

Exercise 11.4b

In each of the following, form an inequality in x and solve it to find the range of values of x.

1. A man gathers x kg each day of outdoor tomatoes, and $(x + 5)$ kg each day of indoor tomatoes. He gathers not less than 10 kg of tomatoes each day.

2. A boy scored $5x$ marks in the first of two exam papers, and $(x + 10)$ marks in the second. He came second in the exams, the first boy scoring a total of 118 marks.

3. When I bought a packet of mints with a £1 note, I received x 10p pieces and $(x - 2)$ 2p pieces change.

4. A girl rides x km at 10 km/h, then walks $\frac{1}{2}x$ km at 3 km/h. She is away from home less than 4 hours.

5. A car travels x km in 3 hours, then $(x + 120)$ km in 5 hours. Its average speed does not exceed 100 km/h.

Simultaneous equations

Sometimes it is more convenient if we introduce a second letter to represent another unknown. One equation in the unknowns is satisfied by as many pairs of values as we wish, but if we have two equations in two unknowns, they usually have only one pair of solutions.

Example 9. Solve the equations:
$$3x + 2y = 12, \quad 4x - 3y = -1.$$

$$3x + 2y = 12 \qquad \text{(i)}$$
$$4x - 3y = -1 \qquad \text{(ii)}$$

Multiply (i) by 4 and (ii) by 3. This is to produce equations in which the coefficients of x are equal. Sometimes it is easier to choose multipliers so that the coefficients of y become equal.

$$12x + 8y = 48 \qquad \text{(iii)}$$
and $$12x - 9y = -3 \qquad \text{(iv)}$$

Subtract (iv) from (iii), so that the term in x disappears.

$$17y = 51.$$
$$\therefore y = 3.$$

Substitute in (i): $\quad 3x + 6 = 12.$
$$\therefore 3x = 6 \quad \text{and} \quad x = 2.$$

Check. L.H.S. of (i) $= 3(2) + 2(3) = 12.$
L.H.S. of (ii) $= 4(2) - 3(3) = -1.$

There follows an example of a type of problem which is common and which is solved by the same method.

Example 10. The expression $(ax + by)$ is equal to 8 when $x = 1$ and $y = 2$; it is equal to 13 when $x = 2$ and $y = 3$. Find its value when $x = 3$ and $y = 2$.

Put $x = 1$ and $y = 2$. $a + 2b = 8$ (i)
Put $x = 2$ and $y = 3$. $2a + 3b = 13$ (ii)
Multiply (i) by 2: $2a + 4b = 16$ (iii)
Do not alter (ii): $2a + 3b = 13$ (iv)
Subtract (iv) from (iii) $b = 3$.
Substitute in (i): $a + 6 = 8$
                                 $\therefore a = 2$.

The expression is therefore $2x + 3y$.
When $x = 3$ and $y = 2$, its value is
$$2(3) + 3(2) = 12.$$

Exercise 11.5a

Solve the pairs of equations in questions **1** to **12**.

1. $x + y = 8$, $x - y = 4$.

2. $2x + y = 7$, $2x - y = 3$.

3. $2x + 3y = 8$, $2x - 3y = 2$.

4. $3y + z = 7$, $2y - z = 3$.

5. $2p + 3q = 5$, $3p + 4q = 7$.

6. $3p + 2q = 4$, $2p + 3q = 7$.

7. $5p + 4q = 22$, $3p + 5q = 21$.

8. $\dfrac{c + d}{c - d} = \dfrac{1}{2}$, $\dfrac{c + 1}{d + 1} = 2$.

9. $\dfrac{u + 1}{2} = \dfrac{v + 2}{3} = \dfrac{v + 3}{4}$.

10. $\dfrac{h + k + 1}{4} = \dfrac{h}{2} = k$.

11. $\dfrac{h + 2k}{4} = \dfrac{h + k + 1}{7} = \dfrac{2h + k}{10}$.

12. $\dfrac{x}{2} + \dfrac{y}{3} = 2$, $2x + 3y = 13$.

13. If $y = ax + b$, find the values of a and b, given that $y = 5$ when $x = 1$ and that $y = 7$ when $x = 2$.

14. If $y = ax + b$, find the value of y when $x = 4$, given that $y = 4$ when $x = 1$ and that $y = 7$ when $x = 2$.

15. If $y = ax^2 + b$, find the values of a and b, given that $y = -3$ when $x = 1$ and that $y = 5$ when $x = 2$.

16. If $s = ut + \frac{1}{2}at^2$, find the value of s when $t = 3$, given that $s = 8$ when $t = 1$ and that $s = 20$ when $t = 2$.

17. Find x and y from the equations: $\dfrac{1}{x} + \dfrac{1}{y} = 3$, $\dfrac{2}{x} + \dfrac{3}{y} = 7$.

18. If the solutions of the pair $x + y = a$, $3x + 4y = b$ are $x = 2$, $y = 3$, find a and b.

19. Find x and y from the equations: $2x^2 + y^2 = 18$, $3x^2 + 2y^2 = 35$.

20. The value of the expression $(ax^2 + bx)$ is 6 when $x = 1$ and 10 when $x = 2$. Find its value when $x = 2$.

Exercise 11.5b

Solve the pairs of equations in questions **1** to **12**.

1. $x + y = 6$, $x - y = 2$.

2. $3x + y = 8$, $3x - y = 2$.

3. $3x + 4y = 7, 3x - 4y = -1.$　　　4. $2y + z = 8, 5y - z = 6.$

5. $3p + 4q = 7, 5p + 6q = 11.$　　　6. $2p - 3q = 1, 3p + 2q = 8.$

7. $3p + 2q = 13, 2p + 3q = 12.$　　8. $\dfrac{a+1}{b+1} = 2, \dfrac{2a+1}{2b+1} = \dfrac{1}{3}.$

9. $\dfrac{u+1}{2} = \dfrac{2u+1}{3} = \dfrac{v+3}{4}.$　　10. $\dfrac{c+d+2}{7} = \dfrac{c}{3} = \dfrac{d}{2}.$

11. $\dfrac{y + 2z + 1}{4} = \dfrac{3y + z + 1}{8} = \dfrac{2y + 3z + 2}{9}.$

12. $\dfrac{x}{3} + \dfrac{y}{4} = 2, 3x + 4y = 25.$

13. If $y = ax + b$, find the values of a and b, given that $y = 6$ when $x = 1$ and $y = 10$ when $x = 2$.

14. If $y = ax + b$, find the value of y when $x = 0$, given that $y = -1$ when $x = 1$ and that $y = 4$ when $x = 3$.

15. If $y = ax^2 + b$, find the values of a and b, given that $y = 7$ when $x = 1$ and that $y = 13$ when $x = 2$.

16. If $P = aW + b$, find the value of P when $W = 4$, given that $P = 8$ when $W = 2$ and that $P = 20$ when $W = 3$.

17. Solve the equations: $\dfrac{3}{x} + \dfrac{4}{y} = 2, \dfrac{4}{x} - \dfrac{1}{y} = 3.$

18. If the solutions of the pair $2x + 3y = a, 3x - y = b$ are $x = -1, y = 2$, find a and b.

19. Solve the equations: $x^2 + y^2 = 13, 3x^2 - 2y^2 = -6.$

20. The value of the expression $(ax^2 + bx)$ is 8 when $x = 2$ and 27 when $x = 3$. Find its value when $x = -1$.

Problems leading to simultaneous equations

Most problems which can be solved by using two unknowns and simultaneous equations may also be solved by using one unknown and a simple equation. In the following worked examples, the method of simultaneous equations is chosen.

Example 11. *A tobacconist bought a certain number of pipes at £1.50 each and others at £1.70 each. Had he bought half as many at £1.50 and twice as many at £1.70, his bill would have been £196 instead of the £188 which he actually paid. How many pipes did he buy altogether?*

Suppose he bought x pipes at £1.50 and y pipes at £1.70.
The cost of x pipes at £1.50 each is £1.5x.
The cost of y pipes at £1.70 each is £1.7y.

$$1.5x + 1.7y = 188$$
or　　　　　　　$$15x + 17y = 1880. \qquad\qquad\text{(i)}$$

The cost of $\frac{1}{2}x$ pipes at £1.50 is

$$\tfrac{1}{2}x(1.5) = £0.75x.$$

The cost of $2y$ pipes at £1.70 is

$$2y(1.7) = £3.4y$$
$$\therefore 0.75x + 3.4y = 196$$

or $\qquad\qquad 3.75x + 17y = 980.$ $\qquad$ (ii)

Subtract (ii) from (i), $\qquad 11.25x = 900.$

Multiply by 4, $\qquad\qquad 45x = 3600$
$$x = 80.$$

Substitute in (i), $\qquad 1200 + 17y = 1880$
$$17y = 680$$

or $\qquad\qquad\qquad\qquad y = 40.$

He bought 80 pipes at £1.50 and 40 at £1.70, i.e. 120 in all.

Check.   80 pipes at £1.50 each cost £120.
40 pipes at £1.70 each cost £68.
Total cost is £188.
40 pipes at £1.50 each cost £60.
80 pipes at £1.70 each cost £136.
Total cost is £196.

Example 12. *A motorist travels 15 km to the litre of petrol and 600 km to the litre of oil. He estimates that an annual distance of 6000 km will cost him £102 in petrol and oil. In fact he used twice as much oil as he estimated and the cost was £108. Find the cost of a litre of petrol.*

Suppose the cost of a litre of petrol is x pence and the cost of a litre of oil is y pence.

In travelling 6000 km, he estimates to use 400 litres of petrol and 10 litres of oil.

The cost of these is $(400x + 10y)$ pence.

$$\therefore 400x + 10y = 10\ 200$$ $\qquad$ (i)

He actually used 20 litres of oil.

$$\therefore 400x + 20y = 10\ 800$$ $\qquad$ (ii)

Subtract (i) from (ii), $\qquad 10y = 600$
$$y = 60$$

Substitute in (i), $\qquad 400x + 600 = 10\ 200$
$$400x = 9600$$
$$\therefore \qquad x = 24$$

The cost of a litre of petrol is 24 pence.

Check.   400 litres of petrol at 24p cost £96.
10 litres of oil at 60p cost £6.
20 litres of oil at 60p cost £12.
The estimated cost is £102.
The actual cost is £108.

Exercise 11.6a

1. When petrol cost 10 pence a litre and oil 25 pence a litre, a motorist found that his cost in petrol and oil was £10.50 for each 1000 km. When petrol was increased to $10\frac{1}{2}$ pence a litre and oil to 26p a litre, the cost was £11.02. Find how many litres of petrol and oil he used to travel 1000 km.

2. At a concert, tickets were 70p or 50p each. Three-quarters of those who paid 70p bought programmes and half the 50p seat-holders bought programmes. The receipts from the tickets were £580 and from the programmes £60. If the programmes cost 10 pence each, how many paid 70p for the concert?

3. I invest £x at 4% and £y at 5%. My annual income is £230. Had I invested £x at 5% and £y at 4%, my annual income would have been £220. Find x and y.

4. The total cost of 12 kg of apples and 24 kg of plums is £21.60. The cost of 24 kg of apples and 12 kg of plums is £18. Find the cost of apples per kg.

5. A number of two digits is increased by 54 when the digits are reversed. The sum of the digits is 12. Find the number.

6. A bottle and a cork together cost 8 pence. The bottle costs 6 pence more than the cork. Find the cost of the cork.

7. A gramophone with 12 records cost £10. The same gramophone with 18 similar records costs £11.50. What is the cost of the gramophone?

8. 100 cigarettes and 4 cigars cost £9.80; 50 cigarettes and 8 cigars cost £12.10. What is the price of a cigar?

9. A man walks at 8 km/h and runs at 12 km/h. To get to the station takes him 20 min. Had he run twice as far, it would have taken him 17½ min. How far is the station?

10. A man travels x km at 8 km/h and y km at 20 km/h. His total time is 3½ hours. The total time taken to travel $(2x + 4)$ km at 8 km/h and $y/3$ km at 20 km/h is 5 hours. Find x and y.

Exercise 11.6b

1. To travel 1000 km a motorist estimated that he would need 100 litres of petrol and 2 litres of oil, and that the cost would be £8.40. He actually used 95 litres of petrol and 3 litres of oil and the cost was £8.20. Find the cost of a litre of petrol.

2. At an entertainment, the price of the first three rows of chairs was £1 each and the other seats cost 60p each. The takings were £204. On the second night of the entertainment, the price of the fourth row of chairs was increased to £1 and the takings were £212. How many chairs were there in the hall, assuming it was full on each occasion and that there were the same number of chairs in each row?

3. If I invest £2000 at x% and £2500 at y%, my annual income is £160. Had I invested £2500 at x% and 2000 at y%, my income would have been £155. Find x and y.

4. 10 kg of apples and 20 kg of plums cost, in all, £15.20; 20 kg of apples and 10 kg of plums cost £11.20. What is the cost of plums per kg?

5. A number of two digits is such that twice the ten digit is 6 greater than the unit digit. When the digits are reversed, the number is increased by 9. What is the number?

6. The sum of the ages of a father and a son is 52 years. Eight years ago, the father was eight times as old as his son. How old is the father now?

7. The charge for electricity is 3 pence per unit for lighting and $\frac{1}{2}$ penny per unit for heating. A man's bill for a quarter should have been £8. By mistake, his lighting was charged at $\frac{1}{2}$ penny per unit and his heating at 3 pence per unit. The amount of the bill was £13. How many units of electricity did he use altogether?

8. 20 cigars and 200 cigarettes cost £17: 16 cigars and 100 cigarettes cost £11.50. Find the cost of a cigar.

9. To go to the station in the morning, I first walk to the garage at 8 km/h and motor the rest of the way at 40 km/h. It normally takes me 21 min in all. One morning when I am late, I run to the garage at 16 km/h and motor at 60 km/h. I complete the journey in $11\frac{1}{2}$ min. How far is the garage from the station?

10. The total time taken to travel x km at 30 km/h and y km at 60 km/h is $2\frac{1}{2}$ hours. The time taken to travel x km at 15 km/h and y km at 45 km/h is 4 hours 20 minutes. Find x and y.

Identities

The equation $4x = 8$ is satisfied only by $x = 2$, but an equation like $2(x + 4) = 2x + 8$ is satisfied by all values of x; such an equation is called an **identity**. Strictly, all the equations that we have written when factorising expressions are identities, for they are satisfied by all values of x. We usually denote an identity by $\equiv$, so that we perhaps ought to write

$$x^2 - 1 \equiv (x - 1)(x + 1)$$

but we shall only use $\equiv$ when we wish to emphasise that we are considering an identity and not just an equation.

If two expressions are identically equal, then they must have the same coefficients for each different power of x on both sides of the identity, i.e. if $Ax^2 + Bx + C \equiv 3x^2 + 4x + 5$, then $A = 3$, $B = 4$, $C = 5$, and the identity is true for all values of x. We can make use of whichever property is the more convenient.

Example 13. Find the constants A, B and C if
$$Ax(x + 1) + Bx + C \equiv x^2 + 2x + 3.$$
Comparing coefficients of x^2; $A = 1$.
Comparing coefficients of x; $A + B = 2$, $\therefore B = 1$.
Comparing the constants; $C = 3$.

Example 14. Prove $(x - 1)^3 + (x + 1)^3 \equiv 2x(x^2 + 3)$.
Always work the two sides of the identity separately.
$$\text{L.H.S.} = (x-1)^3 + (x+1)^3 = (x^3 - 3x^2 + 3x - 1) + (x^3 + 3x + 3x + 1)$$
$$= 2x^3 + 6x$$
$$= 2x(x^2 + 3) = \text{R.H.S.}$$

Alternative method.

If $x = 0$,	L.H.S. $= -1 + 1 = 0$;	R.H.S. $= 0$.
If $x = 1$,	L.H.S. $= 0 + 8 = 8$;	R.H.S. $= 2(4) = 8$.
If $x = -1$,	L.H.S. $= -8 + 0 = -8$;	R.H.S. $= -2(4) = -8$.
If $x = 2$,	L.H.S. $= 1 + 27 = 28$;	R.H.S. $= 4(7) = 28$.

So there are four values of x which satisfy the equation. But the equation is of the third degree and so should be satisfied by three values of x only. Therefore the equation is an identity and is satisfied by all values of x.

Exercise 11.7a

1. Find constants A, B and C if
$$3x^2 - 2x + 1 \equiv Ax(x + 1) + Bx + C.$$

2. Find constants A, B and C if
$$2x^2 + 4x + 3 \equiv Ax(x - 1) + Bx + C.$$

3. Show that the values 0, 1, -1 for x all satisfy $(x - 1)^2 + (x + 1)^2 = 2(x^2 + 1)$. Is this equation an identity?

4. Show that the values 1, 2 for x satisfy $(x - 1)^2 + (x + 1)^2 \equiv x^2 + 3x$. Is this equation an identity?

5. Prove that $\left(x + \dfrac{1}{x} \right)^2 - \left(x - \dfrac{1}{x} \right)^2 \equiv 4.$

6. Find the values of A and B if
$$x \equiv A(x - 1) + B(x - 2).$$

7. Find the values of A and B if
$$2x \equiv A(x + 1) + B(x - 1).$$

8. Find the values of A, B and C if
$$x^2 \equiv A(x - 1)(x - 2) + B(x - 2)(x - 3) + C(x - 3)(x - 1).$$
(Hint: substitute $x = 1$.)

9. Find the values of A, B and C if

10. Show that $x^4 \equiv (x + 1)(x^3 - x^2 + x - 1) + 1$.
Deduce that $\dfrac{x^4}{x + 1} \equiv x^3 - x^2 + x - 1 + \dfrac{1}{x + 1}.$

Exercise 11.7b

1. Find constants A, B and C if
$$3 + 4x + 5x^2 \equiv A + B(x + 1) + Cx(x + 1).$$

2. Find constants A, B and C if
$$1 + x + x^2 \equiv A + B(x + 1) + Cx(x + 1).$$

3. Show that the values 0, 2, -2 for x all satisfy $(x - 2)^2 + (x + 2)^2 = 2(x^2 + 4)$. Is this equation an identity?

4. Is $x^3 = (x + 1)(x^2 - x + 1)$ an equation or an identity?
5. Find the values of A and B if $x \equiv A(x - 2) + B(x - 3)$.
6. Find the values of A and B if $3x \equiv A(2x + 1) + B(2x - 1)$.
7. Show that $(x - y)^2 \equiv (x + y)^2 - 4xy$.
8. Find the values of A, B and C if
$$x^2 \equiv A(x - 1)(x + 1) + B(x + 1)(x + 2) + C(x + 2)(x - 1).$$
9. Find the values of A, B and C if
$$4x^2 \equiv A(2x - 1)(2x + 1) + B(2x + 1)(x - 1) + C(x - 1)(2x - 1).$$
10. Find A, B and C if
$$x^2 \equiv (Ax + B)(2x - 1) + C.$$
Hence find the quotient and remainder when x^2 is divided by $2x - 1$.

Exercise 11.8: Miscellaneous

1. Solve the equation $\dfrac{2}{x} = \dfrac{9}{2x} - 1$.

2. Solve the equation $\dfrac{2}{x} - \dfrac{1}{6} = \dfrac{9}{2x} - 1$.

3. Solve the equation $\dfrac{2}{x} = \dfrac{x}{8}$.

4. Solve the equation $\dfrac{3}{x + 1} = \dfrac{x + 1}{12}$.

5. Solve the inequality $4 - 3x > 1$.

6. Solve the inequality $\dfrac{4}{x} > 3$.

7. Solve the inequality $\dfrac{4}{x} > -2$.

8. Solve the inequality $\dfrac{4}{x - 1} > -1$.

9. Solve the simultaneous equations
$$0.5x - 0.3y = 1.65$$
$$0.7x + 0.2y = 0.14.$$

10. Solve the simultaneous equations.
$$\frac{x + y + 4}{4} = \frac{2x - y}{3}$$
$$\frac{3x - y - 2}{3} = \frac{2x - y}{4}.$$

11. A cyclist leaves home at 10.00 hours to cycle to church 7 km away. He cycles at 10 km/h until he has a puncture, then he has to push his bicycle the rest of the way at 3km/h. He arrives at church at 11.10 hours. Find how far he walked.

12. A woman walks to the shops at 5 km/h, spends 20 minutes shopping and then takes the bus, which averages 30 km/h, home. She is away from home 55 minutes. How far are the shops from her home?

13. A mother is three times as old as her daughter. Six years ago, she was five times as old. How old is the daughter now?

14. I can buy y articles for £2 or $(y + 2)$ similar articles for £2.20. Find y.

15. I am thinking of a number. I double the number, add 6, and multiply the result by 10. I now divide by 20, then subtract the number I first thought of. What is the result?

16. If 7 is added to both the numerator and the denominator of a fraction, the fraction becomes equal to $\frac{5}{6}$. If instead, 5 is added to the numerator and 7 to the denominator, the fraction becomes equal to $\frac{3}{4}$. Find the original fraction.

17. Half the sum of two numbers is 51; one quarter of their difference is 13. Find the numbers.

18. A number with three digits has the hundred digit three times the unit digit and the sum of the digits is 19. If the digits are written in reverse order, the value of the number is decreased by 594. Find the number.

19. The cost £C of a journey on a certain railway is given by the formula $C = a + bx$, where x is the number of kilometres travelled and a and b are constants. A journey of 50 km costs £2, and a journey of 70 km costs £2.50. Find the cost of a journey of 90 km.

20. Show that $(x + y + z)^2 - (x + 2y)(x + 2z) \equiv (y - z)^2$.

12 Quadratic Equations and Inequalities

Quadratic equations

If the product of two expressions is equal to zero, then one or other of those expressions must be zero. A quadratic expression is of the form $ax^2 + bx + c$ (though b or c may be zero), and if we can factorise it into two linear expressions, then the only values of x for which the quadratic is zero are the two values which make one or other linear factor zero.

Example 1. Solve $x^2 - 2x - 3 = 0$.

Factorising, $(x - 3)(x + 1) = 0$
either $x - 3 = 0$ or $x + 1 = 0$
i.e. $x = $ either 3 or -1.

N.B. x is equal to either one or the other of these values. It cannot be equal to both simultaneously.

Example 2. Solve $\dfrac{3}{2x + 1} + \dfrac{4}{5x - 1} = 2$.

Multiply both sides by $(2x + 1)(5x - 1)$, the L.C.M.
$$3(5x - 1) + 4(2x + 1) = 2(2x + 1)(5x - 1),$$
i.e. $15x - 3 + 8x + 4 = 20x^2 + 6x - 2,$
$$20x^2 - 17x - 3 = 0.$$
Factorise $(20x + 3)(x - 1) = 0,$
$$x = 1 \quad \text{or} \quad -\tfrac{3}{20}$$

Exercise 12.1a

Solve the following equations:

1. $x^2 - 4x + 3 = 0$.
2. $x^2 - 5x + 6 = 0$.
3. $x^2 - 5x - 6 = 0$.
4. $x^2 + 7x + 10 = 0$.
5. $x^2 - 3x - 10 = 0$.
6. $x^2 - 4x - 12 = 0$.
7. $x^2 - 2x - 8 = 0$.
8. $x^2 - 10x + 9 = 0$.
9. $x^2 - 7x - 18 = 0$.
10. $2x^2 - 5x + 2 = 0$.
11. $6x^2 - 5x + 1 = 0$.
12. $6x^2 - 7x + 2 = 0$.
13. $3x^2 + 14x + 8 = 0$.
14. $x(2x + 1) = 10$.
15. $x(x + 1) + (x + 2)(x + 3) = 42$.
16. $\dfrac{1}{x + 1} + \dfrac{4}{3x + 6} = \dfrac{2}{3}$.

17. $\dfrac{1}{x} + \dfrac{1}{x+1} = \dfrac{9}{20}.$

18. $\dfrac{x+2}{x+3} = \dfrac{2x-3}{3x-7}.$

19. $\dfrac{3}{x-2} + \dfrac{8}{x+3} = 2.$

20. $\dfrac{x+1}{x-2} + \dfrac{x+11}{x+3} = 4.$

Exercise 12.1b

Solve the following equations:

1. $x^2 - 6x - 7 = 0.$

2. $x^2 - 6x - 16 = 0.$

3. $t^2 - 14t + 24 = 0.$

4. $x^2 - 4x - 21 = 0.$

5. $u^2 - u - 20 = 0.$

6. $v^2 + 11v + 18 = 0.$

7. $y^2 + 9y + 18 = 0.$

8. $x^2 + 8x + 16 = 0.$

9. $2z^2 - 7z + 3 = 0.$

10. $6x^2 - x - 2 = 0.$

11. $6p^2 - 17p + 12 = 0.$

12. $3x^2 - 13x - 10 = 0.$

13. $2x^2 + 35 = 19x.$

14. $(k+1)(2k+1) = 15.$

15. $x(x+2) + (x+1)(2x-1) = 17.$

16. $\dfrac{1}{x+2} + \dfrac{1}{x+3} = \dfrac{7}{12}.$

17. $\dfrac{3}{q} + \dfrac{4}{q+1} = 2.$

18. $\dfrac{x+3}{x+2} = \dfrac{x+4}{x+1}.$

19. $\dfrac{6}{z+1} + \dfrac{5}{2z+1} = 3.$

20. $\dfrac{x+7}{x+1} + \dfrac{2x+6}{2x+1} = 5.$

Inequalities

If $(x-1)(x-2)$ is positive, then either both factors are positive, or both are negative. $(x-1)$ is positive if and only if $x > 1$; $x - 2$ is positive if and only if $x > 2$, so both are positive if and only if $x > 2$.

$$x-2<0 \qquad 2 \quad x-2>0$$

$$x-1<0 \quad 1 \qquad x-1>0$$

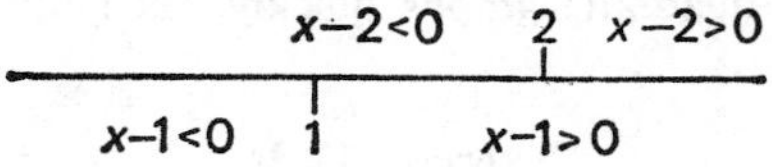

Fig. 12.1

Marking the range of values for which they are positive on a number-line often helps, or alternatively we can sketch the graph of the function $(x-1)(x-2)$.

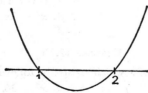

Fig. 12.2

From Fig. 12.1, or from Fig. 12.2, we see that $(x - 1)(x - 2)$ is positive if and only if $x > 2$ or $x < 1$.

Example 3. *Find the range of values of x for which* $x^2 < 1$.

If $$x^2 < 1, x^2 - 1 < 0,$$
i.e. $$(x - 1)(x + 1) < 0.$$

From the number line we see that $x + 1 < 0$ when $x < -1$ and $x - 1 < 0$ when $x < +1$, so that one or other but not both is negative only when $-1 < x < 1$.

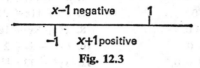

Fig. 12.3

Alternative method

Sketch the graph of $y = x^2$.

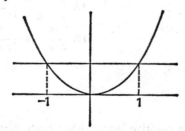

Fig. 12.4

$x^2 < 1$ only if $-1 < x < 1$.

If we can sketch quadratic graphs quickly, then the second method is quicker and clearer.

Exercise 12.2a

Find the range of values of x which satisfy the following inequalities:

1. $x^2 < 4$.
2. $x^2 > 25$.
3. $4x^2 < 9$.
4. $(x - 2)(x - 3) > 0$.
5. $(x - 1)(x + 2) < 0$.
6. $(2x - 1)(x + 2) > 0$.
7. $(2x + 1)(x + 3) < 0$.
8. $x^2 - 3x + 2 > 0$.
9. $x^2 - 3x - 4 < 0$.
10. $x^2 - 2x + 1 > 0$.

Exercise 12.2b

Find the range of values of x which satisfy the following inequalities:

1. $x^2 > 16$.
2. $x^2 < 36$.
3. $9x^2 > 4$.
4. $(x - 3)(x - 4) < 0$.
5. $(x + 1)(x - 2) > 0$.

6. $(2x + 3)(x + 2) < 0.$ **7.** $(2x + 5)(x + 3) > 0.$

8. $x^2 - 2x - 3 < 0.$ **9.** $x^2 - 4x + 3 > 0.$

10. $x^2 + 4x + 4 < 0.$

Problems leading to quadratic equations

Form the equation as when solving linear equations, then factorise the resulting quadratic equation. We obtain two solutions (sometimes they are the same), and often one will not be acceptable, usually because it is negative. Check both answers to see if they are acceptable.

Example 4. The sum of two numbers is 17; their product is 60. Find the numbers.

Let one number be x; then the other is $(17 - x)$, since their sum is 17. Since their product is 60,

$$x(17 - x) = 60,$$
$$x^2 - 17x + 60 = 0,$$
$$(x - 5)(x - 12) = 0$$
$$x = 5 \text{ or } 12,$$

the two numbers are 5 and 12. Notice that if $x = 5$, then $17 - x = 12$, whereas if $x = 12$, $17 - x = 5$.

Example 5. The perimeter of a rectangle is 30 cm; the area of the rectangle is 50 cm². Find the length of the rectangle.

Let the length of the rectangle be x cm. Then since the perimeter is 30 cm, the breadth is $(15 - x)$ cm. Since the area is 50 cm²,

i.e.
$$x(15 - x) = 50,$$
$$x^2 - 15x + 50 = 0,$$
$$(x - 5)(x - 10) = 0$$
$$x = 5 \text{ or } 10$$

But the length must be more than 7.5 cm (one quarter of the perimeter) so the length is 10 cm.

Exercise 12.3a

1. Two numbers differ by 7; their product is 60. Find the numbers.
2. The sum of two numbers is 13; their product is 36. Find the numbers.
3. The length of a rectangle is 3 m greater than its width; the area of the rectangle is 108 m². Find the length of the rectangle.
4. The lengths of the sides of a right-angled triangle are $(2x + 1)$ cm, $2x$ cm and $(x - 1)$ cm. Find x.
5. Two numbers differ by 2: the sum of their squares is 650. Find the numbers.
6. The sum of the lengths of the two sides containing the right angle in a right-angled triangle is 15 cm; the area of the triangle is 28 cm². Find the lengths of these two sides.

7. A man can row at 5 km/h in still water. He rows upstream 3 km then back to his starting point. If the total time taken is 1 hour 15 minutes, find the speed of the current.

8. x articles cost $(3x + 20)$ pence; $(x + 4)$ similar articles cost $(5x - 4)$ pence. Find x.

9. The perimeter of a rectangle is 28 cm; the area must be greater than 45 cm². Within what limits does the length of the rectangle lie?

10. The sides of an obtuse-angled triangle are $(2x + 3)$ cm, $(2x + 2)$ cm and x cm. Within what limits does x lie?

Exercise 12.3b

1. Two numbers differ by 10; their product is 39. Find the numbers.

2. The sum of two numbers is 12; their product is 32. Find the numbers.

3. The length of a rectangle is 10 cm more than its breadth: the area of the rectangle is 56 cm². Find the breadth of the rectangle.

4. The lengths of the sides of a right-angled triangle are $(2x + 1)$ cm, $(2x - 1)$ cm and x cm. Find x.

5. Two numbers differ by 5; the difference of their squares is 45. Find the numbers.

6. The lengths of the sides containing the right angle in a right-angled triangle are x cm and $(x + 3)$ cm; the area of the triangle is 20 cm². Find x.

7. A boat travels at 6 km/h in still water. A man sails 3 km upstream, then back to his starting point. If the total time he takes is 1 hour $7\frac{1}{2}$ minutes, find the speed of the current.

8. $3x$ articles cost $(7x + 2)$ pence; $4(x + 1)$ similar articles cost $(12x + 2)$ pence. Find x.

9. The perimeter of a rectangle is 19 cm; the area must be less than 78 cm². Within what limits does the breadth of the rectangle lie?

10. The lengths of the sides of an acute-angled triangle are $(10x + 1)$ cm, $10x$ cm and $(2x + 1)$ cm. Within what limits does x lie?

Quadratic equations by completing the square and by formula

If a quadratic equation cannot be solved by factorisation, one of the following methods should be used.

(1) *Completing the square*

Example 6. Solve the equation: $3x^2 - 4x - 5 = 0$.

Arrange with terms containing x^2 and x on L.H.S.:
$$3x^2 - 4x = 5.$$
Divide through by 3 (the coefficient of x^2):
$$x^2 - \tfrac{4}{3}x = \tfrac{5}{3}$$

Add to each side (half the coefficient of x)²:

$$x^2 - \tfrac{4}{3}x + (\tfrac{2}{3})^2 = \tfrac{5}{3} + \tfrac{4}{9}.$$
$$\therefore (x - \tfrac{2}{3})^2 = \tfrac{19}{9}.$$

Take the square root of each side:

$$x - \tfrac{2}{3} = \pm \sqrt{\frac{19}{9}} = \pm \frac{4.359}{3}.$$

Solve for x,
$$x = \frac{2 \pm 4.359}{3},$$

$$x = \frac{6.359}{3} \quad \text{or} \quad -\frac{2.359}{3},$$

$$x = 2.12 \quad \text{or} \quad -0.79,$$

each correct to two places of decimals.

(2) *Formula*

The solutions of the equation $ax^2 + bx + c = 0$ are

$$x = \frac{-b \pm \sqrt{b^2 - 4ac}}{2a}.$$

This formula should be memorised. The expression $(ax^2 + bx + c)$ may be made identical to any given quadratic expression, by giving a, b and c suitable values. To make $(ax^2 + bx + c)$ identical to $(5x^2 + 2x - 6)$ for example, put $a = 5$, $b = 2$ and $c = -6$.

The formula may be found by the method of completing the square, applied to the equation $ax^2 + bx + c = 0$.

Rearrange:
$$ax^2 + bx = -c.$$

Divide throughout by a:
$$x^2 + \frac{b}{a} = -\frac{c}{a}.$$

Add to each side (half the coefficient of x)²:

$$x^2 + \frac{b}{a}x + \left(\frac{b}{2a}\right)^2 = -\frac{c}{a} + \frac{b^2}{4a^2}$$

$$\left(x + \frac{b}{2a}\right)^2 = \frac{b^2 - 4ac}{4a^2}.$$

Take the square root of both sides:

$$x + \frac{b}{2a} = \pm \frac{\sqrt{b^2 - 4ac}}{2a}.$$

Subtract $\dfrac{b}{2a}$ from each side:

$$x = -\frac{b}{2a} \pm \frac{\sqrt{b^2 - 4ac}}{2a}.$$

$$x = \frac{-b \pm \sqrt{b^2 - 4ac}}{2a}.$$

Notice that the sum of the roots is $-\dfrac{b}{a}$; this is a useful check when solving quadratics.

Example 7. *Solve the equation:* $3x^2 - 4x - 5 = 0$.
 For this equation $a = 3$, $b = -4$, $c = -5$.

$$\text{Substitute in } x = \frac{-b \pm \sqrt{b^2 - 4ac}}{2a}$$

$$x = \frac{4 \pm \sqrt{(-4)^2 - 4(3)(-5)}}{6}$$

$$x = \frac{4 \pm \sqrt{76}}{6}$$

$$= \frac{4 \pm 8.718}{6}$$

$$= \frac{12.718}{6} \quad \text{or} \quad \frac{-4.718}{6}$$

$$= 2.12 \quad \text{or} \quad -0.79,$$

each correct to two places of decimals.

Check: $2.12 + (-0.79) = 1.33$; $-\dfrac{b}{a} = \tfrac{4}{3} = 1.33$, to 2 d.p.

Exercise 12.4a

Solve the following equations, giving your answers correct to two places of decimals:

1. $2x^2 - 3x - 7 = 0$.
2. $3x^2 - x - 1 = 0$.
3. $u^2 - 3u = 8$.
4. $5x^2 - 8x + 2 = 0$.
5. $7z^2 - 2z - 3 = 0$.
6. $3x^2 + 5x + 1 = 0$.
7. $4y^2 + y - 3 = 0$.
8. $6x^2 + 10x + 3 = 0$.
9. $z^2 + z - 8 = 0$.
10. $u + \dfrac{1}{u} = 3$.
11. $v + \dfrac{1}{2v} = 4$.
12. $\dfrac{1}{x+1} + \dfrac{1}{x+2} = \dfrac{2}{3}$.
13. $(p - 1)(2p + 4) = 9$.
14. $\dfrac{c+1}{c-1} = \dfrac{2c-4}{3c-2}$.
15. $\dfrac{1}{x} + \dfrac{2}{x+1} = \dfrac{4}{x-1}$.
16. $(z - 1)(z + 1) + (2z - 1)(z + 2) = 6$.

17. $z + 1 + \dfrac{1}{z+1} = 3.$

18. $y(y+1) + (y+2)(y+3) = 4.$

19. $\dfrac{1}{2x+1} + \dfrac{1}{3x-1} = 1.$

20. $p(p-1)(p+2) = p(p+2)(p-4).$

Exercise 12.4b

Solve the following equations, giving your answers correct to two decimal places:

1. $x^2 - 2x - 7 = 0.$ **2.** $3x^2 - 5x - 1 = 0.$

3. $u^2 - u - 3 = 0.$ **4.** $5v^2 + 2v - 1 = 0.$

5. $3z^2 - z - 1 = 0.$ **6.** $7z^2 - 3z - 8 = 0.$

7. $2x^2 + 5x + 1 = 0.$ **8.** $4y^2 + 2y - 5 = 0.$

9. $z^2 + z = 5.$ **10.** $u + \dfrac{1}{u} = 5.$

11. $v + \dfrac{1}{2v} = 6.$ **12.** $\dfrac{1}{x+2} + \dfrac{1}{x+3} = \dfrac{2}{3}.$

13. $(2p-1)(p+3) = 8.$ **14.** $\dfrac{x+1}{x-2} = \dfrac{3x+1}{2x-1}.$

15. $\dfrac{1}{x+1} + \dfrac{2}{x+3} = \dfrac{2}{3}.$ **16.** $(q-2)(2q+6) = 7.$

17. $(z-1)(z+3) + (2z+1)(z+4) = 6.$

18. $z + 1 + \dfrac{1}{z+1} = 5.$

19. $(y+1)(y+2) + (y+3)(y+4) = 4.$

20. $p(p+2)(p+3) = (p-1)(p+4)(p+5).$

Simultaneous equations—one linear and one quadratic

Two simultaneous equations, one linear and one quadratic, have in general two pairs of solutions. Graphically, the equations give a straight line and a curve as shown in Fig. 12.5. The values of x and y at the points of intersection are the solutions of the equations.

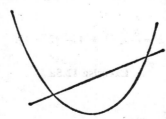

Fig. 12.5

Example 8. Solve the equations: $x + y = 2$, $2x^2 + y^2 = 3$.

Find either x or y in terms of the other from the linear equation. We do whichever we think will make our working easier.

From $x + y = 2$,

$$y = 2 - x.$$

Substitute in the quadratic,

$$2x^2 + (2 - x)^2 = 3,$$

or $\qquad 2x^2 + 4 - 4x + x^2 = 3.$

$$\therefore 3x^2 - 4x + 1 = 0 \quad \text{or} \quad (x - 1)(3x - 1) = 0.$$
$$\therefore x = 1 \quad \text{or} \quad \tfrac{1}{3}.$$

When $x = 1$,

$$y = 2 - 1 = 1.$$

When $x = \tfrac{1}{3}$,

$$y = 2 - \tfrac{1}{3} = 1\tfrac{2}{3}.$$

The solutions are $x = 1$, $y = 1$ or $x = \tfrac{1}{3}$, $y = 1\tfrac{2}{3}$.

N.B. Make it clear in the answer that $x = 1$ must pair with $y = 1$. *Do not write $x = 1$ or $\tfrac{1}{3}$, $y = 1$ or $1\tfrac{2}{3}$.*

Example 9. Solve the equations: $2x + 3y = 8$, $x^2 - xy + y^2 = 3$.

Find x in terms of y:

$$x = \frac{8 - 3y}{2}.$$

Substitute: $\qquad \left(\dfrac{8 - 3y}{2}\right)^2 - \left(\dfrac{8 - 3y}{2}\right)y + y^2 = 3.$

Remove brackets:

$$\frac{64 - 48y + 9y^2}{4} - \frac{8y - 3y^2}{2} + y^2 = 3.$$

Simplify:

$$64 - 48y + 9y^2 - 16y + 6y^2 + 4y^2 = 12,$$

or $\qquad 19y^2 - 64y + 52 = 0.$

$$\therefore (y - 2)(19y - 26) = 0,$$

and $\qquad\qquad\qquad\qquad y = 2 \quad \text{or} \quad \tfrac{26}{19}.$

When $y = 2$,

$$x = \frac{8 - 6}{2} = 1.$$

When $y = \tfrac{26}{19}$,

$$x = \frac{8 - \tfrac{78}{19}}{2} = 4 - \tfrac{39}{19} = \tfrac{37}{19}.$$

Exercise 12.5a

Solve the equations:

1. $x + y = 5$, $xy = 6$.
2. $2x + y = 7$, $x^2 - xy = 6$.

3. $2x + 3y = 2$, $4x^2 + 9y^2 = 2$.
4. $x + y = 2$, $x^2 - xy + y^2 = 1$.
5. $3x - 5y = 1$, $x^2 - 3xy + 2y^2 = 0$.

Exercise 12.5b

Solve the equations:

1. $x + y = 7$, $xy = 12$.
2. $3x + y = 5$, $x^2 + xy = 3$.
3. $x + 2y = 2$, $x^2 + 4y^2 = 2$.
4. $3x = 2y$, $x^2 + xy + y^2 = 19$.
5. $x + y = 3$, $x^2 + xy = 4$.

Harder problems leading to quadratic equations

Example 10. *A boy bought some packets of biscuits for £1.20. If the biscuits had been 3p a packet cheaper, he would have received 2 more packets for his money. How many packets did he buy?*

Let x be the number of packets he bought for £1.20. Then the price of 1 packet is $\dfrac{120}{x}$ pence.

If the biscuits had been 3p a packet cheaper, each would have cost

$$\left(\frac{120}{x} - 3\right) \text{ pence.}$$

But he would then have received $(x + 2)$ packets, so each would have cost $\dfrac{120}{x + 2}$ pence.

Equating these two prices,

$$\frac{120}{x} - 3 = \frac{120}{x + 2}.$$

Multiplying by the L.C.M., $x(x + 2)$ and dividing by 3,

$$40(x + 2) - x(x + 2) = 40x,$$
$$x^2 + 2x - 80 = 0$$
$$(x - 8)(x + 10) = 0$$
$$x = 8 \quad \text{or} \quad -10.$$

A negative answer is unacceptable, so $x = 8$, he bought 8 packets of biscuits.

Example 11. *The distance from London to Bournemouth is 160 km. If an express train were 16 km/h slower, it would take 20 minutes longer on the journey. Find the average speed of the express.*

Let the average speed of the express be x km/h.

The time taken to travel 160 km is $\dfrac{160}{x}$ hours.

If the express were travelling at 16 km/h slower, then its speed would be $(x - 16)$ km/h.

The train would then take $\dfrac{160}{x - 16}$ hours.

But this is 20 minutes, i.e. $\frac{1}{3}$ hour, more than when travelling at x km/h.

$$\frac{160}{x - 16} = \frac{160}{x} + \frac{1}{3}.$$

Multiply by $3x(x - 16)$, the L.C.M.,

$$480x = 480(x - 16) + x^2 - 16x,$$

i.e.
$$x^2 - 16x - 7680 = 0$$
$$(x - 96)(x + 80) = 0$$
$$x = 96 \quad \text{or} \quad -80.$$

The negative speed is not acceptable, so the average speed of the express is 96 km/h.

Exercise 12.6a

1. A motorist has to travel 160 km. His average speed is 8 km/h slower than he anticipated and he takes 1 hour longer than intended on the journey. Find his actual average speed.

2. The average age of x boys in a form is 14 years 2 months. A boy of 15 years 2 months joins the form and the average age is increased by 1 month. Find x.

3. A man spends £5 on bars of chocolate. Had they been $2\frac{1}{2}$p cheaper, he could have bought 10 more. How many did he buy?

4. The area of a room is 288 m². If the length were increased by 6 m and the breadth decreased by 4 m, the area would be unaltered. Find the length of the room.

5. When the price of an orange is reduced by $\frac{1}{2}$p, I find that by paying an extra 3p I can buy 3 dozen instead of $2\frac{1}{2}$ dozen. How much did an orange cost before the reduction in price?

Exercise 12.6b

1. On a journey of 300 km, the driver of a train calculates that if he reduced his average speed by 5 km/h, he would take 40 minutes longer. Find his average speed.

2. The average of a cricketer is 32 runs per innings. In his last innings of the season he is out for 95 and thereby increases his average by 3. Find how many completed innings he played in the season.

3. A greengrocer buys a number of kg of apples for £4. Had the apples been 4p per kg cheaper, he could have bought 5 kg more for his money. How much did he pay for the apples per kg?

4. The area of a room is 300 m². If the length were decreased by 5 m and the breadth increased by 5 m, the area would be unaltered. Find the length of the room.

5. Sweets are shared equally between 20 children at a party. If there were 1 more sweet and 1 more child, each child would get 1 sweet less. How many sweets were shared at the party?

Exercise 12.7: Miscellaneous

1. Solve the equation $x^2 - 2 + \dfrac{1}{x^2} = 0$.

2. Solve the equation $\dfrac{x + 2}{x - 2} = \dfrac{x + 3}{x - 9}$.

3. Solve the equation $9x^2 - 2x = 10$, giving your answers correct to two places of decimals.

4. Solve for x, the equation $x^2 - 2x = a^2 - 2a$.

5. Solve the equation $\dfrac{1 - x}{x} + \dfrac{x}{1 - x} = \dfrac{5}{2}$.

6. If $a^2 + 4ab + 4b^2 - 9c^2 = 0$, express a in terms of b and c.

7. Solve the equation $\dfrac{5}{(2x - 1)^2} - \dfrac{30}{2x - 1} = 18$, giving your answers correct to two decimal places.

8. A motor-boat travelling at an average speed of v km/h is timed over a distance of 4 km. When its speed is increased to $(v + 3)$ km/h, it takes 4 min less for the journey. Find v.

9. The sum of the ages of x boys in a form is 84 years. When a new boy, aged 8 years 1 month, joins the form, the average age is increased by 1 month. Find x.

10. A batsman scores 450 runs in x completed innings. In his next innings, he is out for 63 and thereby increases his average by 2. Find x.

11. A man leaves Southampton at 11.20h and travels by train to Bournemouth at an average speed of 72 km/h. He spends 2 hours in Bournemouth and returns to Southampton by bus at an average speed of 40 km/h. If the bus route is 2 km longer than the train journey and if he arrives in Southampton at 15.15h, find the distance from Southampton to Bournemouth by train.

12. A greengrocer bought a certain number of kg of strawberries for £72. He was unable to sell 10% of them but made a profit of £14.40 by selling the rest at a profit of 8 pence per kg. How many kg did he buy?

13. A buys a car for £300 and sells it to B at a profit of $x\%$. B sells it to C at a profit of $x\%$. C paid £$(6x + \frac{3}{4})$ more for the car than A paid. Find x.

14. A man allowed £120 for his holiday. He afterwards calculated that if he had spent £1 less a day, he could have extended his holiday by 4 days on the same money. How long a holiday did he have?

15. A certain sum of money is made up of a number of ten penny pieces and twice as many five penny pieces. If the number of ten penny pieces were doubled and the number of five penny pieces decreased by 10, the sum of money would be increased by £1.10. Find the original number of five penny pieces.

13 Graphs

Linear functions

The function $ax + b$, where a and b are constants, is called a linear function of x. It contains no power of x above the first. If we plot such a function against x, we will always get a straight line. We can, of course, draw the line from two pairs of values only. However, it is much safer to consider three pairs of values. The third point will serve as a check for our arithmetic.

Example 1. Draw the graph of $y = 3x - 2$.

Consider the values 1, 2, 3 for x.

x	1	2	3
y	1	4	7

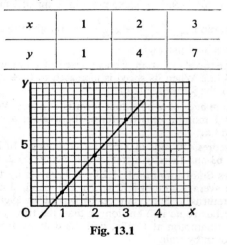

Fig. 13.1

Example 2. Draw the graphs of $5y = 12 - 2x$ and $y = x + 1$, using the same scales and axes. Write down the coordinates of the point of intersection of the two graphs.

x	0	2	3
$\dfrac{12 - 2x}{5}$	2.4	1.6	1.2
$x + 1$	1	3	4

The graphs intersect at the point $(1, 2)$. This gives the solutions $x = 1$, $y = 2$ for the pair of simultaneous equations $5y = 12 - 2x$ and $y = x + 1$.

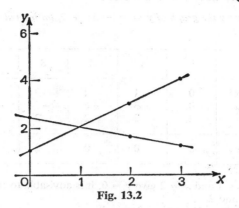

Fig. 13.2

Exercise 13.1a

Draw the graphs of:

1. $y = x + 1$.
2. $y = 3x - 2$.
3. $3y = 4x + 1$.
4. $x/2 + y/3 = 1$.
5. $2x + 3y = 6$.

6. $x = y + 2$ and $2y = x + 1$, using the same scales and axes. Write down the coordinates of the point of intersection of the graphs.

Exercise 13.1b

Draw the graphs of:

1. $y + x = 2$.
2. $y = 2x - 1$.
3. $2y = 3x + 5$.
4. $x/3 + y/4 = 1$.
5. $3x + 4y = 12$.

6. $2x = y - 1$ and $x + y = 4$, using the same scales and axes. Write down the coordinates of the point of intersection of the graphs.

Quadratic functions

The expression $ax^2 + bx + c$, where a, b and c are constants, is called a quadratic function of x. The highest power of the variable x is the second. If such an expression is plotted against x, the resulting curve will have one of the shapes shown.

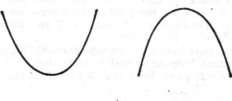

a, positive a, negative

Fig. 13.3

The sign of a determines which way up the curve is.

Example 3. *Draw the graph of* $y = x^2 - 3x + 2$, *taking values of x between 0 and 4.*

x	0	1	2	3	4
x^2	0	1	4	9	16
$-3x$	0	-3	-6	-9	-12
2	2	2	2	2	2
y	2	0	0	2	6

Since both $x = 1$ and $x = 2$ give $y = 0$, it is advisable to consider a value of x between 1 and 2.

When $x = 1\frac{1}{2}$, $y = 2\frac{1}{4} - 4\frac{1}{2} + 2 = -\frac{1}{4}$.

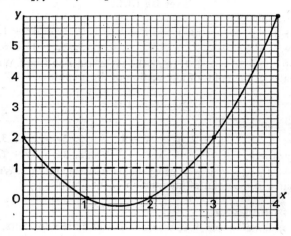

Fig. 13.4

N.B. The points where the curve crosses the x-axis are the points where $x = 1$ and $x = 2$. These are the points on the curve for which $y = 0$, or $x^2 - 3x + 2 = 0$. Therefore $x = 1$ and $x = 2$ are the solutions of the quadratic equation $x^2 - 3x + 2 = 0$.

Similarly, we can solve from the graph any equation of the type $x^2 - 3x = k$, where k is a constant. Suppose we wish to solve $x^2 - 3x + 1 = 0$.

If $x^2 - 3x + 1 = 0$, by adding 1 to each side of the equation,

$$y = x^2 - 3x + 2 = 1.$$

We need the points on the curve for which $y = 1$. Draw the line $y = 1$ to meet the curve. The values of x at the points of intersection are the solutions of the equation.

From the graph, $\qquad x = 2.6 \quad \text{or} \quad 0.4.$

Example 4. Draw the graph of $y = 1 - 2x - 3x^2$ between $x = -3$ and $x = +3$.

x	-3	-2	-1	0	$+1$	$+2$	$+3$
1 $-2x$ $-3x^2$	1 6 -27	1 4 -12	1 2 -3	1 0 0	1 -2 -3	1 -4 -12	1 -6 -27
y	-20	-7	0	1	-4	-15	-32

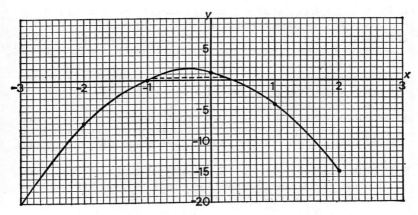

Fig. 13.5

N.B. The solutions of the equation $1 - 2x - 3x^2 = 0$ are given by the values of x at the points where the curve crosses the x-axis. They are $x = -1$ and $x = 0.3$.

To solve from the graph the equation $6x^2 + 4x = 1$.
If $6x^2 + 4x = 1$,

$$3x^2 + 2x = \tfrac{1}{2} \quad \text{and} \quad y = 1 - 2x - 3x^2 = 1 - \tfrac{1}{2} = \tfrac{1}{2}.$$

Find the values of x at the points where the line $y = \tfrac{1}{2}$ cuts the curve. They are 0.2 and -0.85.

Exercise 13.2a

1. Draw the graph of $y = x^2 - 1$, taking values of x between -2 and $+2$. Solve from your graph the equations (i) $x^2 - 1 = 0$, (ii) $x^2 - 4 = 0$.

2. Draw the graph of $y = x^2 + x - 2$, taking values of x between -3 and $+3$. Solve from your graph the equations (i) $x^2 + x - 2 = 0$, (ii) $x^2 + x - 5 = 0$.

3. Draw the graph of $y = x^2 + 4x + 3$, taking values of x between 0 and -5. Solve from your graph the equations (i) $x^2 + 4x + 3 = 0$, (ii) $x^2 + 4x + 2 = 0$.

4. Draw the graph of $y = x^2 + x + 1$, taking values of x from -3 to $+3$. Solve from your graph the equations (i) $x^2 + x - 2 = 0$, (ii) $x^2 + x - 1 = 0$. What can you say about the solutions of the equation $x^2 + x + 1 = 0$?

5. Draw the graph of $y = x^2 + x - 3$, taking values of x between -3 and $+3$. Solve from your graph the equations (i) $x^2 + x - 3 = 0$, (ii) $x^2 + x - 5 = 0$.

Exercise 13.2b

1. Draw the graph of $y = x^2 - 4$, taking values of x between -3 and $+3$. Solve from your graph the equations (i) $x^2 - 4 = 0$, (ii) $x^2 - 6 = 0$.

2. Draw the graph of $y = x^2 + 2x - 3$, taking values of x between -3 and $+3$. Solve from your graph the equations (i) $x^2 + 2x - 3 = 0$, (ii) $x^2 = -2x$.

3. Draw the graph of $y = x^2 - 4x + 3$, taking values of x between 0 and 5. From your graph, solve the equations (i) $x^2 - 4x + 3 = 0$, (ii) $x^2 - 4x + 1 = 0$.

4. Draw the graph of $y = 2x^2 + x + 1$, taking values of x between -3 and $+3$. Write down the coordinates of the lowest point of your graph. What is the least possible value of y? From your graph, solve the equation $2x^2 + x - 1 = 0$.

5. Draw the graph of $y = 1 + x - 2x^2$, taking values of x between -3 and $+3$. What is the greatest possible value of y? From your graph solve the equations (i) $1 + x - 2x^2 = 0$, (ii) $4x^2 - 2x = 5$.

Intersecting graphs

Equations may also be solved graphically by drawing intersecting graphs. An example is given.

Example 5. Draw, using the same scales and axes, the graphs of $y = x^2 - 4x + 7$ and $y = x + 1$. Write down the x coordinates of the points of intersection of the graphs. Find the equation which has these values as roots.

Consider values of x from 0 to 5.

x	0	1	2	3	4	5
x^2	0	1	4	9	16	25
$-4x$	0	-4	-8	-12	-16	-20
7	7	7	7	7	7	7
$x^2 - 4x + 7$	7	4	3	4	7	12
$x + 1$	1	2	3	4	5	6

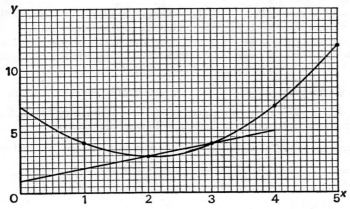

Fig. 13.6

The values of x at the points of intersection are 2 and 3. At these points, the value of y on one graph is equal to the value of y on the other graph.

Therefore $\qquad x^2 - 4x + 7 = x + 1$

or $\qquad x^2 - 5x + 6 = 0$.

The solutions of the equation $x^2 - 5x + 6 = 0$ are $x = 2$ and $x = 3$.

Exercise 13.3a

1. Draw, using the same scales and axes, the graphs of $y = x$ and $y = x^2 - 4$. Write down the values of x at the points of intersection of the graphs and find the equation which has these values as roots.

2. Draw, using the same scales and axes, the graphs of $y = 2x + 1$ and $y = x^2 + x + 1$. Write down the values of x at the points of intersection of the graphs and find the equation having these values as roots.

3. Draw, using the same scales and axes, the graphs of $y = x + 3$ and $y = x^2 - x + 1$. From your graphs, solve the equations
(i) $x^2 - 2x - 2 = 0$; (ii) $x^2 - x - 2 = 0$.

4. Draw the graph of $y = \dfrac{3 - x^2}{5 + x}$ for values of x between -3 and $+3$. From your graph, solve the equation $2x^2 + x - 1 = 0$.

5. Draw, using the same scales and axes, the graphs of
$$y = (x - 1)(x + 2) \quad \text{and} \quad y = x.$$
From your graphs, write down an approximate value for the square root of 2.

Exercise 13.3b

1. Draw, using the same scales and axes, the graphs of $y = 2x$ and $y = x^2 + 1$. Explain why the line must be a tangent to the curve.

2. Draw, using the same scales and axes, the graphs of $y = 2x - 1$ and

$y = x^2 - x + 1$. Write down the values of x at the points of intersection of the graphs and find the equation having these values as roots.

3. Draw, using the same scales and axes, the graphs of $y = x + 2$ and $y = 2x^2 - x - 1$. From your graphs solve the equations

$$\text{(i) } 2x^2 - 2x - 3 = 0; \text{ (ii) } 2x^2 - x = 0.$$

4. Draw the graph of $y = \dfrac{x^2}{x^2 + 2}$ for values of x between -4 and $+4$.

On the same figure, draw the graph of $4y = x$. Write down the equation which has the values of x at the points of intersection as roots.

5. Draw the graph of $y = \dfrac{x^2 + 3}{x + 4}$ for values of x between -3 and $+3$.

From your graph, solve the equation $x^2 - x - 1 = 0$.

Travel graphs

A completely different type of problem may often be solved by a graphical method. If the horizontal axis represents the time and the vertical axis the distance travelled, the position of a train or cyclist may be represented by a point.

Example 6. A man starts from Southampton at 09.00 hours walking to Winchester at 6 km/h. After an hour he meets the bus which runs from Winchester to Southampton. The bus waits a quarter of an hour at Southampton and then returns to Winchester. The bus travels at 24 km/h. Find when and where the man is overtaken by the returning bus.

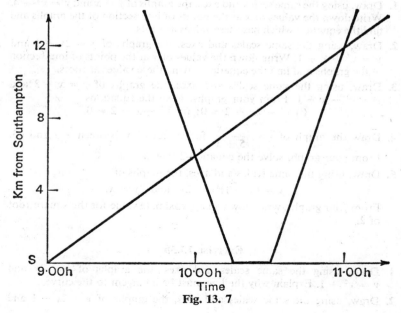

Fig. 13. 7

Take suitable scales, time horizontally and distance from Southampton in km, vertically. The man will be 6 km from Southampton at 10.00 hours and the straight line joining the point represented by this to the origin represents the progress of the man. The progress of the bus is represented by the three straight lines shown in the diagram. The second point of intersection with the graph of the man shows where and when he is overtaken.

The bus overtakes the man at 11.00 hours, 12 km from Southampton.

Exercise 13.4a

1. A slow train starts from London for Carlisle at 10.00 hours travelling at 60 km/h. An hour later, a faster train leaves London for Carlisle at 96 km/h. Find graphically how far from London the faster train overtakes the slower.

2. A train leaves Waterloo for Bournemouth West at 10.12 hours travelling at 60 km/h. Another train leaves Bournemouth West for Waterloo at 10.32 hours travelling at 80 km/h. If the distance between stations is 160 km, find graphically when the trains pass each other.

3. A man starts from Bedford at noon to walk to Eaton Socon, 20 km away. A cyclist from Eaton Socon to Bedford passes him at 12.40 hours, goes on to Bedford where he waits 10 minutes before returning to Eaton Socon by the same route. Find graphically when the cyclist overtakes the man, assuming that he walks steadily at 6 km/h and that the cyclist travels at 16 km/h.

4. A man pays £40 Income Tax on a salary of £1000 and £280 tax when his salary is £1500. Assuming there is a linear relation between his salary and the tax due, draw a straight line graph to enable you to read the tax due on all salaries from £1000 to £2000. Use your graph to find the tax due on a salary of £1250.

5. When I visit a friend's house, I can either go by car at an average speed of 60 km/h, or by train at an average speed of 90 km/h. It takes me 15 min to get from my house to the station and 20 min to get to my friend's house from the station. Assuming the distance is the same by train and by road, find this distance graphically, assuming the two journeys take exactly the same time.

Exercise 13.4b

1. A bus from London to Penzance leaves London at 13.00 hours and travels at 60 km/h. A car averaging 80 km/h leaves London by the same route an hour later. Find graphically when it overtakes the bus.

2. A train travels from X to Y at 80 km/h. Half an hour later another train leaves Y for X travelling at 90 km/h. If the distance from X to Y is 210 km, find how far from X the trains meet.

3. To travel by train costs me 5p per km. To travel by car costs an annual outlay of £80 together with running costs of 2.5p per km. Calculate the cost (i) by train, (ii) by car, of annual distances of 1200, 2400, 3600 and so on up to 12 000 km. Hence determine diagrammatically the least annual distance for which the car is the more economical.

4. A cyclist who averages 24 km/h leaves a town X at 07.00 hours. A car averaging 72 km/h leaves X by the same road half an hour later. He goes to Y, 90 km away, stays there 15 min and returns by the same road. How far from X does he meet the cyclist for the second time?

5. A train running between two stations arrives at its destination 50 min late when it averages 80 km/h and 20 min late when it averages 100 km/h. Find graphically the distance between the stations.

Exercise 13.5 Miscellaneous graphs

1. The distance from Oxford to Bicester is 24 km. A man leaves Oxford at 12.00 hours and walks to Bicester at 8 km/h. At 13.00 hours a cyclist leaves Bicester for Oxford at 16 km/h. Find when the two meet.

2. A cyclist intends to ride from Southampton to Lyndhurst, 20 km away. He rides at 16 km/h until his bicycle breaks down and, after wasting 10 minutes, he walks the rest of the way at 8 km/h. If the whole journey takes 2 hours 10 minutes, find how far he walks.

3. A cyclist leaves A for B at 09.00 hours and travels at 24 km/h. The distance between A and B is 72 km. After 45 minutes he meets a car coming from B to A, where it waits 10 minutes then returns to B. The car travels at 60 km/h. Find when the car overtakes the cyclist.

4. The following table gives the height of the barometer in cm at various heights, measured in metres, above sea level.

Barometer reading	76.2	63.5	52.1	43.2	36.8	30.5
Altitude	0	1500	3000	4500	6000	7500

Represent these readings graphically and from your graph estimate the barometric pressure at a height of 750 m above sea level, and the altitude at which the barometer reads 50 cm.

5. The table gives the distance it is possible to see on a clear day in km from different heights in metres.

Distance	4.0	5.6	7.0	8.1	9.2	10.2
Height	1.5	3	4.5	6	7.5	9

Plot these values and from your graph estimate the distance to the horizon from a height of 4 m.

6. If a stone is thrown vertically upwards with a velocity of 49 m/s, the height above the ground after t seconds is given in feet by the formula $s = 49t - 4.9t^2$. Draw a graph showing the height for various times and from your graph find (i) the greatest height reached, (ii) the time taken for the stone to return to the ground.

(The scales given in the following questions will be found suitable for paper 20 cm by 28 cm; the scale on the x-axis is given first.)

7. (2 cm to 1 unit, 0.5 cm to 1 unit.) Draw the graph of $y = x^3 - 2x$ for values of x from -3 to $+3$, plotting points for which $x = -\frac{1}{2}$, $x = \frac{1}{2}$. Draw also the graph of $y = \frac{3}{2}x + 1$. Find from your graph

(i) the value of $x^3 - 2x$ when $x = 1.5$,

(ii) the solution of the equation $x^3 - 2x = 2$,

(iii) the range of values of x for which $x^3 - 2x < \frac{3}{2}x + 1$.

Shade on your graph the region described by $x > 0$, $y > x^3 - 2x$, $y < \frac{3}{2}x + 1$.

8. (2 cm, 1 cm.) Draw the graph of $y = (x + 1)(x - 1)(x - 2)$ for values of x from -1.5 to 3. Draw suitable to straight line graphs to solve

(i) $(x + 1)(x - 1)(x - 2) = 4 - x$,

(ii) $(x + 1)(x - 1)(x - 2) = 4 - 2x$,

(iii) $(x + 1)(x - 1)(x - 2) = x$.

9. (2 cm, 5 cm.) Draw the graph of $y = \dfrac{x^2}{x + 2}$ for values of x from -1 to 4.

Use your graph to find the range of values of x for which $\dfrac{x^2}{x + 2} < 0.5$.

10. (2 cm, 0.5 cm.) Draw the graph of $y = x^2 (2 - x)$ for values of x from -2 to 3. Draw a suitable straight line to solve $x^2(2 - x) = 2 - 2x$.

11. (4 cm, 2 cm.) Draw the graphs of $y = x^2$, $y = x$ and $y = \dfrac{x + 1}{x + 2}$ plotting points for which $x = -1.5, -1, -0.5 \ldots 2$. Use your graphs to find the values of x between -1.5 and 2 which satisfy

(i) $\dfrac{x + 1}{x + 2} = x$,

(ii) $\dfrac{x + 1}{x + 2} = x^2$.

12. (2 cm, 4 cm.) Draw the graph of $y = \frac{1}{4}(2 + x)(4 - x)$ for values of x from -3 to 5. Draw also the straight line graph $y = \frac{1}{3}(3 - x)$. Hence find the range of values of x for which

$$\tfrac{1}{4}(2 + x)(4 - x) > \tfrac{1}{3}(3 - x).$$

13. (2 cm, 2 cm.) Draw the graph of $y = \dfrac{2x + 3}{x + 4}$ for values of x from -3 to 3. Draw also the graph of $y = x + \frac{1}{2}$, and find the values of x at the points of intersection of the graphs. Find, in its simplest form, the quadratic equation whose roots will be the x coordinates of the points of intersection of the graphs.

14. (2 cm, 2 cm.) Draw the graphs of $y = (x - 2)(x + 1)$ and $y = \dfrac{4}{x + 5}$ for values of x from -2 to 3. From your graphs estimate two roots of the equation $(x + 5)(x - 2)(x + 1) = 4$.

15. (2 cm, 2 cm.) Draw the graph of $y = \sqrt{(25 - x^2)}$, for values of x from -5 to 5. Draw a suitable straight line graph and estimate solutions to the equation $\sqrt{(25 - x^2)} = \frac{1}{2}(8 - x)$.

16. (2 cm, 2 cm.) Draw the graph of $y = (2 - x)/(1 + x)$ for values of x from $-\frac{1}{2}$ to 4. Draw also the graphs of $y = \frac{1}{2}x$ and $y = 1/2x$. Use your graphs to find solutions between $-\frac{1}{2}$ and 4 of

    (i) $2(2 - x)x = 1 + x$,
    (ii) $2(2 - x) = x(1 + x)$.

17. (4 cm, 2 cm.) Draw the graph of $y = x - 2/x$ for values of x from 1 to 5. Use your graph to estimate a value for $\sqrt{2}$.

18. (4 cm, 1 cm.) Draw the graph of $y = x^2 - 2/x$ for values of $x = 1$, $1.5, \ldots 3.5$. Use your graph to estimate

    (i) a value of $\sqrt[3]{2}$,
    (ii) a solution to $x^3 - 7x = 2$.

19. (2 cm, 1 cm.) Draw the graph of $y = \frac{1}{2}x(x - 1)(x - 4)$ for values of x from -1 to 4.5. Use your graph to estimate the range of values of x for which $x(x - 1)(x - 4) > 0.8$.

20. (2 cm, 2 cm.) The graph of $y = A + B/x$ passes through the points $(1,7)$ and $(8,0)$. Calculate the values of A and B. Using your values of A and B, draw the graph of $y = A + B/x$, plotting points for which $x = 1,2,3, \ldots 8$. Show that your graph does pass through $(1,7)$ and $(8,0)$.

14 Graphical Inequalities; Linear Programming

Inequalities

The line $x = 2$ passes through all the points whose x coordinate is 2.

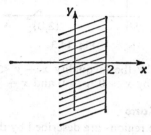

Fig. 14.1

The shaded half-plane to the left of $x = 2$ in Fig. 14.1 contains all the points whose x coordinate is less than 2, and so is described by $x < 2$. The unshaded half-plane to the right of $x = 2$ likewise is described by $x > 2$.

The line $y = x$ passes through all the points whose x and y coordi-

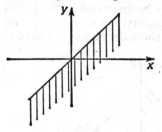

Fig. 14.2

nates are equal. The shaded half-plane in Fig. 14.2 contains all the points whose y coordinate is less than their x coordinate, and so is described by $y < x$. Likewise, the unshaded half-plane is described by $y > x$.

Description of region

Consider the region shown in Fig. 14.3, bounded by the x-axis, the y-axis, and the line $x + y = 2$. All the points in that region are such that $x > 0$, and $y > 0$. Taking the coordinates of any one point in that region, say $(1,\frac{1}{2})$, we see that $1 + \frac{1}{2} < 2$, so that the point lies in

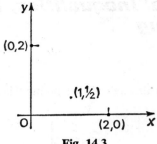

Fig. 14.3

the region described by the inequality $x + y < 2$. Thus the region is described completely by $x > 0$, $y > 0$ and $x + y < 2$.

Boundaries of regions

The boundaries of the regions are described by the equations, so that if we wish to include the boundary in the region, we have to write $\geqslant$ or $\leqslant$. Thus if the boundaries are included in the region in Fig. 14.3, the region will be described by

$$x \geqslant 0, y \geqslant 0, x + y \leqslant 2.$$

Shading of regions

Since we shall usually be interested in the points inside the region, we shade the outer boundary, so that we do not obscure the region. Shading Fig. 14.3 to show the region, we have Fig. 14.4.

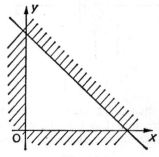

Fig. 14.4

Drawing straight-line graphs

Most of the boundaries of the regions we meet at this stage are straight lines, and it helps if we can draw these quickly and accurately. We saw in Chapter 13 that two points are sufficient to determine a straight line; a third point checks. Find the value of x for which $y = 0$, and the value of y for which $x = 0$. (These give the intercepts on the coordinate axes.) A third point for checking can often be found by inspection.

Example 1. Draw the straight line whose equation is $3x + 4y = 12$. Shade the outer boundary of the half-plane described by $3x + 4y < 12$.

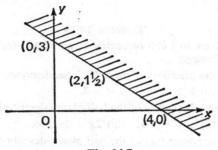

Fig. 14.5

When $y = 0$, $3x = 12$ so $x = 4$.
When $x = 0$, $4y = 12$ so $y = 3$.
The line meets the coordinate axes in $(4,0)$ and $(0,3)$. We see that $(2,1\frac{1}{2})$ also satisfies $3x + 4y = 12$, and that the line we have drawn passes through $(2,1\frac{1}{2})$.
The origin $(0,0)$ is often the easiest point to use to determine which half-plane we want, and $3 \times 0 + 4 \times 0 < 12$, so that $3x + 4y < 12$ describes the half-plane containing the origin, and we shade the boundary of $3x + 4y = 12$ away from the origin.

Example 2. Shade the outer boundary of the region described by $y > 0$, $y < x$, $3x + 4y > 12$ and $3x + 4y < 18$.

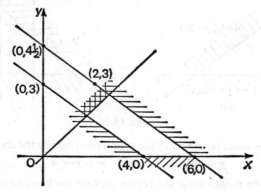

Fig. 14.6

The line $y = x$ passes through (0,0), (1,1) and (2,2). The line $3x + 4y = 12$ we drew in Fig. 14.5; the line $3x + 4y = 18$ we see meets the coordinate axes in (6,0) and (0,4½), and also passes through (2,3). We can now draw the four straight lines bounding the region.

Testing (0,0), we see that we want the region containing the origin for $3x + 4y < 18$, but not containing the origin for $3x + 4y > 12$.

Points such that $y < x$ lie to the right of $y = x$, and those for which $y > 0$ lie above the x-axis, We can now shade the outer boundaries of the region required.

Exercise 14.1a

Take a scale of 2 cm to 1 unit on each axis, and draw small sketch-graphs to illustrate the following:

1. Shade the outer boundary of the half-planes described by
   (i) $x < 0$, (ii) $x > 3$, (iii) $y < -2$.

2. Shade the outer boundary of the half-planes described by
   (i) $y > x$, (ii) $x + y > 1$, (iii) $2x + 3y > 6$.

3. Shade the outer boundary of the half-planes described by
   (i) $x + y < 1$, (ii) $x + y < 3$, (iii) $x + y < -1$.

4. Shade the outer boundary of the region described by the three inequalities
   $$x > 0, y > 0 \quad \text{and} \quad x + y < 4.$$

5. Shade the outer boundary of the region described by the three inequalities
   $$x < 2, y < 4 \quad \text{and} \quad x + y > 3.$$

6. Shade the outer boundary of the region described by the four inequalities
   $$y > 0, y < x, 2x + 3y > 6, \text{ and } 2x + 3y < 12.$$

7. Write down three inequalities describing each of the regions in Fig. 14.7 (a) and (b).

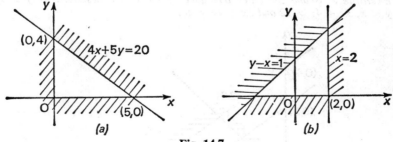

Fig. 14.7

8. Shade the outer boundary of the region described by the three inequalities
   $$y > 0, y < 3x \quad \text{and} \quad x + y < 4.$$

List all the points *inside* the region (not on the boundaries) which have integer coordinates.

9. Shade the outer boundary of the region described by the three inequalities

$$x > 0, 2y > x, x + y < 8.$$

List all the points inside the region whose y coordinate is an integer and whose x coordinate is an even integer.

10. Shade the outer boundary of the region described by

$$x > 0, y > 2x, x + y \leqslant 12.$$

Which point on the boundary of this region has the greatest x coordinate? Which point has the greatest y coordinate?

Exercise 14.1b

Take a scale of 2 cm to 1 unit on each axis, and draw small sketch graphs to illustrate the following:

1. Shade the outer boundary of the half-planes described by

   (i) $y < 0$, (ii) $x > -3$, (iii) $y < 3$.

2. Shade the outer boundary of the half-planes described by

   (i) $y + x < 0$, (ii) $x + y < -1$, (iii) $2x + 3y < -6$.

3. Shade the outer boundaries of the half-planes described by

   (i) $x - y < 1$, (ii) $x - y < 3$, (iii) $x - y < -3$.

4. Shade the outer boundary of the region described by the three inequalities

   $$x > 0, y > 0 \quad \text{and} \quad 2x + y < 4.$$

5. Shade the outer boundary of the region described by the three inequalities

   $$x < 3, y < 4 \quad \text{and} \quad 4x + 3y > 12.$$

6. Shade the outer boundary of the region described by the four inequalities

   $$x > 0, y > \tfrac{1}{2}x, x + y > 1 \quad \text{and} \quad x + y < 3.$$

7. Write down three inequalities describing each of the regions in Fig. 14.8 (a) and (b).

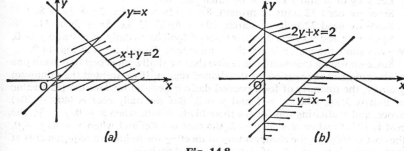

Fig. 14.8

8. Shade the outer boundary of the region described by the three inequalities

   $$x > 0, y > 2x \quad \text{and} \quad x + y < 6.$$

List all the points inside the region which have integer coordinates.

9. Shade the outer boundary of the region described by the four inequalities

$$x > 0, y > 0, 2x + y < 12, x + 2y < 12.$$

List all the points inside the region whose x and y coordinates are both even integers.

10. Shade the outer boundary of the region described by the four inequalities

$$x > 0, y > 0, 3x + y \leqslant 6, x + 3y \leqslant 6.$$

Which point on the boundary of this region is such that the sum of its x and its y coordinate is greatest?

Linear programming

In some of the later questions in Exercise 14.1 a and b we saw that we could find the greatest value of an unknown, say x, subject to the given conditions. This often enables us to solve problems in which we want to maximize a certain quantity, say profit, or minimize another quantity, say costs. This branch of mathematics has been developed only since about 1940, and is of great importance to commerce and industry. Of course their problems have more than two unknowns, and the constraints are not necessarily linear, but these examples do suggest some of the harder problems that can be solved by linear programming.

Example 3. The number of units of protein and starch contained by each of two foods A and B are shown in the table.

	Protein	Starch	Cost per kg
A	8	10	40p
B	12	6	50p
Minimum daily requirement	32	22	

What is the cheapest way of satisfying the minimum daily requirements?

Let x kg of A and y kg of B be eaten each day.

Since we need 32 units of protein, $8x + 12y > 32$, i.e. $2x + 3y > 8$.

Since we need 22 units of starch, $10x + 6y > 22$, i.e. $5x + 3y > 11$.

Since x kg and y kg are the quantities of food eaten daily, $x > 0$ and $y > 0$. We can show the region in which we must look for solutions in Fig. 14.9.

Since we want a minimum, we realise that we shall almost certainly want one of the points at the vertices of the shaded region. If we wanted the minimum *quantity*, the quantity of food needed daily is $(x + y)$ kg, and the least value of that is 3 kg, when $x = 1$ and $y = 2$. But the daily cost is $(40x + 50y)$ pence, and evaluating this for the three likely points, when $x = 0, y = 3\frac{2}{3}$, the cost is $183\frac{1}{3}$p; when $x = 1, y = 2$, the cost is 140p and when $x = 4, y = 0$, the cost is 160p, so the cheapest way of meeting the minimum requirements at these prices is to eat 1 kg of A and 2 kg of B each day.

Variations on this problem

Suppose that we can buy A more cheaply, say at 30p a kg. The constraints are the same, so that we can still look at Fig. 14.9 for likely solutions. Finding the cost corresponding to solutions at each of the

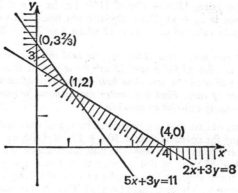

Fig. 14.9

vertices, when $x = 0$, $y = 3\frac{2}{3}$, the cost is unchanged at $183\frac{1}{3}$p: when $x = 1$, $y = 2$, the cost is now 130p, and when $x = 4$, $y = 0$, the cost is now 120p, so the cheapest way of meeting these requirements now is to eat only 4 kg of A and none of B.

Suppose instead that the price of B rises to 60p a kg, the price of A remaining at 40p a kg. The costs at the critical points are 220p at $(0,3\frac{2}{3})$, 160p at $(1,2)$ and 160p at $(4,0)$, so that it will be just as cheap whether we eat 1 kg of A and 2 kg of B, or whether we eat 4 kg of A and none of B. This may be surprising, and if we consider the cost corresponding to any other point on $2x + 3y = 8$, say $(2\frac{1}{2},1)$, we find the cost is also 160p. Thus any value of x between 1 and 4 and the corresponding value of y between 2 and 0, will give equally good solutions to the problem.

We can see that we could easily introduce additional constraints. We might insist on at least 2 kg of A each day, so that $x \geqslant 2$; we might only have enough of B to be able to have $1\frac{1}{2}$ kg a day, so $y \leqslant 1\frac{1}{2}$, or we might wish to consider also other nutrients in A and B.

When solving linear programming problems, choose the unknowns x and y so that they represent numbers which describe the quantity required of each of the variables. The following examples obtain the inequalities, and leave the reader to draw the graphs to obtain the solutions.

Example 4. *A gardener wishes to plant not more than 23 m² of ground with blackcurrant and gooseberry bushes. He allows 1 m² of ground for a blackcurrant bush and 2 m² for a gooseberry bush. A blackcurrant bush costs £1.20 and a gooseberry bush costs 40p; he does not wish to spend more than £11.60. What is the greatest number of bushes he can plant?*

Let the number of blackcurrant bushes be x, of gooseberry bushes be y. Then from the constraint of the ground available, $1x + 2y < 23$. From the constraint of the price, $120x + 40y < 1160$, i.e. $3x + y < 29$. The number of bushes planted is $(x + y)$. Draw the straight lines and we show that the greatest acceptable value of $(x + y)$ is 15 when $x = 7$, $y = 8$.

Example 5. *A car park, area 1400 m^2, is laid out for cars and for vans. 10 m^2 of space is allowed for a car, 15 m^2 for a van. It is estimated that the number of vans will never be less than half the number of cars, nor more than twice the number of cars. Find how many spaces must be marked for cars, in order to park as many vehicles as possible.*

Let the number of cars that can be parked be x, and the number of vans be y. Then, considering the space available, $10x + 15y < 1400$, i.e. $2x + 3y < 280$. Since the number of vans is not less than half the number of cars, $y > \frac{1}{2}x$, and since the number of vans is not greater than twice the number of cars, $y < 2x$. Draw three straight line graphs to describe a region, and find the point in that region which maximizes $(x + y)$. The greatest number of vehicles that can be parked is 120.

Example 6. *A smaller car park is to be laid out, subject to the same restrictions as in Example 5, but with only 500 m^2 space available.*

Use two of the inequalities from Example 5, and find a third inequality. Notice the difficulty when we try to find the largest number of cars that can be parked—we have to have a whole number of cars. The 'best' points with integer coordinates are (28,14), (27,15), (26,16), all of which enable the car park to hold 42 vehicles.

Exercise 14.2a

1. A factory manager is planning to buy two types of machines. Type X needs 3 m^2 of floor space, type Y needs 2 m^2 and he has 40 m^2 available. The cost of type X is £20, of type Y £100 and he can spend up to £900. Find the greatest number of machines he can buy.

2. A farmer is going to sow oats and barley. He estimates that oats need 4 men per hectare, barley 6 men, and he has 26 men available. Oats cost £12 per hectare, barley £8 per hectare, and he is prepared to spend up to £48. Find the greatest possible area of land he can sow.

3. A dealer is going to buy radio and television sets. He intends to buy a total of 100 sets. A radio set costs £40, a television set £120 and he is prepared to spend £10 400. The profit on a radio set is £16, on a television £32. Find the maximum profit he can make.

4. Two foods A and B contain respectively 4 and 6 units of protein, and 5 and 3 units of starch. The cost of A is 40p a kg, of B is 50p a kg. If the minimum daily intake is to be at least 16 units of protein and 11 units of starch, what is the cheapest way of meeting those conditions?

5. A shepherd is planning a sheepfold using an existing wall as one side of the sheepfold. He will use hurdles 2 m in length and there will be no overlap of hurdles. The breadth of the sheepfold must be greater than 20m, and the length greater than 40 m. He has 45 hurdles. List the possible lengths and breadths of the sheepfold, and find which has the greatest area.

Exercise 14.2b

1. A tobacconist proposes to buy up to 500 pipes. He has the choice of two kinds, one at £1 each and the other at £3 each, and he can spend up to £1100. The profit on a £3 pipe is twice that on a £1 pipe. How many of each kind should he buy to make as large a profit as possible?

2. The number of units of vitamins A and B per kg of two breakfast cereals X and Y are shown in the table.

	Vitamin A	Vitamin B
X	16	8
Y	18	4

The minimum daily intake required is 60 units of A and 40 units of B. What is the least total weight of breakfast cereal a man must eat to have enough of these vitamins?

3. A man has 7 guests in his house and is prepared to spend £1 on papers for them. *The Financial Times* costs 10p and *The Sporting Record* costs 12½p. He wishes each guest to have at least one of these papers, but obviously no guest wants more than one copy of either paper. Six guests he knows insist on *The Financial Times*. List all possible solutions to his problem of how many of each paper to buy.

4. A factory manager wishes to install two types of machines, small and large. Small machines need 2 operators and occupy 4 m² of floorspace; large machines need 3 operators and 8 m² of floorspace. There are up to 56 operators available and 136 m² of floorspace. The profit per week is £3 on a small machine and £5 on a large. Find the greatest weekly profit.

5. The Post Office is planning to sell books of stamps containing only 6p and 8p stamps. The cost of the book must be less than £2, and must must contain at least 30 stamps. What is the greatest number of 8p stamps there could be in the book? If the stamps are printed in multiples of four, and there must be some 8p stamps in the book, how many of each can be included in the book?

Cost lines, profit lines

It may have seemed lucky that the points which gave the least cost in Example 3 were the ones we found by inspection at the vertices of the region, and we may have wondered if we could find a systematic method of finding the best points.

The daily cost is $(40x + 50y)$p; if the daily cost is 200p, x and y can have any values such that $40x + 50y = 200$, i.e. $4x + 5y = 20$. If the daily cost is 300p, x and y are such that $4x + 5y = 30$; if the daily cost is 400p, then x and y satisfy $4x + 5y = 40$. Draw the graphs of these lines and we see they are parallel (Fig. 14.10), and the one corresponding to the least cost is that nearest to the origin O.

Any cost line for this problem must be parallel to $4x + 5y = 0$, so to find the least cost graphically we return to the original graph and draw on it the line $4x + 5y = 0$. We then draw parallel lines until we

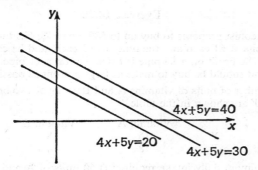

Fig. 14.10

find one which just includes an acceptable point. That is the line of least cost.

We should want to *maximize* profits, in which case we draw the line parallel to a given 'profit' line which is as far as possible away from the origin.

To draw a 'cost' or 'profit' line like $4x + 5y = 0$, we can draw the straight line through the origin and $(5, -4)$ or we can take any easy

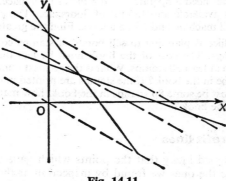

Fig. 14.11

multiple of the coefficients, here 4 and 5, and draw the straight line $4x + 5y = 20$. This has the advantage of avoiding negative values of x and y, and we may have taken our coordinate axes near the edge of our graph paper.

Transport problem

The hardest type of problem we consider at this stage is that facing a mythical transport manager. He has two depots D_1 and D_2, holding

120 tonnes of cement and 40 tonnes of cement respectively. He has two customers C_1 and C_2 who have ordered 80 and 50 tonnes of cement respectively. C_1 is 20 km from D_1 and 40 km from D_2; C_2 is 15 km from D_1 and 30 km from D_2 and delivery costs are proportional to the distance travelled. How should he supply his customers to minimize the transport costs?

Let him supply C_1 with x tonnes from D_1; then C_1 must receive $(80 - x)$ tonnes from D_2; if he supplies C_2 with y tonnes from D_1, C_2 must have $(50 - y)$ tonnes from D_2. It often helps if we set this out in a matrix, and use another matrix to show the distances between customers and depot.

	D_1	D_2	D_1	D_2
C_1	x	$80 - x$	20	40
C_2	y	$50 - y$	15	30
Maximum available	120	40		

Obviously $x \geqslant 0$ and $y \geqslant 0$. Since there is only 120 tonnes in D_1 $x + y \leqslant 120$; since there is only 40 tonnes in D_2, $(80 - x) + (50 - y) \leqslant 40$, i.e. $x + y \geqslant 90$. C_1 and C_2 must each receive positive quantities from the depots, so $(80 - x) \geqslant 0$ and $(50 - y) \geqslant 0$, i.e. $x \leqslant 80$ and $y \leqslant 50$. These inequalities enable us to draw the region in which we must look for solutions.

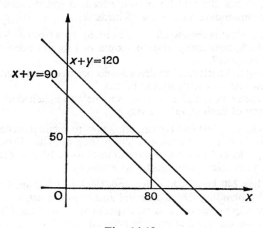

Fig. 14.12

The transport costs are proportional to the distance travelled, and the distance travelled is

$$20x + 15y + 40(80 - x) + 30(50 - y) \text{ km}$$

i.e.
$$(4700 - 20x - 15y) \text{ km}$$

Draw lines parallel to $20x + 15y = 300$, and we see that the best solution is at (40,50). He must supply his customers as shown in this matrix.

	D_1	D_2
C_1	40	40
C_2	50	0

Exercise 14.3

1. A car park is to be built with spaces for cars or small lorries. Cars are allowed 15 m², lorries 30 m², and up to 6000 m² of land is available. There must be space for at least 50 lorries, and the car park must have at least twice as many spaces for cars as lorries. What is the largest number of vehicles it could hold? If lorries pay 50p an hour and cars 20p, and the car park is always full, what number of car spaces and lorry spaces will bring in most money?

2. A rectangular metal sheet is to be made from metal squares of side 1 cm, so that its length and breadth must each be a whole number of centimetres. The perimeter of the sheet must be greater than 10 cm but less than 18 cm. The length must be greater than the breadth but less than twice the breadth. List all possible lengths and breadths of the sheet.

3. A small farmer wishes to keep a few chickens and ducks. Chickens cost 40p each, ducks 20p and he cannot afford to spend more than £8. He cannot accommodate more than 25 birds.

   (i) If he obtains each week twice as many eggs from a chicken as from a duck, how many of each should he buy to obtain as many eggs as possible?

   (ii) If instead each chicken gives 5 eggs per week and each duck 4 eggs, how many of each should he buy?

   (iii) If each chicken gives 4 eggs per week and each duck 5 eggs, how many of each should he buy?

4. A boy has £2 to spend and is wondering how to spend it on coke and buns. Coke costs 5p a can and buns 4p each. A shop only has 40 buns, and coke is sold in packs of 12 cans and buns in packets of 10, and these cannot be split. List all possible solutions to his problem.

5. A grocer has 240 kg of China tea and 250 kg of Indian. He blends two mixtures by blending China and Indian tea in the ratios 4 : 1 and 1 : 5. The first mixture gives twice as much profit as the second. Find how many of each he should blend to give the greatest profit.

6. A dealer buys three oils A, B, C and blends them to make two brands X and Y. Brand X contains A, B and C in the ratio 1 : 1 : 3, and brand Y contains A, B and C in the ratio 5 : 1 : 1. The profit on X is twice that on Y. The dealer has 100 tonnes of A, 80 tonnes of B and 104 tonnes of C. How many tonnes of X and Y should he make to give the maximum profit?

7. The quantities available of three chemicals X, Y and Z are 120 tonnes,

160 tonnes and 170 tonnes respectively. Three blends A, B and C are made from them by mixing in the following ratios:

	X	Y	Z
A	2	2	1
B	1	1	3
C	1	2	2

How much of A, B and C will be made if all the chemicals are to be used? If the profit on A is £10 per tonne, on B £20 per tonne and on C £15 per tonne, find the total profit. Is this the greatest profit possible?

8. A gravel dealer has two quarries Q_1 and Q_2 which produce each week 3000 m^3 and 1500 m^3 respectively. Three builders B_1, B_2 and B_3 require each week 2000 m^3, 1500 m^3 and 1000 m^3 of gravel respectively. The distances in km between quarries and builders are given below:

	B_1	B_2	B_3
Q_1	7	4	2
Q_2	3	2	2

How should the builders be supplied, to keep the transport costs, proportional to the distance travelled, to a minimum?

15 Indices and Logarithms

Indices

We write $a \times a \times a \times a$ as a^4, and more generally, $a \times a \times a \ldots \times a$, the product of n as, as a^n. Thus

$$a^m \times a^n = (a \times a \times \ldots) \times (a \times a \times \ldots)$$

where the first bracket contains m as and the second bracket n as, so their product is the product of $(m + n)$ as, thus

$$a^m \times a^n = a^{m+n} \qquad \text{(i)}$$

Hence

$$\frac{a^{m+n}}{a^n} = a^m = a^{(m+n)-n}$$

$$\therefore \ a^p \div a^q = a^{p-q} \qquad \text{(ii)}$$

so that when multiplying we add the indices, when dividing we subtract the indices. Extending our multiplication, we have also

$$(a^m)^{\,n} = a^m \times a^m \ldots n \text{ times,}$$
$$= a^{mn} \qquad \text{(iii)}$$

Negative and fractional indices

$a^m \div a^m = 1$, for all m, yet when we subtract the indices we have $a^{m-m} = a^0$,

$$\therefore \qquad\qquad a^0 = 1.$$

Also $\qquad\qquad a^m \times a^{-m} = a^0 = 1$

$$\therefore \qquad\qquad a^{-m} = \frac{1}{a^m}.$$

If $\qquad\qquad (a^m)^n = a^p$, where $p = mn$

$\qquad\qquad\qquad a^m$ is the nth root of a^p,

i.e. $\qquad\qquad a^{p/n}$ is the nth root of a^p.

Since strictly a number may have more than one nth root, we say that $a^{p/n}$ is *one* value of the nth root.

Example 1. $\qquad$ (i) $\quad 3^0 = 1.$

$\qquad\qquad\qquad$ (ii) $\quad 3^{-2} = \dfrac{1}{3^2} = \dfrac{1}{9}.$

(iii) $27^{\frac{2}{3}} = (\sqrt[3]{27})^2 = 9.$

(iv) $27^{-\frac{2}{3}} = \dfrac{1}{27^{\frac{2}{3}}} = \dfrac{1}{9}.$

Exercise 15.1a

Write down the values of the following:

1. 3^{-3}.   2. $8^{\frac{2}{3}}$.   3. $8^{-\frac{1}{3}}$.   4. $16^{\frac{3}{2}}$.   5. $16^{-\frac{3}{2}}$.

6. $4^{\frac{3}{2}}$.   7. $27^{-\frac{1}{3}}$.   8. $100^{-\frac{3}{2}}$.   9. $(\sqrt[3]{2^6})$.   10. $\left(\dfrac{144}{169}\right)^{-\frac{1}{2}}$.

Exercise 15.1b

Write down the values of the following:

1. 2^{-3}.   2. $8^{\frac{1}{3}}$.   3. $8^{-\frac{1}{3}}$.   4. $25^{-\frac{1}{2}}$.   5. $125^{-\frac{1}{3}}$.

6. $9^{-\frac{1}{2}}$.   7. $64^{-\frac{1}{3}}$.   8. $16^{-\frac{3}{2}}$.   9. $(\frac{3}{8})^{-1}$.   10. $(\frac{16}{25})^{-\frac{1}{2}}$.

LOGARITHMS

The logarithm of a number to the base 10 is the power to which 10 must be raised to give the number.

Example 2.   (i) $\log_{10} 1000 = \log_{10} 10^3 = 3.$

(ii) $\log_{10} \frac{1}{100} = \log_{10} 10^{-2} = -2.$

(iii) $\log_{10} \sqrt{10} = \log_{10} 10^{\frac{1}{2}} = \frac{1}{2}.$

To every logarithmic equation, there is a corresponding index equation.

Example 3.   (i) *If* $\log_{10} 1000 = 3$, *then* $1000 = 10^3$.

(ii) *If* $\log_{10}x = y$, *then* $x = 10^y$.

The base 10 is the most familiar because 10 is our accepted unit for counting and logarithms to this base are most helpful in arithmetic calculations. Logarithms, however, may be found to any base.

The logarithm of x to the base a is written $\log_a x$.

If the base is not given, it is understood to be 10.

The logarithm of a number to the base a is the power to which a must be raised to give the number.

For example, if $\log_a x = y$, then $a^y = x$.

When a number is expressed as a power of a, then its logarithm to the base a is equal to the index.

Example 4.   (i) $\log_a a^3 = 3.$

(ii) $\log_a \sqrt{a} = \frac{1}{2}.$

(iii) $\log_2 8 = \log_2 2^3 = 3.$

(iv) $\log_5 5\sqrt{5} = \log_5 5^{\frac{3}{2}} = \frac{3}{2}.$

Exercise 15.2a

Write down the values of the following:

1. $\log_a a^4$.
2. $\log_2 16$.
3. $\log_6 6\sqrt{6}$.
4. $\log_2 8$.
5. $\log_4 2$.
6. $\log_4 8$.
7. $\log_{12} \sqrt{12}$.
8. $\log_7 \sqrt[3]{7}$.
9. $\log_5 25\sqrt{5}$.
10. $\log_5 \{(\sqrt[3]{5})(\sqrt{5})\}$.
11. $\log_{10} 100$.
12. $\log_2 8$.
13. $\log_5 5$.
14. $\log_4 16$.
15. $\log_8 1$.

Find x in questions 16 to 20.

16. $\log_{10} x = 2$.
17. $\log_{10} x = -2$.
18. $\log_3 x = 3$.
19. $\log_4 x = \frac{1}{2}$.
20. $\log_2 x = 1$.

Exercise 15.2b

Write down the values of the following:

1. $\log_a \sqrt{a}$.
2. $\log_3 27$.
3. $\log_3 \frac{1}{3}$.
4. $\log_3 \frac{1}{3^2}$.
5. $\log_3 9\sqrt{3}$.
6. $\log_3 \sqrt{27}$.
7. $\log_9 27$.
8. $\log_{27} 3$.
9. $\log_{27} 9$.
10. $\log_{10} \frac{10}{\sqrt{10}}$.
11. $\log_{10} 1$.
12. $\log_2 32$.
13. $\log_3 81$.
14. $\log_3 \frac{1}{27}$.
15. $\log_{16} \frac{1}{4}$.

Find x in questions 16 to 20.

16. $\log_{10} x = 4$.
17. $\log_{10} x = -1$.
18. $\log_4 x = 2$.
19. $\log_9 x = \frac{1}{2}$.
20. $\log_8 x = 2$.

Formulae connecting logarithms

There are three formulae which are true for logarithms to any base and which correspond to the three index formulae already given.
    These are

$$\log_a x + \log_a y = \log_a xy \qquad \text{(iv)}$$

$$\log_a x - \log_a y = \log_a \frac{x}{y} \qquad \text{(v)}$$

$$\log_a x^n = n \log_a x \qquad \text{(vi)}$$

They are proved from the corresponding index formulae in the following way.

Let $\qquad \log_a x = u$ and $\log_a y = v$.

Then $\qquad x = a^u$ and $y = a^v$.

$$\therefore xy = a^u \times a^v = a^{u+v}.$$

So $$\log_a xy = u + v = \log_a x + \log_a y \qquad \text{(iv)}$$

Also $$\frac{x}{y} = \frac{a^u}{a^v} = a^{u-v}$$

and, therefore, $$\log_a \left(\frac{x}{y}\right) = u - v = \log_a x - \log_a y \qquad \text{(v)}$$

Again $$x^n = (a^u)^n = a^{un}$$

and, therefore, $$\log_a x^n = un = n \log_a x \qquad \text{(vi)}$$

The first of these formulae tells us how to add logarithms. For example,

$$\log 2 + \log 5 = \log (2)(5) = \log 10 = 1,$$

and $\log_{12} 9 + \log_{12} 16 = \log_{12} (9)(16) = \log_{12} 144 = \log_{12} 12^2 = 2.$
The second tells us how to subtract logarithms.
For example,

$$\log 70 - \log 7 = \log \frac{70}{7} = \log 10 = 1,$$

and $\log_8 56 - \log_8 \left(\frac{7}{8}\right) = \log_8 \left(\frac{56}{7/8}\right) = \log_8 \frac{56 \times 8}{7} = \log_8 8^2 = 2.$

The third tells us how to find the logarithm of a power.

For example, $\log 8 = \log 2^3 = 3 \log 2. \quad \therefore \frac{\log 8}{\log 2} = 3.$

Also $$\log_7 \sqrt[3]{17} = \log_7 17^{\frac{1}{3}} = \tfrac{1}{3} \log_7 17.$$

These formulae are, of course, equivalent to the laws we apply in numerical calculation.

(1) To multiply two numbers, add their logarithms and find the antilog of the result.

(2) To divide two numbers, subtract their logarithms and find the antilog of the result.

(3) To raise a number to a given power, multiply the logarithm of the number by the power and antilog the result.

Exercise 15.3a

Simplify the following without using tables or calculator:

1. $\log 4 + \log 25.$ 2. $\log 20 - \log 2.$ 3. $\log_2 14 - \log_2 7.$

4. $\log_3 8.1 + \log_3 10.$ 5. $\dfrac{\log 16}{\log 2}.$ 6. $\log 2^2 + \log 5^2.$

7. $\dfrac{\log 27}{\log 3}.$ 8. $\dfrac{\log a^3}{\log a}.$ 9. $\dfrac{\log \sqrt{5}}{\log 5}.$

10. $\log_a \sqrt{a} + \log_a a^2.$

Exercise 15.3b

Simplify the following without using tables or calculator:

1. $\log 2 + \log 50$.   **2.** $\log_2 32 - \log_2 4$.   **3.** $\log_3 21 - \log_3 7$.

4. $\log_4 3.2 + \log_4 20$.   **5.** $\dfrac{\log 25}{\log 5}$.   **6.** $\log 3^2 + \log \frac{10}{9}$.

7. $\dfrac{\log 64}{\log 4}$.   **8.** $\dfrac{\log \sqrt{a}}{\log a}$.   **9.** $\dfrac{\log \sqrt[3]{5}}{\log \sqrt{5}}$.

10. $\dfrac{\log a}{\log \sqrt{a}}$.

Change of base

The formula which tells us how to change from one base to another is

$$\log_b x = \frac{\log_a x}{\log_a b}.$$

Suppose that $\log_b x = u$, so that

$$b^u = x.$$

Take logarithms to base a,

$$\log_a b^u = \log_a x.$$

or $u \log_a b = \log_a x.$

$$\therefore u = \frac{\log_a x}{\log_a b}.$$

Therefore $\log_b x = \dfrac{\log_a x}{\log_a b}.$

Worked Examples

There follow some worked examples illustrating how logarithm and index formulae are applied.

Example 5. Evaluate $\log_7 17$.

$$\log_7 17 = \frac{\log_{10} 17}{\log_{10} 7} = \frac{1.2304}{0.8451} = \frac{10^{0.0899}}{10^{\overline{1}.9270}} \quad \text{(using log tables)}$$

$$= 10^{0.1629}$$

$$= 1.455 \quad \text{(antilog tables).}$$

Example 6. Given that $x = a^2 b$ *and that* $y = a^3 \sqrt{b}$, *express* b *in terms of* x *and* y.

Eliminate a from the equations.

$$a^2 = \frac{x}{b} \quad \text{and} \quad a^3 = \frac{y}{\sqrt{b}}.$$

Therefore $\qquad a^6 = \dfrac{x^3}{b^3}$ or $\dfrac{y^2}{b}$,

and so $\qquad \dfrac{x^3}{b^3} = \dfrac{y^2}{b}$.

$$\therefore b^2 = \frac{x^3}{y^2} = x^3 y^{-2} \quad \text{and} \quad b = x^{\frac{3}{2}} y^{-1}.$$

Example 7. *Evaluate, without tables, 2 log 5 + log 36 − log 9.*

$$2 \log 5 + \log 36 - \log 9 = \log 25 + \log 36 - \log 9$$
$$= \log \frac{25 \times 36}{9}$$
$$= \log 100$$
$$= 2.$$

Example 8. *If $3^x = 8$, find x.*

Whenever the unknown is an index, take logs.

$$3^x = 8.$$
$$\therefore \log 3^x = \log 8.$$

or $\qquad x \log 3 = \log 8.$

$$x = \frac{\log 8}{\log 3} = \frac{0.9031}{0.4771}.$$
$$= \frac{10^{\bar{1}.9657}}{10^{\bar{1}.6786}}$$
$$= 10^{0.2771}$$
$$= 1.892.$$

Example 9. *If log y + 3 log x = 2, express y in terms of x.*

$$\log y + 3 \log x = \log y + \log x^3 = \log yx^3.$$
$$\therefore \log yx^3 = \log 100$$

or $\qquad yx^3 = 100 \quad \text{and} \quad y = \dfrac{100}{x^3}.$

Exercise 15.4: Miscellaneous

1. Find x from the formula $e^x = 10$, given that $e = 2.718$.
2. Find y in terms of x from the equation $\log y + 2 \log x = 2$.
3. Find y in terms of x from the equation $\log_6 y + 2 \log_6 x = 3$.
4. Express $\sqrt[3]{(8\bar{x}^{-3})^{-2}}$ as simply as you can.
5. Simplify the expression $\sqrt{\dfrac{25x^4}{16}} \times \left(\dfrac{x}{2}\right)^{-3}$.
6. Without using tables, write down the values of:

   (i) $\sqrt[3]{\dfrac{x^9}{27}}$, (ii) $3^0 \times 2^{-2}$, (iii) $(4x^{-8})^{\frac{3}{2}}$.

7. Given that $\log y - \frac{3}{4} \log x = 2$, find y in terms of x.

8. Without using tables, write down the values of:

   (i) $27^{-\frac{2}{3}}$,   (ii) $16^{\frac{3}{4}} \times 16^{\frac{1}{8}}$,   (iii) $\dfrac{\log 27}{\log 9}$.

9. Calculate the value of $\log_3 8$.

10. Calculate the value of $\log_{\sqrt{2}} 1.7$.

11. Calculate the value of $\log_{12} 14$.

12. Find the value of x if $2^x = \frac{1}{8}$.

13. Find the value of x^{-2y} if $x = 4.2$ and $y = 0.6$.

14. If $x = 5$ and $y = 6$, find the value of $(8x + 4y)^{x/y}$.

15. Evaluate, without tables, $4 \log 2 + \log 5 - \log 8$.

16. Given that $\log 12 = 1.0792$ and that $\log 24 = 1.3802$ each correct to 4 d.p., deduce the values of $\log 2$ and $\log 6$, without using tables.

17. If $a^{-2} = b^3$ and $a^{\frac{1}{2}}b = c^{\frac{1}{2}}$, express c in terms of a.

18. If $2^{2x} - 4(2^x) + 3 = 0$, find the possible values of x.

19. Evaluate, without tables, $3 \log 2 + \log 20 - \log 1.6$.

20. Evaluate, without tables, $2 \log_6 3 + \log_6 12 + \log_6 8 - \log_6 24$.

21. If $a^{\frac{1}{2}} = b^{\frac{1}{3}}c^2$ and $a^{\frac{1}{3}} = b^{\frac{1}{2}}c^3$, find a.

22. If $3^x = 4^2$, find x.

23. If $4^{x+1} = 2^{x-1}$, find x.

24. If $3^{x+1} = 4^{x-1}$, find x.

25. If $(3^x - 1)(3^x - 2) = 0$, what are the possible values of x?

26. If $3^{2x} - 4(3^x) + 3 = 0$, what are the possible values of x?

27. Simplify $2^x \times 4^{x-1} \times 8^{x+1}$.

28. Simplify $\dfrac{3^x \times 9^{x+1}}{27^{x-1}}$.

29. If $\log x + 2 \log y = 1$, find x in terms of y.

30. If $\log \sqrt{x} + \log y = 2 \log z$, express x in terms of y and z.

31. Write down the values of (i) $\log_2 8$, (ii) $\log_4 8$, (iii) $\log_{\sqrt{2}} 8$.

32. Simplify $2p^3 \times 3p^2 \times \frac{1}{4}p^{-1}$.

33. Simplify $2z^3y^2 \times 3z^2y \times (\frac{1}{4}zy)^{-1}$.

34. If $17^x = 35$, find x.

35. If $8^{x-1} = 16$, find x.

16 Variation

Direct proportion

If the ratio of y to x is always constant, then y is said to vary directly as x. The equation connecting the two quantities is $\frac{y}{x} = k$ or $y = kx$. The relationship may also be written, $y \propto x$, which reads 'y is proportional to x'. The graph of y plotted against x is a straight line through the origin. It is obvious that if x is doubled, y must also be doubled. If y is halved, x must be halved and so on.

Examples of direct proportion are common and a few are given.

1. The distance gone by a car moving at constant speed is directly proportional to the time taken.

2. The circumference of a circle is directly proportional to the radius.

3. The volume of a cylinder of given radius is directly proportional to the height.

Other cases of direct proportion

The volume of a cylinder is given by the equation $V = \pi r^2 h$. From this we see that the volume varies directly as the height but not as the radius. For a given height, $\frac{V}{r^2}$ is constant and V is said to vary directly as the square of the radius.

If y varies directly as the square of x, then $\frac{y}{x^2}$ is constant. This may be written $y \propto x^2$ and the algebraic relation is $y = kx^2$. If y varies directly as x^2 and x is trebled, y is 9 times greater than its previous value.

Again, the formula for the period of a simple pendulum is

$$T = 2\pi \sqrt{\frac{l}{g}}.$$

From this, we see that the periodic time varies directly as the square root of the length.

Example 1. The extension of a stretched string is directly proportional to its tension. If the extension produced by a tension of 8 newtons is 2 cm, find the extension produced by a tension of 12 newtons.

Method (i). $\dfrac{\text{Tension}}{\text{Extension}}$ is constant.

Let x cm be the extension produced. Then
$$\frac{8}{2} = \frac{12}{x}.$$
$$\therefore \quad x = 3.$$

Method (ii). Let $T = kx$, where T is the tension in newtons and x the extension in centimetres. (A different choice of units will merely alter the value of k.)

When $T = 8$, $x = 2$,
$$\therefore \quad 8 = 2k \quad \text{and} \quad k = 4.$$
So $\qquad\qquad T = 4x \quad$ and when $\quad T = 12$, $x = 3$.

Example 2. *The cost of electroplating a square tray varies as the square of its length. The cost of a tray 8 cm square is £15. Find the cost for a tray 12 cm square.*

Method (i). If C is the cost in pounds and x cm the side of the square, $\frac{C}{x^2}$ is constant.

$$\therefore \quad \frac{15}{64} = \frac{C}{144} \quad \text{and} \quad C = \frac{144 \times 15}{64} = 33.75.$$

The cost is £33.75.

Method (ii). The equation connecting C and x is $C = kx^2$.

When $C = 15$, $x = 8$.
$$\therefore \quad 15 = 64k \quad \text{and so} \quad k = \tfrac{15}{64}.$$

Therefore $\qquad\qquad C = \tfrac{15}{64}x^2$.

When $x = 12$, $\qquad\qquad C = \tfrac{15}{64} \times 144 = 33.75$.

The cost is £33.75.

Exercise 16.1a

1. Write down an equation connecting x and y, given that x metres of curtain material cost y pence.

2. A disc of given thickness and radius r cm is of mass m kg. Write down a connection between r and m.

3. Watering cans are made of the same height but of varying diameter. If a can of diameter d cm holds g litres, write down a connection between d and g.

4. The resistance R newtons to the motion of a car varies directly as the square of the speed, v km/h. Write down a connection between R and v.

5. If y varies directly as x and $y = 8$ when $x = 3$, find y when $x = 18$.

6. If y varies directly as the square root of x and $y = 12$ when $x = 4$, find y when $x = 9$.

7. Complete the following statements:
   (i) If $V \propto r^3$, then $r \propto$;
   (ii) If $T \propto \sqrt{l}$ then $l \propto$;
   (iii) If $S = kr^2$, then $r \propto$.

8. Models are made in different sizes of a ship. The mass of a model varies directly as the cube of its length. The mass of a model of length 3 cm is 1 kg. What is the mass of a model of length 12 cm?

9. The surface area of a sphere is proportional to the square of its radius. In order to treble the surface area of the sphere, in what ratio must the radius be altered?

10. The distance it is possible to see on a clear day varies directly as the square root of the height above sea level. At a height of 6 m above sea level, it is possible to see 10 km. What distance can be seen from a height of 54 m?

Exercise 16.1b

1. Squares of area A m^2 are cut from a piece of sheet metal. If the side of a square is x m, write down the equation connecting A and x.

2. If it takes t minutes to cut a circular lawn of radius r metres, write down the connection between t and r.

3. The radius of a sphere varies as V^n where V is its volume. What is the value of n?

4. The radius of a sphere varies as S^n where S is the surface area. What is the value of n?

5. If p varies directly as q and $q = 70$ when $p = 10$, find q when $p = 12$.

6. If y varies directly as the square root of x and y is 10 when $x = 1$, find y when $x = 4$.

7. Complete the following statements:
   (i) If y varies as x^4, then x varies as   ;
   (ii) If A varies as $V^{\frac{2}{3}}$, then V varies as   .

8. If $y - 3$ is directly proportional to x^2 and $y = 5$ when $x = 2$, find y when $x = 6$.

9. A solid sphere of radius 4 cm is of mass 16 kg. Find the mass of a sphere of the same material of radius 6 cm.

10. The time of revolution of a planet round the sun is proportional to $d^{\frac{3}{2}}$ where d is the distance of the planet from the sun. Compare the times of revolution of two planets, one of which is four times the distance of the other from the sun.

Inverse proportion

If $y \propto \dfrac{1}{x}$, y is said to be inversely proportional to x. The equation connecting x and y is $y = \dfrac{k}{x}$ or $xy = k$.

If y is plotted against $\dfrac{1}{x}$, the graph is a straight line through the origin.

If y varies inversely as x, y varies directly as $\dfrac{1}{x}$.

A few common examples of inverse proportion are given.

1. The time for a piece of work is inversely proportional to the number of men employed.

2. The volume of a gas at constant temperature is inversely proportional to the pressure.

3. The length of a rectangle of constant area is inversely proportional to the breadth.

Other cases of inverse proportion

One quantity may vary inversely as some power of another. Three examples are given.

(i) If $y \propto \dfrac{1}{x^2}$, y varies inversely as the square of x. The equation connecting x and y is $y = \dfrac{k}{x^2}$ or $yx^2 = k$.

(ii) If $y \propto \dfrac{1}{x^3}$, y varies inversely as the cube of x. The equation connecting x and y is $y = \dfrac{k}{x^3}$ or $yx^3 = k$.

(iii) If $y \propto \dfrac{1}{\sqrt{x}}$, y varies inversely as the square root of x. The equation connecting x and y is $y = \dfrac{k}{\sqrt{x}}$ or $y\sqrt{x} = k$.

Example 3. The electrical resistance of a wire varies inversely as the square of its radius. Given that the resistance is 0.4 ohms when the radius is 0.3 cm, find the resistance when the radius is 0.45 cm.

If R is the resistance in ohms and r the radius in cm, $R = \dfrac{k}{r^2}$.

When $\qquad R = 0.4,\ r = 0.3.$

$$\therefore\ 0.4 = \frac{k}{(0.3)^2} \quad \text{and} \quad k = 0.036.$$

Therefore the connecting equation is

$$R = \frac{0.036}{r^2}.$$

When $\qquad r = 0.45,\ R = \dfrac{0.036}{(0.45)^2} = 0.18$ ohm approximately.

Joint variation

The formula for the volume of a cone is $V = \frac{1}{3}\pi r^2 h$. Here the volume is a function of two variables, r and h, and the volume varies directly as the height and directly as the square of the radius. This is called joint variation, and V is said to vary jointly as the height and the square of the radius.

Example 4. The volume of a gas of given mass varies directly as the temperature and inversely as the pressure. Write down a formula for the volume.

Here
$$V \propto T \quad \text{and} \quad V \propto \frac{1}{P}.$$

Therefore
$$V = k\frac{T}{P}.$$

Given a set of corresponding values for V, T and P, we can find k.

Variation as the sum of two parts

The function $(ax + bx^2)$ obviously varies with x but it is neither directly proportional to x nor to x^2. The function is described as the sum of two quantities one of which (ax) varies directly as x and the other of which (bx^2) varies directly as the square of x. An alternative description is a function which varies partly as x and partly as the square of x.
N.B. Always read any question carefully to make sure whether we are dealing with joint variation or variation as the sum of two parts.

Example 5. The cost of printing a book varies partly as the number of pages and partly as the square of the number of diagrams. Write down a formula for the cost.

If C is the cost, x the number of pages and y the number of diagrams, then $C = ax + by^2$, where a and b are constants.

Exercise 16.2a

Write down formulae to express the following:

1. The volume of a cylinder varies directly as the square of the radius and directly as the height.
2. The height of a cone varies directly as its volume and inversely as the square of the radius.
3. The electrical resistance of a wire varies directly as the length and inversely as the square of the radius.
4. The resistance to motion of a car at high speeds is the sum of two parts, one of which varies as the velocity and the other as the square of the velocity.
5. The pressure of a given mass of gas varies directly as the temperature and inversely as the volume.

6. The cost of running a car is partly constant and partly varies as the number of miles travelled.

7. The time taken to dig a trench varies as the length of the trench and inversely as the number of men employed.

8. The volume of a tetrahedron varies jointly as the base area and the height.

9. The resistance to the motion of a train is partly constant and partly varies as the square of the velocity.

10. The distance travelled by a particle varies directly as the square of its speed and inversely as its acceleration.

Exercise 16.2b

1. The distance travelled by a particle varies jointly as the acceleration and the square of the time.

2. The time taken by a committee is partly constant and varies partly as the square of the number of members present.

3. The principal is proportional to the interest and inversely proportional to both the rate and the time.

4. The acceleration produced in a body is proportional to the force and inversely proportional to the mass.

5. The depth of water in a tank is proportional to the volume of water and inversely proportional to the area of cross section of the tank.

6. The distance gone by a particle varies partly as the time and partly as the square of the time.

7. The square of the time taken by a planet to go round the sun varies as the cube of its mean distance from the sun.

8. The time of a bus journey varies directly as the distance and inversely as the square root of the number of passengers who board the bus *en route*.

9. The time of swing of a pendulum varies directly as the square root of its length and inversely as the square root of the acceleration due to gravity.

10. The Kinetic Energy of a body is jointly proportional to its mass and the square of its velocity.

Worked Examples

Some worked examples on the different problems on variation we meet are now given.

Example 6. The electrical resistance of a copper wire of circular cross-section varies directly as the length and inversely as the square of the radius. Two wires have equal resistances and one is four times as long as the other. Find the ratio of their radii.

If R is the resistance, l the length and r the radius,

$$R = k\frac{l}{r^2}.$$

Suppose the first wire has resistance R', length l' and radius r'; and that the second wire had resistance R', length $4l'$ and radius x.

Then
$$R' = k\frac{l'}{r'^2} \quad \text{and} \quad R' = k\frac{4l'}{x^2}.$$

$$\therefore \frac{l'}{r'^2} = \frac{4l'}{x^2} \quad \text{or} \quad x^2 = 4r'^2.$$

$$\therefore x = 2r'.$$

$\therefore$ The ratio of their radii is $1 : 2$.

Example 7. *The Kinetic Energy (E) of a body varies directly as its mass m and directly as the square of its velocity v. The momentum M varies jointly as the mass and the velocity. Show that if E is expressed in terms of M and m, then E varies as the square of M and inversely as m.*

$$E = k_1 mv^2 \quad \text{and} \quad M = k_2 mv.$$

(Notice that the constants k_1 and k_2 must not be assumed equal.)
From the second equation,

$$v = \frac{M}{k_2 m}$$

Substitute,
$$E = k_1 m \cdot \frac{M^2}{k_2^2 m^2} = \frac{k_1}{k_2^2}\frac{M^2}{m}.$$

Since $\dfrac{k_1}{k_2^2}$ is a constant, E varies as the square of M and inversely as m.

Example 8. *The cost of making a table is the sum of two parts. One is proportional to the area and the other to the square of the length. If the cost of a table 2 m by 3 m is £50 and the cost of a table 1.5 m by 4 m is £64, find the cost of a table 2.5 m square.*

If l is the length of the table in metres, A the area in m^2 and C the cost in pounds,

$$C = aA + bl^2, \quad \text{where } a \text{ and } b \text{ are constants.}$$

When $l = 3$ and $A = 6$, $C = 50$.

$$\therefore 50 = 6a + 9b \tag{i}$$

When $l = 4$ and $A = 6$, $C = 64$

$$\therefore 64 = 6a + 16b \tag{ii}$$

Subtract (i) from (ii)
$$14 = 7b$$
$$b = 2$$

Substitute:
$$50 = 6a + 18$$
$$\therefore a = 5\tfrac{1}{3}$$

The equation connecting C, A and l is
$$C = \frac{16A + 2l^2}{3}$$

When $l = 2.5$ and $A = 6.25$,

$$C = \frac{100}{3} + \frac{25}{2} = 45\tfrac{5}{6}.$$

The cost is therefore £45.83.

Exercise 16.3: Miscellaneous

1. If y is inversely proportional to x and $y = 2\tfrac{1}{2}$ when $x = 2$, find y when $x = 4$.

2. The greatest mass which can be supported by a beam of given thickness varies directly as the breadth and inversely as the length. If a beam of breadth 2 cm and length 15 m can support a mass of 200 kg, find the mass which can be supported by a beam 3 cm broad and 20 m long.

3. If Q varies inversely as the square of P and if $Q = 8$ when $P = 2$, find Q when $P = 4$.

4. A motorist estimates that his annual expenditure is partly constant and partly varies as the distance travelled. The cost for an annual distance of 5000 km is £250 and for an annual distance of 6000 km is £275. Find the cost for an annual distance of 8000 km.

5. The number of lead shot that can be made from a given weight of metal varies inversely as the cube of the radius of the shot. If 5000 shot of radius 1 mm can be made, find how many shot of radius $\tfrac{1}{2}$ mm can be made from an equal weight of metal.

6. The time taken to sink a well varies partly as the square of the depth and partly as the cube of the depth. If it takes 60 hours to sink a well 10 m deep and $146\tfrac{1}{4}$ hours to sink a well 15 m deep, find the time taken to sink a well 20 m deep.

7. The cost of catering per head is partly constant and partly inversely proportional to the number of people present. If the cost per head for 40 people is 60p and 50 people 50p, find the cost per head for 100 people.

8. If x varies directly as the cube of y and if the square of y varies inversely as the cube of z, prove that x^2z^9 is constant.

9. The safe speed for a train rounding a corner is proportional to the square root of the radius. If the safe speed for a curve of radius 50 m is 25 km/h, what is the safe speed for a curve of radius 98 m?

10. If x varies inversely as y and y varies inversely as z, how does x vary with z?

11. The square of the velocity of a particle varies as the cube root of its distance from a fixed point. When the distance of the particle from this point is 54 m, its velocity is 12 m/s. What is its velocity when it is 16 m from the fixed point?

12. The illumination of a bulb varies inversely as the square of the distance. If the illumination is 4 candle-power at a distance of 4 m, what is the illumination at a distance of 3 m?

13. The attraction between two bodies is directly proportional to the product of their masses and inversely proportional to the square of the distance

between them. If each mass is doubled, how must the distance between them be altered to give the same attraction?

14. The heat generated by a current in a wire varies directly as the time, directly as the square of the voltage and inversely as the resistance. If the voltage is 50 volts and the resistance 40 ohms, the heat generated is 75 units per second. Find the heat generated in $\frac{1}{2}$ minute if the voltage is 40 volts and the resistance 50 ohms.

15. Electrical conductivity in a wire of circular cross section varies directly as the square of the radius and inversely as the length. The lengths of two wires are in the ratio 3 : 4 and their radii in the ratio 1 : 2. Find the ratio of their conductivities.

16. The cost of annealing a length of chain is the sum of two parts, one proportional to the length of chain and the other proportional to the square of the length. If the cost of annealing a chain of length l is £x and the cost of annealing a chain of length $2l$ is £y, find the cost of annealing a chain of length $3l$.

17. A solid sphere of radius 4 cm is of mass 64 kg. Find the mass of a shell of the same metal whose internal and external radii are 3 cm and 2 cm respectively.

18. A solid cone of height 8 cm is of mass 128 kg. A smaller cone of height 2 cm is cut away leaving a frustum. Find the mass of the frustum.

19. If V varies directly as the square of x and inversely as y, and if $V = 18$ when $x = 3$ and $y = 4$, find V when $x = 5$ and $y = 2$.

20. If $P \propto \dfrac{x^2}{z}$ and $z \propto xt$, find (i) how P varies with x and t, (ii) how P varies with z and t.

21. The resistance to the motion of a car is partly constant and partly proportional to the square of the velocity. When the velocity is 20 km/h, the resistance is 50 newtons; when the velocity is 30 km/h, the resistance is 100 newtons. Find for what velocity the resistance is 170 newtons.

22. The effort required to raise a load is partly constant and partly proportional to the load. The effort necessary for a load of 8 newtons is 6 newtons and for a load of 12 newtons is 8 newtons. Find the effort necessary for a load of 20 newtons.

23. The braking force necessary to stop a train is directly proportional to the weight of the train and the square of its velocity and inversely proportional to the distance gone before stopping. If one train is twice as heavy as another and moving twice as fast, find the ratio of the stopping distances, assuming equal braking forces.

24. A model is made of a ship. If the ratio of the displacement of the ship to that of the model is 8000 : 1, find the ratio of the areas of deck space.

17 Arithmetic and Geometric Progressions

ARITHMETIC PROGRESSIONS

If a sequence of terms is such that the difference between any term and the one immediately preceding it is constant, the terms are said to form an arithmetic progression. This difference is called the common difference. Examples of arithmetic progressions are:

(i) 5, 8, 11, 14. . . . ; common difference $+3$.
(ii) 3, -1, -5, -9. . . . ; common difference -4.
(iii) -2, $-\frac{3}{4}$, $+\frac{1}{2}$, $+1\frac{3}{4}$. . . . ; common difference $1\frac{1}{4}$.
(iv) a, $a + d$, $a + 2d$, $a + 3d$. . . . ; common difference d.

The *n*th term

Suppose the first term of an arithmetic progression (in future shortened to A.P.) is 7 and the common difference 3.

The second term is $\quad 7 + 1(3) = 10$.
The third term is $\quad 7 + 2(3) = 13$.
The fourth term is $\quad 7 + 3(3) = 16$.
The nth term is $\quad 7 + (n - 1)3 = 3n + 4$.

Similarly the nth term of an A.P. whose first term is a and whose common difference is d, is $a + (n - 1)d$.

Example. 1. Find which term 383 is of the series $5 + 8 + 11 + \ldots$.
The difference between the first term and 383 is 378 and the common difference is 3. The common difference must therefore be added to the first term (378/3) times or 126 times. Therefore 383 is the 127th term.

The arithmetic mean

If a, b, c are three consecutive terms of an A.P., then the common difference equals $b - a$ or $c - b$.

Therefore $\qquad\qquad b - a = c - b$
or $\qquad\qquad\qquad 2b = a + c$.

b is called the arithmetic mean of a and c. For example, the arithmetic mean of 3 and 15 is $\frac{1}{2}(3 + 15)$ or 9.

The sum of an A.P.

Suppose we wish to find the sum of 50 terms of the A.P.

$$3 + 5 + 7 + \ldots \tag{i}$$

The 50th term is $3 + 49(2) = 101$.
Write the series backwards and it is:

$$101 + 99 + 97 + \ldots \tag{ii}$$

Each term of series (i) added to the corresponding term of series (ii) is 104. There are 50 terms and so the sum of all terms is $50(104) = 5200$.

But we have added two equal series and so the sum of each series is 2600.

Now find in a similar way the sum of n terms of the A.P.

$$a + (a + d) + (a + 2d) + \ldots \tag{i}$$

The nth term is $a + (n - 1)d$. Call this l.
Write the series backwards and it becomes

$$l + (l - d) + (l - 2d) + \ldots \tag{ii}$$

Each term of series (i) added to the corresponding term of series (ii) gives $(a + l)$. There are n terms and so the total sum is $n(a + l)$. This is the sum of two equal series and so the sum of each is $\frac{1}{2}n(a + l)$.

But $l = a + (n - 1)d$ and so $a + l - 2a = (n - 1)d$.

$$\therefore S = \frac{n}{2}(a + l) = \frac{n}{2}\{2a + (n - 1)d\}.$$

Example 2. *Find the sum of 28 terms of the A.P. 3 + 10 + 17 + ...*

Here $a = 3$, $d = 7$ and $n = 28$.

$$S = \frac{n}{2}\{2a + (n - 1)d\} = \frac{28}{2}\{6 + 27(7)\} = 14(195) = 2730.$$

Exercise 17.1a

1. Find the 23rd term of the A.P. $-7, -3, +1, \ldots$.
2. Find the 40th term of the A.P. 8, 5, 2, $\ldots$.
3. Find the value of n given that 77 is the nth term of the A.P. $3\frac{1}{2}, 7, 10\frac{1}{2}, \ldots$.
4. Find the value of n given that -49 is the nth term of the A.P. 11, 8, 5, $\ldots$.
5. If, 3, x, y, 18 are in A.P., find x and y.
6. If -5, p, q, 16 are in A.P., find p and q.
7. Find the sum of 12 terms of the series $7 + 11 + 15 + \ldots$.
8. Find the sum of 18 terms of the series $2\frac{1}{2} + 3\frac{3}{4} + 5 + \ldots$.

9. What is the nth term of the A.P. 4, 9, 14, ...?
10. What is the nth term of the A.P. 10, 7, 4, ...?

Exercise 17.1b

1. Find the 50th term of the A.P. 100, 97, 94, ...
2. The first of the odd numbers is 1 and the second is 3. What is the 50th odd number?
3. Find n given that 697 is the nth term of the A.P. 4, 11, 18, ...
4. Is 280 a term of the A.P. $11 + 15 + 19 + \ldots$?
5. The 5th term of an A.P. is 9 and the 8th term is 27. Find the 6th term.
6. The 2nd term of an A.P. is 2 and the 6th term is -14. What is the first term?
7. Find the sum of the first 50 odd numbers.
8. Find the sum of n terms of the A.P. $3 + 6 + 9 + \ldots$
9. What is the nth term of the A.P. 6, 5.9, 5.8, ...?
10. How many terms are there in the series 1, 4, 7, ... $(6n - 2)$?

GEOMETRIC PROGRESSIONS

If, in a sequence of terms, each term is a constant multiple of the preceding, the terms are said to be in geometric progression (G.P.). This multiple is called the common ratio.

Examples of terms in G.P. are:

(i) 3, 6, 12, 24, ...; common ratio $+2$.
(ii) 8, 4, 2, 1, $\frac{1}{2}$, ...; common ratio $\frac{1}{2}$.
(iii) 2, -10, $+50$, ...; common ratio -5.
(iv) a, ar, ar^2, ar^3, ...; common ratio r.

The *n*th term

Suppose that 3 is the first term of a G.P. whose common ratio is 2.

The second term is $3(2)$.
The third term is $3(2^2)$.
The fourth term is $3(2^3)$.
The nth term is $3(2^{n-1})$.

Similarly the nth term of a G.P. whose first term is a and whose common ratio is r is ar^{n-1}.

The geometric mean

If x, y, z are three terms in G.P., the common ratio is equal to $\frac{y}{x}$ or to $\frac{z}{y}$.

$$\therefore \frac{y}{x} = \frac{z}{y} \quad \text{or} \quad y^2 = xz.$$

This is the condition for x, y, z to be three consecutive terms of a G.P. y is said to be the geometric mean of x and z, so the geometric mean of two numbers is the square root of their product.

For example, the geometric mean of 4 and 9 is $\sqrt{4 \times 9}$ or 6.

The sum of a G.P.

Suppose S denotes the sum of n terms of the G.P. whose first term is a and whose common ratio is r.

Then
$$S = a + ar + ar^2 + ar^3 + \ldots + ar^{n-2} + ar^{n-1} \qquad \text{(i)}$$

Multiply each side by r,

$$rS = ar + ar^2 + ar^3 + \ldots + ar^{n-1} + ar^n \qquad \text{(ii)}$$

Subtracting (ii) from (i).

$$S - rS = a - ar^n$$
$$\therefore \ S(1 - r) = a(1 - r^n)$$

or
$$S = \frac{a(1 - r^n)}{1 - r}.$$

If $r < 1$, both numerator and denominator of the fraction are positive and this is the most convenient form for S. If, however, $r > 1$, both top and bottom are negative and a more convenient form is

$$S = a\frac{r^n - 1}{r - 1}.$$

Example 3. *Find the sum of eight terms of the G.P. 2, 6, 18, ...*

Here $a = 2$, $r = 3$ and $n = 8$.

$$S = a\frac{r^n - 1}{r - 1} = 2\,\frac{3^8 - 1}{3 - 1} = 3^8 - 1 = 6560.$$

Exercise 17.2a

(Leave answers in index form)

1. What is the 40th term of the G.P. 2, 14, 98, ...?
2. What is the 17th term of the G.P. 16, -8, 4, ...?
3. Find n given that 1024 is the nth term of the G.P. 4, 8, 16, ...
4. Find n given that 3^{20} is the nth term of the G.P. 27, 81, 243, ...
5. Find the geometric mean of 18 and 72.
6. If 5, x, y, 40 are in G.P., find x and y.
7. Find the sum of 20 terms of the G.P. 4, 8, 16, ...

8. Find the sum of 17 terms of the G.P. 9, -3, $+1$, $-$. . . .

9. What is the nth term of the G.P. 6, 18, 54, ...?

10. What is the nth term of the G.P. 5, $-\frac{5}{2}$, $\frac{5}{4}$, $-\frac{5}{8}$, ...?

Exercise 17.2b

(Leave answers in index form)

1. What is the 24th term of the G.P. 5, 15, 45, ...?

2. What is the 18th term of the G.P. 3, -1, $\frac{1}{3}$, ...?

3. Find n given that $\dfrac{1}{2^{17}}$ is the nth term of the G.P. 16, 8, 4, ...

4. Is 2^n a term of the G.P. 81, 27, 9, ...?

5. Find the geometric mean of 35 and 140.

6. If 7, x, y, 189 are in G.P., find x and y.

7. Find the sum of 18 terms of the G.P. 3, 6, 12, ...

8. Find the sum of $2n$ terms of the G.P. 2, -6, 18, -54 . . . .

9. What is the 18th term of the G.P. 8, 2, $\frac{1}{2}$, ...?

10. What is the nth term of the G.P. x, $2xy$, $4xy^2$, ...?

Worked Examples

Three examples of typical problems on progressions are now given.

Example 4. Find approximately how many grains of corn are needed to put one on the first square of a chess board, two on the second square, four on the third square, and so on.

There are 64 squares. So the required number is

$$1 + 2 + 2^2 + 2^3 \ldots \text{ to 64 terms.}$$

This sum equals

$$\frac{2^{64} - 1}{2 - 1} = 2^{64} - 1.$$

$$\log 2 = 0.3010.$$
$$\log 2^{64} = 64 \times 0.3010 = 19.264.$$
$$\therefore \ 2^{64} = 1.837 \times 10^{19}.$$

The subtraction of 1 from 2^{64} will not affect this answer, which is correct to three significant figures only.

Example 5. Find a formulae for the sum of the first n odd numbers. Prove that the sum of the odd numbers from 1 to 125 inclusive is equal to the sum of the odd numbers from 169 to 209 inclusive.

$$S = 1 + 3 + 5 + \ldots \text{ to } n \text{ terms.}$$

But $\qquad S = \dfrac{n}{2}\{2a + (n - 1)d\} = \dfrac{n}{2}\{2 + (n - 1)2\} = n^2.$

The number of terms from 1 to 125 is 63.

Therefore $1 + 3 + 5 \ldots + 125 = 63^2.$

Consider

$(169 + 171 + 173 \ldots + 209)$ as

$$(1 + 3 + \ldots + 209) - (1 + 3 + \ldots + 167).$$

There are 105 terms in the first bracket and 84 in the second bracket.

Therefore $\quad 1 + 3 + \ldots + 209 = 105^2$

and $\quad\quad\quad 1 + 3 + \ldots + 167 = 84^2,$

and so $\quad 169 + 171 + \ldots + 209 = 105^2 - 84^2 = (189)(21)$

by the difference of two squares.

But $\quad\quad\quad\quad (189)(21) = 9.21.21 = 63^2.$

Example 6. If x, y, z are in G.P., prove that log x, log y, log z are in A.P.

The condition that x, y, z are in G.P. is $y^2 = xz.$

Take logs: $\quad\quad\quad \log y^2 = \log (xz)$

or $\quad\quad\quad\quad\quad 2 \log y = \log x + \log z.$

But this is the condition that $\log y$ should be the arithmetic mean of $\log x$ and $\log z.$

Exercise 17.3: Miscellaneous

1. Find the sum of n terms of the A.P. $1 + 2 + 3 + \ldots$. How many terms of this series are needed to give a sum greater than 500?

2. Find the sum of 20 terms of the A.P. $1 + 5 + 9 + \ldots$. Prove that the sum of n terms is equal to $n(2n - 1).$

3. A man puts £10 in the bank for his son on each of his birthdays from the first to the twentieth inclusive. If the money accumulates at 3% compound interest, what is the total value on the son's twenty-first birthday?

4. In an A.P., the first term is 6 and the common difference is 3. How many terms are needed to make a sum greater than 600?

5. A man's salary in 1960 was £2000 per annum and it increased by 10% each year. Find how much he earned in the years 1960 to 1969 inclusive.

6. A man is able to save £50 of his salary in a certain year and afterwards saves in any one year £20 more than he saved in the preceding year. How long does it take him to save £4370?

7. Insert 5 arithmetic means between 20 and 92.

8. Insert 3 geometric means between 15 and 1215.

9. Find how many terms of the G.P. $3 + 6 + 12 + \ldots$ are necessary to make the sum exceed 1000.

10. Find the values of x and y if $x, 2, 4\frac{1}{2}, y$ are (i) in A.P.; (ii) in G.P. Find the sum of twenty terms of the A.P.

11. One of these sets of numbers is in A.P. and the other in G.P.

$$\tfrac{1}{6}, \tfrac{1}{2}, \tfrac{5}{6}, \ldots ; \tfrac{1}{6}, \tfrac{1}{2}, \tfrac{3}{2}, \ldots$$

Find the sum of 18 terms of the A.P. and find the 7th term of the G.P.

12. The sum of n terms of a series for all values of n is $(n^2 + 3n)$. Find the first three terms of the series.

13. The third term of a G.P. is 18 and the sixth is 486. Find the first term and write down a formula for the sum to n terms.

14. If the first, second and fifth terms of an A.P. are three consecutive terms of a G.P., find the common ratio.

15. If the 16th term of an A.P. is 3 times the 4th term, prove the 23rd term is 5 times the 3rd term.

16. If the sum of the first $2n$ terms of an A.P. is equal to the sum of the next n terms, prove that the sum of the first $(n + 2)$ terms is $2a(n + 2)$, where a is the first term.

17. If the pth term of an A.P. is x and the qth term is y, find the rth term.

18. Find which is the first negative term of the A.P. 15, $13\frac{1}{2}$, 12, ... What is the sum of all the positive terms?

19. Find the sum of all numbers between 1 and 100 which are divisible by 3.

20. Find the sum of all numbers between 1 and 100 which are divisible by 3 but not by 5.

21. If the first term of a G.P. is a and the nth term b, find the sum of these n terms.

22. Find the sum of all integers between 50 and 100.

23. How many terms of the series $1 + 2 + 3 + \ldots$ are necessary to give a sum greater than 450?

24. The series of positive integers is divided in the following way: $1 + (2 + 3) + (4 + 5 + 6) + (7 + 8 + 9 + 10) + \ldots$ What is the first term of the nth group? What is the sum of the terms in the nth group?

25. The second term of an A.P. is $(x - y)$ and the 5th term is $(x + y)$. Find the third and fourth terms.

26. Find the sum of 20 terms of the series
$$\log 2 + \log 4 + \log 8 + \log 16 + \ldots$$

27. Find the sum of all the numbers which can be made using all the digits 1, 2, 3 and 4.

28. The 3rd, 5th, and 8th terms of an A.P. are consecutive terms of a G.P. Find the common ratio.

29. The arithmetic mean of two numbers is 15 and their geometric mean is 9. Find the numbers.

30. How many terms of the G.P. $2 + 4 + 8 + \ldots$ are necessary to give a sum greater than 10 000?

MATHEMATICAL STRUCTURE

MATHEMATICAL STRUCTURE

Notation

{ }	the set of
$n\{A\}$	the number of elements in the set A
$\{x: \ \}$	the set of elements x such that
$\in$	is an element of
$\notin$	is not an element of
$\mathscr{E}$ (or $\mathscr{U}$)	the universal set
$\emptyset$	the empty (null) set
$\cup$	union
$\cap$	intersection
$\subset$	is a subset of
A'	the complement of the set A
PQ	the operation Q followed by the operation P
$f: x \longmapsto y$	the function mapping the set X (the domain) into the set Y (the range)
f^{-1}	the inverse of the function f
$\mathbb{R}$	the set of all real numbers
$\mathbb{Z}$	the set of all integers
$\mathbb{Z}_+$	the set of all positive integers
$\mathbb{Q}$	the set of all rationals, e.g. $\frac{3}{4}$

SUMMARY OF CHAPTERS 18–21

Set theory

A **set** is a well-defined class of objects, so that we can tell without ambiguity whether any one object does or does not belong to that class.

The **empty** (or **null**) **set** is the set without any elements.

Any element(s) chosen from a set form a **subset** of that set; all subsets except the set itself and the empty set are called **proper** subsets.

Those elements common to two sets A, B, form a set called the **intersection** of A and B, written $A \cap B$; those elements in either A or B or both form a set called the **union** of A and B, written $A \cup B$.

The **complement** of a set A is the set of all elements in the universal set $\mathscr{E}$ which are not in A, and is denoted by A'.

The **cardinal number** of a set A is the number of elements in A, and is written $n\{A\}$; sometimes $n(A)$ is used by other writers.

Binary operations

A binary operation * is defined on two elements of a given set S, e.g. 'add 3 to 4', 'divide 5 by 6'.

A set S is **closed** under an operation * if $a*b \in S$ for all a, $b \in S$, e.g. when any one integer is added to any integer, their sum is an integer, so that the set $\mathbb{Z}$ of all integers is closed under addition.

An operation * is **commutative** over S if $a*b = b*a$ for all a, $b \in S$.

An operation * is **associative** over S if $a* (b*c) = (a*b)*c$ for all a, b, $c \in S$.

The **identity** element e for a given operation * is the element such that $a*e = e*a = a$ for all $a \in S$.

The **inverse** (a^{-1}) of an element a is such that $a*a^{-1} = a^{-1}*a = e$.

Relations, mappings, functions

A **relation** associates an element x of one set (the **domain** D) with one or more elements y of another set (the **range** R). The range can be the same set as the domain.

The element y is the **image** of x under that relation.

A **function** (mapping) is a relation under which every element in D has one and only one image in R, i.e. it is a one-one or a many-one relation.

A composite function fg is one in which first g maps an element x into $g(x)$, then f maps $g(x)$ into $fg(x)$. The inverse function is $g^{-1} f^{-1}$.

Groups

A set S is a **group** under an operation * if

    (i) S is closed under *,
    (ii) the operation * is associative over S,
    (iii) there is an identity element in S,
    (iv) every element $a \in S$ has an inverse $a^{-1} \in S$.

18 Set Theory

Definition

Watching a small child, we may be tempted to think that our own first action must have been to divide things that came within our reach into two classes, those we could eat and those we could not eat! Mathematicians call any well-defined class a **set**, and by well-defined we mean that we must be able to decide definitely whether any one object does or does not belong to that set. Thus the definition 'all edible objects' would not really be satisfactory. Is a button edible?

Nor could we define a set of the 'best ten football teams', or the 'most popular five singers'; even Example 1 (ii) would be unsatisfactory if we used 'main line' railway stations, unless we say how to determine whether a station is 'main line'.

Anything belonging to a set we call an element (or member) of that set. We list each identical element only once, so that the set of all current English postage stamps value less than 5p, at the time of writing, is

$$\tfrac{1}{2}p, \ 1p, \ 1\tfrac{1}{2}p, \ 2p, \ 2\tfrac{1}{2}p, \ 3p, \ 3\tfrac{1}{2}p, \ 4p, \ 4\tfrac{1}{2}p$$

and we do not say that there are many millions of each. By convention, we list the elements in a set between curly brackets,

$$\{\tfrac{1}{2}p, \ 1p, \ 1\tfrac{1}{2}p, \ 2p, \ 2\tfrac{1}{2}p, \ 3p, \ 3\tfrac{1}{2}p, \ 4p, \ 4\tfrac{1}{2}p\}$$

Example 1. List the elements in the following sets:

(i) *coins in current use in England,*
(ii) *'Inter-City' terminus railway stations in London, North of the Thames,*
(iii) *even numbers less than 13.*

At the time of writing, the elements of (i) are

$$\{\tfrac{1}{2}p, \ 1p, \ 2p, \ 5p, \ 10p, \ 50p\}$$

and the elements of (ii) are

{Liverpool Street, King's Cross, St. Pancras, Euston, Paddington, and Charing Cross}

At all times, the elements of (iii) are

$$\{2, \ 4, \ 6, \ 8, \ 10, \ 12\}$$

In each case, we could decide whether any element we considered did or did not belong to the set.

Example 2. List the elements of each of the following sets:

(i) *the prime numbers less than 20,*
(ii) *even numbers less than 15,*
(iii) *multiples of 4 less than 15.*

Call these sets A, B and C respectively, as we shall want to refer to them later. The elements of A are {2, 3, 5, 7, 11, 13, 17, 19},

of B are {2, 4, 6, 8, 10, 12, 14},

and of C are {4, 8, 12}.

Intersection and union

The sets A and B both contain the element 2, and that is the only element common to both. The set of all the elements common to sets A and B is called the intersection of A and B, written $A \cap B$, so that

$$A \cap B = \{2\}.$$

The set of all elements in A or B or both is called the union of A and B, written $A \cup B$, so that

$$A \cup B = \{2, 3, 4, 5, 6, 7, 8, 10, 11, 12, 13, 14, 17, 19\}.$$

Notice that $B \cap C = C$ and $B \cup C = B$.

Empty set

Sets A and C do not have any elements in common, so that the intersection of A and C does not have any elements. A set without any elements is called the empty (or null) set, written $\emptyset$, so that

$$A \cap C = \emptyset.$$

The empty set does not contain any elements at all, not even a zero.

Subsets

We have noticed that every element of C is a member of B, and we call C a **subset** of B. Strictly, we say that C is a **proper** subset of B, for C does not contain every element of B. Every set is a subset of itself but not a **proper** subset, nor is the empty set counted as a proper subset. Thus the proper subsets of C are

{4}, {8}, {12}, {4,8}, {4,12}, {8,12}.

It can be shown that there are 126 proper subsets of B, so don't try to list all of them!

Venn diagrams

The relationship between sets can often be clearly shown by a diagram of a sort first devised by John Venn (1834–1923), an English mathematician. He used ovals to represent sets, and the ovals overlapped if the sets had any element in common. Thus the relation between A and B is illustrated:

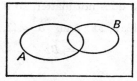

Fig. 18.1

between A and C by:

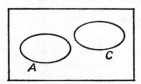

Fig. 18.2

while since B contains every element of C, their relation is illustrated:

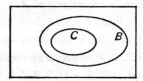

Fig. 18.3

Example 3. *If* $A = \{$*the prime factors of 30*$\}$,
        $B = \{$*the prime factors of 70*$\}$,
and $C = \{$*the prime factors of 42*$\}$,
list the elements of A, B and C and show their relationship in a Venn diagram.

The elements of the sets are

$$A = \{2, 3, 5\}$$
$$B = \{2, 5, 7\}$$
and $$C = \{2, 3, 7\}$$

and their relationship is shown in Fig. 18.4.

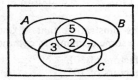

Fig. 18.4

Equal sets

Sets are only said to be equal if they have the same elements. The order in which these are listed does not matter, so if

$$A = \{2, 3, 5\} \quad \text{and} \quad B = \{2, 5, 3\},$$

$$A = B.$$

Exercise 18.1a

List the elements in $A \cap B$ and $A \cup B$ in each of Q.1–4, and draw a Venn diagram to show the relation between A and B.

1. $A = \{1, 2, 4, 5, 6\}$
   $B = \{1, 4, 6\}$

2. $A = \{1, 2, 3, 4, 5\}$
   $B = \{1, 2, 3\}$

3. $A = \{1, 2, 3\}$
   $B = \{1, 2, 3, 5\}$

4. $A = \{1, 3, 5, 7\}$
   $B = \{2, 4, 6, 8\}$

5. If $A = \{3, 4, 5\}$, list the proper subsets of A.

6. If $B = \{3, 4, 5, 6\}$, how many proper subsets has B?

7. If $A = \{7, 14, 21, 28\}$, describe in words the set A.

8. If $B = \{64, 81, 100, 121, 144\}$, describe in words the set B.

9. If $A = \{1, 2, 3, 4\}$ and $A \cup B = \{1, 2, 3, 4, 5, 6\}$, list two elements in the set B.

10. If $A = \{2, 4, 8, 16\}$ and $A \cup B = \{1, 2, 4, 8, 16\}$, find one element in the set B.

11. If $\quad A = \{\text{all even numbers less than } 10\}$,
    and $B = \{\text{all multiples of 3 less than } 10\}$,
    list the elements in A and B and draw a Venn diagram showing the relationship between A and B.

12. If $\quad A = \{\text{all positive numbers less than } 8\}$,
    and $B = \{\text{all primes less than } 10\}$,
    list the elements in the sets $A \cap B$; $A \cup B$.

13. If $\quad A = \{\text{all multiples of 3 less than } 20\}$,
    and $B = \{\text{all multiples of 6 less than } 20\}$,
    list the elements in $A \cap B$, $A \cup B$ and draw a Venn diagram showing the relation between the sets.

14. Draw Venn diagrams illustrating that $A \cup (B \cup C) = (A \cup B) \cup C$. (Assume the sets have some elements in common.)

15. If $\quad A = \{\text{all prime factors of } 30\}$,
    and $B = \{\text{all prime factors of } 105\}$,
    (i) list the elements of A and B,

(ii) draw a Venn diagram to illustrate the relation between A and B, marking each element in the diagram,

(iii) list the elements in $A \cap B$, $A \cup B$.

Find the H.C.F. and L.C.M. of 30 and 105.

(iv) If $C = \{$all prime factors of the H.C.F. of 30 and 105$\}$, and $D = \{$all prime factors of the L.C.M. of 30 and 105$\}$, list the elements of C and D.

(v) What do you notice about the sets $A \cap B$, $A \cup B$, C and D?

Exercise 18.1b

List the elements in $A \cap B$ and $A \cup B$ in each of the following, and draw a Venn diagram to show the relation between A and B.

1. $A = \{2, 4, 6, 8, 10\}$
$B = \{2, 3, 4, 8, 10\}$

2. $A = \{1, 2, 3, 4\}$
$B = \{1, 2, 3, 4, 5\}$

3. If $A = \{1, 2, 3\}$ and $A \cap B = \{2, 3\}$, can you list two elements in set B? Can you list all the elements of set B?

4. If $A = \{$all prime numbers between 10 and 30$\}$
$B = \{$all even numbers between 9 and 29$\}$
$C = \{$all multiples of 3 between 10 and 29$\}$
list the elements in each set and draw a Venn diagram illustrating the relation between the sets.

5. If $A = \{$all natural numbers less than 10 which are not multiples of 3$\}$
and $B = \{$all multiples of 6 less than 20$\}$,
list the elements in $A \cap B$, $A \cup B$.

6. If $A = \{$all multiples of 5 less than 30$\}$
and $B = \{$all multiples of 10 less than 30$\}$
list the elements of $A \cap B$, $A \cup B$ and draw a Venn diagram illustrating the relation between A and B.

7. Draw Venn diagrams illustrating $A \cup (B \cap C) \neq (A \cap B) \cup C$. (Again assume the sets have some elements in common.)

8. If $A = \{$all prime factors of 55$\}$
and $B = \{$all prime factors of 70$\}$
(i) list the elements of A and B,
(ii) draw a Venn diagram to illustrate the relation between A and B, marking each element in the diagram,
(iii) list the elements in $A \cap B$, $A \cup B$.
Find the H.C.F. and L.C.M. of 55 and 70.
(iv) If $C = \{$all prime factors of the H.C.F. of 55 and 70$\}$
and $D = \{$all prime factors of the L.C.M. of 55 and 70$\}$, list the elements of C and D.
(v) What do you notice about the sets $A \cap B$, $A \cup B$, C and D?

9. Repeat Q.8 using the numbers 42 and 110 instead of 55 and 70.

10. Repeat Q.9 using the numbers 24 and 36 instead of 42 and 110. Can you see why $C \neq A \cap B$ and $D \neq A \cup B$?

Number of elements in a set: cardinal number

The number of elements in a set A is called the cardinal number of the set and is written $n\{A\}$, thus

if $\quad\quad A = \{2, 3, 5\}, \quad n\{A\} = 3,$
if $\quad\quad B = \{2, 4, 6, 7\}, \quad n\{B\} = 4,$
and if $\quad C = \varnothing, \quad\quad\quad n\{C\} = 0.$

The universal set

When listing the elements in the sets in Example 2 we knew we had to consider only numbers. No point in wondering whether a railway station was a prime number! When listing the London railway termini on page 185, we considered only railway stations, then decided which were Inter-City termini, then which of these were North of the Thames. This 'background set' is called the universal set, and denoted by $\mathscr{E}$ (occasionally $\mathscr{U}$ or E). Thus in nearly all the examples considered so far the universal set could have been the set of all natural numbers, or even the set of all positive integers less than 200. The elements of any set A must have been chosen from the universal set, i.e. A is a subset of $\mathscr{E}$.

Example 4. If $\mathscr{E} = \{all\ positive\ integers\}$ and
$\quad\quad\quad\quad\quad A = \{all\ perfect\ squares\ less\ than\ 50\},$
then $\quad\quad\quad\quad A = \{1, 4, 9, 16, 25, 36, 49\}$
but if $\quad\quad\quad\quad \mathscr{E} = \{all\ even\ integers\}$
then $\quad\quad\quad\quad A = \{4, 16, 36\}$
as 1, 9, 25, 49 are not members of the universal set, and so cannot be chosen for A.

Example 5. If $\mathscr{E} = \{all\ pupils\ in\ your\ class\}$
and $\quad\quad\quad\quad A = \{all\ pupils\ with\ red\ hair\}$
the elements of A will not, unless $A=\varnothing$, be the same as if $\mathscr{E} = \{all\ children\ in\ the\ next\ class\}$.

It is usual to represent $\mathscr{E}$ in a Venn diagram by a rectangle.

Complement of a set

All the elements of the universal set which are not members of a set A themselves form a set called the complement of A, written A'. Thus if
$\quad\quad\quad\quad \mathscr{E} = \{1, 2, 3, 4, 5, 6\}.$
and $\quad\quad A = \{1, 2, 4, 5\}$
then $\quad\quad A' = \{3, 6\}.$

Example 6. If $\mathscr{E} = \{1, 2, 3, 4, 5, 6, 7, 8\}$
$\quad\quad\quad\quad\quad A = \{2, 4, 6, 8\}$
and $\quad\quad\quad\quad B = \{2, 3, 5\}$
list the elements A' and B'. Show that $(A \cap B)' = A' \cup B'$ and illustrate this by a Venn diagram.

The elements of A' are {1, 3, 5, 7}
and   of B' are {1, 4, 6, 7, 8}.
The elements of $A \cup B$ are {2, 3, 4, 5, 6, 8}
and so $(A \cup B)' = \{1, 7\}$.

The only elements common to A' and B' are 1 and 7, so that

$$A' \cap B' = \{1, 7\} = (A \cup B)'.$$

The area shaded in Fig. 18.5 can be described either as $A' \cap B'$ or as $(A \cup B)'$, illustrating that these sets are equal.

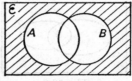

Fig. 18.5

Relation between sets of numbers

We have met various types of numbers, integers, rationals, positive and negative numbers; Venn diagrams illustrate clearly the relation between these sets. The universal set $\mathscr{E}$ we shall take as the set of all real numbers, knowing that all the numbers we have found are in this set. All integers, e.g. 2, are rationals, since they can be written $\frac{2}{1}$, so if

$$A = \{\text{all integers}\}$$
$$B = \{\text{all rationals}\},$$

A is a subset of B and the relation between A and B is shown in Fig. 18.6.

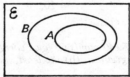

Fig. 18.6

No number is both positive and negative, so if $P = \{\text{all positive numbers}\}$ and $N = \{\text{all negative numbers}\}$, $P \cap N = \emptyset$, as illustrated in Fig. 18.7.

Fig. 18.7

Problems

Example 7. *A survey in a class shows that 15 of the pupils play cricket, 11 play tennis and 6 play both cricket and tennis. How many pupils are there in the class, if everyone plays at least one of these games?*

Draw a Venn diagram showing overlapping sets. Let $T = \{$all who play tennis$\}$, $C = \{$all who play cricket$\}$, $\mathscr{E}$ being the set of all pupils in this class. Then $n\{C \cap T\} = 6$, since 6 pupils play both cricket and tennis.

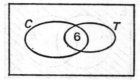

Fig. 18.8

Given that 15 pupils play cricket, there must be 9 of these who do not play tennis, i.e. $n\{C \cap T'\} = 9$. Similarly, there must be 5 pupils who do not play cricket, i.e. $n\{C' \cap T\} = 5$.

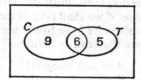

Fig. 18.9

From Fig. 18.9, $\qquad n\{C \cup T\} = 9 + 6 + 5$
$$= 20,$$

there are 20 children in the class.

Example 8. *In a school of 140 pupils 60 have cornflakes for breakfast and 90 have toast. What is* (i) *the greatest,* (ii) *the least, number who have both?*

Let $\qquad\qquad \mathscr{E} = \{$all pupils in the school$\}$,
$\qquad\qquad\qquad C = \{$all who have cornflakes$\}$,
$\qquad\qquad\qquad T = \{$all who have toast$\}$.

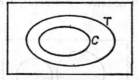

Fig. 18.10

Then if C is a subset of T (Fig. 18.10), $n\{C \cap T\} = 60$, so that 60 pupils have both.

But $n\{C'\} = 140 - 60$

$\qquad = 80,$

so as many as 80 pupils could be in $C' \cap T$ (Fig. 18.11),

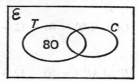

Fig. 18.11

i.e. $\qquad\qquad\qquad n\{C \cap T\} = 10,$

only 10 pupils have both. Thus the greatest number who can have both is 60 and the least is 10.

Example 9. *A newsagent sells three papers, the Echo, the Mail and the Advertiser. 70 customers buy the Echo, 60 the Mail, and 50 the Advertiser. 17 buy both the Echo and the Mail, 15 the Mail and the Advertiser and 16 the Advertiser and the Echo, while 3 customers buy all three papers. How many customers has he?*

Let $\qquad\qquad \mathscr{E} = \{$all who buy newspapers$\}$,

$\qquad\qquad\qquad E = \{$all who buy the *Echo*$\}$,

$\qquad\qquad\qquad M = \{$all who buy the *Mail*$\}$,

$\qquad\qquad\qquad A = \{$all who buy the *Advertiser*$\}$.

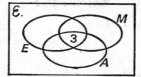

Fig. 18.12

Since 3 customers buy all three papers, $n\{E \cap M \cap A\} = 3$ (Fig. 18.12). There are 17 customers who buy the *Echo* and *Mail*,

so $\qquad\qquad n\{E \cap M \cap A'\} = 17 - 3$

$\qquad\qquad\qquad\qquad\qquad = 14.$

Similarly $\qquad\quad n\{E' \cap M \cap A\} = 15 - 3$

$\qquad\qquad\qquad\qquad\qquad = 12.$

and $\qquad\qquad n\{E \cap M' \cap A\} = 16 - 3$

$\qquad\qquad\qquad\qquad\qquad = 13$ (Fig. 18.13.)

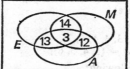

Fig. 18.13

Since 70 buy the *Echo*,

$$n\{E \cap (M' \cup A')\} = 70 - 14 - 13 - 3$$
$$= 40,$$

i.e. 40 buy only the *Echo*,

$$n\{M \cap (A' \cup E')\} = 60 - 12 - 14 - 3$$
$$= 31$$

i.e. 31 buy only the *Mail*,

and

$$n\{A \cap (E' \cup M')\} = 50 - 12 - 13 - 3$$
$$= 22,$$

i.e. 22 buy only the *Advertiser*.

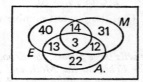

Fig. 18.14

From Fig. 18.14 we see that the newsagent has

$$40 + 14 + 31 + 12 + 3 + 13 + 22 \text{ customers,}$$

that is, 135 customers.

Exercise 18.2

1. If $\mathscr{E} = \{1, 2, 3, 4, 5, 6, 7, 8\}$
   $A = \{1, 2, 4, 8\}$
   $B = \{1, 3, 5, 7\}$
   $C = \{2, 4, 6, 8\}$
   list the elements in A', B', C'.

2. If $A = \{1, 3, 5, 7, 9\}$
   and $A' = \{2, 4, 8, 10\}$
   list the elements in $\mathscr{E}$.

3. If $\mathscr{E} = \{$John, Jill, James, Jane, Jack$\}$
   and $A = \{$John, Jill, Jane$\}$
   list the elements in A'.

4. If $\mathscr{E} = \{$oak, ash, elm, beech$\}$
   and $A' = \{$oak, beech$\}$
   list the elements in A.

5. If $A = \{$red, green, blue$\}$
   and $A' = \{$black, white$\}$
   list the elements in $\mathscr{E}$.

6. Given that $\mathscr{E} = \{1, 2, 3, 4, 5, 6, 7\}$,
   $A = \{1, 2, 3, 4\}$, $B = \{3, 4, 5, 6\}$,
   correct each of the following:

   (i) $A \cup B = \{1, 2, 4, 5, 6\}$

    (ii) $A \cap B = \{3, 4, 3, 4\}$

    (iii) $A' = \{4, 5, 6, 7, 8\}$

    (iv) $B' = \{1, 2\}$

    (v) $n(A \cup B) = 7$.

7. Given that $\mathscr{E} = \{0, 1, 2, 4, 8, 16\}$
   $A = \{1, 2, 4\}$ and $B = \{8, 16\}$
   correct each of the following:

    (i) $A \cup B = \mathscr{E}$

    (ii) $A \cap \mathscr{E} = \{0\}$

    (iii) $n\{A' \cup B'\} = 7$

    (iv) $A' = B$

    (v) $(A \cup B)' = \emptyset$.

8. Given that $\mathscr{E} = \{0, 1, 2, 3, 4, 5, 6, 7\}$
   and $A = \{0, 1, 2, 3\}$ and $B = \{5, 6, 7\}$,
   only one of the following is correct. Say which is correct, and give correct
   versions of each of the other statements:

    (i) $A' = B$

    (ii) $A' \cup B' = \mathscr{E}$

    (iii) $A' \cap B' = \emptyset$

    (iv) $n\{A \cap B\} = 1$

    (v) $n\{A' \cup B'\} = 9$.

9. Assuming that A and B have some elements in common, draw a Venn
   diagram, shading $A' \cap B$. Draw a second Venn diagram shading $A' \cup B$.
   Draw two more Venn diagrams to illustrate

   (i) $A \cup (A' \cap B) = A \cup B$, (ii) $A \cap (A' \cup B) = A \cap B$.

10. Repeat Q.9, if A and B do not have any elements in common.

11. Assuming A, B and C have elements in common, draw a Venn diagram
    shading the area representing $A' \cup B' \cup C'$. Hence verify that
    $A' \cup B' \cup C' = (A \cap B \cap C)'$.

12. Verify in a similar manner that $A' \cap B' \cap C' = (A \cup B \cup C)'$.

13. Verify in a similar manner that $(A \cup B') \cap (A \cup B) = A$.

14. If $A = \{$all even numbers$\}$
    and $B = \{$all primes$\}$
    draw a Venn diagram illustrating the relation between A and B.

15. If $A = \{$all primes$\}$
    and $B = \{$all perfect squares$\}$,
    draw a Venn diagram illustrating the relation between A and B.

16. All the 120 pupils in a certain school learn French or German or both.
    75 learn French and 60 learn German. How many learn both?

17. Every man in a certain club owns either a Rolls-Royce or an Aston
    Martin. 23 own Rolls-Royces, 14 own Aston Martins, and 5 own both.
    How many men are there in the club?

18. A survey of 150 householders, all of whom kept a cat or a dog or both,
    showed that 85 kept cats and 70 kept dogs. How many kept both?

19. A survey of 70 men who had spent a day playing cricket or golf showed
    that 40 played cricket and 8 played both cricket and golf. How many
    played golf?

20. All the pupils in a certain school win prizes for Regular Attendance or Good Conduct. 55 win prizes for Regular Attendance, 45 win prizes for Good Conduct, and 25 win prizes for both. How many pupils are there in the school?

21. Given that $n\{X\} = 18$, $n\{Y\} = 24$ and $n\{X \cup Y\} = 40$, find $n\{X \cap Y\}$.

22. Given that $n\{X\} = 17$, $n\{Y\} = 13$ and $n\{X \cap Y\} = 5$, find $n\{X \cup Y\}$.

23. Given that $n\{X\} = 18$, $n\{X \cap Y\} = 7$, $n\{X \cup Y\} = 40$, find $n\{Y\}$.

24. Given that $n\{Y\} = 10$, $n\{X \cup Y\} = 26$, $n\{X \cap Y\} = 5$, find $n\{X\}$.

25. Given that $n\{X\} = 19$, $n\{Y\} = 20$ and $n\{X \cup Y\} = 39$, find $n\{X \cap Y\}$.

26. A boy has 50 marbles. 35 had some red markings, 20 some blue markings and 12 both red and blue. How many had neither red nor blue markings?

27. There are 22 masters at a certain school. 8 read *The Financial Times*, 10 read *The Racing World*. Find
    (i) the greatest,   (ii) the least
    number who read neither.

28. In a certain road of 100 houses 70 have doorbells and 45 have knockers. What can you say about the number n that have both?

29. The captain of a cricket eleven knows that 7 of his team can bat and 5 can bowl. What can you say about the number n who can both bat and bowl?

30. In a certain examination, 72 candidates offered Mathematics, 64 English and 62 French. 18 offered both Mathematics and English, 24 Mathematics and French and 20 English and French. 8 candidates offered all three subjects. How many candidates were there for the examination?

31. In a survey at an airport, 55 travellers said that last year they had been to Spain, 53 to France and 79 to Germany. 18 had been to Spain and France, 17 to Spain and Germany, and 25 to France and Germany, while 10 had been to all three countries. How many travellers took part in the survey?

32. Of pupils surveyed at the end of a term, 118 had applied for Oxford University, 98 for Cambridge and 94 for London. 42 had applied for Oxford and Cambridge, 24 for London and Cambridge and 34 for London and Oxford, while 8 had applied for all three. How many pupils took part in the survey?

19 Binary Operations

Binary operations

'Divide 6 by 3', 'Add 5 to 2'. These are examples of binary operations, in which two elements of a set are combined according to some clearly-defined rule. The rule may be stated in words, as in the examples above, or symbolically, e.g. '6 ÷ 3', '5 + 2'. A binary operation is often denoted by *, and different definitions given of the operation.

*Example 1. A binary operation * is defined over $\mathbb{R}$ the set of all real numbers, that $x * y = x + 2y$. Find $x * y$ if (i) $x = 2$, $y = 1$, (ii) $x = 4$, $y = -2$*

Since $\qquad x * y = x + 2y$ in this example,
$$2 * 1 = 2 + 2 \times 1 = 4,$$
and $\qquad 4 * (-2) = 4 + 2(-2) = 0.$

Common binary operations on the set of real numbers $\mathbb{R}$ are addition, subtraction, multiplication and division. 'Finding the square root' is not a binary operation, as the operation is only performed on one element of $\mathbb{R}$.

Closure

When a binary operation is defined over a set S, the set S is said to be *closed* under the operation * if $a*b$ belongs to S for all $a, b \in S$. Thus, the set $\mathbb{R}$ is closed under the binary operation 'add', for the sum of any two real numbers is a real number. It is closed over the set of integers $\mathbb{Z}$, and over the set of rationals $\mathbb{Q}$, for the sum of two integers is always an integer, and the sum of two rationals is also a rational. But $\mathbb{Z}$ is not closed under the operation 'divide'; for, although $18 \div 3 = 6$, an integer, $18 \div 5$ is not an integer, and the set has to be closed for *all* elements.

Example 2. (i) *The set of all rationals $\mathbb{Q}$ is closed under the operations 'addition', 'subtraction', 'multiplication' and 'division', for if a, b are any rationals,*
$$a + b, a - b, a \times b, a \div b \text{ are all rationals.}$$

(ii) The set of all even numbers is closed under addition, subtraction and multiplication, but not under division, for $12 \div 4 = 3$, which is not even.

(iii) The set of all positive integers, $\mathbb{Z}_+$, is closed under addition and multiplication, but not under subtraction (for $3 - 5 = -2$, which is not a member of $\mathbb{Z}_+$), nor under division, as shown above.

Note that one counter-example is sufficient to show that a set is not closed under a particular operation.

Commutative

If the order in which the elements are combined does not affect the result, the operation is called commutative. Thus addition is commutative, for $x + y = y + x$, but subtraction is not commutative, for $x - y \neq y - x$.

Associative

If we wish to combine three elements under a binary operation, we have to combine them in pairs. Thus to find $4 + 5 + 7$ we first decide whether to add 4 and 5, then add 9 to 7, or whether to add 5 and 7, then add 4 to 12. If the order is immaterial, as in this example, the operation is said to be associative. Expressed algebraically, the operation * is associative if

$$a * (b * c) = (a * b) * c$$

for all a, b, c in the set S on which the operation is defined.

Example 3. (i) *Addition is both associative and commutative over* $\mathbb{R}$ *for* $a + (b + c) = (a + b) + c$ *for all a, b, c in* $\mathbb{R}$, *and* $a + b = b + a$ *for all a, b in* $\mathbb{R}$.

(ii) Subtraction is neither associative nor commutative over $\mathbb{R}$, for $a - (b - c) \neq (a - b) - c$ for all a, b, c nor is $a - b = b - a$, for all a, b.

Notice that the operation has to commute all elements: it is not sufficient that $a - b = b - a$ when $a = b$.

(iii) Addition mod 5 is associative and commutative over the set S {0, 1, 2, 3, 4}, and the set S is closed under addition mod 5.

*Example 4. The operation * is defined over* $\mathbb{R}$ *by* $x * y = \frac{1}{2}(x + y)$. *Investigate whether* $\mathbb{R}$ *is closed under *, and whether * is associative or commutative.*

If x and y are any real numbers, $\frac{1}{2}(x + y)$ is also a real number, so $\mathbb{R}$ is closed under *. But notice that the set $\mathbb{Z}$ of integers is not closed under *, for $\frac{1}{2}(2 + 1)$ is not an integer.

Since $\frac{1}{2}(x + y) = \frac{1}{2}(y + x)$, $x * y = y * x$, so the operation is commutative. But $x * (y * z) = x * \frac{1}{2}(y + z) = \frac{1}{2}x + \frac{1}{4}(y + z)$, whereas $(x * y) * z = \frac{1}{4}(x + y) + \frac{1}{2}z$, so the operation is not associative.

Example 5. Investigate whether the set S of all 2×2 matrices of the form $\begin{pmatrix} a & 0 \\ 0 & b \end{pmatrix}$ *whose elements a, b belong to* $\mathbb{R}$ *is closed under matrix multiplication, and whether matrix multiplication is associative or commutative or both over S.*

Let $\begin{pmatrix} a & 0 \\ 0 & b \end{pmatrix}$, $\begin{pmatrix} l & 0 \\ 0 & m \end{pmatrix}$, $\begin{pmatrix} x & 0 \\ 0 & y \end{pmatrix}$ be any three 2×2 matrices $\in S$. Then

$\begin{pmatrix} a & 0 \\ 0 & b \end{pmatrix}\begin{pmatrix} l & 0 \\ 0 & m \end{pmatrix} = \begin{pmatrix} al & 0 \\ 0 & bm \end{pmatrix}$, $\in S$ so that the set S is closed under matrix multiplication. Further $\begin{pmatrix} a & 0 \\ 0 & b \end{pmatrix}\begin{pmatrix} l & 0 \\ 0 & m \end{pmatrix} = \begin{pmatrix} al & 0 \\ 0 & bm \end{pmatrix} = \begin{pmatrix} l & 0 \\ 0 & m \end{pmatrix}\begin{pmatrix} a & 0 \\ 0 & b \end{pmatrix}$ for all a, b, l, m, so that for matrices of this type, matrix multiplication is commutative. Also

$$\left[\begin{pmatrix} a & 0 \\ 0 & b \end{pmatrix}\begin{pmatrix} l & 0 \\ 0 & m \end{pmatrix}\right]\begin{pmatrix} x & 0 \\ 0 & y \end{pmatrix} = \begin{pmatrix} alx & 0 \\ 0 & bmy \end{pmatrix} = \begin{pmatrix} a & 0 \\ 0 & b \end{pmatrix}\left[\begin{pmatrix} l & 0 \\ 0 & m \end{pmatrix}\begin{pmatrix} x & 0 \\ 0 & y \end{pmatrix}\right]$$

so that matrix multiplication is associative for these matrices.

Notice that if we had wished to investigate matrix multiplication over the set of all 2×2 matrices, we should have had to consider matrices of the form $\begin{pmatrix} a & b \\ c & d \end{pmatrix}$. It would be sufficient to find just one counter example to show that matrix multiplication was not commutative over all 2×2 matrices,

e.g. $\begin{pmatrix} 3 & 1 \\ 2 & 2 \end{pmatrix}\begin{pmatrix} 2 & 1 \\ 0 & 1 \end{pmatrix} = \begin{pmatrix} 6 & 4 \\ 4 & 4 \end{pmatrix}$ whereas $\begin{pmatrix} 2 & 1 \\ 0 & 1 \end{pmatrix}\begin{pmatrix} 3 & 1 \\ 2 & 2 \end{pmatrix} = \begin{pmatrix} 8 & 4 \\ 2 & 2 \end{pmatrix}$

but the algebra required to show that

$$\left[\begin{pmatrix} a & b \\ c & d \end{pmatrix}\begin{pmatrix} l & m \\ n & p \end{pmatrix}\right]\begin{pmatrix} x & y \\ z & w \end{pmatrix} = \begin{pmatrix} a & b \\ c & d \end{pmatrix}\left[\begin{pmatrix} l & m \\ n & p \end{pmatrix}\begin{pmatrix} x & y \\ z & w \end{pmatrix}\right]$$

would be more difficult than in the example chosen.

Exercise 19.1a

1. The operation $*$ is defined over $\mathbb{R}$ by $x * y = x + y^2$. Find (i) $2 * 3$, (ii) $3 * 2$, (iii) $3 * 5$, (iv) $2 * (3 * 5)$, (v) $(2 * 3) * 5$.
   Is $\mathbb{R}$ closed under this operation? Is the operation associative or commutative?

2. If $x * y = \sqrt{(xy)}$, where the positive square root is taken, find (i) $2 * 18$, (ii) $18 * 2$, (iii) $2 * (12 * 27)$, (iv) $(2 * 12) * 27$.
   Is the operation $*$ associative or commutative?
   Is the set $\mathbb{R}$ closed under $*$?

3. If $*$ is multiplication mod 6, find (i) $3 * 3$, (ii) $3 * 4$, (iii) $3 * 5$.

4. Find which of the following sets are closed under the operation given:
       (i) {integers}   under multiplication
       (ii) {powers of 2} under multiplication
       (iii) {powers of 2} under subtraction
       (iv) {1, 3, 5}   under multiplication mod 6
       (v) {1, 3, 5}   under addition mod 6
       (vi) {0, 2, 4}   under addition mod 6
       (vii) {all primes}   under addition
       (viii) {all positive numbers less than 1} under multiplication.

5. If $x * y = xy + x + y$, find the value of $3 * 5$. Show that the operation is both commutative and associative over the set of real numbers.

6. If $x * y = xy + x + y$, solve the equations

  (i) $x * 3 = 19$,

  (ii) $(x * 3) + (2 * x) = 40$,

  (iii) $x * x = 48$.

Exercise 19.1b

1. The operation $*$ is defined over $\mathbb{R}$ by $x * y = (x - y)^2$. Find (i) $2 * 3$ (ii) $3 * 2$,   (iii) $2 * (3 * 5)$,   (iv) $(2 * 3) * 5$.
   Is $\mathbb{R}$ closed under this operation? Is the operation associative or commutative?

2. The operation $*$ is defined over the set of integers $\mathbb{Z}$ by 'the highest common factor of', i.e. $18 * 27 = 9$. Find (i) $6 * 8$,   (ii) $6 * (8 * 18)$.
   Is the operation associative or commutative or both?

3. The operation $*$ is defined over the set S {2, 3, 4} as multiplication mod 5. Find (i) $3 * 4$,   (ii) $2 * 4$,   (iii) $2 * 3$.
   Is S closed under this operation?

4. The operation $*$ is defined over the set of all sets as union, i.e. $A * B = A \cup B$. Draw a pair of Venn diagrams, suitably shaded, to illustrate that $A * (B * C) = (A * B) * C$.
   If the operation is defined as intersection, draw another pair of Venn diagrams, suitably shaded, to show that $A * (B * C) = (A * B) * C$.

5. The operation $*$ is defined over the set of all integers thus: if x, y are any two integers, $x * y$ is the quotient when x is divided by y, e.g. $13 * 4 = 3$, $23 * 4 = 5$.
   (i) Find $33 * 4$, $35 * 5$, $100 * 7$, $100 * 8$, $100 * 9$.
   (ii) Solve the equation $100 * x = 9$.

6. The set S consists of all 2×2 matrices of the form

$$\begin{pmatrix} a & 1 - b \\ 1 - a & b \end{pmatrix}$$

That is, the sum of the elements in each column is 1. Show that S is closed under matrix multiplication.

Identity element

When zero is added to any number, the value of the number is not altered: when any number is multiplied by 1, that number is not altered. Zero is called the identity element under addition: 1 is the identity element under multiplication. The identity element (usually denoted by e) has the property that $a * e = e * a = a$ for all a.

Example 6. Show that $\begin{pmatrix} 1 & 0 \\ 0 & 1 \end{pmatrix}$ *is the identity matrix under matrix multi-plication.*

   Let $\begin{pmatrix} x & y \\ z & w \end{pmatrix}$ be any 2×2 matrix.

Then
$$\begin{pmatrix} 1 & 0 \\ 0 & 1 \end{pmatrix}\begin{pmatrix} x & y \\ z & w \end{pmatrix} = \begin{pmatrix} x & y \\ z & w \end{pmatrix}$$

and
$$\begin{pmatrix} x & y \\ z & w \end{pmatrix}\begin{pmatrix} 1 & 0 \\ 0 & 1 \end{pmatrix} = \begin{pmatrix} x & y \\ z & w \end{pmatrix}$$

so $\begin{pmatrix} 1 & 0 \\ 0 & 1 \end{pmatrix}$ is the identity 2×2 matrix for multiplication.

Inverse element

$3 \times \frac{1}{3} = 1$ and $\frac{1}{3} \times 3 = 1$. 1 is the identity element for multiplication, and $\frac{1}{3}$ is called the **multiplicative inverse** of 3. The inverse of an element a is the element (usually denoted by a^{-1}) such that $a * a^{-1} = e = a^{-1} * a$. Thus the additive inverse of 3 is -3, since $3 + (-3) = 0$, the identity element under addition.

Example 7. Show that $\begin{pmatrix} -2 & 3 \\ 3 & -4 \end{pmatrix}$ is the inverse under matrix multiplication of $\begin{pmatrix} 4 & 3 \\ 3 & 2 \end{pmatrix}$.

By matrix multiplication, $\begin{pmatrix} -2 & 3 \\ 3 & -4 \end{pmatrix}\begin{pmatrix} 4 & 3 \\ 3 & 2 \end{pmatrix} = \begin{pmatrix} 1 & 0 \\ 0 & 1 \end{pmatrix}$

and $\begin{pmatrix} 4 & 3 \\ 3 & 2 \end{pmatrix}\begin{pmatrix} -2 & 3 \\ 3 & -4 \end{pmatrix} = \begin{pmatrix} 1 & 0 \\ 0 & 1 \end{pmatrix}$

so that $\begin{pmatrix} -2 & 3 \\ 3 & -4 \end{pmatrix}$ is the inverse of $\begin{pmatrix} 4 & 3 \\ 3 & 2 \end{pmatrix}$

Strictly we should say under what operation the element is the inverse, but 'matrix multiplication' is usually understood in this case.

Example 8. A binary operation $*$ is defined on $\mathbb{R}$ by $x * y = \sqrt{xy}$ for all $x, y \in \mathbb{R}$. Show that there is not an identity element for this operation.

Suppose that e is the identity element. Then $x * e = e * x = x$ for all x,

i.e. $\sqrt{xe} = x$

i.e. $e = x$.

But we have to have the same element as identity element for *all* members of the set, so we do not have identity element for this operation.

Example 9. A binary operation $*$ is defined on $\mathbb{R}$ by $x * y = xy + x + y$ for all $x, y \in \mathbb{R}$. Find the identity element and the inverse of 5 under this operation.

Suppose that e is the identity element.

Then
$$x * e = x,$$
$$xe + x + e = x,$$
$$x(1 + e) = 0,$$
$$e = -1.$$

Let the inverse of 5 be x.
Then
$$5 * x = -1$$
$$5x + 5 + x = -1$$
$$6x = -6.$$
$$x = -1.$$

When this operation is defined over $\mathbb{R}$, the inverse of every element is -1.

Exercise 19.2

1. Find (a) the identity element,
         (b) the inverse of the element 2,
   for each of the operations defined below on $\mathbb{R}$.
   (i) $a*b = a + b$,
   (ii) $a*b = ab$,
   (iii) $a*b = ab + a + b$,
   (iv) $a*b = a + b - \frac{1}{2}ab$.

2. Find the identity element, if it exists, when each of the operations below is defined on $\mathbb{R}$.
   (i) $a*b = ab + 2a + 2b$,
   (ii) $a*b = \sqrt{ab}$,
   (iii) $a*b = \dfrac{a}{b} + \dfrac{b}{a}$.

3. Find the identity element and the inverse of $\begin{pmatrix} 2 \\ 1 \end{pmatrix}$ under matrix addition defined on the set of all 2 by 1 matrices.

4. The operation 'rotate in the plane through an angle θ in an anti-clockwise sense' is defined on the set of all plane geometrical figures. Find the identity element and the inverse of the element $\theta = 90°$.

Distributive law

If x, y and z are any real numbers, then $x(y + z) = xy + xz$ but $x \div (y + z) \neq x \div y + x \div z$. We say that multiplication is distributive over addition, but that division is not distributive over addition. We need to have two binary operations defined before we can determine whether the distributive law is satisfied.

Example 10. (i) *The operation of matrix multiplication is distributive over matrix addition, i.e.*

$$A(B + C) = A.B + A.C$$

for any matrices A, B, C.

(ii) *Defined on the set of all sets, union and intersection are distributive over each other.*

i.e. $X \cup (Y \cap Z) = (X \cup Y) \cap (X \cup Z)$

and $X \cap (Y \cup Z) = (X \cap Y) \cup (X \cap Z)$.

Many of the more interesting applications of the distributive law, e.g. to vector multiplication and to Boolean Algebra are outside the scope of this book.

Exercise 19.3

1. Draw Venn diagrams illustrating
   (i) $\qquad X \cap (Y \cup Z) = (X \cap Y) \cup (X \cap Z)$
   and (ii) $\qquad X \cup (Y \cap Z) = (X \cup Y) \cap (X \cup Z)$

2. If $*$ denotes multiplication mod 6 and $\oplus$ denotes addition mod 6, consider any two sets of numbers to show $*$ is distributive over $\oplus$. (e.g. $2 * (3 \oplus 4) = 2 * 3 \oplus 2 * 4$).

3. Is multiplication mod 5 distributive over addition mod 5?

4. If $*$ denotes 'select the greater' and $\oplus$ denotes 'select the smaller', give an example to show that $*$ is not distributive over $\oplus$, and another example to show that $\oplus$ is not distributive over $*$.

5. If $*$ denotes 'the lowest common multiple of' and $\oplus$ denotes 'the highest common factor of', is $*$ distributive over $\oplus$? (e.g. $18 \oplus 12 = 6$ and $12 * 15 = 60$).

20 Relations, Mappings and Functions

Relations

'3 is greater than 2', '4 is the square of 2', 'Leonardo da Vinci painted the Mona Lisa'. These statements are examples of relations. In the first, the relation 'is greater than' is defined over the set $\mathbb{R}$ of all real numbers, relating some elements in $\mathbb{R}$ to others. In the second, the relation 'is the square of' is also defined over $\mathbb{R}$, whereas in the third, an element in the set of all painters is associated with an element in another set, the set of all paintings.

Not all statements are relations. 'I am hungry' is a statement that is either true or false; '$\pi = 22/7$' is certainly false, and '5 is happy' appears to be meaningless, except in some fantasy world. '2 is a prime number' is not a relation, as it does not relate '2' to any other number.

It will not always be possible to find a relation between elements in any pair of given sets. If $A = \{$yellow, blue, green, red$\}$ and $B = \{$London, Paris, Madrid$\}$, there does not appear to be any relation associating the elements of set A with those of set B.

Domain, codomain; image, range

A relation associates an element of one set (the **domain**) with one or more elements of another set (the **codomain**), which may be the same set as the first. Those elements in the second set which are associated with elements in the domain are called the **images** of the elements in the domain, thus 2 is the image of 1 under the relation 'is half of'; the set of all images is called the **range**.

Into, onto relations

If the relation 'is a square root of' is defined over $\mathbb{R}$, then the domain and the codomain are $\mathbb{R}$ but the range is the set of all positive numbers, $\mathbb{R}_+$ (Fig. 20.1(a)). Whereas if the relation 'is twice' is defined over $\mathbb{R}$, the range is the same as the codomain, the set $\mathbb{R}$ of all real numbers (Fig. 20.1(b)).

The first relation is called 'into'; the second, 'onto'. Relations can be illustrated by Papy graphs, as in Fig. 20.1.

One-one; one-many

If each element in the domain has only one image in the range then the relation is said to be 'one-one'; if there is more than one image, the

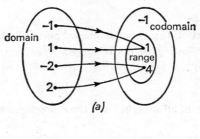

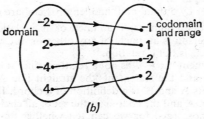

Fig. 20.1

relation is 'one-many'. Thus the relation 'is one half of' defined on $\mathbb{R}$ is one-one; the relation 'is the square of' is one-many, since 4 is the square of both 2 and -2.

Non-algebraic relations

Although many relations are algebraic, that is not essential. The relation 'is a friend of' could be defined on a set S, where S is the set of all members of an exclusive club, Mr A, Mrs B, Miss C and Lord D, and the relation between them could be illustrated by Fig. 20.2, which shows who is a friend of whom.

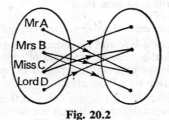

Fig. 20.2

Examples. If the domain and codomain are $\mathbb{R}$, x an element in the domain, and y an element in the codomain, the following are relations:

'is equal to'
'is greater than'
'is greater by 5 than'

'is the square of'
'is the reciprocal of'
'is the integral part of' (e.g. 7 is the integral part of 7.4).

Defined on ℝ, the following are not relations:

'is equal in area to'
'is a perfect square'
'is stronger than'.

Defined on the set 𝕃 of all straight lines, the following are relations:

'is parallel to'
'is not perpendicular to'

whereas 'is greater by 5 than' and 'had length 3 cm' are not relations on 𝕃.

Defined on the set of all humans, 'is an uncle of' is a relation, since we can tell whether any one person is an uncle of any other or not, whereas on this set 'is a good footballer' is not a relation.

If $x \in \mathbb{R}$, the statement '$3x = 7$' is not a relation, since it is a property of just one element in ℝ, and the statement 'the straight line AB is a chord of a circle through points A and B' is a definition of a chord. If the domain is the set of all straight lines, and the codomain the set of all circles, then 'intersects' is a relation on those sets, for we can tell whether any one straight line intersects any one circle.

Exercise 20.1

1. A relation is defined with domain {1, 2, 3} and codomain ℝ. Find the range in each case when the relation is

    (i) 'is one half of'
    (ii) 'is the square of'
    (iii) 'is greater than'
    (iv) 'is less than or equal to'.

2. A relation is defined on ℝ. Find whether the relation is 'into' or 'onto' if the relation is

    (i) 'is one half of'
    (ii) 'is the logarithm to base 10 of'
    (iii) 'is the cube of'
    (iv) 'is the integer next below' (e.g. the image of 2.7 is 2).

3. The set A consists of a boy and his sister: the set B of their parents. If A is the domain, is the relation 'is a child of' one-one or one-many? If B is the domain, is the relation 'is a parent of' one-one or one-many? Illustrate each relation by a Papy graph.

4. The domain is the set of all integers between 0 and 9 inclusive, and a relation is defined whereby each element is associated with the last digit of its square, e.g. the image of 7 is 9, since $7^2 = 49$. Find the range under this mapping, and illustrate by a Papy graph. Which of the following does this show *not* to be perfect squares:

        28177,   431444,   281918,   500000?

Notice that this method does *not* show that any of these numbers are perfect squares.

5. The domain is the set {George IV, William IV}, the codomain {George III, George IV} and the relation defined is 'was the son of'. Find the range, and illustrate by a Papy graph.

Symmetric

Friendship (we hope) is a **symmetric** relation; if Mrs B is a friend of Miss C, then Miss C is a friend of Mrs B. Not all relations are symmetric. If the relation 'is greater than' is defined on $\mathbb{R}$ and if $a, b \in \mathbb{R}$, then if $a > b$, $b \not> a$. If the relation between elements a in the domain and b in the range is denoted by $a\,R\,b$, then a relation is symmetric if and only if

$$a\,R\,b \text{ implies and is implied by } b\,R\,a.$$

Transitive

Although Mrs B is a friend of Miss C and Miss C is a friend of Lord D, it may not follow that Mrs B is a friend of Lord D. But with the relation 'is greater than' defined on $\mathbb{R}$, if a is greater than b and b is greater than c, then a is greater than c. A relation which has this property is called a **transitive** relation, and the condition that a relation is transitive can be written

$$a\,R\,b \text{ and } b\,R\,c \text{ implies } a\,R\,c.$$

Unfortunately friendship is not transitive!

Reflexive

The third property some relations possess is that they are **reflexive**. 'Is equal to' defined on $\mathbb{R}$ is one such relation. For every element in $\mathbb{R}$ is equal to itself. The relation 'is greater than' is not reflexive, for no element is greater than itself. The property that a relation R is reflexive is written $a\,R\,a$.

Reflexive, symmetric, transitive; equivalence relations

These are usually listed in alphabetical order, and may be summarized.
Reflexive: a relation R is reflexive if $a\,R\,a$ for all a in the domain.
Symmetric: a relation R is symmetric if $a\,R\,b \Leftrightarrow b\,R\,a$.
Transitive: a relation R is transitive if $a\,R\,b$ and $b\,R\,c$ implies $a\,R\,c$.
A relation which is reflexive, symmetric and transitive is called an equivalence relation.
In the following examples, a relation R is defined in a set S.

Example 1. If $S = \mathbb{Z}$, i.e. the set of all integers, and the relation R is 'is a factor of', then R is reflexive (for every integer is a factor of itself), transitive (for if a is a factor of b and b is a factor of c, then a is a factor of c) but not symmetric (for if a is a factor of b, b need not be a factor of a).

Example 2. If S is the set of all rectangles and R is 'is equal in area to', then R is reflexive, symmetric and transitive. But if R is 'is greater in area than', R is transitive, but neither symmetric nor reflexive.

Example 3. If S is the set of all people and R is 'is a cousin of', then R is not reflexive (for no one is their own cousin), nor transitive (for if a is a cousin of b and b is a cousin of c, a and c need not be cousins) but it is symmetric. But if R is 'is a brother of' then R is now both symmetric and transitive.

Example 4. If S is the set of all straight lines in three-dimensional space and R is 'is parallel to', then R is reflexive, symmetric and transitive. But if R is 'is perpendicular to', then R is symmetric, but neither reflexive nor transitive.

Exercise 20.2

Find whether the following relations are reflexive, symmetric and/or transitive:

1. The relation 'is a brother of' defined over the set of all people.
2. The relation 'is a brother of' defined over the set of all men.
3. The relation 'is taller than' defined over the set of all people.
4. The relation 'is a fan of the same pop star as' defined over the set of all young persons.
5. The relation 'is richer than' defined over the set of all oil sheikhs.
6. The relation 'runs as fast as' defined over the set of all Olympic sprinters.
7. The relation '$x + y = 0$' defined over $\mathbb{R}$ the set of all real numbers, x and y being elements of that set.
8. The relation '$x^2 = y^2$' defined over $\mathbb{R}$.
9. The relation '$x^2 > y^2$' defined over $\mathbb{R}$.
10. The relation '$x > y$' defined over $\mathbb{R}$.
11. The relation 'x is the square of y' defined over $\mathbb{R}$.
12. The relation 'is equal in area to' defined over the set of all circles.

Functions

'Choose a number, double it, then subtract 3'. If we chose the number 5, we obtained the result 7; if we chose the number -1, we obtained the result -5, assuming we carried out the instructions correctly. Once the instruction had been given there was only one result we could obtain. This is an example of a **function** *, which associates one member of the domain with one and only one member of another set, the range. In this example, we were told to think of a number; this defined the domain, and may have restricted the range. If we were told to think of a positive number, we should have a different domain, and a different range. If we had not had the domain defined, and we had thought of, say, a circle, the instruction to double it then subtract 3 would have been meaningless.

*sometimes called a mapping: the terms are synonymous.

We say that a function (or mapping) maps every element in the domain into a corresponding element (the image) in the range. The function can often be expressed algebraically, so that if x denotes any element in the domain, $2x - 3$ will denote the corresponding element in the range, and the function is written

$$f : x \mapsto 2x - 3.$$

If the domain is the set $\{-1, 0, 1, 2, 3, 4, 5\}$, the association of some elements in the domain with an element in the range is illustrated by the Papy graph in Fig. 20.3.

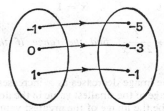

Fig. 20.3

One-one, many-one mappings

It is not necessary for one and only one element in the domain to correspond to one and only one element in the range, as illustrated by Fig. 20.3. (Such a function is called a one-one mapping.) If the domain is the set $\mathbb{R}$ of all real numbers, the function $f : x \mapsto x^2$ maps two elements in the domain into one element of the range in most cases (it would be a one-one mapping if the domain was restricted to the element 0), and is an example of a many-one mapping (Fig. 20.4).

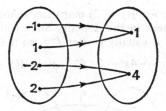

Fig. 20.4

The important property of a function is that for every element in the domain there is one and only one image in the range, so that the relation $x \mapsto \sqrt{x}$ is not a function if the range is the set $\mathbb{R}$, but is a function if the range is restricted to $\mathbb{R}_{+}$.

Example 5. Find the images under the function f: $x \mapsto 4x - 2$ of $-1, 0, \frac{1}{2}$.

Substituting in $4x - 2$, we see that -1 is mapped into $-6, 0$ into -2 and $\frac{1}{2}$ into 0.

The image of an element x under a function f is often written $f(x)$, e.g. here $f(-1) = -6$.

Example 6. What element in the domain has image of 2 under the function f: $x \rightarrow 4x - 2$?

Call this element x. Then its image is $4x - 2$, so that

$$4x - 2 = 2,$$

i.e. $$x = 1$$

Example 7. If $f : x \mapsto \dfrac{1}{2 + x}$, find the range if the domain is the set $\{x : 1 \leqslant x \leqslant 5\}$.

Since the value of the image decreases as x increases, the greatest value in the range will be the image of the smallest value in the domain, and the smallest value in the range will be the image of the greatest value in the domain. Since f maps 1 into $\frac{1}{3}$ and 5 into $\frac{1}{7}$, the range is the set $\{y : \frac{1}{7} \leqslant y \leqslant \frac{1}{3}\}$.

Inverse functions

In Example 6 we found which element in the domain had image 2 in the range under function $f: x \mapsto 4x - 2$. We can extend this to find which element has been mapped into any given image in the range. Denote the image of an element x in the domain by y. Then since $f: x \mapsto 4x - 2$,

$$y = 4x - 2,$$

i.e. $$x = \tfrac{1}{4}(y + 2)$$

and the function $F: y \mapsto \frac{1}{4}(y + 2)$ maps any element in the range back into its original element in the domain. Such a function is called the inverse function, and is usually written f^{-1}, so that

if $$f: x \mapsto 4x - 2, \quad f^{-1} : y \mapsto \tfrac{1}{4}(y + 2).$$

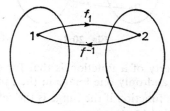

Fig. 20.5

Since a function maps every element in the domain into a unique image in the range, only one-one functions possess an inverse function.

Restricted domain

One important way whereby a many-one function can have an inverse is to restrict the domain, so that the function becomes a one-one mapping. This can be done with the function $f: x \longmapsto x^2$, by restricting the domain to the set of positive numbers as illustrated in Fig. 20.6.

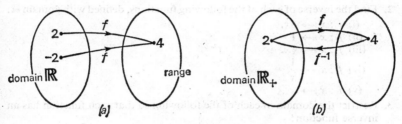

Fig. 20.6

(a) Domain set of all real numbers, many one mapping; no inverse function.

(b) Domain set of all positive real numbers, one-one mapping; inverse functions $f^{-1} : x \longmapsto \sqrt{x}$.

A particularly important example of this occurs with the trigonometric functions. Consider the function $f: x \longmapsto \sin x°$. Then f maps 30 into $\sin 30°$, i.e. $\frac{1}{2}$, and maps 150 into $\sin 150°$, also $\frac{1}{2}$. As it is a many-one mapping, there is not an inverse function. But if we restrict the domain to the set $\{x: -90 \leqslant x \leqslant 90\}$, then the mapping is one-one and so there is an inverse function. The set of values forming the domain in this case is called the set of principal values, and is used extensively later in Mathematics.

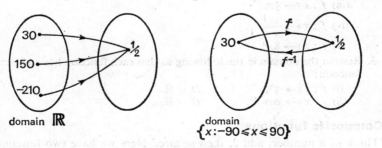

Fig. 20.7

Exercise 20.3a

1. Find the range in each of the following functions defined over the domain D:

   (i) $f : x \mapsto 3x,$ $D = \{-1, 0, 1, 2\}.$

   (ii) $f : x \mapsto 1 - x,$ $D = \{-2, -1, 0, 1, 2\}.$

   (iii) $f : x \mapsto 2^x,$ $D = \{-2, -1, 0, 1, 2\}.$

   (iv) $f : x \mapsto \dfrac{1}{x},$ $D = \{\text{all positive integers}\}.$

   (v) $f : x \mapsto \dfrac{1}{x^2},$ $D = \{x : 1 < x < 5\}.$

2. Find the inverse of each of the following functions, defined with domain $\mathbb{R}$:

   (i) $f : x \mapsto 3x.$

   (ii) $f : x \mapsto 1 - x.$

   (iii) $f : x \mapsto \tfrac{1}{4}x.$

   (iv) $f : x \mapsto \dfrac{1}{x}.$

   (v) $f : x \mapsto x^3.$

3. Restrict the domain in each of the following so that each function has an inverse function:

   (i) $f : x \mapsto x^2,$ $D = \{-2, -1, 0, 1, 2\}.$

   (ii) $f : x \mapsto 1 - x^2,$ $D = \{x : -1 < x < 1\}.$

Exercise 20.3b

1. Find the range of each of the following functions defined over the domain D:

   (i) $f : x \mapsto 4x,$ $D = \{0, \tfrac{1}{4}, \tfrac{1}{2}, \tfrac{3}{4}\}.$

   (ii) $f : x \mapsto -x,$ $D = \{-2, -1, 0, 1, 2\}.$

   (iii) $f : x \mapsto x^3,$ $D = \{x : 1 < x < 2\}.$

   (iv) $f : x \mapsto -x,$ $D = \mathbb{Z}_+.$

   (v) $f : x \mapsto \log_{10} x,$ $D = \{x : 1 < x < 10\}.$

2. Find the inverse of each of the following functions, defined over the domain $\mathbb{R}$:

   (i) $f : x \mapsto 4x.$

   (ii) $f : x \mapsto -x.$

   (iii) $f : x \mapsto \tfrac{1}{2}x.$

   (iv) $f : x \mapsto \dfrac{1}{x + 1}.$

   (v) $f : x \mapsto 8x^3.$

3. Restrict the domain in the following so that each function has an inverse function:

   (i) $f : x \mapsto x^2,$ $D = \mathbb{R}.$

   (ii) $f : x \mapsto \cos x°,$ $D = \mathbb{R}_+.$

Composite functions

'Think of a number, add 2, then square.' Here we have two functions mapping first an element, say 3, into 5, then 5 into 25. If $f : x \rightarrow x + 2$

and $g : x \rightarrow x^2$, we can describe first f then g by the composite function gf. Thus

$$gf(3) = g(5) = 25, \text{ since } f \text{ maps 3 into 5 then } g \text{ maps 5 into 25.}$$

Notice that the order in which the functions operate usually matters. $fg(3) = f(9) = 11 \neq gf$ (3). When finding the inverse functions, we also have to note the order in which they are applied, as in this example.

Example 8. If $f : x \longmapsto 2x$ and $g : x \longmapsto x - 3$, find

    (i) gf (3) *and* fg (3),
    (ii) $g^{-1}f^{-1}(3)$ *and* $f^{-1}g^{-1}(3)$,
    (iii) $g^{-1}f^{-1}(0)$ *and* $f^{-1}g^{-1}(0)$.

(i) f maps 3 into 6, then g maps 6 into 3, so gf (3) $= g(6) = 3$.
    But g maps 3 into 0 and f maps 0 into 0, so $fg(3) = f(0) = 0$.

(ii) $f^{-1} : x \longmapsto \frac{1}{2}x$ and $g^{-1} : x \longmapsto x + 3$,
    so $g^{-1}f^{-1}(3) = g^{-1}(\frac{3}{2}) = \frac{9}{2}$,
    and $f^{-1}g^{-1}(3) = f^{-1}(6) = 3$.

(iii) $g^{-1}f^{-1}(0) = g^{-1}(0) = 3$.
    and $f^{-1}g^{-1}(0) = f^{-1}(3) = \frac{3}{2}$.

Notice that gf maps 3 into 3 and $f^{-1}g^{-1}$ maps 3 back into 3, whereas fg maps 3 into 0 and $g^{-1}f^{-1}$ maps 0 back into 3.

Exercise 20.4a

1. If $f : x \longmapsto 3x$ and $g : x \longmapsto x + 2$, find

    (i) gf (2) and fg(2),
    (ii) $g^{-1}f^{-1}(8)$ and $f^{-1}g^{-1}(8)$,
    (iii) $g^{-1}f^{-1}(12)$ and $f^{-1}g^{-1}(12)$.

2. If $f : x \longmapsto 2x$, $g : x \longmapsto x + 3$, $h : x \longmapsto x^2$ find

    (i) hgf (2) and fgh(2),
    (ii) $f^{-1}g^{-1}h^{-1}(49)$.

3. If $f : x \longmapsto x^2$ and $g : x \longmapsto \dfrac{1}{x}$, show that $gf = fg$.

4. If $f : x \longmapsto x + 2$ and $g : x \longmapsto x^2 + 1$, show that $fg : x \longmapsto x^2 + 3$, and find the mapping gf.

5. If $f : x \longmapsto 2x - 3$ and $g : x \longmapsto 4 - x$, show there is no value of x for which $gf(x) = fg(x)$.

Exercise 20.4b

1. If $f : x \longmapsto \frac{1}{2}x$ and $g : x \longmapsto x - 2$, find

    (i) gf (4) and fg(4),
    (ii) $g^{-1}f^{-1}(0)$ and $f^{-1}g^{-1}(0)$,
    (iii) $g^{-1}f^{-1}(1)$ and $f^{-1}g^{-1}(1)$.

2. If $f : x \longmapsto \frac{1}{2}x$, $g : x \longmapsto x + 4$ and $h : x \longmapsto \frac{1}{x}$, find

(i) $hgf(2)$ and $fgh(2)$,

(ii) $h^{-1}g^{-1}f^{-1}(0.2)$ and $f^{-1}g^{-1}h^{-1}(0.2)$.

3. If $f : x \longmapsto 3x + 4$ and $g : x \longmapsto 2(x + 1)$, show that $fg = gf$.

4. If $f : x \longmapsto x^2$ and $g : x \longmapsto x + 4$, show that $gf : x \longmapsto x^2 + 4$ and find the mapping fg.

5. If $f : x \longmapsto x^2 + 2$ and $g : x \longmapsto 2x + 3$, find the two values of x for which $gf(x) = fg(x)$.

Exercise 20.5

1. If the domain is $\{1, 4, 9, 16, 25\}$, which of the following define a relation?

(i) 'is greater than'

(ii) 'is positive'

(iii) 'is one half of'

(iv) 'is a perfect square'

(v) 'has the same number of digits as'.

2. The triangle ABC is isosceles with $AB = BC$. If the domain is the set of the sides of this triangle {AB, BC, CA}, which of the following define a relation?

(i) 'intersects'

(ii) 'is equal in length to'

(iii) 'is equal in area to'

(iv) 'is parallel to'

(v) 'is a straight line'.

3. If the domain is the set $\mathbb{Z}$ of all integers, explain why 'is a multiple of' is a relation, whereas 'is a multiple of 2' is not a relation.

4. If the domain is 'un, deux, trois', is the relation 'is French for' a function?

If the domain is 'one, two, three', why is the relation which maps English into French NOT a function?

5. If the domain is the set of all dresses in a fashion shop, is the relation 'is the same colour as' reflexive, symmetric and transitive?

6. The table below is part of that issued by the Post Office in 1976 to show the relation between the mass of a parcel in kg and the postal charge:

Not over	1 kg	55p
1 kg and not over	2 kg	70p
2 kg and not over	3 kg	85p
etc. *until*		
9 kg and not over	10 kg	160p

(i) Is the relation many-one, one-one, or one-many?

(ii) Find the domain x and the range y.

7. A certain bus company calculates its fares as 5p for the first 1 km or part thereof, then 4p for each subsequent km, so that the fare for a journey of 0.8 km is 5p, for a journey of 4.4 km is 21p, etc.

    (i) Is the relation between distance travelled in km to the fare in pence
      many-one, one-one or one-many?

    (ii) Find the domain x and the range y, if the longest journey possible
      on these buses is 25 km.

8. If $f : x \longmapsto \sin x°$ and the domain is $\{x : 0 < x < 30\}$, find the range.

9. If $f : x \longmapsto \log_{10} x$, and the range is $\{y : 1 < y < 3\}$, find the domain.

10. If $f : x \longmapsto$ integer next below or equal to x (e.g. $f(3.7) = 3$), and the
domain is $\mathbb{R}$, find the range.

11. If $f : x \longmapsto x^2 + 1$ and the domain is $\mathbb{R}$, find the range.

12. If $f : x \longmapsto x + 2$ and $g : x \longmapsto x^2$, find the two values of x for which
$f(x) = g(x)$.

13. If $f : x \longmapsto x^2 - 4$ and $g : x \longmapsto 3x$, find the two values of x for which
$f(x) = g(x)$.

14. If $f : x \longmapsto 2x$, $g : x \longmapsto x^2$, and $h : x \longmapsto 3x + 1$, find the two values of x
for which $gf(x) = h(x)$. Illustrate this by a Papy graph.

15. If $f : x \longmapsto x + 5$ and $g : x \longmapsto x - 7$, show that $fg(x) = gf(x)$ for all x.

16. If $f : x \longmapsto x^2$ and $g : x \longmapsto 3x - 2$, find the two values of x for which
$f(x) = g(x)$.

17. If $f : x \longmapsto 2x^2$ and $g : x \longmapsto x + 1$, find the two elements in the domain
$\mathbb{R}$ which have the same image in the range $\mathbb{R}$.
If the domain is restricted to $\mathbb{R}_+$, how many elements in the domain
now have the same image in the range?

18. If $f : x \longmapsto 20/x$ and $g : x \longmapsto x - 1$, find the elements x in the domain $\mathbb{R}$
such that $gf(x) = x$.

19. If $f : x \longmapsto x + 1$, $g : x \longmapsto x^2$ and $h : x \longmapsto 3x$, show that $hgf : x \longmapsto$
$3x^2 + 6x + 3$, and find fgh.

20. Find functions f, g, h as in Q.19 such that $fgh : x \longmapsto \dfrac{1}{(x + 2)^2}$.

21 Groups

Problem : to solve *a**x = *b*, for all operations *.

Let us consider the steps we take when solving $3y = 7$. We have one operation, multiplication, defined in $3y$, so we multiply both sides of the equation by the inverse of 3, i.e. $\frac{1}{3}$

so
$$\tfrac{1}{3}(3y) = \tfrac{1}{3}(7).$$

Next we use the associative property of multiplication to deduce

$$(\tfrac{1}{3} \times 3)y = \tfrac{1}{3}(7)$$

and then if we are looking for a solution in the set $\mathbb{R}$, we say that

$$y = \tfrac{7}{3}.$$

If we require a solution in the set of integers, $\mathbb{Z}$, we say there is not a solution to the equation.

First, the element 3 has to have an inverse; then the operation has to be associative, and finally the set from which the elements are chosen, 7 and $\frac{1}{3}$, has to be closed. If there is to be an inverse element for 3, there has to be a unit element, so that we see four conditions are necessary and sufficient if we are to be able to solve any equation $a * x = b$. Any set S which satisfies these conditions under an operation * is called a **group**.

Group : definition

A set of elements S is a group under an operation * if

(i) S is closed under *,
(ii) the operation is associative over S, i.e. $a *(b*c) = (a*b)*c$ for all $a, b, c \in S$,
(iii) there is an identity element $e \in S$, i.e. an element such that $a*e = e*a = a$ for all $a \in S$,
(iv) every element $a \in S$ has an inverse a^{-1} in S, i.e. $a*a^{-1} = a^{-1}*a = e$.

If in addition the operation * is commutative, then the group is called a commutative (or Abelian) group.

Modulo arithmetic

The tables opposite show multiplication mod 5 defined over two sets. Looking at Table I, we see that the set $\{1, 2, 3, 4\}$ is closed under multiplication mod 5, that there is an identity element (1), that each element has an inverse, the inverse of 1 being 1, of 2 being 3, of 3 being 2 and of 4 being 4. It is tedious to check that the operation is associative but that is so. Thus the set {1, 2, 3, 4} is a group under multiplication mod 5.

	1	2	3	4
1	1	2	3	4
2	2	4	1	3
3	3	1	4	2
4	4	3	2	1

Table I

	0	1	2	3	4
0	0	0	0	0	0
1	0	1	2	3	4
2	0	2	4	1	3
3	0	3	1	4	2
4	0	4	3	2	1

Table II

If we add the element 0 to the set (Table II), we see that 0 has no inverse, so that the set {0, 1, 2, 3, 4} does not form a group under multiplication mod 5. As we expect, we can solve the equation $a*x = b$ if * is multiplication mod 5, *providing a is not zero*. (If a and b are both zero, then all values of x satisfy the equation.)

	1	2	3
1	1	2	3
2	2	0	2
3	3	2	1

Consider now the set {1, 2, 3} under multiplication mod 4. Although there is an identity element (1), the element 2 does not have an inverse, so the set {1, 2, 3} does not form a group under multiplication mod 4. Again, we see that we cannot solve $2*x = 1$ in multiplication mod 4.

Group of rotations

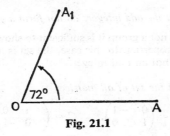

Fig. 21.1

Consider the rotation of a line OA about the point O. Let us call the operation which rotates OA through 72° anticlockwise to OA_1, P; the operation that rotates OA through 144°, Q; that through 216°, R; and that through 288°, S. The operation which rotates through 0° we shall call I, the identity operation; a rotation through 360° returns OA to its original position, and is equivalent to a rotation through 0°.

Our rule of combination $x*y$ is to perform first one operation y then another x, e.g. $P*Q$ means we first rotate through 216°, then through 72°, so that $P*Q = R$. Similarly $Q*Q = S$ and our operation table is displayed below.

	I	P	Q	R	S
I	I	P	Q	R	S
P	P	Q	R	S	I
Q	Q	R	S	I	P
R	R	S	I	P	Q
S	S	I	P	Q	R

Table III

	0	1	2	3	4
0	0	1	2	3	4
1	1	2	3	4	0
2	2	3	4	0	1
3	3	4	0	1	2
4	4	0	1	2	3

Table IV

The set of rotations I, P, Q, R, S forms a group under the law of combination $x*y$, first perform y, then x, and we notice that the table has exactly the same lay-out as Table IV, the set 0, 1, 2, 3, 4 under addition mod 5. Two groups which have the same structure like this are called isomorphic. (A more precise definition of isomorphism is needed when the idea is considered in advanced mathematics.)

Example 1. Show that the integers $\mathbb{Z}$ form a group under addition.

(1) The sum of any two integers is an integer, so that $\mathbb{Z}$ is closed under addition.

(2) The element 0 has the property that $a + 0 = 0 + a = a$ for all $a \in \mathbb{Z}$.

(3) Each element has an inverse, the additive inverse of 2 being -2, of 3 being -3, of a being $-a$.

(4) The operation is associative, the order in which three integers are added not affecting the sum.

Example 2. Show that the odd integers do not form a group under addition.

To show that a set is not a group it is sufficient to show that it does not have one of the necessary properties. In this case, the set is not closed, as the sum of two odd integers is not an odd integer.

Example 3. Show that the set of all matrices

$$A = \begin{pmatrix} 1 & 0 \\ 0 & 1 \end{pmatrix}, \quad B = \begin{pmatrix} 0 & -1 \\ 1 & 0 \end{pmatrix}, \quad C = \begin{pmatrix} -1 & 0 \\ 0 & -1 \end{pmatrix}, \quad D = \begin{pmatrix} 0 & 1 \\ -1 & 0 \end{pmatrix}$$

is a group under matrix multiplication.

The table below shows that the set is closed, that the matrix A is the unit element, and that each matrix has an inverse. It can be shown that matrix multiplication is associative, so the set satisfies the four necessary conditions, and is a group under matrix multiplication.

	A	B	C	D
A	A	B	C	D
B	B	C	D	A
C	C	D	A	B
D	D	A	B	C

The table is exactly the same as the table for {0, 1, 2, 3} under addition mod 4, and for the set of rotations through 0°, 90°, 180°, 270° combined as in the example under the law 'first y, then x'. It is not the same as the table in Example 4.

Example 4. Let the operation of reflecting in the x-axis be called X, of reflecting in the y-axis be Y, of reflecting in the origin O be O and of leaving unchanged be I. The law of combination x * y is 'first y then x', so that $X(Y(P)) = O(P)$ is illustrated in Fig. 21.2, where O(P) is the image of P under O. Show that the set {I, X, Y, O} is a group under this operation.

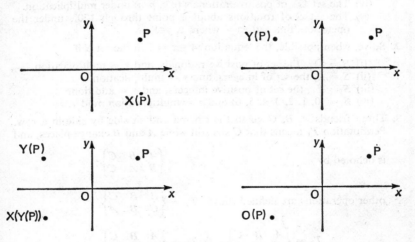

Fig. 21.2

	I	X	Y	O
I	I	X	Y	O
X	X	I	O	Y
Y	Y	O	I	X
O	O	Y	X	I

The table shows that the set is closed under *, that I is the unit element, and that each element is its own inverse. It can be shown that the operation is associative. Thus $\{I, X, Y, O\}$ is a group under *. Groups isomorphic to this group are called Klein four-groups, and it can be shown that all groups of order four are isomorphic either to the Klein four-group or to that in Example 3 (called the cyclic group, since one member is the group of cyclic rotations through 90°).

We have seen that the addition of numbers, multiplication and addition in different moduli, rotations about a point, reflections, and matrix multiplication can all form groups. It is this generalization which explains the reason for the study of groups, especially the groups of symmetries, in later mathematics.

In the following exercises, assume that the operations are associative.

Exercise 21a

1. Which of the following sets are groups under the operations given?
   - (i) The set $\mathbb{Z}$ of integers under multiplication.
   - (ii) The set of even integers under multiplication.
   - (iii) The set $\mathbb{Z}_+$ of positive integers under addition.
   - (iv) The set $\mathbb{Q}_+$ of positive rationals (e.g. p/q) under multiplication.
   - (v) The set S of rotations about a point through 120°, under the operation 'first x then y' where x, $y \in S$.

2. Solve, when possible, the 'equation' $4 * x = 5$ in the set S if
   - (i) $S = \mathbb{Q}_+$, the set of positive rationals, and $* = $ multiplication.
   - (ii) $S = \mathbb{Z}$, the set of integers, and $* = $ multiplication.
   - (iii) $S = \mathbb{Z}_+$, the set of positive integers, and $* = $ addition.
   - (iv) $S = \{0, 1, 2, 3, 4, 5, 6\}$ and $* = $ multiplication mod 7.

3. Three friends A, B, C go to the cinema and sit side by side in a row. Permutation T_1 means that C sits still while A and B change places, and

   is denoted by
   $$T_1 = \left\{ \begin{matrix} A & B & C \\ B & A & C \end{matrix} \right\};$$

   other operations are defined thus:
   $$T_2 = \left\{ \begin{matrix} A & B & C \\ C & B & A \end{matrix} \right\},$$

   $$T_3 = \left\{ \begin{matrix} A & B & C \\ A & C & B \end{matrix} \right\}, \qquad T_4 = \left\{ \begin{matrix} A & B & C \\ B & C & A \end{matrix} \right\},$$

   $$T_5 = \left\{ \begin{matrix} A & B & C \\ C & A & B \end{matrix} \right\}, \qquad T_6 = \left\{ \begin{matrix} A & B & C \\ A & B & C \end{matrix} \right\},$$

   so that under T_6 each returns to his original place. An operation $*$ is defined on the set of T_s where $T_x * T_y$ means 'first T_y then T_x', thus

   $$T_2(T_4) = T_2 \left(\begin{matrix} A & B & C \\ B & C & A \end{matrix} \right) = \left(\begin{matrix} A & B & C \\ A & C & B \end{matrix} \right) = T_3.$$

(i) Find the identity element.

(ii) Find the inverse of each element.

(iii) Show that $\{T_1, T_2, T_3, T_4, T_5, T_6\}$ is a group under *.

(Assume associative property.)

4. a, b are two distinct elements of a universal set $\mathscr{E}$. Subsets $S_0 = \varnothing$, $S_1 = \{a\}$, $S_2 = \{b\}$ and $S_3 = \{a, b\}$ are defined. Find, if possible, the identity element and the inverse of each element under (i) union, (ii) intersection. Is the set $\{S_0, S_1, S_2, S_3\}$ a group under either union or intersection?

5. All groups of order two, three, four and five are commutative. Copy the table below, and try to complete it to obtain a non-commutative group. (The table then will not be symmetrical about its leading diagonal.)

	e	a	b	c
e	e	a	b	c
a	a			
b	b			
c	c			

Exercise 21b

1. Which of the following sets are groups under the operation given?

   (i) The set $\{0, 1, 2\}$ under addition mod 3,

   (ii) The set $\{0, 1, 2\}$ under multiplication mod 3,

   (iii) The set $\{1, 2, 3, 4, 5\}$ under multiplication mod 6,

   (iv) Positive, negative and zero powers of 2 (e.g. 2^4, 2^{-2}) under multiplication,

   (v) The set $\{1, 3, 5, 7\}$ under multiplication mod 8.

2. Solve, when possible, the 'equation' $a*x = b$, in the set S if $S = \{0, 1, 2, 3, 4, 5\}$, * = multiplication mod 6 and

   (i) $a = 3, b = 5$,   (ii) $a = 5, b = 3$,

   (iii) $a = 4, b = 2$,   (iv) $a = 2, b = 4$.

3. A woman has a rectangular table-cloth $ABCD$ which just covers a rectangular table. To save washing the table-cloth, she can turn it over so that

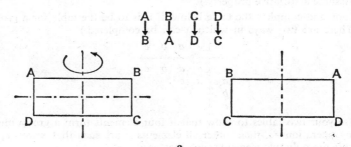

a

she can turn it over so that

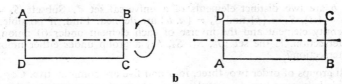

b

or she can turn it so that

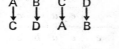

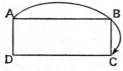

c

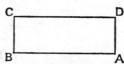

Fig. 21.3

Call these changes T_1, T_2, T_3. Define a suitable T_0 to serve as an identity element, and show that $\{T_0, T_1, T_2, T_3\}$ form a group under $*$ if $x*y$ means 'first y then x'. (This is the group of symmetries of a rectangle, and is isomorphic to the Klein four-group.)

4. S is the set of all matrices of the form $\begin{pmatrix} a & 1-b \\ 1-a & b \end{pmatrix}$
that is, whose columns total 1. Let $X = \begin{pmatrix} a & 1-b \\ 1-a & b \end{pmatrix}$
and $Y = \begin{pmatrix} c & 1-d \\ 1-c & d \end{pmatrix}$. Show that $X.Y \in S$.
Find the inverse of X, and show that $X^{-1} \in S$. Find the unit element, and hence show that S forms a group under matrix multiplication. (Assume associative property.)

5. Copy and complete the table below, if it is to be the table for a group. (There are two ways in which it can be completed.)

	e	a	b	c
e	e	a	b	c
a	a	e		
b				
c				

Use your two tables to show that if four elements form a group under any operation $*$, then either all elements x are such that $x*x = e$, or there are only two elements such that $x*x = e$.

22 Averages; Mean, Mode, Median

Average

'The average number of goals scored in First Division football matches last Saturday was 2.3.' 'The average number of children in a family is 2.7.' We certainly do not think that all teams scored 2.3 goals last Saturday, nor that most families have 2.7 children. The numbers 2.3 and 2.7 are statistics which, when interpreted correctly, give us some information about the goals scored or the size of a family.

Mean

The commonest way of determining a statistic to give information about the 'average' of a sample is to add all the elements together, then divide by the number of elements. This is used in finding the batting average of cricketers, in finding the average number of goals scored in football, and so on. This statistic is called the **mean**.

Example 1. *Find the mean of the numbers 1, 3, 5, 6, 7.*

$1 + 3 + 5 + 6 + 7 = 22$.
There are five numbers, so the mean is $\frac{22}{5}$, i.e. 4.4.

Notice that the mean is not one of the numbers in the sample. It may happen to be, but generally it is not.

Example 2. *Find the mean of the numbers 1, 3, 5, 6, 107.*

The mean is $\dfrac{1 + 3 + 5 + 6 + 107}{5} = 24.4$.

The mean is now very large, because of the 107, and does not describe very well the majority of scores in the sample.

Median

A statistic which is not affected by a few very unusual extreme scores is the **median**. This is found by ranking the scores in order of size, and taking the middle: if there is an even number of scores, we take the mean of the two middle scores.

Example 3. *Find the median of 4, 3, 5, 2, 11.*

Rank the scores in ascending order,
$$2, 3, 4, 5, 11.$$
The median is the third score, i.e. 4.

Example 4. Find the median of 4, 3, 5, 2, 6, 11.

Rank the scores in ascending order,

$$2, 3, 4, 5, 6, 11.$$

Since there are six scores, we take $\frac{1}{2}(4 + 5)$,
the median is $4\frac{1}{2}$.

Strictly, we could use any score between 4 and 5, but a widely-observed convention is to use the mean of the two middle scores. The median is defined to be that score which has as many scores greater than itself as there are less than itself.

Mode

When making a very exotic shirt, or a very fancy pair of ladies shoes, we shall be most certain of selling our product if we make it in the commonest size, the one size that occurs most frequently. This is called the mode.

Example 5. Find the mode of 1, 2, 2, 2, 3, 1, 3, 0.

The score 2 occurs most frequently and is the mode.

Exercise 22.1a

1. Find the mean, median, and mode of each of the following sets of scores:

    (i) 1, 1, 2, 3, 4.   (ii) 2, 1, 5, 4, 5.

    (iii) 0, 1, 1, 2, 3, 4.   (iv) 1, 0, 3, 0, 6, 8.

2. Find the mean of the following scores:

    (i) 1, 2, 4, 5, 8.   (ii) 11, 12, 14, 15, 18.

    (iii) 101, 102, 104, 105, 108.   (iv) 20, 40, 80, 100, 160.

3. Find the mean of the following scores:

    (i) 1, 3, 4, 4, 8.   (ii) 1.1, 3.1, 4.1, 4.1, 8.1.

    (iii) 0.9, 2.9, 3.9, 3.9, 7.9.   (iv) 5, 15, 20, 20, 40.

4. Find the mean of the following scores:

    (i) 4, 4, 4, 4, 4, 4, 5, 5, 5, 5.   (ii) 1, 2, 2, 2, 2, 2, 2, 2, 2, 2.

    (iii) 2, 2, 2, 3, 3, 3, 3, 3, 4, 4.   (iv) 13, 13, 13, 14, 14.

5. Find x if the mean of 1, 3, 6, 8, x is 6.

Exercise 22.1b

1. Find the mean, median and mode of each of the following sets of scores:

    (i) 2, 3, 3, 3, 5.   (ii) 4, 0, 2, 9, 0.

    (iii) 0, 1, 4, 5, 9, 9.   (iv) 5, 0, 1, 2, 0, 7.

2. Find the mean of the following scores:

    (i) −1, 0, 1, 3, 7.   (ii) 9, 10, 11, 13, 17.

    (iii) 49, 50, 51, 53, 57.   (iv) −4, 0, 4, 12, 28.

3. Find the mean of the following scores:

    (i) $-2, -1, 1, 3, 4.$ (ii) $-1.9, -0.9, 1.1, 3.1, 4.1.$

    (iii) $17, 18, 20, 22, 23.$ (iv) $-10, -5, 5, 15, 20.$

4. Find the mean of the following scores:

    (i) $-1, -1, -1, 3, 3, 3, 3, 3, 3, 3.$ (ii) $2, 3, 3, 3, 3, 3, 2, 3, 2, 3.$

    (iii) $3, 4, 5, 4, 5, 4, 5, 3, 4, 4.$ (iv) $23, 24, 25, 25, 25.$

5. Find x if the mode of $1, 2, 2, 3, x, 5, 5, 1, 4,$ is 2.

Frequency

In Ex. 22.1a, Q.4 and Ex. 22.1b, Q.4, many of the scores were repeated, and we probably found that instead of adding say six 4's it
was easier to multiply 4 by 6; that instead of adding $13 + 13 + 13$ it
was easier to multiply 13 by 3. The **frequency** of 4 is 6; the frequency
of 13 is 3. This can be used to reduce working, especially when there
are many scores.

Example 6. *Find the mean of the following scores:*

$$21, 21, 21, 22, 22, 22, 22, 23, 23, 24,$$
$$24, 24, 24, 24, 24, 25, 26, 27, 27, 27.$$

The scores were tabulated and we can see there are 20 scores. Display the
scores x and frequencies f in a table, and add a third column fx showing the
product of each score and the corresponding frequency.

x	f	fx
21	3	63
22	4	88
23	2	46
24	6	144
25	1	25
26	1	26
27	3	81
Total:	20	473

Since the sum of 20 scores is 473, the mean is $\frac{473}{20}$ i.e. 23.65.

Use of working zero

In Qs 2 and 3 of each of the previous exercises we may have noticed that
we could deduce all later parts of each question from the first part.
Thus when we have found the mean of 1, 2, 3, 4 to be 2.5, we can see
that the mean of 1.2, 2.2, 3.2 and 4.2 is $2.5 + 0.2$, i.e. 2.7; similarly
when we have found the mean of 1, 2, 2, 4, 5 to be 2.8, we can deduce
the mean of 0.9, 1.9, 1.9, 3.9, 4.9 to be $2.8 - 0.1$, i.e. 2.7. Conversely,
when we wished to find the mean of the scores in Example 6, we could
have subtracted 20 from each score, found the mean of the new scores
to be 3.65, then added 20 to obtain 23.65.

Example 7. *Find the mean of the following scores:*
      9.81, 9.81, 9.82, 9.82, 9.82, 9.83, 9.83, 9.84, 9.84, 9.84.

The scores will be simplified if we subtract 9.8 from each score. 9.8 is called the working zero, denoted by w. (Some writers call w the working mean or the 'estimated mean', but it is certainly NOT an estimated mean, as the mean must be greater than 9.8.) Tabulate the data:

x	$x - w$	f	$f(x - w)$
9.81	0.01	2	0.02
9.82	0.02	3	0.06
9.83	0.03	2	0.06
9.84	0.04	3	0.12
Total		10	0.26

The mean of the scores $(x - w)$ is $\dfrac{0.26}{10}$ i.e., 0.026, so the mean of the original scores is 9.826.

Scaling

In some of the earlier problems, the scores were easy multiples of numbers whose mean we had already found. Thus when we know the mean of 2, 4, 7, 8, 9 to be 6, we can deduce that the mean of 20, 40, 70, 80, 90 is 60; of 0.02, 0.04, 0.07, 0.08, 0.09 is 0.06. Scaling and the use of working zeros can often simplify very awkward data.

Example 8. *Find the mean of 475, 515, 535, 545, 585.*

Use 535 as a working zero: the new scores are -60, -20, 0, 10 and 50.
Scale by dividing by 10; -6, -2, 0, 1, 5.
The mean of these scores is $-\frac{2}{5}$ i.e. -0.4.
Multiplying by 10, -4.
Adding to 535,   531.
The mean of the original scores is   531.

Exercise 22.2a

1. Find the mean of the following:
      17, 17, 17, 17, 21, 21, 21, 25, 25, 26.

2. Find the mean of each of the following sets of scores, given with their corresponding frequencies:

(i) Score (x)	frequency (f)	(ii) x	f
4	4	15	4
5	7	17	3
6	9	19	3
		20	4
		29	6

3. Use a working zero to find the mean of each of the following sets of scores, given with their corresponding frequencies:

(i)	x	f		(ii)	x	f
	71	4			99	4
	72	6			101	5
	73	7			103	3
	74	3			104	4
					106.5	4

4. Scale the data, then find the mean of each of the following scores:

   (i) 120, 120, 240, 240, 480.

   (ii) 0.01, 0.03, 0.04, 0.07, 0.09.

5. The table below gives the percentage distribution of income in the United States in 1957, so that 7% of the population had an income of less than $1000, etc.

Income in dollars	Percentage frequency
Under 1000	7
1000–1999	12
2000–2999	10
3000–3999	10
4000–4999	12
5000–7499	26
7500–10 000	23
Total:	100

Estimate that the 7% with an income of under $1000 had an income of $500; that the 12% with an income $1000–1999 had an income of $1500, and so on, and hence estimate the mean income of this group of people. (These are called *mid-interval values*; we do not know that the incomes are evenly distributed over the intervals, so we can only *estimate* the mean, using these estimates in our calculations.)

Exercise 22.2b

1. Find the mean of the following:

   17, 17, 17, 13, 13, 17, 17, 13, 13, 17.

2. Find the mean of each of the following sets of scores, given with their corresponding frequencies:

(i)	Score (x)	frequency (f)		(ii)	x	f
	1	43			3	31
	2	47			4	33
	3	57			5	31
	4	53			6	35
					8	20

3. Use a working zero to find the mean of each of the following sets of scores, given with their corresponding frequencies:

(i) x	f		(ii) x	f
81	17		93	6
82	27		94	17
83	40		97	27
84	16		101	34
			102	10
			110	6

4. Scale the data, then find the mean of the following sets of scores.

   (i) 1000, 1000, 2500, 4000, 5000.
   (ii) 0.003, 0.013, 0.023, 0.021, 0.007.

5. The table below gives the heights correct to the nearest cm, of the pupils in a certain school and corresponding frequencies.

Height h (in cm)	Frequency f	h	f
155–164	10	173	84
165–170	41	174	42
171	37	175	27
172	38	176–180	12

By taking mid-interval estimates where appropriate, estimate the mean height of these pupils.

Weighted mean

If a boy scored 20 and 30 marks in two examinations, then his mean ('average') score is 25. But if the first examination was twice as long as the second, it would be fair to weight it by a factor of 2. Thus his weighted mean would be

$$\frac{20 \times 2 + 30 \times 1}{2 + 1} = \frac{70}{3} = 23\tfrac{1}{3}.$$

If the first examination was in English and the second in Maths, then for some purposes Mathematics might be considered twice as important as English, and we would weight the Maths marks by a factor of 2. This time the weighted mean is

$$\frac{20 \times 1 + 30 \times 2}{1 + 2} = \frac{80}{3} = 26\tfrac{2}{3}.$$

Notice that we always divide by the sum of the weightings.

Example 9. Two ladies are choosing a new car. They judge the cars by index numbers as below, and their own preferences are shown by the weightings they give to each of the qualities of the car. Which car would each prefer?

	Car		Purchaser	
	Mini	Maxi	Mrs Go	Miss Stop
Petrol consumption	108	92	3	2
Speed	104	107	1	3
Comfort	82	110	1	5

These figures show that the petrol consumption of the Mini is much better than that of the Maxi, and that Mrs Go wants a car with a good petrol consumption. The Maxi is clearly the more comfortable car, and Miss Stop attaches great importance to comfort. The weighted means are:

Mrs Go: Mini: $\dfrac{108 \times 3 + 104 \times 1 + 82 \times 1}{3 + 1 + 1} = 102.$

Maxi: $\dfrac{92 \times 3 + 107 \times 1 + 110 \times 1}{3 + 1 + 1} = 98.6.$

Miss Stop: Mini: $\dfrac{108 \times 2 + 104 \times 3 + 82 \times 5}{2 + 3 + 5} = 93.8.$

Maxi: $\dfrac{92 \times 2 + 107 \times 3 + 110 \times 5}{2 + 3 + 5} = 105.5$

Moving averages

Some quantities such as share prices, unemployment figures, trade figures, vary considerably and a better indication of the underlying trend can sometimes be seen by taking averages over a period. If the period varies, this is called a moving average.

Example 10. The figures below show the number of unemployed in a certain holiday resort during six consecutive months. Find the three-month moving average for that period.

Number unemployed	Mar	Apr	May	June	July	Aug
(in thousands)	84	50	64	53	45	21

The average for March, April and May is $\dfrac{84 + 50 + 64}{3} = 66,$

for April, May, June is $\dfrac{50 + 64 + 53}{3} = 55\frac{2}{3},$

for May, June, July is $\dfrac{64 + 53 + 45}{3} = 54,$

and for June, July, August is $\dfrac{53 + 45 + 21}{3} = 39\frac{2}{3}.$

Exercise 22.3a

1. Find the weighted mean of 20, 25, 30, 35 if they are assigned weightings of
   (i) 1, 2, 3, 4,
   (ii) 1, 3, 7, 9,
   (iii) 5, 2, 2, 1, respectively.

2. The weighted mean of 15 and 25 is 21. Find the ratio of the weightings assigned to each score.

3. If the designer of the Maxi (see Example 9) is determined that Mrs Go shall prefer his car to the Mini, what is the least amount by which he must increase the figure for petrol consumption, the other figures remaining constant?

4. The annual rate of inflation in a certain country, calculated each month over a period of eight consecutive months, was:

$$24\%; \quad 28\%; \quad 30\%; \quad 27\%; \quad 20\%; \quad 17\%; \quad 20\%; \quad 25\%.$$

Calculate (i) the six-month moving average for each of the three periods of six months,

(ii) the three-month moving average for each of the six periods of three months.

Which moving average would you quote if you were a politician claiming to have cured inflation?

5. The monthly rainfall in Cambridge in cm during a period of twelve consecutive months was:

9.32, 2.03, 0.13, 3.48, 3.76, 7.64, 5.06, 2.77, 4.06, 0.72, 2.08, 5.33.

Calculate each of the three-month moving averages, and draw a graph to illustrate them.

Exercise 22.3b

1. Find the weighted mean of 3, 5, 6, 9 if they are assigned weightings of

(i) 1, 1, 1, 2,   (ii) 2, 2, 3, 3,   (iii) 4, 3, 2, 1.

2. The following table gives the index and weighting for each of four commodities used in forming a price index:

	Index no.	Weighting
Food	112	6
Rent	125	3
Fuel	106	1
Clothing	107	2

Calculate the price index.

3. A firm testing artificial fibres assigns indices to certain qualities it wishes to obtain in its fibres, strength, comfort in wear and ease of dyeing. Three fibres X, Y and Z are tested and given the following scores:

	X	Y	Z
Strength	120	107	130
Comfort	94	102	59
Dyeing	82	98	120

For a certain material, weightings 5, 4 and 3 are given to strength, comfort and dyeing respectively. Which fibre has the highest weighted mean?

Material to be used for a different purpose has weightings 1, 3, 1 respectively. Which fibre now has the highest weighted mean?

4. The total number of working days lost by strikes each year over 11 years were, in thousands:

   1389, 1694, 1792, 2184, 2457, 3794, 2083, 8415, 3462, 5270, 3008.

   Calculate the five-year moving average, and illustrate this moving average graphically.

5. The price of a certain share on the first working day of each of 12 months was:

   61p, 77p, 66p, 85p, 83p, 90p, 76p, 80p, 72p, 61p, 74p, 63p.

   Calculate the ten three-month moving averages and illustrate them graphically.

Exercise 22.4 Miscellaneous

1. The mean of seven numbers is 15: when another number is added, the mean of the eight numbers is 16. Find the eighth number.

2. The mean of seven numbers is 15; when three more numbers are added, the mean of the ten numbers is 12. Find the mean of the three numbers.

3. Write down five consecutive integers whose median is m. Show that the mean of any five consecutive integers is always equal to the median.

4. Find the median and mode of the following scores: 0, 1, 1, 2, 0, 3, 0, 5.

5. Find the two modes of the following scores: 0, 1, 1, 2, 0, 3, 0, 5, 1. (Such scores are said to be bimodal, with two modes.)

6. Find the median of the following scores:
   (i) 0, 1, 1, 0, 2, 5, 7, 8, 9, 11.
   (ii) 0, 11, 1, 2, 0, 5, 0, 3, 0.

7. The number of children per family given as a percentage of the number of families sampled in a census is shown below:

Number of children	Percentage		
x	f	x	f
0	2.4	4	5.4
1	29.5	5	0.9
2	39.5	6	0.6
3	21.4	7	0.1

   Calculate the mean number of children in a family in those sampled.

8. 100 oranges were tested for their vitamin C content, with the following results:

Vitamin C in milligrams, correct to the nearest milligram	30	31	32	33	34	35
Number of oranges	11	18	23	19	17	12

Find the mean vitamin C content of these oranges.

9. The number of passengers per compartment in an Inter-City express was recorded as below:

Number of passengers	0	1	2	3	4	5	6
Number of compartments	2	2	5	25	24	17	5

Find the mean number of passengers per compartment.

10. The marks scored by 500 candidates in an examination in which the maximum mark was 50 were:

Mark range	frequency	mark range	frequency
1–5	10	26–30	81
6–10	41	31–35	71
11–15	72	36–40	27
16–20	83	41–45	13
21–25	94	46–50	8

Use the mid-interval values to estimate a mean mark for these candidates.

23 Representation of Data

Collection of data

Statistics depend on the fair collection of data, and we have probably collected data of various forms and made simple diagrams to display this data. Fair sampling is difficult to achieve; what might be wrong with these methods for sampling?

(i) Sending a representative to call at houses during the morning, to ask if the occupiers had washing machines?

(ii) Selecting telephone numbers at random from a directory and asking whoever answers the phone if they ate porridge for breakfast?

(iii) Telephoning as in (ii), but asking if they owned two cars?

In (i), our sample might be distorted because many wives who own washing machines might be out working; the answers in (ii) might depend on whether the telephone directory was for part of Scotland or England; the answers for (iii) would probably be distorted because in general people with telephones are more wealthy than those without, and so more likely to own two cars.

Representation of data

We assume that the data has been collected fairly. Suppose a housewife finds that of her £30 per week housekeeping, she spends £15 on groceries, £5 on meat, £5 on fruit and vegetables, £3 on bread and £2 on milk. This information could be represented by a pictogram, using a picture of a £1 note to represent each £ she spends.

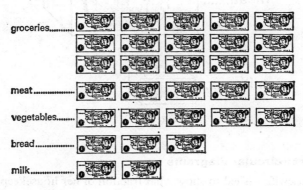

Fig. 23.1

Misrepresentation of data

It would be misleading to make the edge of each £1 note proportional to the amount spent, for our eye compares areas:

meat........

bread........

milk..........

Fig. 23.2

and worse still if we try to give a three-dimensional effect! Yet these are occasionally used in newspapers.

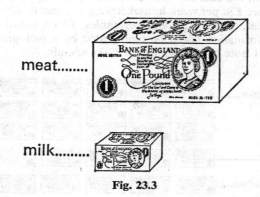

meat........

milk.........

Fig. 23.3

Pie-chart: circular diagrams

If the housewife wished to show what fraction of her housekeeping was spent on each type of purchase, she would probably use a pie-chart (sometimes called a circular diagram). In this, the angle at the centre of the circle is proportional to the quantity the sector represents. Since the total expenditure is £30, groceries are represented by a sector angle $\frac{15}{30} \times 360°$, i.e. 180°, meat and vegetables each by a sector angle $\frac{5}{30} \times 360°$ i.e. 60°, bread by a sector $\frac{3}{30} \times 360°$, i.e. 36° and milk by an sector angle 24° (Fig. 23.4).

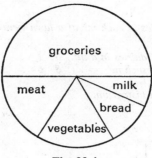

Fig. 23.4

Bar chart

The housewife could, however, use a bar chart in which the bars are drawn of equal width and the length of the bar is proportional to the quantity it represents (Fig. 23.5).

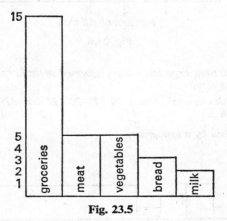

Fig. 23.5

Histogram

A histogram often looks like a bar chart and the two are easily confused. The essential characteristics of a histogram are that it represents frequency by area and that there is a numerical relationship between quantities whose frequencies it represents. Thus a histogram could be used to represent the marks scored by candidates in an examination after they have been grouped (or even ungrouped marks, though that would be misleading unless many marks are being considered), because there is a numerical relation between the intervals 1–5, 6–10, etc.; it could not be used to describe Fig. 23.5, even though the areas represent frequencies, because there is not a numerical relation between say, bread and milk.

Example 1. *The daisies in each m^2 of a lawn were counted and the number recorded below:*

No. of daisies in a m^2	0	1	2	3	4	5
No. of such m^2	8	6	5	3	2	1

Represent this data by a histogram.

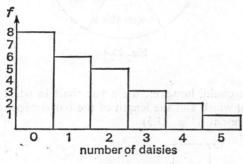

Fig. 23.6

Example 2. *The same experiment was performed on the lawn of a neighbouring garden with these results:*

No. of daisies in a m^2	Under 10	11–20	21–30	31–40	41–100
No. of such m^2	1	3	2	1	18

Represent this data by a histogram.

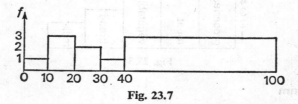

Fig. 23.7

Notice the width of the last class-interval is six times that of the others, so the area is six times that of the interval 11–20, which has the same vertical height.

Exercise 23.1a

1. In a certain school, the lessons each week are allocated as below:

   English   4   Maths   5   Science   6
   French   4   German   4   Others   13

   Draw a pie-chart illustrating this data.

2. A survey of the occupations of the men in a certain club produced the following information:

   Accountants   1   Bankers   3   Doctors   3
   Engineers   7   Lawyers   4   Salesmen   5

   Draw a bar chart to display this data.

3. The number of attempts a certain darts-player needed one evening before he hit a 'bull's eye' were recorded:

No. of attempts	1	2	3	4	5	6	7	8
frequencies	0	0	1	0	3	4	5	2

   Draw a histogram to display this data.

4. The distance travelled to school each day by the pupils in several classes were recorded as below:

Distance in km (to the nearest $\frac{1}{10}$ km)	Under 2	2–2.9	3–3.9	4–4.9	5–5.9
No. of pupils	100	40	20	30	10

   Draw a histogram to display this date.

5. The ages of agricultural workers in a certain country were recorded as below:

Age in years	18 and under 20	20 and under 60	60 and under 70
No. in thousands	20	400	100

   Draw a histogram to display this data.

Exercise 23.1b

1. The pupils at a certain school recorded the main dish in their dinners over a period of 36 days:

   Roast meat   1   Fish   7   Sausages   8
   Pies   6   Stew   10   Cheese flan   4

   Draw a pie chart to display this data.

2. The teachers at a certain school were asked how they travelled to school one day, and their replies recorded below:

   Walked   13   Cycled   8   By car   12
   By train   2   On Horse 1

   Draw a bar chart to display this information.

3. 100 cars were tested to see how far they travelled on 10 litres of a certain petrol and their results recorded below:

Distance in km (to the nearest km)	100–109	110–119	120–129	130–139	140–149
Number of cars	5	15	25	35	20

Draw a histogram to display this data.

4. A sample of 50 cooks were asked how long they boiled cabbage, and their replies recorded below:

Time in minutes (to the nearest minute)	Under 5	5–9	10–14	15–19	20–59
Number of cooks	5	20	5	10	10

Draw a histogram to display this data.

5. The goals scored one Saturday by a sample of 54 football teams were recorded as below:

Number of goals	0	1	2	3	4	5	6	7
Frequency	8	12	14	10	5	4	0	1

Draw a histogram to display this data.

Cumulative frequency curve

Looking at the data of Ex.23.1b, Q. 3, we see that 5 cars travelled less than 110 km, 20 cars travelled less than 120 km, 45 cars travelled less than

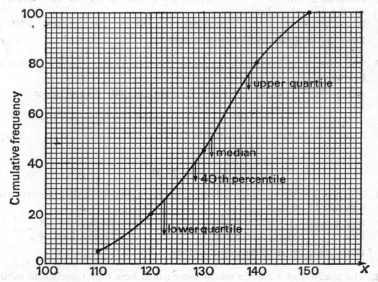

130 km, 80 cars less than 140 km and all 100 cars less than 150 km. We can draw a graph to display this, which shows the cumulative frequency of the data, and so is called a cumulative frequency curve. (Sometimes called an ogive in the past; the term is little heard today.)

If the cars are arranged in order of distance travelled, how far would the 50th car have travelled? Read on the x-axis the distance corresponding to a cumulative frequency of 50, and we have 131.5, so we estimate that the 50th car would have travelled 131.5 km. 40 of the cars we estimate would have travelled less than 128 km; reading against the 70, we estimate that 30 of the cars would have travelled *more* than 137 km (since 70 of the cars would have travelled less than 137 km).

Median, quartiles, percentiles

The median of the distances covered by these cars is found by reading against the cumulative frequency of 50*, and is 131.5 km; reading against cumulative frequencies of 25 and 75 we have scores of 122.5 km and 138.5 km, and these are called the lower and upper quartiles respectively. The distance between these quartiles is the inter-quartile range, and the **semi-inter-quartile range** is sometimes used as a measure of dispersion, or spread, of the readings. Percentiles are defined in a similar manner, so that the 25th percentile is the lower quartile.

Estimates only

Readings from a cumulative frequency curve are only estimates. When used unwisely, they can be very misleading. (Try drawing, then interpreting, the cumulative frequency curve for Ex. 23.1b, Q. 5; how many goals did the 40th team score?) They should never be used for small samples.

Exercise 23.3

1. Draw a cumulative frequency curve illustrating the data in Ex. 22.2a, Q.5. Estimate
    (i) the income exceeded by 50% of the group studied,
    (ii) what percentage of the group had an income in excess of $3500 per annum.

2. Draw a cumulative frequency curve illustrating the data in Ex. 22.2b, Q.5. Estimate
    (i) the height exceeded by 45 of the pupils,
    (ii) the number of pupils taller than 174.5 cm,
    (iii) the semi-inter-quartile range.

3. Draw a cumulative frequency curve illustrating the data in Ex. 22.4, Q.8. Use your curve to estimate
    (i) how many oranges exceeded 32.5 units vitamin C content,
    (ii) the vitamin C content exceeded by 40% of the oranges,
    (iii) the semi-inter-quartile range.

*Strictly, the median of 100 scores should be read against $50\frac{1}{2}$, and the quartiles against $25\frac{1}{4}$ and $75\frac{3}{4}$. With a large number of scores, the difference is negligible.

4. Draw a cumulative frequency curve illustrating the data in Ex. 22.4, Q.10. Use your curve to estimate

    (i) the percentage of candidates who fail if the pass mark is 22,
    (ii) the pass mark if 40% of the candidates pass.

5. Draw a cumulative frequency curve illustrating the data in Ex. 23.1a, Q.4. Use your curve to estimate the distance travelled exceeded by 60% of the pupils.

Exercise 23.4

1. An ornithologist recorded the number of eggs in 100 bird nests as below:

1	egg	3 nests	4 eggs	43 nests
2	eggs	15 nests	5 eggs	9 nests
3	eggs	25 nests	6 eggs	5 nests

    (i) Draw a histogram to display this data.
    (ii) Find the mean number of eggs per nest.

2. The pocket money received each week by a group of children was recorded as below:

50p	13 children	90p	11 children
60p	25 children	£1	6 children
70p	24 children	£1.50	3 children
80p	15 children	£2	3 children

    (i) Draw a histogram to display this data.
    (ii) Find the mean weekly pocket money each child received.

3. A charity recorded donations received in a certain year as below:

£1	160 donations	£10	25 donations
£2	80 donations	£20	15 donations
£3	40 donations	£40	2 donations
£4	25 donations	£50	2 donations
£5	25 donations	£100	1 donation

Find the mean of this data, and say why this is not typical of the donations received.

4. The Hire Purchase debts, correct to the nearest £50, of 80 families were recorded below:

£100	12 families	£300	7 families
£150	23 families	£350	3 families
£200	17 families	£400	4 families
£250	12 families	£500	2 families

How many families had Hire Purchase debts of less than £125?
Draw a cumulative frequency curve and use this to estimate

    (i) the median Hire Purchase debt of these 80 families,
    (ii) the Hire Purchase debt exceeded by 30% of the families.

5. A survey was made of the weekly earnings of 1000 men, and the results, correct to the nearest £5, recorded below:

£30	£35	£40	£45	£50	£55	£60	£65
40	60	100	140	220	240	140	60

(i) Find the mean wage from this data; why may it differ from the true mean?

(ii) Draw a cumulative frequency curve; from it estimate the median wage, and the number of men who earned between £40 and £50 per week.

24 Probability

If an unbiased coin is tossed, it is just as likely to come down 'Heads' as 'Tails'. We may say that there is a '50–50 chance that it will land showing Heads', or we say 'the odds are evens that it will land showing Heads,' but when studying the relative likelihood of particular outcomes, we assign numbers between 0 and 1 to the probability of events happening. Thus the probability of a fair coin showing 'Heads' is 1/2; if a fair die is thrown, the probability of it showing a '4' is 1/6, as also for any one other score. If an event will not happen, we assign to it the probability 0: to an event which is certain to happen we give the probability 1.

Use of symmetry

These probabilities were found by studying the symmetry of the problem. Two faces to the coin, each just as likely as the other to be showing. So we need two equal numbers whose sum is 1, since one or other of the faces must be showing. A fair die, six faces, each equally likely to be uppermost, the probability of any one face being uppermost is 1/6. A pack of 52 cards, each equally likely to be drawn, the probability of the Ace of Spades being drawn is 1/52.

Definition

If we have a number of equiprobable events, such as drawing cards from a well-shuffled pack, the probability of a particular outcome is defined as

$$\frac{\text{the number of equiprobable favourable outcomes}}{\text{total number of equiprobable outcomes}}$$

Thus the probability of drawing an Ace from a pack of 52 cards is 4/52; the probability of drawing an Ace from a pack if the two jokers are included is 4/54. Note that it does not mean that in 52 draws from a fair pack without jokers we must have 4 Aces.

Importance of equiprobable events

To find the probability that, if three coins are thrown, two show 'Heads', the favourable throws are T, H, H; H, T, H, and H, H, T, whereas the list of all possible outcomes is

$$\text{H, H, H;} \quad \text{T, H, H: H, T, H: H, H, T;}$$
$$\text{T, T, H: T, H, T: H, T, T;} \quad \text{T, T, T}$$

so that the probability of two 'Heads' showing is 3/8.

Notice that if we listed the possible outcomes as

0 Heads, 1 Head, 2 Heads, and 3 Heads,

and so claimed that the probability of two Heads was 1/4, this would be incorrect. These outcomes are not equiprobable, and so we cannot use our definition in this manner.

To find the probability of an event happening, decide how many equiprobable outcomes are favourable, how many equiprobable outcomes there are, and the probability is the ratio of these two. The following examples illustrate the use of the definition.

Example 1. *In a well-shuffled pack of 52 playing cards, find the probability that a card drawn at random is a heart.*

There are 52 possible outcomes, and since there are 13 hearts, we have 13 'favourable' outcomes. Thus the probability of drawing a heart is 13/52, i.e. $\frac{1}{4}$.

Example 2. *From a pack of playing cards, the King of Hearts is removed. The remaining cards are well shuffled. What is the probability that a card drawn at random is a heart?*

There are now 51 equiprobable outcomes, 12 of which are favourable. Thus the probability that a card drawn now is a heart is 12/51, i.e. 4/17.

Example 3. *From a pack of playing cards, two cards are taken, which are not hearts. They are not replaced, and the remaining cards are then well shuffled. What is the probability that the next card drawn is a heart?*

There are now 50 possible outcomes, and of these 13 are favourable, since there are still 13 hearts in the pack. The probability that a card drawn at random is a heart is therefore 13/50.

Example 4. *A man has 4 blue socks in his drawer, 5 brown socks and 6 green socks. Find the probability that the first sock he selects one dark morning is (i) blue, (ii) brown, (iii) green.*

Since there are 15 socks in the drawer and of these 4 are blue, the probability that a sock selected at random is blue is 4/15. Similarly the probability that the first sock is brown is 5/15, and the probability that the sock is green is 6/15.

If the first sock he selects is blue, there are then 3 blue socks remaining out of 14 socks, so the probability that the next sock selected is blue is 3/14. If however the first sock selected is brown, the probability that the second sock selected is brown is 4/14; likewise is the first sock selected is green, the probability that the second selected is green is 5/14.

Example 5. A number is selected at random from those numbers between 50 and 60 inclusive. Find the probability that it is (i) prime, (ii) a multiple of 8.

There are 11 numbers between 50 and 60 inclusive. Of these, only 53 and 59 are primes. Thus the probability that a number chosen at random is prime is 2/11.

Only one number in this range is a multiple of 8, that is 56, so the probability that a number chosen at random is a multiple of 8 is 1/11.

Example 6. Three athletes A, B and C are to run a race. B and C have equal chances of winning, but A is twice as likely to win as either. Find the probability of each athlete winning.

We wish to find three numbers to assign as the probabilities that each of these athletes will win the race. Since one runner is certain to win (ignoring ties), the probabilities must total 1. Since B and C are equally likely to win, their probabilities must be the same, say x.

Since A is twice as likely to win as either, his probability must be $2x$.

$$x + x + 2x = 1$$

i.e.

$$x = \tfrac{1}{4}$$

The probability that A wins is $\tfrac{1}{2}$, that B wins is $\tfrac{1}{4}$, that C wins is $\tfrac{1}{4}$.

Exercise 24.1a

1. From a well-shuffled pack of 52 cards, one card is drawn. Find the probability that it is

    (i) a King,
    (ii) the Queen of Hearts,
    (iii) a Diamond,
    (iv) either the Queen of Hearts or the Jack of Spades,
    (v) either a two or a three,
    (vi) either a two or a Spade,
    (vii) not the Ace of Spades,
    (viii) not a Club,
    (ix) not a Diamond,
    (x) either a King, a Queen or a Jack.

2. A man writes down at random a whole number larger than 1 and smaller than 11. Find the probability that it is

    (i) odd,   (ii) even,   (iii) prime,   (iv) a factor of 12,
    (v) a perfect square,   (vi) a power of 2,   (vii) a perfect cube,
    (viii) a square root of a number less than 50.

3. A bag contains initially 20 marbles, 5 of which are chipped.

    (i) Find the probability that a marble drawn at random is chipped.
    (ii) A chipped marble is drawn and not replaced. Find the probability that the next marble drawn is also chipped.

(iii) 8 marbles, of which three are chipped, are drawn and not replaced. Find the probability that a marble now drawn at random is chipped.

(iv) 8 marbles, of which five are chipped, are drawn and not replaced. Find the probablity that a marble now drawn is not chipped.

4. List the prime numbers less than 50. (Remember that 1 is not a prime.) Find the probability that a number chosen from these at random

   (i) is less than 25,

  (ii) contains a digit 5,

 (iii) contains two digits 3,

 (iv) is a power of 2,

  (v) is between 10 and 20,

 (vi) has its square greater than 100.

5. A desk contains 3 pens, 7 biros and 2 pencils. What is the probability that one object selected at random is a pen?

6. A box of chocolates contains 5 with soft centres and 7 with hard centres. What is the probability that, if one is selected at random and eaten, it is a 'soft centre'? If the first one eaten is 'soft centred', what is the probability that the next one eaten is also 'soft centred'?

7. What is the probability that a letter of the alphabet selected at random is a vowel?

8. List the possible outcomes of throwing a pair of dice, e.g. (1, 1) to denote that both dice show ones, (1, 2) to denote that the first shows a 1 and the second a 2. There should be 36 such outcomes, all equiprobable.

   (i) Find the probability that the total shown on the two dice is 2.

  (ii) List the probability of all totals from 2 to 12. Verify that these add up to 1.

 (iii) What is the most likely total obtained from throwing two fair dice?

Exercise 24.1b

1. A woman has 10 tins of baked beans in her larder and 8 tins of tomatoes. The tins have all lost their labels and are identical in appearance, so that every selection the woman makes is at random. Find the probability that

    (i) the first tin selected is of baked beans,

   (ii) the first tin selected is of tomatoes,

  (iii) after she has opened a tin of baked beans, the next one opened also contains baked beans,

  (iv) after she has opened a tin of tomatoes, the next one opened will contain baked beans,

   (v) after she has opened two tins of baked beans, the next one she opens contains baked beans,

  (vi) after she has opened two tins of tomatoes, the next one opened is baked beans,

 (vii) after she has opened a tin of tomatoes and a tin of baked beans, the next one opened contains baked beans.

2. Using only numerals between 1 and 5 inclusive, a man writes down at

random a proper fraction, not necessarily in its lowest terms. List all possible fractions, and so find the probability that his fraction

    (i) is less than $\frac{1}{2}$,
    (ii) is greater than $\frac{1}{2}$,
    (iii) is in its lowest terms,
    (iv) is a perfect square,
    (v) is equal to a recurring decimal.

3. A man has 3 pairs of black socks and 2 pairs of brown socks. If he dresses hurriedly in the dark, find the probability that

    (i) the first sock he puts on is brown,
    (ii) the first sock he puts on is black,
    (iii) after he has put on a black sock, he will then put on another black sock
    (iv) that after he has first put on a brown sock, the next sock will also be brown.

4. A man estimates gloomily that the bus to work each day is twice as likely to be full as it is to have room for him. If his estimate is correct, what is the probability that the bus will be full?

5. Towards the end of a game of 'Scrabble' a player finds that he has the following letters:

$$A, B, B, F, G, J, J, K, M, X, Z, Z.$$

Find the probability that a letter drawn at random from these is

    (i) B,
    (ii) B or J,
    (iii) F, G, or J,
    (iv) K, M, X, or Z.

6. A man writes down two different letters of the alphabet in a random order. Given that the first letter is not Z, what is the probability that they are consecutive letters in alphabetical order?

7. A bookshelf contains 10 detective stories, 9 historical novels and 7 books on sport. A man selects one at random. What is the probability that it is a book on sport?

    What is the probability that the next person to pick from this shelf (before the first book has been replaced) selects at random

    (i) a book on sport,
    (ii) a detective story,
    (iii) a historical novel?

8. The faces of a fair 6-sided die are marked so that only one shows a '1', two faces show a '2' and three faces show a '3'.

    Denote the two faces that show '2's by 2′, 2″, the three faces that show a '3' by 3′, 3″, 3‴. List all the possible outcomes when the die is thrown twice, e.g. (1, 1), (2′, 1), (2″, 1), (3′, 1), (3″, 1), (3‴, 1). (There should be 36 outcomes, all equi-probable.)

    (i) What is the probability that the total shown on the die is 2?
    (ii) List the probability of each total, from 2 to 6 inclusive.
    (iii) What is the most likely total?

Theoretical probability, experimental probability

What is the probability that a fair coin lands showing 'Heads' when tossed? $\frac{1}{2}$, by symmetry. This is called a theoretical (expected) probability. These theoretical probabilities are dependent on the symmetry of a problem, e.g. the probability of a fair die showing '5' is 1/6, the probability of a card drawn at random from a full pack being a Heart is 1/4.

What is the probability that a word chosen at random from those on this page has five letters? We count the words on this page, count those with five letters, and divide. An 'experiment' has been 'performed', the page was written, and probability calculated on the basis of that experiment is called experimental probability.

The theoretical probability that a 3-digit number chosen at random ends in a '5' is 1/10. Find the experimental probability by asking a group of friends to write down a three-digit number at random. We expect that with a large number of trials the experimental probability will be near to 1/10. Sometimes strange results are obtained when people are asked to choose a single-digit number—the experimental probability that 7 is chosen is often much higher than 1/10. Similarly, if a bag contains 10 marbles, 3 of which are red, the expected probability that a marble drawn is red is 3/10; the experimental probability could only be found by making many draws from such a bag.

If we have n trials, a of which were 'successful', we defined the experimental probability of success as a/n. The theoretical probability we define in a slightly different manner.

Suppose we wish to find the probability that a number chosen at random from 1, 2, 3, 4, 5, 6, 7, 8, 9 is a prime. The universal set $\mathscr{E}$ of all possible outcomes (sometimes called the outcome space or possibility space) is

$$\{1, 2, 3, 4, 5, 6, 7, 8, 9\}$$

and each of these nine equiprobable outcomes has a probability of $\dfrac{1}{n\{\mathscr{E}\}}$, i.e. $\frac{1}{9}$. If a subset A is 'favourable' for a particular problem, the probability of an event associated with an element of A is defined as

$$\frac{n\{A\}}{n\{\mathscr{E}\}}$$

In this example, $A = \{2, 3, 5, 7\}$ so that $n\{A\} = 4$ and the probability of a number drawn at random from $\{1, 2, 3, 4, 5, 6, 7, 8, 9\}$ being prime is $\frac{4}{9}$.

Example 7. *Two numbers are chosen at random from 1, 2, 3. What is the probability that their sum is odd?*

If (1, 2) means that first 1 is chosen, then 2, the universal set $\mathscr{E}$ is $\{(1, 2),$ (1, 3), (2, 1), (2, 3), (3, 1), (3, 2)\} and $n\{\mathscr{E}\} = 6$.

The set A of all favourable outcomes is $\{(1, 2), (2, 1), (2, 3), (3, 2)\}$ and $n\{A\} = 4$.

∴ The probability that the sum of the two numbers is odd $= \frac{4}{6} = \frac{2}{3}$.
We shall abbreviate this to read,

$$\text{Pr(sum of the numbers is odd)} = \tfrac{2}{3}.$$

Addition of probabilities

Example 8. *Find the probability that a number chosen at random from the integers between 10 and 20 inclusive is either a prime or a multiple of 5.*

The universal set $\mathscr{E} = \{10, 11, 12, 13, \ldots 20\}$ and $n\{\mathscr{E}\} = 11$.
The subset A of primes $= \{11, 13, 17, 19\}$ so $n\{A\} = 4$.
The subset B of multiples of 5 $= \{10, 15, 20\}$ so $n\{B\} = 3$.
The favourable subset $C = A \cup B$, since any element in C will be in either A or B or both, and here $n\{C\} = n\{A\} + n\{B\}$,

i.e. $$\text{Pr} = \frac{4+3}{11} = \frac{7}{11}.$$

Notice that $n\{A \cup B\} = n\{A\} + n\{B\}$ only because $A \cap B = \emptyset$, as there is no number between 10 and 20 which is a prime and a multiple of 5. This is illustrated by the Venn diagram (Fig. 24.1).

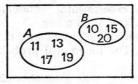   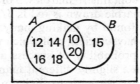

Fig. 24.1   **Fig. 24.2**

Example 9. *Find the probability that a number chosen at random from the integers between 10 and 20 inclusive is a multiple of 5 or a multiple of 2.*

As in the last example,

$\mathscr{E} = \{10, 11, 12, 13 \ldots 20\}$ and $n\{\mathscr{E}\} = 11$. The subset A of multiples of 2 is $\{10, 12, 14, 16, 18, 20\}$ and $n\{A\} = 6$. The subset B of multiples of 5 is $\{10, 15, 20\}$, and $n\{B\} = 3$. But $A \cap B = \{10, 20\}$ so that

$$n\{A \cup B\} = n\{A\} + n\{B\} - n\{A \cap B\},$$
$$= 6 + 3 - 2$$
$$= 7$$

$$\text{Pr} = \frac{7}{11}$$

The elements in each subset are shown in the Venn diagram in Fig. 24.2.

Harder example on experimental probability

Example 10. *Of 24 children on a tennis coaching course, 14 are boys and 8 are left handed, including 5 of the boys. No child is ambidextrous. Find the probability that a child selected at random is*

(i) a left-handed girl, (ii) a right-handed girl, (iii) a left-handed boy, (iv) a right-handed boy.

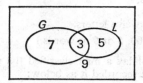

Fig. 24.3

Since there are 24 children, 14 of whom are boys, there must be 10 girls. Since 8 are left handed, including 5 boys, there must be 3 left-handed girls. If G is the set of girls, L the set of left-handed children, the Venn diagram in Fig. 24.3 shows the relation between G and L, and we can put into the diagram the number of elements in each subset. Thus

$$n\{G \cap L\} = 3, \; n\{G \cap L'\} = 7,$$
$$n\{G' \cap L\} = 5, \; n\{G' \cap L'\} = 9.$$

But these are respectively the number of left-handed girls, right-handed girls, left-handed boys and right-handed boys, so that the probability that a child selected at random from among those on this course is

(i) a left-handed girl is 3/24, i.e. $\frac{1}{8}$,
(ii) a right-handed girl is 7/24,
(iii) a left-handed boy is 5/24,
(iv) a right-handed boy is 9/24, i.e. $\frac{3}{8}$.

Exercise 24.2

Find the theoretical probability in Q.1–3, and the experimental probability in Q.4–8.

1. Find the probability that an integer selected at random from those between 10 and 100 inclusive is a multiple of 5 or of 9.

2. An Egyptian fraction has numerator 1, though the denominator can be any integer. (The Egyptians did also use 2/3, but that does not affect this problem.) Find the probability that an Egyptian fraction of value between 1/10 and 1/20 inclusive can be expressed as a terminating decimal. (E.g. $\frac{1}{2}$ = 0.5, a terminating decimal, whereas $\frac{1}{3}$ is equal to a recurring decimal.)

3. Three numbers are selected at random from 1, 2, 3 and 4, repetitions being allowed. List all twenty different selections. Find the probability that these numbers could be the lengths in cm of the sides of

(i) an equilateral triangle,
(ii) a triangle that is isosceles but not equilateral,
(iii) any triangle.

4. Find the probability that a word selected at random from those in Q.1

    (i) has five letters,
    (ii) has three letters,
    (iii) has eight letters,
    (iv) has more than twelve letters.

(Do not count 10, 100, 5 or 9 as words.)

5. On a fortnight's holiday a man classified the weather each day as wet (W) or fine (F), and recorded it as below:

$$\text{F, F, W, W, W, F, W ; W, W, F, F, W, F, F.}$$

Every evening of a wet day, the man said to himself 'It was wet today, it will be fine tomorrow.' If one of his forecasts was noted at random, what is the probability that forecast was a correct one?

6. A man recorded his scores in a game of darts one evening:

$$0, 1, 0, 8, 20, 0, 7, 9, 12, 0.$$

What is the experimental probability that he scores

    (i) more than 5,   (ii) an 8,   (iii) less than 3?

7. The probability of carrots for lunch today, is 3/4, the probability of rice pudding for lunch is 1/10 and the probability of both is 3/40. Find the probability of

    (i) carrots but not rice,
    (ii) rice but not carrots,
    (iii) neither rice nor carrots.

8. A hundred men were measured correct to the nearest cm and their heights h cm recorded, grouped and displayed in the histogram below:

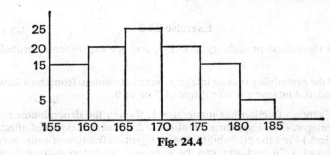

Fig. 24.4

Find the probability that a man selected at random has a height of

    (i) less than 1.60 m,
    (ii) between 1.75 m and 1.85 m,
    (iii) more than 1.70 m.

25 Conditional Probability

Independent events

If two events are such that one has no effect on the other, then they are independent events. It is likely that the probability that Tottenham Hotspur wins the F.A. Cup is independent of the probability that England is holding the Ashes for cricket. It might not, however, be independent on whether it is a fine day for the Cup Final, or even on whether it has been a hard winter. We shall assume events are independent unless there is a clear connection between them, as in the examples in the last chapter where, say, the probability of the outcome of a second draw clearly depended on the outcome of the first draw.

In mathematical terms, we can define independence by saying that two events A, B are independent if and only if

$$\text{Pr(both } A \text{ and } B) = \text{Pr}(A) \cdot \text{Pr}(B)$$

For example, with the data of Exercise 24.2, Q.7, the probability of having carrots was 3/4, the probability of having rice pudding was 1/10, and we were given that the probability of having both carrots and rice pudding was 3/40.

Mutually exclusive events

Some events are such that one outcome automatically excludes another. A number cannot be both odd and even: a coin cannot show Heads and Tails simultaneously. Such events are called mutually exclusive.

Multiplication Law

Suppose we have a fair die, faces numbered 1 to 6, and another fair die, faces coloured blue (B), green (G), orange (O), red (R), white (W) and yellow (Y). When they are both thrown what is the probability that they will show yellow and an even number?

The set of all equiprobable outcomes can be represented by crosses on a grid.

	1	2	3	4	5	6
Y	×	⊗	×	⊗	×	⊗
W	×	×	×	×	×	×
R	×	×	×	×	×	×
O	×	×	×	×	×	×
G	×	×	×	×	×	×
B	×	×	×	×	×	×

The favourable outcomes we can ring, so that the elements in the favourable subset A are (2, Y), (4, Y), (6, Y) and $n\{A\} = 3$. Since $n\{\mathscr{E}\} = 36$,

Pr(one die showing yellow and the other an even number)
$$= 3/36,$$
i.e. $$= 1/12.$$

It is often tedious to list all possible outcomes, so we devise a shorter method. We know that the probability that an even number is shown is $\frac{1}{2}$, so that when the coloured die is also thrown the probability that it is accompanied by an even number is equal to the probability that the same colour is accompanied by an odd number.

$\therefore$ Pr(even and yellow) $= 1/2$ Pr(yellow)
$$= 1/2 \times 1/6.$$
$$= 1/12.$$

Pr(even and yellow) $=$ Pr(even) $\times$ Pr(yellow).

These outcomes, even, yellow, are independent.

Tree diagram

This method can be illustrated by a tree diagram.

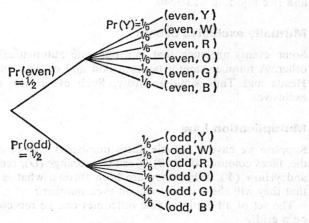

Fig. 25.1

We see that we have twelve equiprobable outcomes, only one of which is acceptable, so that its probability is 1/12.

If we wished to find the probability that the dice showed either an even number and yellow or an odd number and red, there are two acceptable outcomes, so that the probability is 2/12, i.e. 1/6.

Example 1. *A bag contains 10 marbles, 7 are red and 3 black. A marble is drawn at random, and then replaced. Another draw is made. What is the probability that both marbles drawn are red?*

$$\begin{aligned}
\text{Pr(first marble is red)} &= 7/10, \\
\text{Pr(second marble is red)} &= 7/10, \\
\text{Pr(both are red)} &= (7/10)\,(7/10) \\
&= 49/100.
\end{aligned}$$

Similarly, Pr(both marbles are black) $= (3/10)\,(3/10)$
$= 9/100.$

To find the probability that one is red and the other black, we can either say that the only possible outcomes are that both are red, that one is red and the other black, or that both are black. Since the sum of all the probabilities is 1,

$$\begin{aligned}
\text{Pr(one is red and the other black)} &= 1 - 49/100 - 9/100 \\
&= 42/100,
\end{aligned}$$

or we can say that

$$\begin{aligned}
\text{Pr(first is red)} &= 7/10, \\
\text{Pr(second is black)} &= 3/10, \\
\text{Pr(red then black)} &= \tfrac{7}{10} \times \tfrac{3}{10} = \tfrac{21}{100}.
\end{aligned}$$

But 'black then red' is equally acceptable,

and $\qquad$ Pr(black then red) $= \tfrac{21}{100}.$

$\therefore \qquad$ Pr(one red and the other black) $= \tfrac{21}{100} + \tfrac{21}{100}$
$= \tfrac{42}{100}.$

again this is illustrated clearly by a tree diagram.

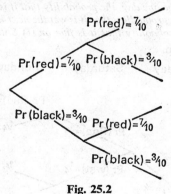

Fig. 25.2

Example 2. *A coin is tossed four times. Find the probability that four heads are shown.*

$$\begin{aligned}
\text{Pr(first throw shows heads)} &= \tfrac{1}{2}, \\
\text{Pr(second throw shows heads)} &= \tfrac{1}{2}, \\
\text{Pr(third throw shows heads)} &= \tfrac{1}{2},
\end{aligned}$$

and $\qquad$ Pr(fourth throw shows heads) $= \tfrac{1}{2}.$

$\therefore \qquad$ Pr(all four throws show heads) $= \tfrac{1}{2} \times \tfrac{1}{2} \times \tfrac{1}{2} \times \tfrac{1}{2} = \tfrac{1}{16}.$

Example 3. *The probability that it is raining at 8.30 a.m. on any one day is $\frac{1}{4}$.* *The probability that a boy wears a coat as he leaves for school at 8.30 a.m. is* *2/3 if it is raining at that time; if it is not raining at that time, the probability* *that he wears a coat is 1/10. What is the probability that he wears a coat on any* *one day?*

Draw a tree diagram.

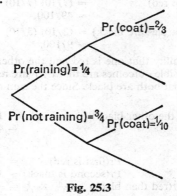

Fig. 25.3

From this we see that the probability that it is raining and that he wears a coat is $\frac{1}{4} \times \frac{2}{3}$, i.e. $\frac{1}{6}$, and that the probability that it was not raining yet he was wearing a coat is $\frac{3}{4} \times \frac{1}{10}$, i.e. $\frac{3}{40}$. Thus the probability that he is wearing a coat is $\frac{1}{6} + \frac{3}{40}$, i.e. $\frac{29}{120}$.

Example 4. *If it is fine one day, the probability that it is fine the next day is $\frac{3}{4}$:* *if it is wet one day, the probability that it is wet the next is 2/3. If it is fine today* *(Thursday), find the probability that it is fine on* (i) *Saturday,* (ii) *Sunday.*

Draw a tree diagram.

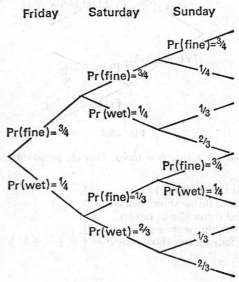

Fig. 25.4

The probability that it is fine on Saturday is $\frac{3}{4} \times \frac{3}{4} + \frac{1}{4} \times \frac{1}{3}$, i.e. $\frac{31}{48}$; we can check that the probability that it is wet on Saturday is $\frac{1}{4} \times \frac{2}{3} + \frac{3}{4} \times \frac{1}{4}$, i.e. $\frac{17}{48}$, and $\frac{31}{48} + \frac{17}{48} = 1$.

The probability that it is fine on Sunday can be found by reading along the upper of each pair of branches of the tree, so the probability is $\frac{3}{4} \times \frac{3}{4} \times \frac{3}{4} + \frac{3}{4} \times \frac{1}{4} \times \frac{1}{3} + \frac{1}{4} \times \frac{1}{3} \times \frac{3}{4} + \frac{1}{4} \times \frac{2}{3} \times \frac{1}{4}$, i.e. $\frac{347}{576}$. Again, the probability that it is wet on Sunday gives us a check. Reading along the lower of each pair of branches, the probability that it is wet on Sunday is $\frac{3}{4} \times \frac{3}{4} \times \frac{1}{4}$ $+ \frac{3}{4} \times \frac{1}{4} \times \frac{2}{3} + \frac{1}{4} \times \frac{1}{3} \times \frac{1}{4} + \frac{1}{4} \times \frac{2}{3} \times \frac{2}{3}$, i.e. $\frac{229}{576}$. Notice that by ordering the way in which we label the branches of the tree we make it easier to see where to find acceptable outcomes.

Exercise 25a

1. The probability that a plant from a certain packet of wallflower seeds will produce red flowers is 2/5. Find the probability that

    (i) two plants from this packet are both red,
    (ii) neither of two plants from this packet is red,
    (iii) exactly one of two plants from this packet is red.

2. The probability that a biased coin will show 'Heads' is 2/3. Find the probability that when tossed twice it shows

    (i) two heads, (ii) two tails, (iii) exactly one head.

3. The probability that a person in a certain area owns a car is $\frac{1}{4}$. Find the probability that

    (i) two persons selected at random both own cars,
    (ii) of two persons selected at random, only one owns a car,
    (iii) three persons selected at random all own cars,
    (iv) of three persons selected at random, none owns a car,
    (v) of three persons selected at random, exactly one owns a car.

4. From a pack of 52 cards, those below 7 are discarded, keeping only 7, 8, 9, 10, J, Q, K, A. Cards are drawn at random, examined but not replaced. Find the probability that

    (i) when one is drawn, it is an 8,
    (ii) that the first card drawn is an '8' and the next a '9',
    (iii) the first card is an '8' and the next another '8',
    (iv) the first card is a 'K' the next a 'Q' and the third a 'J'.

5. A man travels to work by either car or by train. There is a probability of 2/3 that he travels by train on Monday. If he travels by train on any one day, there is a probability of 3/4 that he will travel by car the next day. If he travels by car on any one day, there is a probability of 5/6 that he will travel by train the next day. Find the probability that he travels

    (i) by car on Tuesday,
    (ii) by train on Tuesday,
    (iii) by car on Wednesday,
    (iv) by train on Wednesday.

6. The probability of England beating Wales at Rugby Football is 9/10; of beating Scotland is 3/4 and of beating Ireland is 1/3. Find the probability that

    (i) England wins all three matches,
    (ii) England beats Wales and Scotland but not Ireland,
    (iii) England only beats Wales,
    (iv) England does not win any of the three matches.

(The probabilities in this question, unfortunately, are fictitious.)

7. Each evening a man either watches television or reads a book; the probability that he watches television is 4/5. If he watches television, there is a probability of 3/4 that he will fall asleep; if he reads a book, there is a probability of 1/4 that he will fall asleep. Find the probability that he falls asleep.

8. The man in Q.7 is not very truthful. When asked if he has been asleep there is a probability of only 1/5 that he will admit he has been to sleep and a probability of 3/5 that he will claim to have been asleep when he has not been asleep. Find the probability that

    (i) he goes to sleep and admits it,
    (ii) he goes to sleep and does not admit it,
    (iii) he does not go to sleep but claims that he has been asleep,
    (iv) he does not go to sleep and says that he has not been asleep.

Compare the answers to (i) and (iii) to find the probability that if he says he has been to sleep one evening then he is, in fact, telling the truth.

Exercise 25b

1. Two numbers are selected at random from the integers 1 to 10 inclusive, repetitions being allowed. Find the probability that

    (i) both are prime,
    (ii) neither is prime,
    (iii) both are powers of 2,
    (iv) one is a prime and the other a power of 2,
    (v) one is a multiple of 2 and the other is a prime.

2. In a certain town the probability that a man selected at random on a Saturday afternoon is carrying more than 30 kg of shopping ('heavily-laden') is 0.7. Find the probability that

    (i) two men selected at random are both heavily-laden,
    (ii) three men selected at random are all heavily-laden,
    (iii) two out of three men selected at random are heavily-laden,
    (iv) four men selected at random are all heavily-laden.

3. The probability that the centre-forward of a certain football team will score a goal at any one attempt is constant, and is 0.01. Find the probability that he will score

    (i) on each of two attempts,
    (ii) on each of three attempts,
    (iii) on two out of three attempts,

Find also the probability that he will not score in 10 attempts.

4. A man knows from long experience that when he catches a fish the probability it is too small to take home is 0.9. Find the probability that

    (i) when he catches two fish, both are too small,
    (ii) three out of four fish caught are all too small,
    (iii) five fish out of five caught are large enough to take home,
    (iv) of 10 fish caught, only one is large enough to take home.

5. A woman has either coffee or tea for her midmorning break. There is a probability of 0.9 that she has coffee on Mondays; if she has coffee any one day, there is a probability of 0.4 that she has coffee the next day. If she has tea one day, the probability she has tea the next day is 0.3. Find the probability that she has

    (i) coffee on Tuesday,   (ii) tea on Tuesday,
    (iii) coffee on Wednesday,   (iv) tea on Wednesday.

6. The probability that a certain girl wakes up late in the morning is 0.6. When she wakes up late, the probability that she is late arriving at the office is 0.85; when she is not late waking up, the probability that she arrives late is 0.4. Find the probability that she is late arriving on any one morning.

7. Seeing the girl in Q.6 arriving late, the 'boss' asks her if she overslept. What is the probability that she ought to answer 'Yes'?

8.

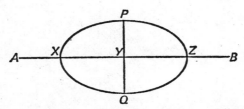

Fig. 25.5

Figure 25.5 shows a plan of the streets of a small village. A man wishes to drive from A to B. The probability that he will keep straight on at any road junction is 0.4. When there is a choice, he is equally likely to turn left or right. He never turns through 180°. Find the probability that he takes the route

    (i) $AXPZB$,   (ii) $AXQZB$,   (iii) $AXYZB$.

Find the probability that he drives from A to B without passing through any junction twice.

(Use the symmetry of the plan to reduce the number of calculations.)

Fig. 24.2

GEOMETRY

GEOMETRY

Some definitions

A triangle with no sides equal is called a **scalene** triangle.

A triangle with two (or more) sides equal is called an **isosceles** triangle.

A triangle with all three sides equal is called an **equilateral** triangle.

A quadrilateral with both pairs of opposite sides parallel is called a **parallelogram**.

A parallelogram with two adjacent sides equal is called a **rhombus**.

A parallelogram with one angle a right angle is called a **rectangle**.

A rectangle with adjacent sides equal is called a **square**.

Some properties of parallel lines

Corresponding angles and **alternate** angles are equal.

Adjacent angles are supplementary.

Some commonly-used theorems

1. The sum of the angles of a triangle is two right angles. (Corollary: the sum of the angles of an n-sided polygon is $(2n - 4)$ right angles.)

2. In any right-angled triangle, the square on the hypotenuse is equal to the sum of the squares on the other two sides (Pythagoras' Theorem).

3. The angle subtended by an arc of a circle at the centre is double the angle subtended by that arc at any other point on the remaining part of the circumference. (Corollaries: angles in the same segment of a circle are equal; the opposite angles of a cyclic quadrilateral are supplementary; the angle in a semi-circle is a right angle.)

4. If two chords AB, CD of a circle intersect at a point X, $AX . XB = CX . XD$.

5. Equiangular triangles have their corresponding sides proportional.

26 Axioms, Parallel lines, Triangles and Polygons

Geometry was developed by the Egyptians more than 1000 years before Christ to help them mark out again their fields after the floods from the Nile, but was abstracted by the Greeks into a logical system of proofs many centuries later. For measurement, the length of lines and sizes of angles were needed: for a logical system of proofs, basic postulates or axioms were necessary.

Measurement of angle

The measurement of a straight line presented no difficulty; the only problem was that of the unit of measurement, now based on the metre and suitable fractions and multiples. The measurement of an angle was in fractions and multiples of complete rotations. Since the Babylonians found that many fractions could be added easily if they were expressed with 360 as a common denominator, a complete rotation was divided into 360 smaller angles (called a degree), and those degrees into 60 smaller parts. Thus a quarter-turn (right angle) is equal to 90°.

The straight line

A line can be defined as a **straight line** if, when X, Y and Z are *any* three points on the line, angle $XYZ = 180°$.

Axioms

Euclid's (c. 330–275 BC) axioms may be stated:

(i) A straight line may be drawn from any one point to any other point.

(ii) A finite straight line may be extended at each end in a straight line.

(iii) A circle can always be drawn with any given centre and any given radius.

(iv) All right angles are equal one to another.

(v) If a straight line meets two other straight lines so that the two adjacent angles on one side of it are together less than two right angles, the other straight lines, when extended, will meet on that side of the first straight line. (See Fig. 26.1. Angles a, b are such that $a + b < 180°$ so that AB and DC meet on the same side of XY as angles a and b.)

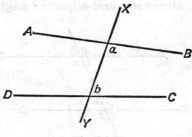

Fig. 26.1

Constructions and proofs are based on these axioms.

Definitions

An **acute** angle is less than 90°; an **obtuse** angle is between 90° and 180°; a **reflex** angle is between 180° and 360°.

Two angles whose sum is 90° are called complementary: two angles whose sum is 180° are called supplementary.

Vertically opposite angles

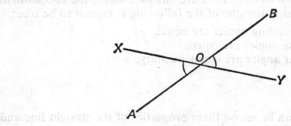

Fig. 26.2

When two straight lines *XOY*, *AOB* meet at *O*, angle *XOA* = angle *BOY*, each being supplementary to angle *AOY* (and to angle *XOB*). Such angles are called vertically opposite. *XOB* and *AOY* are also vertically opposite.

Parallel lines

From Euclid's fifth axiom, we can deduce that if angles *a* and *b* in Fig. 26.1 total 180°, *AB* and *CD* do not meet either to the left or to the right of *XY*, *AB* and *CD* are then called **parallel**.

Corresponding, alternate, and adjacent angles

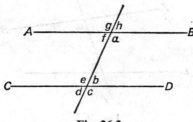

Fig. 26.3

Pairs of corresponding angles in Fig. 26.3 are a and c, h and b, g and e and f and d. If AB is parallel to CD, corresponding angles are equal, e.g. $a = c$, each being supplementary to b.

Pairs of alternate angles are f and b, a and e. If AB is parallel to CD alternate angles are equal, e.g. $a = e$, each being supplementary to b.

Pairs of adjacent angles are a and b, f and e. If AB is parallel to CD, adjacent angles are supplementary, from the definition of parallel lines. These angles are sometimes called interior angles.

Conversely, if two straight lines are cut by a third, the two straight lines will be parallel if any one of the following is known to be true:

(i) two corresponding angles are equal,
(ii) two alternate angles are equal,
(iii) two adjacent angles are supplementary.

Problems

Simple problems can be set on these properties of the straight line and parallel straight lines.

Example 1. In Fig. 26.4, AOB is a straight line. Find x.

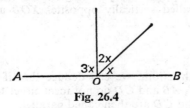

Fig. 26.4

Since AOB is a straight line, angle $AOB = 180°$,
$$\therefore x + 2x + 3x = 180°,$$
$$x = 30°.$$

Example 2. In Fig. 26.5, AB is parallel to DE. Find x.

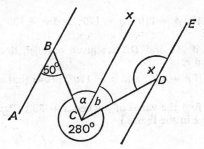

Fig. 26.5

Draw a line *CX* through *C* parallel to *AB*.
Then

$$a = 50° \text{ (alternate, } AB \text{ parallel to } CX)$$
$$b = 360° - 280° - 50° = 30°$$
$$b + x = 180° \text{ (adjacent, } CX \text{ parallel to } DE)$$
$$\therefore \quad x = 150°.$$

The symbol ‖ is often used to denote 'is parallel to', e.g. here *AB* ‖ *CX*.

Exercise 26.1a

1. In Fig. 26.6, given that *OA* is perpendicular to *OB*, find *x*.

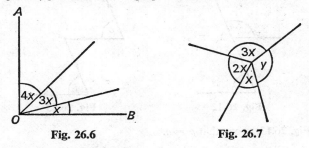

Fig. 26.6   **Fig. 26.7**

2. Find a relationship between *x* and *y* from Fig. 26.7.

3. Given that *AB* is parallel to *CD* in Fig. 26.8, find *x*, *y* and *z*.

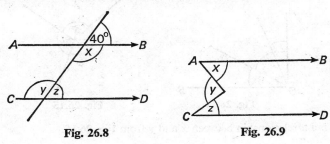

Fig. 26.8   **Fig. 26.9**

4. In Fig. 26.9, given that AB is parallel to CD, find an equation connecting x, y and z.

5. If, in Fig. 26.10, $p = 110°$, $q = 120°$ and $r = 130°$, prove that BA is parallel to DE.

6. In Fig. 26.10, if BA and DE are given parallel, find an equation connecting p, q and r.

7. In Fig. 26.10, if $p = 140°$ and $q = 110°$, prove that when BA is parallel to DE, $q = r$.

8. In Fig. 26.11, find the value of y when $x = 30°$. (Parallel lines are indicated by arrows in the figure.)

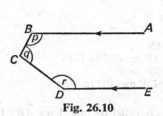

Fig. 26.10

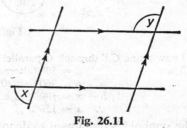

Fig. 26.11

9. Find a relationship between x and y from Fig. 26.12.

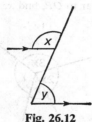

Fig. 26.12

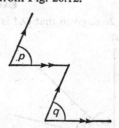

Fig. 26.13

10. In Fig. 26.13, prove that $p = q$.

Exercise 26.1b

1. In Fig. 26.14, given that AO is perpendicular to OB, find x.

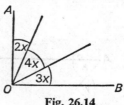

Fig. 26.14

Fig. 26.15

2. Find a relationship between x and y from Fig. 26.15.

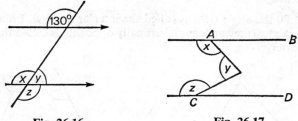

| Fig. 26.16 | Fig. 26.17 |

3. Find x, y and z from Fig. 26.16.

4. Given in Fig. 26.17, that $x = 120°$, $y = 100°$, $z = 140°$, prove that AB is parallel to CD.

5. If in Fig. 26.17, AB and CD are parallel, find an equation connecting x, y and z.

6. If in Fig. 26.17, $x = 100°$, $y = 140°$, find z if AB and CD are parallel.

7. In Fig. 26.18, if $y = 40°$ and $z = 80°$, find x.

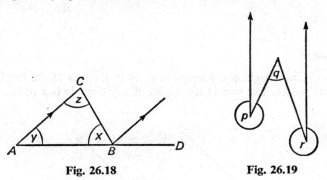

| Fig. 26.18 | Fig. 26.19 |

8. Find a relationship between x, y and z from Fig. 26.18.

9. If, in Fig. 26.18, $y = 50°$ and $z = 80°$, prove that $x = y$.

10. From Fig. 26.19, find an equation connecting p, q and r.

THE TRIANGLE

A figure bounded by three straight lines is called a triangle.

A scalene triangle is one in which no two sides are equal in length.

An isosceles triangle is one in which two (or more) sides are equal in length.

An equilateral triangle is one in which all three sides are equal in length.

An obtuse-angled triangle is a triangle having one obtuse angle.

Exterior angle

In Fig. 26.20 the angle *CBA* is called the interior angle *B*. Either of the angles marked *x* (they are equal; vertically opposite) is called the corresponding exterior angle.

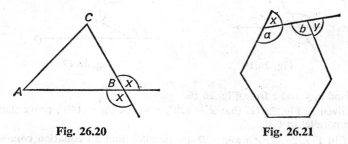

Fig. 26.20   Fig. 26.21

Similarly, in the 6-sided figure (hexagon) shown in Fig. 26.21, *x* is the exterior angle corresponding to *a*; *y* is the exterior angle corresponding to *b*.

Theorem

The exterior angle of a triangle is equal to the sum of the two interior opposite angles. The sum of the angles of any triangle is 2 right angles.

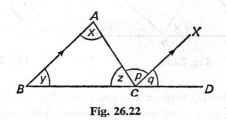

Fig. 26.22

Given the triangle *ABC* with *BC* produced to *D*.
To prove that the angle $ACD = x + y$, and that $x + y + z = 180°$.
Construction: Draw *CX* parallel to *BA*.
Proof:

$$x = p \text{ (alternate; } BA \parallel CX).$$
$$y = q \text{ (corresponding; } BA \parallel CX).$$

By addition, $p + q = x + y,$
or   the angle $ACD = x + y.$
Also $x + y + z = p + q + z = 180°$ (straight line).

THE POLYGON

Any plane closed figure bounded by straight lines is called a polygon.

A **convex polygon** is one which no interior angle is greater than 180°.

A **re-entrant polygon** is one in which at least one angle is greater than 180° (see Fig. 26.23).

A **regular polygon** has all its sides equal and all its angles equal.

A **quadrilateral** is a polygon of 4 sides.

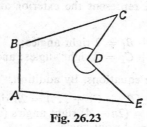

Fig. 26.23

A **pentagon** is a polygon of 5 sides.

A **hexagon** is a polygon of 6 sides.

An **octagon** is a polygon of 8 sides.

Theorem

In a convex polygon of n sides, the sum of the interior angles is $(2n - 4)$ right angles. The sum of the exterior angles is 4 right angles, whatever the value of n.

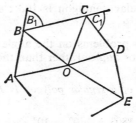

Fig. 26.24

Given a convex polygon $ABCDE$ . . . of n sides.

To prove that (i) $\widehat{ABC} + \widehat{BCD} + \ldots = (2n - 4)$ right angles;

(ii) the sum of the exterior angles is 4 right angles.

Construction: Join the vertices A, B, C . . . to any point O inside the polygon.

Proof: (i) There are n triangles such as AOB, BOC, etc.

Therefore the sum of all the angles of all these triangles is $2n$ right angles.

These angles include $\widehat{AOB} + \widehat{BOC} + \widehat{COD} + \ldots$, i.e. 4 right angles (a complete revolution).

They also include $\widehat{ABC} + \widehat{BCD} + \widehat{CDE} + \ldots$, i.e. the sum of the interior angles of the polygon.

∴ the sum of the interior angles + 4 right angles = $2n$ right angles.

∴ the sum of the interior angles = $(2n - 4)$ right angles.

(ii) Let B_1, C_1, etc. represent the exterior angles corresponding to B, C, etc.

Then $B + B_1 = 2$ right angles;
        $C + C_1 = 2$ right angles; and so on.

There are n of these equations. By addition,

$(B + C + \ldots) + (B_1 + C_1 + \ldots) = 2n$ right angles.

But $(B + C + \ldots) = (2n - 4)$ right angles (already proved).

$(B_1 + C_1 + \ldots) = 4$ right angles.

Particular cases

Putting $n = 4$, we see that the sum of the interior angles of a quadrilateral is 4 right angles.

Putting $n = 5$, we see that the sum of the interior angles of a pentagon is 6 right angles.

Putting $n = 6$, we see that the sum of the interior angles of a hexagon is 8 right angles.

Each angle of a regular pentagon is 108°; each angle of a regular hexagon is 120°.

N.B. In dealing with problems on the angles of a polygon, work if possible with the exterior angles rather than the interior.

Example 3. Each angle of a regular polygon is 170°. Find the number of sides of the polygon.

Each exterior angle $= 180° - 170° = 10°$.

The sum of the exterior angles is 360° and so there must be 36 exterior angles.

The number of sides is 36.

Example 4. One angle of a hexagon is 140°. The other five angles are equal to each other. Find them.

One exterior angle is $180° - 140° = 40°$.

The sum of the other five equal exterior angles is 320°.

Each of these exterior angles must be 64°.

Each of the equal interior angles is $(180° - 64°) = 116°$.

Exercise 26.2a

1. Each angle of a regular polygon is 168°. How many sides has it?

2. One angle of a pentagon is 140°. Find each of the other angles, given that they are all equal to each other.

3. State for which of the following exterior angles a regular polygon is possible; (i) 20°, (ii) 25°, (iii) 30°.

4. State for which of the following interior angles a regular polygon is possible; (i) 165°, (ii) 160°, (iii) 155°.

5. Three of the angles of a hexagon are each $x°$. The other three are each $2x°$. Find x.

6. Write down the number of degrees in each angle of a regular 15-sided polygon.

7. In a 7-sided figure, three of the angles are equal and each of the other four angles is 15° greater than each of the first three. Find the angles.

8. If the angles of a pentagon are $y°$, $(y + 10)°$, $(y + 20)°$, $(y + 40)°$ and $(y + 50)°$, find y.

9. If the angles of a hexagon are $x°$, $(x + 10)°$, $(x + 20)°$, $(x + 30)°$, $(x + 40)°$ and $(x + 50)°$, find x.

10. Find the sixth angle of a hexagon when each of the others is 118°.

Exercise 26.2b

1. Each angle of a regular polygon is 162°. How many sides has it?

2. One angle of a pentagon is 160°. Find each of the other equal angles.

3. State for which of the following exterior angles a regular polygon is possible: (i) 24°, (ii) 28°, (iii) 36°.

4. State for which of the following interior angles a regular polygon is possible: (i) 170°, (ii) 168°, (iii) 164°.

5. Three of the angles of a hexagon are each $2x°$ and each of the others is $3x°$. Find x.

6. Write down the number of degrees in each angle of a regular 16-sided figure.

7. In an octagon (8-sided figure), four of the angles are equal and each of the others is 20° greater than each of the first four. Find the angles.

8. If the angles of a quadrilateral are $(p + 10)°$, $(2p − 30)°$, $(3p + 20)°$ and $4p°$, find p.

9. If four of the angles of a heptagon (7-sided figure) are equal and each of the other three is 20° greater than each of the first four, find the angles.

10. Find the fifth angle of a pentagon when each of the other angles is 120°.

Exercise 26.3: Miscellaneous

1. If three angles of a quadrilateral are 40°, 100° and 150°, find the fourth angle.

2. If three angles of a quadrilateral are $x°$, $y°$ and $z°$, express the fourth angle in terms of x, y and z.

3. Find the third angle of a triangle when the other two are $(2x - 30)°$ and $(3x + 20)°$.

4. Find x when the three angles of a triangle are $(x + 20)°$, $(2x + 30)°$ and $(3x + 40)°$.

5. Find x when the four angles of a quadrilateral are $x°$, $2x°$, $(x + 10)°$ and $(x + 20)°$.

6. Find x from Fig. 26.25.

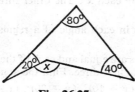

Fig. 26.25

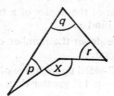

Fig. 26.26

7. Find x in terms of p, q, r from Fig. 26.26.

8. From Fig. 26.27, prove that $z = 90°$.

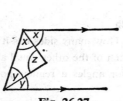

Fig. 26.27

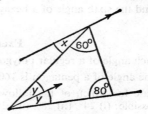

Fig. 26.28

9. Find x in Fig. 26.28.

10. Find y and z in Fig. 26.29.

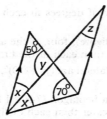

Fig. 26.29

11. If the exterior angles of a quadrilateral are $x°$, $(x + 5)°$, $(x + 10)°$ and $(x + 25)°$, find x.

12. If three of the exterior angles of a hexagon are each $x°$ and each of the other three is $2x°$, find x.

27 Congruent Triangles; Triangles and Quadrilaterals; Inequalities

Congruent triangles

Two triangles are called congruent if one can be superimposed on the other. Thus in Fig. 27.1, triangles ABC, XYZ will be congruent if

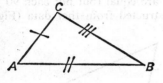

Fig. 27.1

$AB = XY$, $BC = YZ$ and $CA = ZX$, as these lengths determine a triangle uniquely; this case is abbreviated SSS. In Fig. 27.2, triangles

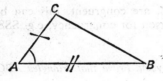

Fig. 27.2

ABC, XYZ will be congruent if $AB = XY$, $CA = ZX$ and angle CAB = angle ZXY; again, a triangle is determined uniquely; this case

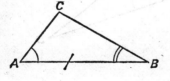

Fig. 27.3

is abbreviated SAS. In Fig. 27.3, triangles ABC, XYZ will be congruent if $AB = XY$, angle ABC = angle XYZ and angle CAB = angle ZXY; (ASA). Notice that this can be extended, for if any two pairs of angles are equal then the third pair must be equal, so that triangles are congruent if ASA, AAS or SAA. The final case of congruency is when $AB = XY$, $AC = XZ$ and angle ACB = angle $XYZ = 90°$ (RHS). (See Fig. 27.4.)

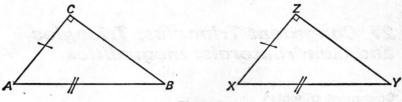

Fig. 27.4

Notice that two triangles are *not* congruent if two pairs of sides are equal and a pair of not-included angles are equal (but not each 90°), as two different triangles could be constructed from that data (Fig. 27.5).

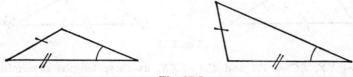

Fig. 27.5

When two triangles, e.g. *ABC*, *XYZ*, are congruent, this can be abbreviated △*ABC* ≡ △*XYZ*. The reason for congruency, e.g. SSS, should always be given.

Example 1. ABCD is part of a regular polygon. Prove that triangles ABC, BCD are congruent, and that AC = BD.

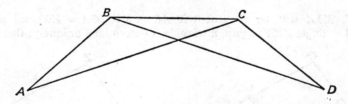

Fig. 27.6

In triangles *ABC*, *BCD*

$$AB = BC \text{ (equal sides of regular polygon)}$$
$$BC = CD \text{ (equal sides of regular polygon)}$$

angle *ABC* = angle *BCD* (angles of a regular polygon)
∴ △ *ABC* ≡ △ *BCD* (SAS).

Since the triangles are congruent, the third sides are equal

i.e. $AC = BD.$

For a different method of proof, see page 342.

Exercise 27.1

A quadrilateral *ABCD* in which *AB* and *DC* are parallel, and *BC* and *AD* are parallel, is called a parallelogram. Using this definition, prove the following:

1. The opposite angles of a parallelogram are equal. (Hint: join *AC*, and prove $\triangle ABC \equiv \triangle CDA$.)

2. The opposite sides of a parallelogram are equal.

3. The diagonals of a parallelogram bisect each other. (Hint: let the diagonals meet at *X*. Prove $\triangle ABX \equiv \triangle DCX$.)

4. If a quadrilateral is such that one pair of opposite sides are equal and parallel, then the other pair of opposite sides are also equal and parallel.

These properties of the parallelogram are considered again on page 342, when we study the symmetry of the parallelogram.

A quadrilateral *ABCD* in which $AB = BC$ and *BD* bisects angle *ABC* is called a kite. The symmetry of the kite is considered on page 342. Use this definition of a kite to prove the following:

5. *Two* pairs of adjacent sides of a kite are equal.

6. The diagonals of a kite intersect at right angles.

7. One of the diagonals of a kite is bisected by the other diagonal.

8. Given that two straight lines *AB*, *CD* bisect each other, prove that $AC = BD$.

9. In the isosceles triangle *ABC*, $AB = BC$. Points *L*, *M* are taken *AB*, *BC* respectively such that $AL : LB = 1 : 2$ and $BM : MC = 2 : 1$. Prove that triangles *AMB*, *CLB* are congruent, and hence that $MA = LC$.

10. In the quadrilateral *ABCD*, $AD = BC$ and $BD = AC$. Prove that angles *DAB*, *ABC* are equal.

The isosceles triangle

In Fig. 27.7, triangle *ABC* is isosceles, i.e. $AB = AC$. There are many different proofs that the base angles of an isosceles triangle are equal. One of the most elegant is, nevertheless, difficult to appreciate.

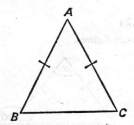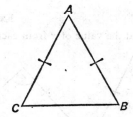

Fig. 27.7

To prove the triangle ABC is congruent to the triangle ACB. (They are, in fact, the same triangles, but orientated differently.)

$$AB = AC$$
and $$AC = AB \text{ (given)}$$
angle $$BAC = \text{angle } CAB \text{ (same angle)}$$

so triangles ABC, ACB are congruent. Since the triangles are congruent, the other pairs of corresponding angles are equal,

i.e. angle $ABC = \text{angle } ACB$

Alternative proof

Draw the bisector of the angle ABC, to meet AC at D. Then in triangles ABD, CBD,

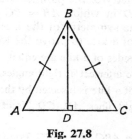

Fig. 27.8

$$AB = BC$$
$$BD = BD \text{ (common)}$$
and angle $ABD = \text{angle } DBC \text{ (construction)}$
$$\therefore \triangle ABD = \triangle DBC \text{ (SAS)}$$
$$\therefore \text{ angle } BAD = \text{angle } BCD.$$

Another proof is given on page 341, using the symmetry of triangle ABC. Many geometrical proofs are shortened and made more elegant using symmetry, but at an early stage it is important to notice that appeals to symmetry can always be justified by proving suitable triangles congruent.

Exercise 27.2a

1. Find the value of x from each of the given figures.

 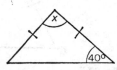 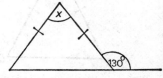

Fig. 27.9

2. *ABCDE* is a regular pentagon. Calculate the angle *BAC*.

3. *ABCDE* is a regular pentagon. Prove that *AC* is parallel to *DE*.

4. A chord *AB* subtends an angle of 80° at the centre *O* of a circle. Find the angle *OAB*.

5. The mid points of the equal sides *AB*, *AC* of an isosceles triangle are *X*, *Y* respectively. Prove that *BY* = *XC*.

6. In Fig. 27.10, *ABCD* is a square and *DEC* is an equilateral triangle. Find the angle *DAE*.

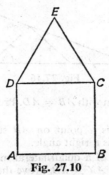

Fig. 27.10

7. In Fig. 27.10, prove that *EA* = *EB*.

8. In Fig. 27.10, calculate the angle *AEB*.

9. From Fig. 27.11, give *y* in terms of *x*.

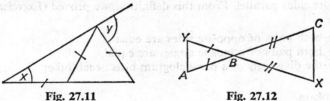

Fig. 27.11   **Fig. 27.12**

10. In Fig. 27.12, given that *AB* = *BY* and that *BX* = *BC*, prove that *AY* is parallel to *XC*.

Exercise 27.2b

1. Find the value of *x* from each of the given figures.

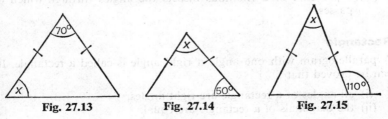

Fig. 27.13   **Fig. 27.14**   **Fig. 27.15**

2. *ABCDEF* is a regular hexagon. Find the angle *FAE*.

3. *ABCDEF* is a regular hexagon. Prove that *BD* is parallel to *AE*.

4. *ABCDEF* is a regular hexagon. Prove that *CF* is parallel to *AB*.

5. In the regular hexagon *ABCDEF*, calculate the angle *ADB*.

6. A chord *XY* subtends an angle of 100° at the centre *O* of a circle. Calculate the angle *OXY*.

7. Find *x* from Fig. 27.16.

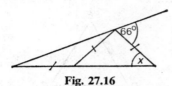

Fig. 27.16

8. *ABCD* is a parallelogram with *AB* = *AD*. Prove that *CA* bisects the angle *BAD*.

9. *ABC* is a triangle. *X* is a point on *AB* such that *AX* = *XB* = *XC*. Prove that the angle *C* is a right angle.

10. The diagonals *AC*, *BD* of a quadrilateral meet at *X*. Given that *AB* is parallel to *DC* and that *AX* = *XB*, prove that *CX* = *XD*.

Parallelogram

We defined a parallelogram as a quadrilateral with both pairs of opposite sides parallel. From this definition, we proved (Exercise 27.1) that

   (i) both pairs of opposite sides are equal,

   (ii) both pairs of opposite angles are equal,

   (iii) the diagonals of a parallelogram bisect each other.

Rhombus

A parallelogram with a pair of adjacent sides equal is called a rhombus. It can be proved that

   (i) all sides of a rhombus are equal,

   (ii) the diagonals of a rhombus bisect each other at right angles,

   (iii) a diagonal of a rhombus bisects the angles through which it passes.

Rectangle

A parallelogram with one angle a right angle is called a rectangle. It can be proved that

   (i) all angles of a rectangle are right angles,

   (ii) the diagonals of a rectangle are equal.

Square

A rectangle with two adjacent sides equal is called a square. A square has all the properties of the parallelogram, rhombus and rectangle.

The relation between these quadrilaterals is illustrated in the Venn diagram in Fig. 27.17.

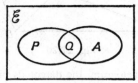

Fig. 27.17

If $\mathscr{E} = \{$all quadrilaterals$\}$
$P = \{$all parallelograms$\}$
$A = \{$all quadrilaterals with at least one pair of adjacent sides equal$\}$
and $Q = \{$all rhombuses$\}$
$P \cap A = Q.$

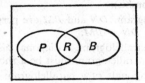

Fig. 27.18

If $B = \{$all quadrilaterals with at least one right angle$\}$
and $R = \{$all rectangles$\}$
$P \cap B = R.$

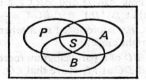

Fig. 27.19

If $S = \{$all squares$\}$
$P \cap A \cap B = S.$

Example 2. In the parallelogram ABCD, points L, M are taken in AB, CD respectively such that AL = CM. Prove that LM and BD bisect each other.

Since $ABCD$ is a parallelogram, $AB \parallel CD$, i.e. $AL \parallel CM$.
Since $AL = CM$ (given), and $AL \parallel CM$, $ALCM$ is a parallelogram.

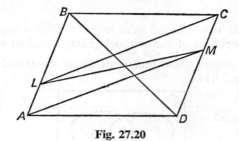

Fig. 27.20

Since *ALCM* is a parallelogram, the diagonals bisect each other, i.e. *LM* and *AC* bisect each other.

But since *ABCD* is a parallelogram, *AC* and *BD* bisect each other, ∴ *LM* and *BD* bisect each other.

Exercise 27.3

1. One angle of a parallelogram is 62°. Find the other angles.

2. The angle between a diagonal of a rhombus and a side is 20°. Calculate the angles of the rhombus.

3. *ABCD* is a parallelogram. *DN* and *BM* are perpendiculars from *D* and *B* to *AC*. Prove that *DN* = *BM*.

4. Prove that if, in a parallelogram, a diagonal bisects the angles through which it passes, the parallelogram must be a rhombus.

5. Prove that if the diagonals of a parallelogram are equal, it must be a rectangle.

6. The median *AD* of a triangle *ABC* is produced to *X* where *AD* = *DX*. Prove that *ABXC* is a parallelogram.

7. *ABCD*, *CEFD* and *ABXY* are all parallelograms. Prove that *FX* and *EY* bisect each other.

8. The two parallelograms *ABCD* and *ABXY* are such that *CDXY* is a straight line. Prove that △*ADY* ≡ △*BCX*.

9. *ABCD* is a rhombus. *BA* is produced to *X* where *BA* = *AX*. Prove that the angle *XDB* is a right angle.

10. If the diagonals *AC*, *BD* of a parallelogram meet at *O* and a line through *O* meets *AB* at *L* and *DC* at *M*, prove that *LO* = *OM*.

Inequalities

It may be assumed that the straight line as we defined it on page 262 is the shortest path between two points. We can now prove three other inequalities.

1. In any triangle, the greatest angle is opposite the greatest side. The proof is left to the reader. Hint: if in triangle *ABC*, *AC* > *AB*, let *X* be the point in *AC* such that *AX* = *AB*. Join *BX*.

2. In any triangle, the greatest side is opposite the greatest angle. Hint: either $AB < AC$ or $AB = AC$ or $AB > AC$. Use (1).

3. The shortest distance from a point to a straight line is the perpendicular distance.

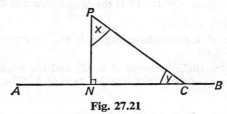

Fig. 27.21

Given a straight line AB and a point P not on AB. Let PN be the line through P perpendicular to AB, and C a point in AB not in PN.

Since $\qquad P\hat{N}C = 90°$ (construction),
$$x + y = 90°$$
$$\therefore y < 90°$$
But $\qquad$ angle $PNC = 90°$
$$\therefore y < P\hat{N}C, \text{ i.e. } PN < PC \text{ by (2) above.}$$

Exercise 27.4

1. In Fig. 27.22, is $AB > BC$?

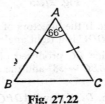

Fig. 27.22

2. In Fig. 27.23, place BC, CA, CD in order of magnitude.

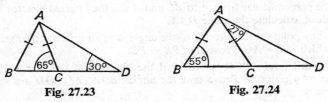

Fig. 27.23 $\qquad\qquad$ **Fig. 27.24**

3. In Fig. 27.24, place BC, CA, CD in order of magnitude.

4. In the triangle ABC, $AB > AC$. The internal bisectors of the angles B and C meet at X. Prove $XB > XC$.

5. X is any point on the side AB of an equilateral triangle ABC. Prove that $AC > CX > AX$.

6. *X* is a point on the side *AB* of an isosceles triangle *ABC* in which *CA = CB*. Prove that *CA > CX*.

7. A point *P* is taken on the side *CA* produced of an isosceles triangle in which *AB = AC*. Prove that *PB > AC*, given the angle *A* is acute.

8. In the quadrilateral *ABCD*, *AB* is the largest side and *CD* the shortest. Prove that $\widehat{ADC} > \widehat{ABC}$.

9. *ABCD* is a parallelogram whose diagonals *AC*, *BD* meet at *O*. Prove that *AB + AD > 2AO*.

10. In the triangle *ABC*, the angle *A* = 72° and the angle *B* = 55°. The internal bisectors of the angles *B* and *C* meet at *X*. Which is the greater, *BX* or *CX*?

Exercise 27.5: Miscellaneous

1. In Fig. 27.25, *ABCD* is a square and *ABP* an equilateral triangle. Calculate the angle *DPC*.

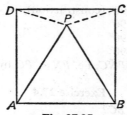

Fig. 27.25

2. *ABCD* is a parallelogram. If the bisectors of the angles *C* and *D* meet at *O*, prove that the angle *DOC* is a right angle.

3. *D* is the point on the side *BC* produced of a triangle *ABC* such that *CD = CA*. If the angle *ABC* = 58° and the angle *ACB* = 46°, calculate the angle *BAD*.

4. *ABCD* is a parallelogram. *DCXY* and *ADPQ* are squares drawn outside the parallelogram. Prove that *AY = CP*.

5. In the triangle *ABC*, the angle *B* = 66° and the angle *C* = 48°. If *AD* is the perpendicular from *A* to *BC* and *AX* is the internal bisector of the angle *A*, calculate the angle *DAX*.

6. *P* is any point on the bisector of the angle *A* of an isosceles triangle *ABC* in which *AB = AC*. Prove that *PB = PC*.

7. A quadrilateral *ABCD* is such that the bisectors of the four angles all meet at a point *O*. Prove that the angles *BOA* and *COD* are supplementary.

8. *ABCDEFGH* is a regular octagon. Calculate the angle *BDC*.

9. *BCD* are consecutive vertices of a regular polygon. If the angle *BDC* is 10°, how many sides has the polygon?

10. In the triangle *ABC*, the angles at *A* and *C* are 40° and 95° respectively. If *N* is the foot of the perpendicular from *C* to *AB*, prove that *CN = BN*.

28 Ratio; Similar Figures, Especially Triangles

Ratio

If the lengths of two straight lines are p cm and q cm respectively, the ratio of their lengths is $p:q$, or p/q. Ratios compare two *numbers*; we cannot compare a length of 5 cm with an area of 5 cm².

Internal and external division

If the point X in AB lies between A and B, X is said to divide AB **internally** in the ratio $AX:XB$ (Fig. 28.1). If the point Y in AB produced

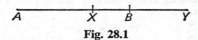

Fig. 28.1

does not lie between A and B, it is said to divide AB **externally** in the ratio $AY:YB$. Since these two line-segments are in opposite directions, such a ratio is often written with a negative number, e.g. $1:-2$.

Example 1. AB is a straight line length 16 cm. The point X divides AB internally in the ratio 5:3; the point Y divides AB externally in the ratio 5:3. Find the lengths of AX and XB, AY and YB.

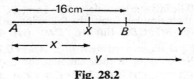

Fig. 28.2

Let the length of AX be x cm. Then the length of XB is $(16-x)$ cm,

and
$$\frac{x}{16-x}=\frac{5}{3},$$

i.e.
$$x=10,$$

the length of AX is 10 cm, of XB is 6 cm.

Similarly, if the length of AY is y cm, the length of YB is $(y-16)$ cm,

so
$$\frac{y}{y-16}=\frac{5}{3},$$
$$y=40,$$

the length of AY is 40 cm, of YB is 24 cm.

Exercise 28.1a

1. Points A, B, C, D lie in order in a straight line. $AB = 2$ cm, $BC = 3$ cm, $CD = 4$ cm. Write down in their simplest forms the following ratios:

   (i) $AB:BC$,   (ii) $BC:CD$,   (iii) $AC:CD$,   (iv) $AB:CD$.

2. Points A, B, C, D lie in a straight line. $AB = 1.2$ cm, $BC = 0.8$ cm, $CD = 1.6$ cm. Find each of the following ratios in its simplest form:

   (i) the ratio in which B divides AC,
   (ii) the ratio in which C divides BD,
   (iii) the ratio in which C divides AD.

3. Points P and Q divide the straight line AB, length 10 cm, internally and externally respectively in the ratio $p:q$. Find the lengths AP, PB and AQ, QB when

   (i) $p = 2, q = 1$,   (ii) $p = 3, q = 1$,
   (iii) $p = 3, q = 2$,   (iv) $p = 4, q = 1$.

4. Points P and Q divide the straight line AB, length 20 cm internally and externally in the ratio $p:q$. Find the lengths AP, PB and AQ, QB in each of the following cases, marking the relative positions of A, P, B and Q on a diagram.

   (i) $p = 1, q = 3$,   (ii) $p = 3, q = 1$,
   (iii) $p = 2, q = 3$,   (iv) $p = 3, q = 2$.

5. In the triangle ABC, angle $ABC = 90°$, $AB = 5$ cm, $BC = 10$ cm. Points X and Y lie in AB, BC respectively such that $AX:XB = CY:YB = p:q$. Find the ratios, in their simplest forms, of area triangle ABC: area triangle XBY when $p:q$ is

   (i) $1:4$,   (ii) $2:3$,   (iii) $3:2$,   (iv) $4:1$.

Exercise 28.1b

1. Points A, B, C, D lie in a straight line, $AB = 1$ cm, $BC = 2$ cm, $CD = 3$ cm. Write down in their simplest forms the following ratios:
   (i) $AB:BC$,   (ii) $BC:CD$,   (iii) $AC:CD$,   (iv) $AB:CD$.

2. Points A, B, C, D lie in a straight line. $AB = 0.7$ cm, $BC = 0.8$ cm, $CD = 1$ cm. Write down, in their simplest forms, the following ratios:
   (i) $AB:BC$,   (ii) $BC:CD$,   (iii) $AC:CD$,   (iv) $AB:CD$.

3. Points P and Q divide the straight line AB, length 30 cm, internally and externally respectively in the ratio $p:q$. Find the lengths of AP, PB and AQ, QB when

   (i) $p = 3, q = 1$,   (ii) $p = 3, q = 2$,
   (iii) $p = 4, q = 1$,   (iv) $p = 5, q = 1$.

4. In the triangle ABC, angle $ABC = 90°$, $AB = 6$ cm, $BC = 12$ cm. Points X and Y lie in AB, BC respectively such that $AX:XB = CY:YB = p:q$. Find the ratios, in their simplest forms, of area triangle ABC: area triangle XBY when $p:q$ is (i) $1:5$,   (ii) $1:2$,   (iii) $2:1$,   (iv) $5:1$.

5. (The Golden Ratio.) If the point X in AB, length 1 unit, divides AB internally in the same ratio as that in which A divides BX externally, i.e. $AX:XB = AB:AX$, show that the length of AX is $\frac{1}{2}(\sqrt{5} - 1)$ units.

Similar figures

Similar figures are figures which have the same shape; one is an enlargement of the other. All circles are similar, but not all ellipses are similar. All squares are similar, but not all rectangles are similar. Two polygons are similar if

- (i) the angles of one are respectively equal to the angles of the other, AND
- (ii) corresponding sides are in a constant ratio.

Similar triangles

It can be proved that if two triangles are such that the angles of one are equal to the angles of the other, then the corresponding sides are proportional, and conversely that if the corresponding sides are proportional, then the angles of one will be equal to the angles of the other, so that either condition (i) or condition (ii) is sufficient for a pair of triangles to be similar.

These three theorems are useful when considering similar triangles.

Theorem I

If two triangles are equiangular, their corresponding sides are proportional.

Theorem II

If two triangles have their corresponding sides proportional, they are equiangular.

Theorem III

If two triangles have one angle of the one equal to one angle of the other, and the sides about these equal angles are proportional, the triangles are similar.

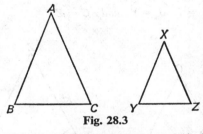

Fig. 28.3

To prove the triangles ABC, XYZ in Fig. 28.3 similar, it is necessary and sufficient to prove any one of the following:

(i) $\widehat{A} = \widehat{X}$ and $\widehat{B} = \widehat{Y}$ (it follows that $\widehat{C} = \widehat{Z}$).

(ii) $\qquad \dfrac{AB}{XY} = \dfrac{AC}{XZ} = \dfrac{BC}{YZ}.$

(iii) $\qquad \widehat{A} = \widehat{X}$ and $\dfrac{AB}{AC} = \dfrac{XY}{XZ}.$

Conversely, if the two triangles are given similar, write the two triangles with the equal angles under each other $\dfrac{ABC}{XYZ}$. Then the equations connecting the lengths of the sides may be obtained by writing down the sides of the first triangle and writing underneath the corresponding letters of the second triangle.

This gives $\qquad \dfrac{AB}{XY} = \dfrac{BC}{YZ} = \dfrac{AC}{XZ}.$

Example 2.

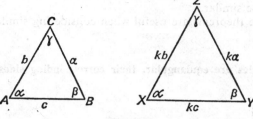

Fig. 28.4

With the data of Fig. 28.4, in which $\widehat{A} = \widehat{X} = \alpha$, $\widehat{B} = \widehat{Y} = \beta$ and $\widehat{C} = \widehat{Z} = \gamma$,

$$\dfrac{AB}{XY} = \dfrac{BC}{YZ} = \dfrac{CA}{ZX} \qquad \qquad (Theorem\ I)$$

Example 3. With the data of Fig. 28.4, if $BC = a$, $YZ = ka$; $CA = b$, $ZX = kb$; $AB = c$, $XY = kc$, then

$$\widehat{A} = \widehat{X},\ \widehat{B} = \widehat{Y} \quad and \quad \widehat{C} = \widehat{Z} \qquad \qquad (Theorem\ II)$$

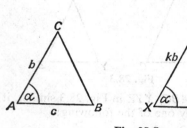

Fig. 28.5

Example 4. *With the data of Fig. 28.5, if* $\widehat{A} = \widehat{X} = \alpha$, $AC = b$ *and* $XZ = kb$, $AB = c$ *and* $XY = kc$, *triangles* ABC, XYZ *are similar. This is most useful when* AB *and* AC *lie along* XY *and* XZ *respectively (Fig. 28.6), showing the enlargement of* BC *into* YZ.

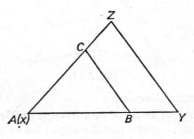

Fig. 28.6

Exercise 28.2a

1. A straight line parallel to the base BC of the triangle ABC meets AB in X, AC in Y. If $AX : XB = 2 : 3$, find the ratios (i) $AY : YC$, (ii) $XY : BC$.

2. A straight line parallel to the base BC of the triangle ABC meets AB in X, AC in Y. If $AY : YC = 3 : 5$, find (i) $AX : XB$, (ii) $XY : BC$.

3. A straight line parallel to the base BC of the triangle ABC meets AB in X, AC in Y. If $XY : BC = 2 : 3$, find the ratio $AX : XB$.

4. $ABCD$ is a parallelogram; X is the point in AB such that $AX : XB = 3 : 2$; XD and AC meet at Y. Find the ratios (i) $AX : CD$, (ii) $AY : YC$.

5. A straight line through the vertex A of a parallelogram $ABCD$ cuts BC at X and DC produced at Y. Prove that $AX : AY = AB : DY$.

Exercise 28.2b

1. A straight line parallel to the base BC of the triangle ABC meets AB in X, AC in Y. If $AX : XB = 4 : 5$, find the ratios (i) $AY : YC$, (ii) $XY : BC$.

2. A straight line parallel to the base BC of the triangle ABC meets AB in X, AC in Y. If $XY : BC = 4 : 5$, find the ratio $AX : XB$.

3. In the parallelogram $ABCD$, H and K are points in AD and BC respectively such that $AH : HD = 3 : 1$ and $BK : KC = 1 : 4$. Lines through H and K parallel to AB meet AC in X and Y respectively. Find the ratios (i) $AX : XC$ (ii) $AY : YC$, (iii) $AX : XY$, (iv) $XY : YC$.

4. In the triangle ABC, angle $ABC = 90°$, and the line through B perpendicular to AC meets AC in D. Prove that triangles ABC and BDC are equiangular.

5. In Fig. 28.7, AC is parallel to BD and CX is parallel to DY. Prove that AX is parallel to BY.

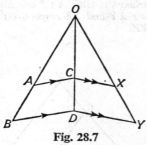

Fig. 28.7

The angle bisector theorems

The internal bisector of an angle of a triangle divides the opposite side in the ratio of the sides containing the angle; the external bisector of an angle divides the opposite sides externally in the ratio of the sides containing the angle.

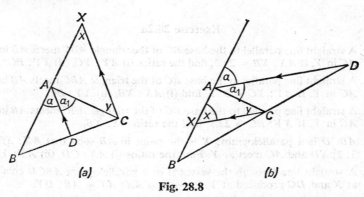

Fig. 28.8

AD bisects angle A internally in Fig. 28.8(a), externally in Fig. 28.8(b). Draw a line through C parallel to DA to meet BA produced (or BA in (b)) in X.

Then $\qquad a = x$ (corresponding; $AD \parallel XC$)

$\qquad\qquad a_1 = y$ (alternate; $AD \parallel XC$).

Since $\qquad a = a_1$ (given)

$\qquad\qquad x = y$.

$\qquad \therefore AX = AC$ (triangle AXC isosceles).

But $\qquad \dfrac{BD}{DC} = \dfrac{BA}{AX}$ ($AD \parallel XC$)

$\qquad \therefore \dfrac{BD}{DC} = \dfrac{BA}{AC}$.

The converse of this theorem is also true.

Exercise 28.3

1. The internal bisector of the angle A of a triangle ABC meets BC at D. If $AB = 7$ cm, $AC = 8$ cm and $BC = 9$ cm, calculate BD.

2. The external bisector of the angle A of a triangle ABC meets BC at X. If $AB = 5$ cm, $AC = 8$ cm, $BC = 6$ cm, calculate BX.

3. D is a point on the side BC of a triangle ABC. Given that $AB = 4$ cm, $AC = 6$ cm, $BD = 2$ cm, $BC = 5$ cm, prove that AD bisects the angle BAC.

4. The internal bisector of the angle A of a triangle meets BC at D; the external bisector meets CB produced at E. If $AB = 4$ cm, $AC = 6$ cm and $BC = 7$ cm, calculate DE.

5. The internal bisector of the angle A of a triangle ABC meets BC at D. Given that $AB : AC = 3 : 4$, find the ratio of the area of the triangle ABD to that of ABC.

6. If AX is a median of the triangle ABC and the internal bisectors of the angles AXB, AXC meet AB, AC respectively at P, Q, prove that PQ is parallel to BC.

7. ABC is an equilateral triangle. The internal bisector of the angle B meets the median AD at G. Prove that G is a point of trisection of AD.

8. In the quadrilateral $ABCD$, it is given that the internal bisectors of the angles B and D meet on AC. Prove that the internal bisectors of the angles A and C meet on DB.

9. D is a point on the side AB of the triangle ABC. It is given that $AB = 4$ cm, $BC = 4$ cm and $BD = 2$ cm. A line through D parallel to BC meets AC at E. Prove that BE bisects the angle B.

10. The bisectors of the angle A of the triangle ABC meet BC and BC produced at X and Y. If F is the mid point of XY, prove that $FA = FX$.

Areas of similar figures

If the radii of two circles are r and kr, their areas are πr^2 and $\pi k^2 r^2$, so they are in the ratio $1 : k^2$. If two rectangles are similar, their sides a, b and ka, kb, then their areas are also in the ratio $1 : k^2$. Any two similar

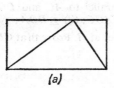
(a)

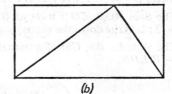
(b)

Fig. 28.9

triangles can be inscribed in two similar rectangles (Fig. 28.9. This statement can be proved; draw the altitudes of the triangles.) so the areas of similar triangles are in the ratio of the squares of the corresponding sides. Any similar figures bounded by straight lines can be divided into triangles, and we can generalize this to say that 'the areas of any two similar figures are proportional to the squares of the ratio of any linear dimension. This is hard to prove—what do we mean by the area of the region enclosed by curved lines?

It is also true that the volumes of similar solids are in the ratio of the cube of their linear dimensions, so that if two bottles are such that any linear dimension of one is twice the corresponding linear dimension of the other, the surface area of one will be four times the surface area of the other, and the volume of one will be eight times the volume of the other.

Exercise 28.4a

1. The sides of a triangle are 6 cm, 7 cm and 8 cm. The shortest side of a similar triangle is 2 cm. Find the lengths of the other two sides, and the ratio of the areas of the two triangles.

2. The sides of a triangle are 5 cm, 6 cm and 7 cm. The longest side of a similar triangle is 21 cm. Find the lengths of the other sides and the ratio of the two areas.

3. $ABCD$ is a trapezium with AB parallel to DC; O is the intersection of the diagonals. If $DO : OB = 3 : 4$, write down the ratios (i) $CO : OA$, (ii) $\triangle DOC : \triangle AOD$, (iii) $\triangle DOC : \triangle AOB$.

4. In the parallelogram $ABCD$, H is the point in AB such that $AH : HB = 2 : 3$, and K is the point in CD such that $CK : KD = 4 : 5$. HD and AK meet at O. Find the ratio of $\triangle AHO : \triangle KDO$.

5. D and E are points on the sides AB, AC respectively of a triangle ABC such that $AD : DB = 2 : 3$ and $AE : EC = 2 : 3$. Find the ratio $\triangle ADE : \triangle ABC$.

6. The medians BD, CE of a triangle ABC meet at G. Prove that the triangles EGD and CGB are similar. Hence prove that $CG = 2GE$.

7. Using the notation of Q.6, prove that $\triangle EGD : \triangle ABC = 1 : 12$.

8. Y is a point on the side BC produced of a parallelogram $ABCD$. AY meets DC at X. Prove that $AY : AX = BY : AD$.

9. The lines AOB, COD meet at O. DB is parallel to AC and $DO : OC = 3 : 2$. Write down the ratios (i) $\triangle DOB : \triangle AOC$; (ii) $\triangle DOA : \triangle BOC$.

10. The altitudes BN, CM of a triangle ABC meet at H. Prove that $CH.HM = BH.HN$.

Exercise 28.4b

1. The sides of a triangle are 2 cm, 3 cm and 4 cm. The length of the shortest side of a similar triangle is 2.5 cm. Find the lengths of the other sides and the ratio of the areas of the two triangles.

2. The sides of a triangle are 2 cm, 3 cm and 4 cm. Show that the longest side of any similar triangle is twice the shortest side of that triangle.

3. The sides of a triangle are 2 cm, 3 cm and 4 cm. The longest side of a similar triangle is 5 cm more than its shortest side. Find the lengths of the three sides of the similar triangle, and the ratio of the areas of the two triangles.

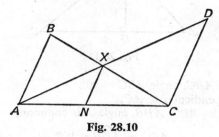

Fig. 28.10

4. In Fig. 28.10, *AB*, *NX* and *CD* are parallel. If *AB* = 3 cm, *CD* = 5 cm, calculate the length *NX*, and the ratios $\triangle AXN : \triangle ACD$ and $\triangle CNX : \triangle CAB$.

5. A cone height 8 m has a circular base radius 4 m. Find the radius of the circular section cut from the cone by a plane parallel to the base and 2 m from it.

6. *ABC* is a triangle. *X* is the point of trisection of *BA* nearer *B* and *Y* is the point of trisection of *CA* nearer *C*. Given that *XC*, *YB* meet at *Z*, find $\triangle XYZ : \triangle ABC$.

7. The triangle *ABC* is right-angled at *A* and *AD* is an altitude. Prove that *AB.AD* = *BD.AC*.

8. The internal bisectors of the angles *B* and *C* of a triangle *ABC* meet the opposite sides at *D* and *E*. If *DE* is parallel to *CB*, prove that *AB* = *AC*.

9. In the parallelogram *ABCD*, *X* is the mid point of *AB* and *Q* is the mid point of *CD*. The point *P* divides the side *CB* in the ratio 1 : 2, the point *Y* divides the side *AD* in the ratio 1 : 2, prove that the triangles *AXY*, *CQP* are similar. Find the ratio of the area of the hexagon *XBPQDY* to that of the parallelogram.

10. A line is drawn through *D*, the mid point of the side *BC* of a triangle *ABC*, to meet *BA* at *P*, *CA* produced at *Q* and the parallel to *AB* through *C* at *K*. If *AB* = *AC*, prove that $\dfrac{AP}{PB} = \dfrac{QP}{QK}$.

Pythagoras' theorem

The Egyptians knew that every triangle whose sides were in the ratio 3 : 4 : 5 contained a right angle. This was generalized, in many stages, until Pythagoras (c. 569–500BC) stated and proved the theorem that now bears his name. **In a right-angled triangle, the square on the hypotenuse is equal to the sum of the squares on the other two sides.** There are many proofs of the theorem. We give a proof using similar triangles.

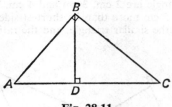

Fig. 28.11

In the triangle ABC, angle $ABC = 90°$.
Draw BD perpendicular to AC.
Then in triangles ABC, ADB, angle A is common,

$$\widehat{ABC} = \widehat{ADB} = 90°$$

∴ triangles ABC, ADB are similar,

$$\therefore \frac{AB}{AD} = \frac{AC}{AB}$$

i.e. $$AB^2 = AC.AD.$$

Likewise, triangles ABC, BDC are similar so

$$BC^2 = AC.DC.$$

Adding $$AB^2 + BC^2 = AC.AD + AC.DC$$
$$= AC(AD + DC)$$
$$= AC^2.$$

The converse of the theorem, that if the square of one side of a triangle is equal to the sum of the squares on the other two sides, then the angle between those sides is a right angle, is also true. This converse is widely used, and we should, of course, distinguish when we are using the theorem, and when we are using the converse of the theorem.

Example 5. *The base BDC of a triangle is such that BD = 4 cm, DC = 25 cm. The line AD is perpendicular to BC, and AD = 10 cm. Prove that the triangle ABC is right angled.*

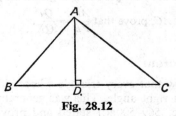

Fig. 28.12

Since $$\widehat{BDA} = 90°, \quad AB^2 = 4^2 + 10^2 \quad \text{(Pythagoras' Theorem)}$$
$$= 116.$$

Since $\qquad C\widehat{D}A = 90°, AC^2 = 10^2 + 25^2$ (Pythagoras' Theorem)
$$= 725.$$
$$\therefore AB^2 + AC^2 = 116 + 725$$
$$= 841$$
$$= 29^2$$
i.e. $\qquad\qquad\qquad AB^2 + AC^2 = BC^2$

$\therefore$ angle BAC is a right angle (converse of Pythagoras' Theorem).

Exercise 28.5a

1. What is the length of the hypotenuse of a right-angled triangle if the other two sides are 5 cm and 12 cm long?

2. The sides of a rectangle are 8 cm and 6 cm long. What is the length of the diagonal?

3. The diagonals of a rhombus are 8 cm and 6 cm long. What is the side of the rhombus?

4. Calculate the length of the altitude of an isosceles triangle whose base is 10 cm long and whose equal sides are 13 cm long.

5. Is the triangle whose sides are 17 cm, 15 cm and 8 cm long right-angled?

6. The altitude AD of a triangle ABC is 12 cm long. If $BD = 8$ cm, $DC = 18$ cm, prove that the angle BAC is a right angle.

7. AD is an altitude of the triangle ABC. Prove that $BA^2 - CA^2 = BD^2 - CD^2$.

8. The side of an equilateral triangle is a units. The length of an altitude is x units. Prove that $4x^2 = 3a^2$.

9. The angle B of a triangle ABC is a right angle. AD is a median of the triangle. Prove that $AD^2 = AC^2 - \frac{3}{4}BC^2$.

10. In the triangle ABC, the angle $B = 90°$. The internal bisector of the angle B meets AC at D. Given that $AB = 3$ cm, $BC = 4$ cm, find AD.

Exercise 28.5b

1. If the hypotenuse of a right-angled triangle is 37 cm long and a second side is 35 cm long, find the length of the third side.

2. The diagonals of a square $ABCD$ meet at O. Prove that $AB^2 = 2AO^2$.

3. The triangle ABC has a right angle at B and $AB = 5$ cm, $BC = 12$ cm. If BD is an altitude find AD and DC.

4. Is the triangle whose sides are 53 cm, 45 cm and 24 cm right angled?

5. ABC is a triangle in which the angle B is 60° and the angle C is 30°. Prove that $BC = 2AB$.

6. ABC is an isosceles triangle in which $AB = AC = 2BC$. AD is an altitude. Prove that $4AD^2 = 15BC^2$.

7. Calculate the length of BC in Fig. 28.13.

8. If X is a point inside the rectangle $ABCD$, prove that $XA^2 + XC^2 = XB^2 + XD^2$.

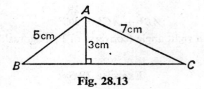

Fig. 28.13

9. ABC is a triangle in which $AB = BC$ and the angle $ABC = 120°$. Prove that $AC^2 = 3AB^2$.

10. Calculate the length of the diagonal of a rectangular box whose edges are 3 cm, 5 cm and 8 cm.

29 The Circle

Definitions

A point in a plane whose distance from a fixed point in that plane is constant lies on the **circumference** of a circle.

The fixed point is called the centre of the circle and the constant distance is called the **radius**.

A **chord** of a circle is a straight line joining two points on the circumference.

A chord through the centre is called a **diameter**.

Concentric circles are circles having the same centre.

Theorem

A straight line drawn from the centre of a circle to bisect a chord which is not a diameter, is at right angles to the chord.

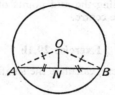

Fig. 29.1

Given AB is a chord of a circle centre O. N is the mid point of AB.

To prove that ON is perpendicular to AB.

Construction: Join OA, OB.

The proof is left as an exercise.

The converse theorem is also true.

Theorems

The perpendicular drawn from the centre of a circle to a chord bisects that chord.

There is one circle, and only one circle, which passes through three points not in a straight line.

Equal chords of a circle are equidistant from the centre.

Chords which are equidistant from the centre of a circle, are equal.

The proofs of these theorems are left as exercises, if desired.

Exercise 29.1a

1. In a circle of radius 5 cm, calculate the length of a chord which is 3 cm from the centre.

2. In a circle of radius 13 cm, calculate the distance from the centre of a chord which is 24 cm long.

3. An isosceles triangle whose sides are 13 cm, 13 cm and 10 cm is inscribed in a circle. Find the radius of the circle.

4. Two parallel chords of a circle are of lengths 6 cm and 8 cm and lie on the same side of the centre. If the radius of the circle is 5 cm, find the distance between the chords.

5. A chord of a circle of radius x cm is y cm distant from the centre. Find its length.

6. Show that in a circle the greater of two chords is nearer the centre.

7. A line cuts two concentric circles at points A, B, C, D. Prove that $AB = CD$.

8. An equilateral triangle is inscribed in a circle of radius r. Find the area of the triangle.

9. An equilateral triangle of side 4 cm is inscribed in a circle. Find the radius of the circle.

10. Prove that the line joining the mid points of two parallel chords of a circle passes through the centre.

Exercise 29.1b

1. In a circle of radius 13 cm, calculate the length of a chord which is 5 cm distant from the centre.

2. In a circle of radius 5 cm, calculate the distance from the centre of a chord which is 8 cm long.

3. An isosceles triangle whose sides are 5 cm, 5 cm and 6 cm is inscribed in a circle. Find the radius of the circle.

4. Two parallel chords of a circle are of lengths 10 cm and 24 cm and lie on opposite sides of the centre. If the radius of the circle is 13 cm, find the distance between the chords.

5. A chord of a circle is x cm long. If it is distant y cm from the centre, find the radius of the circle.

6. AB, CD are chords of a circle, meeting at O. If O is the mid point of each of the chords, prove that AB and CD must be diameters of the circle.

7. The triangle ABC has a right angle at B. Prove that the circle which passes through A, B and C must have the mid point of AC as centre.

8. A square of side x cm is inscribed in a circle. Find the radius of the circle.

9. An equilateral triangle of side x cm is inscribed in a circle. Find the radius of the circle.

10. What is the locus of the centre of the circle which passes through two fixed points A and B?

Angle in a segment

The chord AB divides the circumference of the circle into two arcs, APB, called the major arc, and AQB, the minor arc.

The region APB is called a segment of a circle. Any angle subtended by AB at a point on the arc APB is called the angle in the segment APB. Such an angle is also called the angle subtended by the arc AQB at the circumference.

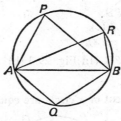

Fig. 29.2

$A\widehat{P}B$ and $A\widehat{R}B$ are called angles in the same segment.

The angle AQB is the angle subtended by the major arc at the circumference.

Theorem

The angle which an arc of a circle subtends at the centre is double that which it subtends at any other point on the remaining part of the circumference.

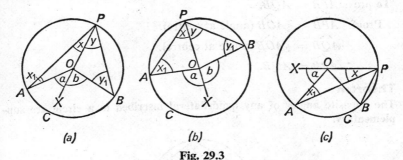

Fig. 29.3

Given ACB an arc of a circle O. P is any point on the remaining arc.

To prove $A\widehat{O}B = 2A\widehat{P}B$.

Construction: Join PO and produce to X.

Proof Since $AO = OP$ (radii),

$$x = x_1.$$

Also $a = x + x_1$ (exterior angle)
                         $= 2x.$

Similarly $b = 2y.$

(In Fig. 29.3 (c),

                $b = X\widehat{O}B$ and $y = X\widehat{P}B.$)

Adding in Figs. 29.3 (a) and (b),

                $a + b = 2(x + y),$

or $A\widehat{O}B = 2A\widehat{P}B.$

   Subtracting in Fig. 29.3 (c),

                $b - a = 2(y - x),$

or $A\widehat{O}B = 2A\widehat{P}B.$ (In Fig. 29.3 (b), reflex $A\widehat{O}B = 2A\widehat{P}B.$)

Theorem

Angles in the same segment of a circle are equal.

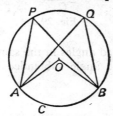

Fig. 29.4

Given an arc ACB of a circle centre O. Any two points P and Q are on the remaining arc.

To prove $A\widehat{P}B = A\widehat{Q}B.$

 Proof $A\widehat{P}B = \frac{1}{2}A\widehat{O}B$ (angle at centre).

          $A\widehat{Q}B = \frac{1}{2}A\widehat{O}B$ (angle at centre).

   $\therefore A\widehat{P}B = A\widehat{Q}B.$

Theorem

The opposite angles of any quadrilateral inscribed in a circle are supplementary.

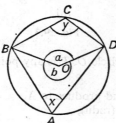

Fig. 29.5

Given a quadrilateral $ABCD$ inscribed in a circle centre O.
To prove $x + y = 180°$.
Construction: Join OB, OD.
Proof: $a = 2x$ (angles on arc BCD),
$\qquad\quad b = 2y$ (angles on arc BAD).
$\therefore a + b = 2(x + y)$.

Since $a + b = 360°$, $x + y = 180°$.

N.B. If BOD is a straight line, i.e. a diameter, $a = 180°$ and therefore $x = 90°$.

This is the important theorem, that **the angle in a semicircle is a right angle.**

Corollary. If the side AB of the cyclic quadrilateral $ABCD$ is produced to X, $C\widehat{B}X$ is supplementary to $C\widehat{B}A$. Since $C\widehat{D}A$ is also supplementary to $C\widehat{B}A$ (opposite angles of cyclic quadrilateral),

$$C\widehat{B}X = A\widehat{D}C.$$

Therefore the exterior angle of a cyclic quadrilateral is equal to the interior opposite angle.

Two converse theorems give useful tests for proving that four points lie on a circle.

1. A quadrilateral in which two opposite angles are supplementary is cyclic.
2. If the line joining two points subtends equal angles at two other points on the same side of it, the four points lie on a circle.

Theorems connecting arcs and chords

The following theorems can be proved, if desired.
In the same circle or in equal circles,

1. Equal chords stand on equal arcs.
2. Equal arcs subtend equal angles at the circumference.
3. Equal arcs are subtended by equal angles at the circumference.

Exercise 29.2a

1. Find the angle subtended at the circumference of a circle by the side of a regular 12-sided figure inscribed in the circle.
2. $ABCD$ is a cyclic quadrilateral. The angle $DAB = 80°$, and the angle $ACB = 50°$. Prove that $AD = AB$.
3. Find the angles of the triangle formed by joining the points representing 03.00 hours, 08.00 hours and 11.00 hours.
4. $ABCDE$ is a regular pentagon inscribed in a circle. Prove that AE is parallel to BD.

5. *ABCD* is a cyclic quadrilateral. The sides *AB, CD* produced meet at *E*; the sides *AD, BC* produced meet at *F*. Given that the angle *CEB* = 20° and that the angle *AFB* = 50°, calculate the angles of the quadrilateral.

6. *ABCD* is a quadrilateral in which the angle *ADB* = 51°, the angle *ABC* = 89° and the angle *BAC* = 40°. Prove that *ABCD* lie on a circle.

7. A circle passes through the points *A*, *B* and *O*. A second circle, centre *O*, also passes through the points *A* and *B*. Any line *AXY* through *A* meets the first circle at *X* and the second circle at *Y*. Prove that *BX* = *XY*.

8. Two chords of a circle *AB* and *CD* meet at a point *O* inside the circle. Prove that the triangles *AOC*, *DOB* are similar and hence prove that *AO . OB* = *CO . OD*.

9. *AB* is a diameter of a circle and *C* is a point on the circle so that $\widehat{BAC}$ = 30°. The internal bisector of the angle *ABC* meets the circle again at *X*. Prove that *CA* is the internal bisector of the angle *BAX*.

10. *BE, CF* are altitudes of a triangle *ABC*. Prove that the angles *AEF* and *ABC* are equal.

Exercise 29.2b

1. Find the angle subtended at the circumference of a circle by the side of a regular 10-sided figure inscribed in that circle.

2. *ABCD* is a cyclic quadrilateral. Given that the angle *DAB* = *x* and that the angle *ACB* = *y*, find a relationship between *x* and *y* if *AB* = *AD*.

3. Find the angles of the triangle formed by joining the points representing 02.00 hours, 06.00 hours and 11.00 hours.

4. *ABCDEFG* is a regular 7-sided figure inscribed in a circle. If *FB* and *AE* meet at *X*, prove that *FE* = *FX*.

5. *ABCD* is a cyclic quadrilateral. The sides *AB, DC* produced meet at *E*; the sides *AD, BC* produced meet at *F*. Given that the angle *CEB* = 40° and that the angle *AFB* = 50°, calculate the angles of the quadrilateral.

6. *ABCD* is a quadrilateral in which the angle *BAC* = *x*, the angle *ADB* = *y* and the angle *ABC* = *z*. Find a relationship between *x*, *y* and *z* if *ABCD* is cyclic.

7. Two chords *AB, CD* of a circle centre *O* are perpendicular. Given that the angle *AOC* = 110°, calculate the angle *BOD*.

8. Two chords of a circle, *AB* and *CD*, meet when produced at a point *O* outside the circle. Prove that the triangles *AOC*, *DOB* are similar and hence prove that *OA . OB* = *OC . OD*.

9. Prove that the internal bisectors of the angles of a cyclic quadrilateral themselves form a cyclic quadrilateral.

10. The altitudes *BE, CF* of a triangle *ABC* meet at *H*. If *X* is the mid point of *AH*, prove that *XF* = *XE*.

Exercise 29.3: Miscellaneous

1. The point *P* inside a circle is such that the shortest chord through it is 8 cm long. The greatest distance from *P* to the circumference is 16 cm. Find the radius of the circle.

2. *XY* and *AB* are diameters of a circle. *P* is any point on the circumference of the circle such that the angle *APY* is acute. Prove that the angles *XPA* and *BPY* are equal or supplementary.

3. In the cyclic quadrilateral *ABCD*, *AB* = *AC*. Given that *X* is any point on *CD* produced, prove that *AD* bisects the angle *BDX*.

4. Two circles intersect at *A* and *B*. Lines *PAQ*, *RBS* are drawn to meet one circle at *P* and *R*, and other circle at *Q* and *S*. Prove that *PR* is parallel to *QS*.

5. *ABCDEF* is a regular hexagon. Prove that *BDF* is an equilateral triangle.

6. *ABCDEF* is a hexagon inscribed in a circle. Prove that the sum of the angles at *A*, *C* and *E* is equal to the sum of the angles at *B*, *D* and *F*.

7. *O* is the centre of a circle and *P* a point distant 2 cm from *O*. If the radius of the circle is 8 cm, calculate the length of the chord through *P* which makes an angle of 45° with *OP*.

8. *PQR* is a triangle inscribed in a circle of radius 4 cm. Given that the angle *PQR* = 30°, calculate the distance of the chord *PR* from the centre of the circle.

9. Two circles intersect at *A* and *B*. Lines *PAQ*, *RAS* are drawn to meet one circle at *P*, *R* and the other circle at *S* and *Q*. Prove that the angles *PBR* and *QBS* are equal.

10. In the figure, *XC* = *CD* = *DA* = *AB*. Prove that *XD* is parallel to *CA*.

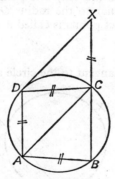

Fig. 29.6

11. *ABC* is an acute-angled triangle inscribed in a circle. Prove that the sum of the angles subtended by *AB*, *BC*, *CA* in their minor segments is 4 right angles.

12. *AB* is a fixed chord of a circle and *P* a variable point on the major arc *AB*. *AP* is produced to *Q* so that *BP* = *PQ*. Prove that the angle *AQB* is constant.

13. *ABC* is a triangle in which *AB* = 2 cm and the angle *ACB* = 30°. Calculate the radius of the circle passing through *A*, *B* and *C*.

14. Two equal circles intersect at *A* and *B*. Any line through *B* meets one circle at *P* and the other at *Q*. Prove that *PAQ* is an isosceles triangle.

15. *ABCD* is a trapezium with *AB* parallel to *DC*. Given that *AC* and *BD* meet at *X* and that *AX* = *XB*, prove that the points *A*, *B*, *C*, *D* are concyclic.

Tangent

The straight line through a point *P* on a circle, perpendicular to the radius through *P*, is called a tangent. It can be proved that the tangent to a circle does not meet the circle at any point other than the point of

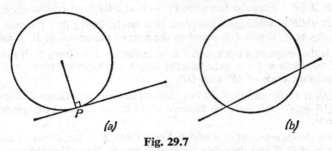

Fig. 29.7

contact. Alternatively, we can define the tangent as a straight line touching the circle at one and only one point *P*, and then prove that the tangent is perpendicular to the radius *OP*. A straight line which cuts a circle in two distinct points is called a secant (Fig. 29.7(b)).

Alternate segment

In Fig. 29.8, *PTQ* is the tangent to the circle at *T* and *TA* is any chord

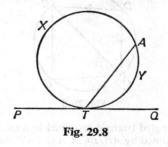

Fig. 29.8

through *T*. If we consider the angle *ATQ*, the segment *TXA* is called the alternate segment: if we consider the angle *PTA*, the segment *AYT* is the alternate segment.

Alternate segment theorem

The angles between a tangent to a circle and a chord through the point of contact of the tangent are equal to the angles in the alternate segments.

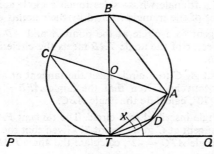

Fig. 29.9

To prove first that $\widehat{QTA} = \widehat{TCA}$.

Draw the diameter through T to meet the circle again at B.

Then if $\widehat{ATQ} = x$, $\widehat{BTA} = 90° - x$, since $\widehat{BTQ} = 90°$.

Since $\widehat{BAT} = 90°$, angle in a semicircle,

$$\widehat{TBA} = 180° - 90° - (90° - x)$$
$$= x.$$

But $\widehat{TBA} = \widehat{TCA}$, angles subtended by the same arc AT

$\therefore \quad \widehat{TCA} = x,$

i.e. $\widehat{ATQ} = \widehat{ACT}$.

Since $\widehat{ATP} = 180° - \widehat{ATQ}$, angles on a straight line,

and $\widehat{ACT} = 180° - \widehat{ADT}$, opposite angles of a cyclic quadrilateral,

$$\widehat{ATP} = \widehat{TDA},$$

i.e. the other angle between the tangent and chord is equal to the angle in the other alternate segment.

There are several other proofs of this theorem. One similar to the above is found joining T to O, O to A, and proving that $\widehat{AOT} = 2\widehat{ATQ}$, etc.

The converse of this theorem is also true: if a straight line is drawn from the extremity of a chord of a circle making with the chord an angle equal to the angle in the alternate segment, then that straight line is a tangent.

Exercise 29.4a

1. Given that the tangents at two points A, B of a circle meet at T, prove that $TA = TB$.

2. Prove that equal chords of a given circle all touch another concentric circle.

3. The angles of a triangle whose sides touch a circle are 50°, 60° and 70°. Find the angles of the triangle formed by their points of contact.

4. AT is the tangent to a circle at the point A and AB is a chord of the circle. The bisector of the angle TAB meets the circle again at X. Prove that $AX = XB$.

5. The tangents at A, C to a circle meet the tangent at another point B at P and Q respectively. Given that the angle $APB = 50°$ and that the angle $CQB = 70°$, calculate the angle ABC.

6. ABC is a triangle inscribed in a circle. The tangent PAT to the circle at A meets the tangent at C in the point T. Given that the angle $BCT = 80°$ and that the angle $ATC = 42°$, calculate the angle PAB.

7. A, B, C are three points on a circle centre O. The tangent at C meets AB produced at T. Given that the angle $AOB = 80°$ and that the angle $CBT = 64°$, calculate the angle TCB.

8. AB is a diameter of a circle and C a point on the circle so that the angle $BAC = 24°$. If the tangent at C meets AB produced at T, calculate the angle CTB.

9. AB is a diameter of a circle. The tangent at a point P is such that the angle $APT = 112°$. Calculate the angle BAP.

10. A, B, C are three points on a circle. The tangent at C meets BA produced at T. Given that the angle $ATC = 36°$ and that the angle $ACT = 48°$, calculate the angle subtended by AB at the centre of the circle.

Exercise 29.4b

1. Given that the tangents at two points A, B of a circle, centre O, meet at T, prove that OT bisects the angle ATB.

2. Given that the tangents at two points A, B of a circle, centre O, meet at T, prove that OT bisects the angle AOB.

3. The angles of a triangle whose sides touch a circle are 48°, 54° and 78°. Find the angles of the triangle formed by their points of contact.

4. AB is a chord of a circle, centre O. The tangent at another point T meets AB produced at C. Given that the angle $BCT = 20°$ and that the angle $BTC = 65°$, prove that the triangle AOB is equilateral.

5. The tangents at A, C to a circle meet the tangent at another point B at P and Q respectively. Given that the angle $APB = x$ and that the angle $CQB = y$, calculate the angle ABC in terms of x and y.

6. ABC is a triangle inscribed in a circle. The tangent PAT to the circle at A meets the tangent at C in the point T. Given that the angle $BCT = x$ and that the angle $ATC = y$, calculate the angle PAB in terms of x and y.

7. A, B, C are three points on a circle centre O. The tangent at C meets AB produced at T. Given that the angle $AOB = x$ and that the angle $CBT = y$ calculate the angle TCB in terms of x and y.

8. AB is a diameter of a circle and C a point on the circle. If the tangent at C meets AB produced at T and the angle $BAC = x$, find the angle CTB in terms of x.

9. *AB* is a diameter of a circle. The tangent *PT* at a point *P* is such that the angle *APT* = *x*. Find, in terms of *x*, the angle *BAP*.

10. *A*, *B*, *C* are three points on a circle. The tangent at *C* meets *BA* produced at *T*. Given that the angle *ATC* = *x* and that the angle *ACT* = *y*, find the angle subtended by *AB* at the centre of the circle in terms of *x* and *y*.

Internal and external contact

Two circles are said to touch at a point *T* if they have a common tangent at the point *T*.

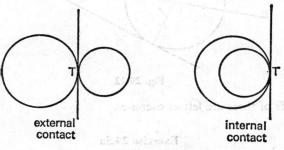

external contact   internal contact

Fig. 29.10

If one circle lies inside the other, the contact is said to be internal; in all other cases, the contact is external.

Two theorems

1. If two circles touch, the point of contact lies on the straight line joining the centres.

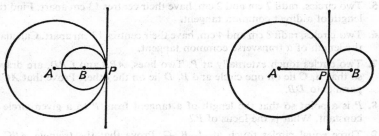

Fig. 29.11

2. If two tangents are drawn to a circle from a point outside the circle, then

    (i) the tangents are equal in length,

    (ii) the angle between the tangents is bisected by the line joining the point of intersection of the tangents to the centre of the circle,

(iii) this line also bisects the angle between the radii drawn to the points of contact.

This can all be summarized by saying that the figure is symmetrical about the line *OP* in Fig. 29.12.

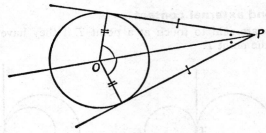

Fig. 29.12

The proofs of these are left as exercises.

Exercise 29.5a

1. Find the length of the tangents drawn to a circle of radius 5 cm from a point 13 cm from the centre.
2. Find the length of the chord of contact in Q.1.
3. A quadrilateral is circumscribed to a circle. Prove that the sums of the lengths of the opposite sides of the quadrilateral are equal.
4. Two circles, radii 9 cm and 4 cm, touch externally. Find the length of a common tangent.
5. Two circles, radii 7 cm and 2 cm, have their centres 13 cm apart. Find the length of a direct common tangent.
6. Two circles, radii 8 cm and 4 cm, have their centres 13 cm apart. Calculate the length of a transverse common tangent.
7. Two circles touch externally at *P*. Two lines, *APB* and *CPD*, are drawn so that *A*, *C* lie on one circle and *B*, *D* lie on the other. Prove that *AC* is parallel to *DB*.
8. *P* is a point so that the length of a tangent from *P* to a given circle is constant. What is the locus of *P*?
9. Three equal circles touch at *A*, *B*, *C*. Prove that the triangle *ABC* is equilateral and that its side is equal in length to the radius of a circle.
10. Two circles touch externally at *P*. A common tangent touches the circles at *R*, *S*. Prove that the angle *RPS* is a right angle.

Exercise 29.5b

1. The length of a tangent from a point distant 5 cm from the centre of a circle is 4 cm. Find the radius of the circle.

2. Find the length of the chord of contact in Q.1.

3. A tangent to a circle centre O meets two parallel tangents at P and Q. Prove that the angle POQ is a right angle.

4. Two circles, radii R and r, touch externally. Find the length of a common tangent.

5. Two circles, radii R and r, have their centres a distance d apart. Find the length of a direct common tangent.

6. Two circles, radii R and r, have their centres a distance d apart. Find the length of a transverse common tangent.

7. What is the locus of the centre of a circle which touches two given equal circles?

8. What is the locus of a point from which tangents to a given circle are perpendicular?

9. The centres of three circles which touch externally form a triangle whose sides are 5 cm, 6 cm and 9 cm. Find the radii of the circles.

10. Two circles touch internally at P. T is any point on the tangent at P. Tangents TX, TY are drawn to the two circles. Prove that $TX = TY$.

Intersecting chord theorem

If two chords of a circle intersect either inside or outside the circle, the product of the segments of one chord is equal to the product of the segment of the other chord.

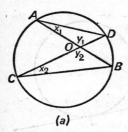

(a)

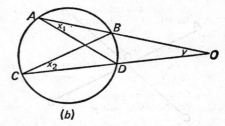

(b)

Fig. 29.13

With Fig. 29.13, to prove $AO.OB = CO.OD$. Join AD, BC.

Proof: In the triangles AOD,

$$x_1 = x_2 \text{ (on same arc)}$$
$$y_1 = y_2 \text{ (vertically opposite in (a))}$$

or   angle y is common, in (b)

$\therefore$ triangles AOD, BOC are similar,

$$\therefore \frac{AO}{CO} = \frac{OD}{OB}$$

i.e. $AO.OB = CO.OD$.

Special case

If, in Fig. 29.13(b), the chord CD becomes a tangent, $AO \cdot OB = OC^2$.

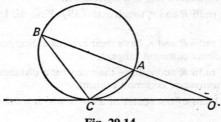

Fig. 29.14

This theorem can also be proved by proving triangles OAC, OCB similar, using the alternate segment theorem.

Converses

The converses of these theorems are true and often used.

1. If two straight lines AB, CD intersecting at a point O are such that $AO.OB = CO.OD$, then A, B, C, D are concyclic.
2. If two straight lines BAO and TO intersecting at O are such that $OA.OB = OT^2$, then OT is a tangent to the circle circumscribing the triangle ABT.

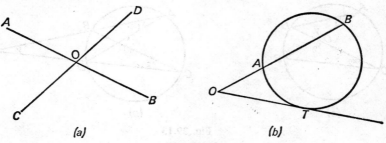

(a)   (b)

Fig. 29.15

Exercise 29.6a

1. The chords AB, CD of a circle meet at a point O outside the circle. Given that $OA = 7$ cm, $OB = 4$ cm and $OD = 2$ cm, find OC.
2. The chords AB, CD of a circle meet at a point O outside the circle. Given that $OA = 8$ cm, $AB = 3$ cm, $OD = 4$ cm, find CD.
3. The chords AB, CD of a circle meet at a point O outside the circle. Given that $OA = 8$ cm, $OB = 3$ cm and $CD = 10$ cm, find OD.
4. The chords AB, CD of a circle meet at a point O inside the circle. If $OA = 5$ cm, $OB = 10$ cm and $OD = 4$ cm, find OC.

5. The chords AB, CD of a circle meet at a point O inside the circle. If $OA = 4$ cm, $AB = 12$ cm and $CD = 18$ cm, find OD, given that it is larger than OC.

6. From a point O outside a circle, a secant OAB is drawn and a tangent OT. If $OT = 6$ cm and $AB = 5$ cm, find OA.

7. OAB and OCD are two straight lines. If $OA = 4$ cm, $AB = 5$ cm, $OC = 3$ cm and $CD = 9$ cm, prove that $ABDC$ is a cyclic quadrilateral.

8. OAB and OT are straight lines. If $OT = 8$ cm, $OA = 4$ cm and $AB = 12$ cm, prove that the angles OTA and OBT are equal.

9. The altitudes BE, CF of a triangle ABC meet at H. Prove that $CH \cdot HF = BH \cdot HE$.

10. Prove that the common chord of two intersecting circles when produced bisects their common tangents.

Exercise 29.6b

1. The chords AB, CD of a circle meet at a point O outside the circle. Given that $OA = 12$ cm, $OB = 4$ cm and $OD = 6$ cm, find OC.

2. The chords AB, CD of a circle meet at a point O outside the circle. Given that $OB = 4$ cm, $BA = 8$ cm and $OD = 6$ cm, find CD.

3. The chords AB, CD of a circle meet at a point O outside the circle. Given that $OA = 9$ cm, $OB = 4$ cm and $CD = 9$ cm, find OC.

4. The chords AB, CD of a circle meet at a point O inside the circle. If $OA = 7$ cm, $OB = 4$ cm and $OD = 3.5$ cm, find OC.

5. The chords AB, CD of a circle meet at a point O inside the circle. If $OA = 6$ cm, $AB = 14$ cm and $CD = 16$ cm, find OD given that it is less than OC.

6. From a point O outside a circle a secant OAB and a tangent OT are drawn. If $OT = 8$ cm, $AB = 12$ cm, find OA.

7. AOB and COD are straight lines. If $OA = 4$ cm, $OB = 9$ cm, $OD = OC = 6$ cm, prove that $ABCD$ is a cyclic quadrilateral.

8. OAB and OT are straight lines. If $OT = 6$ cm, $OA = 4$ cm, $AB = 5$ cm, prove that the angles OTB and OAT are equal.

9. BE, CF are altitudes of a triangle ABC. Prove that $AF \cdot AB = AE \cdot AC$.

10. T is any point on the common chord (produced) of two circles. Prove that the tangents from T to the two circles are equal.

Exercise 29.7: Miscellaneous

1. A circle is inscribed in a quadrilateral. Prove that the angles subtended at the centre of the circle by opposite sides are supplementary.

2. Two circles touch externally at P. Through P a straight line is drawn to cut the circles again at Q and R. Prove that the tangents at Q and R are parallel.

3. Two circles intersect at A and B. Through P, a point on the first circle,

straight lines PAC and PBD are drawn to cut the second circle at C and D. Prove that CD is parallel to the tangent at P.

4. Show that the direct common tangents to two circles intersect on the line of centres produced.

5. The tangent at T to a circle cuts a chord AB produced at C. Prove that the triangles TBC and ATC are similar.

6. The tangent at T to a circle cuts a chord AB produced at C. Prove that $BC:AC = TB^2:TA^2$.

7. Two circles intersect at P and Q. A common tangent touches one circle at A and the other at B. Prove that the angles APB and AQB are supplementary.

8. Two circles touch internally at T. A tangent to the inner circle at P meets the larger circle at X and Y. Prove that TP bisects the angle XTY.

9. X is a point on the side BC of a triangle ABC. The tangents at B and C to the circles through ABX and ACX respectively meet at T. Prove that $ABTC$ is a cyclic quadrilateral.

10. Two circles, centres A and B, touch externally at P. A common tangent of the circles touches them at S and T respectively. Prove that AS touches the circle through S, T and P.

11. Two circles, centres A and B, touch externally at P. A common tangent of the circles touches them at S, T respectively. Prove that the circle on ST as diameter touches AB at P.

12. Two circles, centres A and B, touch externally at P. A common tangent of the circles touches them at S, T respectively. Prove that the circle on AB as diameter touches ST.

13. Prove that, if it is possible to inscribe a circle in a parallelogram, the parallelogram must be a rhombus.

14. A diameter BA of a circle of radius 6 cm is produced to T where $AT = 4$ cm. TX is a tangent from T to the circle and TX produced meets the tangent at B in Y. Calculate the lengths of TX and XY.

15. A diameter AB of a circle bisects a chord CD at X. If $AX = 9$ cm and $CX = 6$ cm, find the radius of the circle.

30 Loci

Definition

A locus is a set of points satisfying a given condition. This can be used to *define* a certain curve, e.g. the set of all points in a plane a constant distance from a fixed point is the usual definition of a circle, or we may need to use geometry to show that a locus takes an already defined curve, e.g. the set of all points in a plane a constant distance from a given straight line can be shown to be a pair of straight lines parallel to the given straight line.

Some plane loci

The locus of all points P in a plane such that the sum of their distances from two fixed points A and B is constant is an **ellipse**, the locus of all

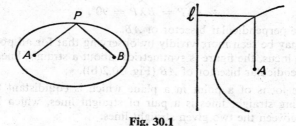

Fig. 30.1

points Q such that their distance from a fixed point A is equal to their distance from a fixed straight line l is a **parabola**.

Some loci in space

A sphere can be defined as the set of all points in three-dimensional space that are a constant distance from a fixed point. The set of all points in space that are a constant distance from a straight line is a cylinder with circular cross-section.

Two important loci

1. The locus of a point in a plane which is equidistant from two fixed points is the perpendicular bisector (sometimes called the mediator) of the straight line joining those two points.

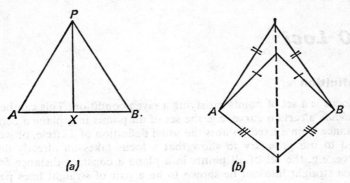

Fig. 30.2

Proof: If P is any point in the locus, join P to the midpoint X of AB. Then triangles APX, BPX are congruent (SSS).

$$\therefore \ A\widehat{X}P = B\widehat{X}P.$$

But $\qquad A\widehat{X}P + B\widehat{X}P = 180°$

$$\therefore \ A\widehat{X}P = B\widehat{X}P = 90°,$$

PX is the perpendicular bisector of AB.

This may be seen more vividly by observing that for all positions of P on the locus, the figure is symmetrical about a straight line, which is the perpendicular bisector of AB (Fig. 30.2(b)).

2. The locus of a point in a plane which is equidistant from two intersecting straight lines is a pair of straight lines, which bisect the angles between the two given straight lines.

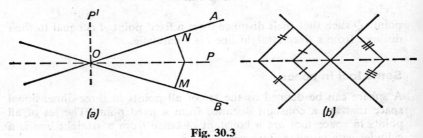

Fig. 30.3

Proof: If P is any point in the locus, draw PN and PM perpendicular to OA and OB respectively. Then triangles ONP, OMP are congruent (RHS).

$$\therefore \ P\widehat{O}N = P\widehat{O}M$$

$\therefore$ P always lies on the internal bisector of angle AOB.

If P lies the other side of OA say at P', then it lies on the other angle bisector.

Again, we see that Fig. 30.3(b) is symmetrical about the angle bisector.

Example 1. *The triangle ABC is such that AB = 4 cm, and the area of the triangle is 6 cm². Find the locus of all possible positions of the vertex C.*

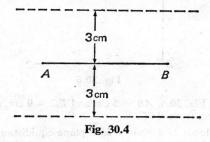

Fig. 30.4

Since the length of the base is 4 cm and the area of the triangle is 6 cm², the perpendicular height of the triangle must be 3 cm. Therefore, the third vertex C lies on one of the two straight lines parallel to AB, 3 cm away from AB.

Constant angle locus

We saw (page 298) that angles subtended by an arc of a circle at the circumference of the circle are equal, so that the set of all points P such that APB is constant is an arc of a circle through AB. Strictly, it is the arcs of two equal circles (Fig. 30.5).

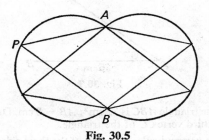

Fig. 30.5

Exercise 30

1. Draw the locus of a point P in a plane a constant distance 3 cm from a straight line.

2. Draw the locus of the point P in a plane a distance of 7 cm from a fixed point O.

3. The straight line AB is of length 6 cm. Draw the locus in a plane through AB of the set of all points 4 cm from AB.

4. In Fig. 30.6, $AB = 4$ cm, $BC = 5$ cm and $ABC = 90°$. Copy the figure and draw the set of all points distant 3 cm from ABC.

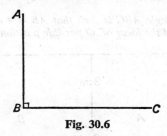

Fig. 30.6

5. If instead, in Fig. 30.6, $AB = 5$ cm and $BC = 9$ cm, and draw the set of all points 6 cm from ABC.

6. What is the locus of a point P in a plane equidistant from two parallel straight lines?

7. The base BC of a triangle is fixed. Find the locus in a plane of the third vertex A if the median AD is of constant length. (The median joins the vertex of a triangle to the midpoint of the opposite side.)

8. Two straight lines AOC, BOD intersect at O. How many points are there in the plane 3 cm from AOC and 4 cm from BOD?

9. $ABCD$ is a parallelogram area 40 cm²; $AB = 8$ cm. Find the locus in a plane of the point of intersection of the diagonals of $ABCD$.

10. In Fig. 30.7, $AB = 5$ cm and angle $BAC = 40°$. Find the point in AC equidistant from A and B.

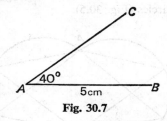

Fig. 30.7

11. The area of the triangle ABC is 32 cm²; $AB = 8$ cm. Draw the locus in the plane of the third vertex C of the triangle.

12. Prove that the perpendicular bisectors of the three sides of a triangle meet at a point.

13. Prove that the internal bisectors of the angles of a triangle meet at a point.

14. What is the locus in space of a point equidistant from two fixed points A, B?

15. What is the locus in space of a point equidistant from two fixed parallel planes?

16. What is the locus in space of a point equidistant from two intersecting straight lines?

17. The line AB is 5 cm long. Find the locus of a point P such that $\widehat{APB} = 90°$.

18. In the triangle ABC, $\widehat{ACB} = 90°$, $AB = 12$ cm and the area of the triangle is 24 cm². Find the four possible positions of C.

19. In the triangle ABC, $\widehat{ACB} = 100°$, $AB = 12$ cm and the area of the triangle is 24 cm². Find the four possible positions of C.

20.

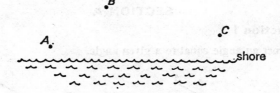

Fig. 30.8

A yachtsman sees three land marks A, B and C (Fig. 30.8). His charts tell him that $AB = 300$ m, $BC = 400$ m and $\widehat{ABC} = 130°$. His instruments show that AB subtends an angle of 35° at his eye, and BC an angle of 55°. Draw Fig. 30.8 to scale and mark his position. How far is he from A?

31 Constructions

Common constructions which should be known are now listed. Proofs are indicated, and can be completed by the reader. The older constructions are given in Section A; those that have recently become popular in Section B.

SECTION A

Construction 1

To construct an angle equal to a given angle.

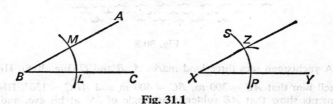

Fig. 31.1

Given an angle *ABC* and a straight line *XY*.

To construct a line through *X* making with *XY* an angle equal to $\widehat{ABC}$.

Construction: With centre *B* and any radius draw a circle to cut *BC* at *L* and *BA* at *M*.

With centre *X* and the same radius draw a circle *S* to cut *XY* at *P*.

With centre *P* and radius equal to *LM* draw a circle to cut *S* at *Z*.

Then *XZ* makes with *XY* the required angle.

Proof: The triangles *LBM* and *PXZ* are congruent. In particular the angles *B* and *X* are equal.

Construction 2

To bisect a given angle.

Given an angle *ABC*.

To construct the internal bisector of the angle *ABC*.

Construction: With centre *B* and any radius, draw an arc to cut *BC* at *X* and *BA* at *Y*. With centres *X* and *Y* and equal radii, greater than $\frac{1}{2}XY$, draw arcs to meet at *P*.

Then *BP* is the required bisector.

Proof: The triangles *PYB* and *PXB* are congruent and in particular the angles *PBX* and *PBY* are equal.

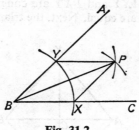

Fig. 31.2

Construction 3

To construct a perpendicular from a given point on a line to that line.

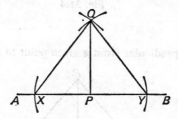

Fig. 31.3

Given a straight line *AB* with a point *P* on it.

To construct a line through *P* perpendicular to *AB*.

Construction: With centre *P* and any radius, draw a circle to cut *AB* at *X* and *Y*.

With centres *X* and *Y* and equal radii greater than *XP*, draw arcs to meet at *Q*.

Then *PQ* is the required perpendicular.

Proof: The triangles *PXQ* and *PYQ* are congruent. In particular, the angles *QPX* and *QPY* are equal.

Construction 4

To draw the perpendicular bisector of a given straight line.

Given a straight line *AB*.

To construct the mid point of *AB*.

Construction: With centres *A* and *B* and equal radii greater than ½*AB*, draw circles to cut at *X* and *Y*. Draw the straight line through *X* and *Y* to meet *AB* at *C*.

Then *C* is the mid point of *AB* and *XY* is the perpendicular bisector of *AB*.

Proof: The triangles AXY and BXY are congruent. In particular, the angles AXC and BXC are equal. Next, the triangles AXC and BXC can be proved congruent.

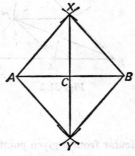

Fig. 31.4

Construction 5

To construct a perpendicular from a given point to a given line.

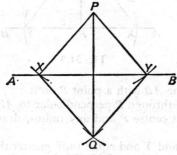

Fig. 31.5

Given a straight line AB and a point P not on the line.
To construct the perpendicular from P to AB.
Construction: Draw a circle with centre P to cut AB at X and Y. With equal radii, and centres X and Y, draw two circles to meet at Q (on the opposite side of AB from P).

Then PQ is the required perpendicular.
Proof: The triangles PXQ and PYQ are congruent, In particular, the angles XPQ and YPQ are equal. Now the triangles XPZ and YPZ may be proved congruent, where Z is the foot of the perpendicular.

Construction 6

To construct an angle of 60°.
Construction: Draw any line AB.

With centre A and any radius, draw a circle S to cut AB at P.
With centre P and the same radius, draw an arc to cut S at Q.
Then the angle $PAQ = 60°$.

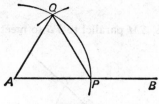

Fig. 31.6

Proof: The triangle AQP is equilateral.

(*N.B.* By bisecting an angle of 60°, an angle equal to 30° may be constructed. An angle of 45° is constructed by bisecting a constructed right angle.)

Construction 7

To construct through a given point a straight line parallel to a given straight line.

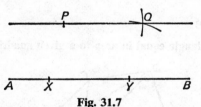

Fig. 31.7

Given a straight line AB and a point P not on the line.
To construct a line through P parallel to AB.
Construction: Take any two points X and Y on AB.
  With centre P and radius equal to XY, draw a circle S.
  With centre Y and radius equal to XP, draw another circle to cut S at Q.
  Then PQ is parallel to XY.
Proof: $XPQY$ is a quadrilateral with its opposite sides equal. It is therefore a parallelogram.

Construction 8

To divide a given straight line into a given number of equal parts.

  Suppose the number of parts to be five. A similar construction holds for any other integral number of parts.
Given a straight line AB.

To construct the points which divide AB into five equal parts.
Construction: Draw any other line through A.

With compasses, mark off any five equal distances AP, PQ, QR, RS, ST on this line.

Join TB.

Draw PH, QK, RL, SM parallel to TB to meet AB at H, K, L, M.

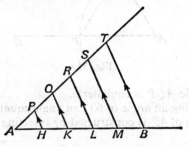

Fig. 31.8

Then H, K, L, M divide AB into five equal parts.
Proof: By the intercept theorem.

Construction 9

To construct a triangle equal in area to a given quadrilateral.

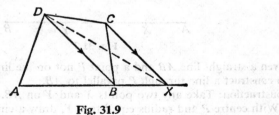

Fig. 31.9

Given a quadrilateral $ABCD$.
To construct a triangle equal in area to $ABCD$.
Construction: Join DB.

Through C draw a line parallel to DB to meet AB produced at X.

Then ADX is a triangle equal in area to $ABCD$.
Proof: The triangles DCB and DXB are equal in area. Add to each the area of ADB.

(*N.B.* By a similar construction, a polygon of $(n-1)$ sides may be constructed equal in area to a given polygon of n sides. By successive reduction, a triangle may be constructed equal in area to any given polygon.)

Construction 10

To construct a tangent to a circle at a given point of the circle.

Fig. 31.10.

Given a point P on a circle centre C.

To construct the tangent at P to the circle.

Construction: Draw through P a line PT so that the angle CPT is a right angle.

PT is the tangent at P.

Proof: The tangent is perpendicular to the radius at the point of contact.

Construction 11

To construct the pair of tangents from a given point to a given circle.

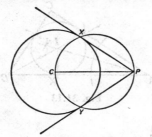

Fig. 31.11

Given a circle centre C and a point P outside the circle.

To construct the tangents from P to the circle.

Construction: Join CP.

Draw the circle on CP as diameter to meet the given circle at X and Y. Then PX and PY are the required tangents.

Proof: The angles CXP and CYP are angles in a semicircle.

Construction 12

To construct the circumcircle of a given triangle.

Given a triangle ABC.

To construct the circumcircle of ABC.

Construction: Construct the perpendicular bisectors of *CA* and *CB* to meet at *O*.

With centre *O* and radius equal to *OC*, draw a circle. This is the circumcircle of the triangle *ABC*.

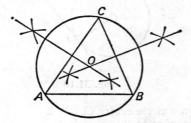

Fig. 31.12

Proof: *O* lies on the perpendicular bisector of *AC* and so is equidistant from *A* and *C*. Similarly *O* is equidistant from *B* and *C*.

Construction 13

To construct the inscribed circle of a given triangle.

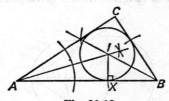

Fig. 31.13

Given a triangle *ABC*.
To construct its incircle.
Construction: Construct the internal bisectors of the angles *A* and *B* to meet at *I*.

Construct *IX*, the perpendicular from *I* to *AB*.

With centre *I* and radius *IX*, draw a circle.

This is the inscribed circle of the triangle *ABC*.

Proof: *I* lies on the internal bisector of the angle *A* and is therefore equidistant from *AB* and *AC*. Similarly *I* is equidistant from *AB* and *BC*.

Construction 14

To construct an escribed circle of a given triangle.

A triangle *ABC* has three escribed circles. These are called the escribed circles opposite A, B and C respectively. The escribed circle opposite A touches BC internally and touches AB and AC produced.

The internal bisector of the angle A and the external bisectors of the

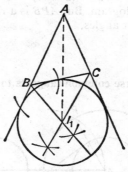

Fig. 31.14

angles B and C all meet in a point I_1, which is the centre of the escribed circle opposite A.

The centres of the escribed circles opposite B and C are conventionally called I_2 and I_3 respectively.

Construction 15

To construct the direct common tangents to two given circles.

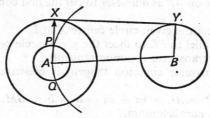

Fig. 31.15

Given two circles centres A and B, radii a and b, where a is greater than b.

To construct their direct common tangents.

Construction: Draw a circle centre A, radius $(a - b)$.

Draw a circle on AB as diameter to cut the first constructed circle at P and Q.

Produce AP to meet the given circle centre A at X.

Through B draw the line parallel to AX to meet the circle centre B at Y.

Then XY is a direct common tangent.

The second direct common tangent is constructed similarly by producing AQ.

Proof: $XP = XA - AP = a - (a - b) = b = BY$.

So $PXYB$ is a parallelogram. But APB is a right angle and therefore AXY and BYX are right angles.

Construction 16

To construct the transverse common tangents to two given circles.

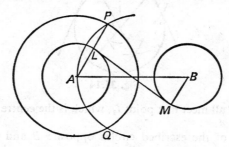

Fig. 31.16

Given two circles centres A and B, radii a and b.

To construct their transverse common tangents.

Construction: Draw a circle centre A, radius $(a + b)$.

Draw the circle on AB as diameter to cut the first constructed circle at P and Q.

Join AP to meet the given circle centre A at L.

Draw BM parallel to PL to meet the given circle centre B at M.

Then LM is a transverse common tangent.

The other transverse common tangent is constructed similarly by joining AQ.

Proof: $PL = AP - AL = (a + b) - a = b = BM$.

So $LPBM$ is a parallelogram.

But $\widehat{APB} = 90°$ and therefore $LPBM$ is a rectangle.

$$\therefore \widehat{ALM} = \widehat{BML} = 90°$$

Construction 17

To construct the locus of a point at which a given straight line subtends a given angle.

Given a fixed straight line AB and an angle x.

To construct the locus of a point P which moves so that $\widehat{APB} = x$.

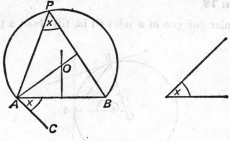

Fig. 31.17

Construction: Draw AC so that $B\widehat{A}C = x$.

Through A draw the perpendicular to AC to cut the perpendicular bisector of AB at O.

With centre O and radius OA draw an arc of a circle terminated by A and B.

This arc is the required locus.

Proof: Since O is on the perpendicular bisector of AB, $OA = OB$.

The arc therefore passes through A and B.

AC is a tangent to the circle and $A\widehat{P}B = x$.

(N.B. The complete locus consists of two equal arcs, one on each side of AB.)

Construction 18

To construct a square equal in area to a given rectangle.

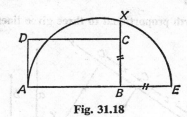

Fig. 31.18

Given a rectangle $ABCD$.

To construct a square equal in area to $ABCD$.

Construction: Produce AB to E so that $BE = BC$.

Draw the circle on AE as diameter to meet BC, produced if necessary, at X.

Then BX is a side of the required square.

Proof: AE bisects any chord of the circle perpendicular to it.

$$\therefore AB . BE = BX^2.$$

Construction 19

To draw a regular polygon of *n* sides (i) in, (ii) about a given circle.

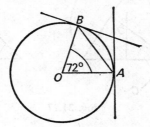

Fig. 31.19

We shall suppose the polygon has five sides. A similar construction holds for any other integral number of sides.

Given a circle centre *O*.

To draw a polygon of 5 sides (i) in, (ii) about the circle.

Construction: (i) Draw an angle at the centre equal to $\dfrac{360°}{5}$, i.e. 72°.

If the arms of this angle meet the circle at *A* and *B*, then *AB* is one side of the required pentagon.

(ii) The tangents at *A* and *B* are two sides of the regular pentagon drawn about the circle.

Proof: The symmetry of the figure.

Construction 20

To construct a fourth proportional to three given lines.

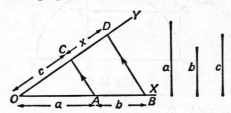

Fig. 31.20

Given three distances *a*, *b* and *c*, the fourth proportional is *x* where $a : b = c : x$.

Given three straight lines *a*, *b* and *c*.

To construct a line *x* such that $a : b = c : x$.

Construction: Draw any two lines OX and OY.

Along OX, mark off $OA = a$ and $AB = b$.

Along OY, mark off $OC = c$.

Join AC.

Through B draw BD parallel to AC to meet OY at D.

Then CD is the fourth proportional.

Proof: By parallels, $\dfrac{OA}{AB} = \dfrac{OC}{CD}$.

Construction 21

To construct a mean proportional to two given lines.

Given two distances a and b, their mean proportional is x where $a : x = x : b$.

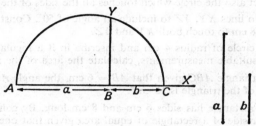

Fig. 31.21

Given two lines a and b.

To construct a mean proportional to a and b.

Construction: Draw any line AX.

Along AX, mark off $AB = a$ and $BC = b$.

Draw the circle on AC as diameter.

Through B draw the perpendicular to AC to cut the circle at Y.

Then BY is the mean proportional to a and b.

Proof: AC bisects any chord perpendicular to it.

$$\therefore AB.BC = BY^2.$$

Exercise 31.1: Miscellaneous Constructions

1. Draw a triangle ABC of sides 6.8 cm, 3 cm and 3.8 cm. Construct its circumcircle.

2. Draw a triangle ABC in which $AB = 5.6$ cm, the angle $B = 80°$ and the angle $C = 50°$. Construct the inscribed circle of the triangle.

3. On a straight line 3.6 cm long describe a segment of a circle to contain an angle of 50°.

4. Construct a circle of radius 2 cm to touch both the circumference and a diameter of a circle of radius 4 cm.

5. Construct a triangle of sides 4 cm, 6 cm and 8 cm. Construct the altitudes of this triangle and verify that they are concurrent.

6. A triangle XYZ is such that $XY = 4$ cm, $YZ = 6$ cm and the angle $Y = 60°$. Construct the triangle and the escribed circle of the triangle opposite Y.

7. Two circles radii 6 cm and 4 cm have their centres 8 cm apart. Construct a direct common tangent and measure its length.

8. Two circles radii 4 cm and 5 cm have their centres 12 cm apart. Construct a transverse common tangent and measure its length.

9. Without using set square or protractor construct a triangle XYZ in which $XY = 2.4$ cm, $YZ = 3$ cm and the angle $Y = 120°$. Find a point P such that $Y\widehat{X}P = 90°$ and $YP = ZP$.

10. Construct a rhombus with sides 6 cm long with one diagonal 8 cm long. Construct also the circle which touches all the sides of the rhombus.

11. Draw two lines XY, YZ to include an angle of 50°. Construct a circle of radius 2.8 cm to touch both XY and YZ.

12. Draw a circle of radius 4 cm and inscribe in it a regular hexagon. By making suitable measurements, calculate the area of the hexagon.

13. Draw a triangle ABC given that $AB = 6$ cm, the angle $ACB = 52°$ and the area of the triangle is 15 cm².

14. A given rectangle has sides 6 cm and 8 cm long. By construction, find the other side of a rectangle of equal area given that one of its sides is 10 cm long.

15. A given rectangle has sides 6 cm and 8 cm. By construction, find and measure the side of a square of equal area.

16. Construct a parallelogram given that one side is 5 cm long and that the diagonals are 8 cm and 6 cm in length.

17. Draw a triangle ABC in which $BC = 5$ cm, $BA = 4$ cm and $AC = 3$ cm. Construct a rectangle $ABXY$ equal in area to the triangle. Hence construct a square equal in area to the triangle.

18. Using ruler and compasses only, construct a trapezium $ABCD$ in which the parallel sides AB and DC are 2 cm apart, given that $AB = 4.8$ cm, the angle $A = 60°$ and $BC = 3.2$ cm. Is it possible to draw more than one such trapezium?

19. Draw a circle centre O of radius 4 cm. P is a point distant 6 cm from O. Construct two chords of the circle, each distant 2 cm from the centre, which pass through P when produced.

20. Using ruler and compasses only, draw a quadrilateral $ABCD$ in which the angle $DAB = 60°$, $AB = 4$ cm, the angle $CDA =$ the angle $ABC = 90°$ and $BC = 3.5$ cm. Construct a triangle on base AB equal in area to the quadrilateral.

21. Construct a triangle XYZ given that XY is greater than XZ and that $YZ = 4$ cm, the angle $X = 60°$ and the altitude through $X = 3$ cm.

22. Draw a triangle ABC in which $BC = 7$ cm, $CA = 5$ cm and $AB = 4$ cm.

Find by construction a point P such that the angle $APB = 40°$ and the angle $APC = 90°$.

23. AB is a diameter of a circle of radius 2 cm. Construct a circle of radius 4 cm to touch the circle externally and also to touch AB produced.

24. Draw a quadrilateral $ABCD$ given that $AB = 4$ cm, $BC = 5$ cm, $CD = 3.4$ cm, $DA = 3.4$ cm and the angle $B = 70°$. Construct a triangle on base BC equal in area to the quadrilateral.

25. Construct a triangle ABC given that $AB = 4$ cm, $BC = 5$ cm and that the length of the median through A is 3.5 cm.

26. Draw a triangle of sides 5 cm, 6 cm and 7 cm and construct the side of a square equal in area to the triangle.

27. The area of an acute-angled triangle is 5.6 cm². Two of the sides are known to be 4.8 cm and 3 cm. Construct the triangle.

28. Draw a circle of radius 6 cm and mark a point P distant 5 cm from the centre. Construct a chord of length 8 cm to pass through P.

29. Draw a triangle ABC in which $BC = 4$ cm, the angle $B = 50°$ and the angle $C = 70°$. Construct a point P which is equidistant from the lines AB and AC and which is also equidistant from B and C.

30. Draw two lines OX and OY inclined at an angle of 60°. On OX, mark points P, Q such that $OP = 6$ cm and $OQ = 10$ cm. Construct a circle to pass through the points P and Q and to touch the line OY.

31. Using ruler and compasses only, construct a triangle PQR in which the angle $P = 45°$, $PQ = 5$ cm and $PR = 6$ cm. Construct a point X inside the triangle which is equidistant from Q and R so that $PX = 4$ cm.

32. Draw a circle of radius 4 cm and mark a point P distant 6 cm from its centre. Construct two circles each of radius 5 cm to touch the given circle externally and also to pass through P.

33. Draw a quadrilateral $ABCD$ given that $AB = 5$ cm, $AD = 6$ cm and the diagonal $BD = 8$ cm. You are also given that the areas of the two triangles BAD and BCD are equal and that the angle at C is a right angle. How many such quadrilaterals are possible?

34. Construct a parallelogram $ABCD$ such that its diagonals are 7.2 cm and 5.6 cm long and such that the angle between the diagonals is 60°. Find a point X on AB produced so that the angle $CXB = 45°$.

35. Draw a triangle ABC in which $BC = 8$ cm, $CA = 6$ cm and $AB = 4$ cm. Construct the circle which touches BC at B and which also passes through A. Find a point P on this circle and inside the triangle equidistant from CA and CB.

36. Draw a circle of radius 4 cm. Construct a regular hexagon to circumscribe the circle. Given that AB and BC are adjacent sides of the hexagon, construct a point O which is equidistant from A and B and also equidistant from A and C. What is the point O?

37. Draw a triangle ABC in which $AC = 5$ cm, $CB = 7$ cm and $AB = 6$ cm. Mark D, the mid point of AB. Draw the circle which touches BC at C and passes through D. Find a point P on this circle such that the angle APB is a right angle.

38. Two circles with centres P and Q intersect at A and B. The radii of the circles are 4 cm and 5 cm and the distance between their centres is 7 cm. Construct a common tangent to the circles and verify that AB produced bisects this common tangent.

39. Draw a parallelogram $ABCD$ in which $AB = 3.6$ cm, $AD = 2.7$ cm and the angle $A = 50°$. Construct an equiangular parallelogram of the same area with one side 3 cm long.

40. Show how to bisect the area of a triangle by a straight line drawn through a given point of one of its sides.

41. A point P is 4 cm from a given line AB. Draw two circles each of radius 3 cm to touch AB and to pass through P.

42. Two parallel lines are cut by a transversal. Show how to draw a circle to touch all three lines.

43. In a circle of radius 4 cm inscribe a rectangle one of whose sides is 6 cm long.

44. Draw a circle of radius 5 cm. Construct two tangents to the circle which include an angle of 80°.

45. Draw a triangle ABC with sides 6 cm, 7 cm and 8 cm long. Find a point P inside the triangle at which AB, BC and CA all subtend equal angles.

46. Draw a circle of radius 5 cm. From a point 8 cm from its centre, construct the two tangents to the circle. Construct a circle to touch these tangents and the given circle.

47. Draw a circle of radius 6 cm and two radii which include an angle of 80°. Construct a circle to touch these two radii and the given circle.

48. Construct a right-angled triangle given that the length of the hypotenuse is 10 cm and that the altitude to the hypotenuse is 4.8 cm long.

49. Draw a regular pentagon of side 4 cm and construct a square of equal area.

50. Draw a circle of radius 6 cm. Construct a chord AB which is 8 cm long. Find a point P on the minor arc AB such that $AP : PB = 4 : 3$.

SECTION B

Construction 22

To reflect a point P in a straight line l.

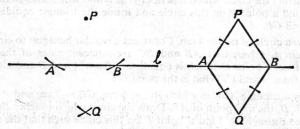

Fig. 31.22

Draw an arc of a circle, radius r, centre P, to cut l in A and B. With centres A, B draw arcs, radii r, to cut at Q. Then Q is the image of P reflected in l.

Proof by showing $\triangle PAB \equiv \triangle QAB$, or by observing that the figure is symmetrical about l.

Construction 23

To reflect a straight line m in a given straight line l, if the point of intersection of m and l is accessible.

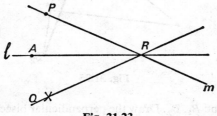

Fig. 31.23

Choose any one point P in m; construct the image of P when reflected in l, by construction 22. If l and m meet in R, join QR. This is the image of m reflected in l.

Proof by showing $\triangle$s PRA, QRA congruent, or by observing that the figure is symmetrical about l.

Construction 24

To reflect a straight line m in a given straight line l, if the point of intersection of l and m is not accessible.

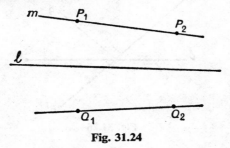

Fig. 31.24

Choose two points P_1 and P_2 in m. Reflect P_1 and P_2 in l to obtain Q_1 and Q_2. (Construction 22.) Draw the straight line through Q_1 and Q_2. This is the image of m reflected in l.

Proof as in Constructions 22, 23.

Construction 25

To find the centre of a rotation under which a line-segment A_1B_1 has been rotated to a position A_2B_2.

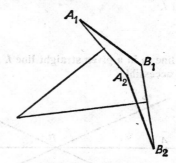

Fig. 31.25

Join A_1, A_2 and B_1, B_2. Draw the perpendicular bisectors of A_1A_2 and B_1B_2. These meet at O, the centre of the rotation.

Proof: Since A_1 has been rotated to A_2, A_1A_2 is the chord of a circle, whose centre lies on the perpendicular bisector of the chord. Similarly the centre of the rotation lies on the perpendicular bisector of B_1B_2.

Construction 26

To draw the shortest path from a point A to a point B, going through some one point on a straight line l not through A or B.

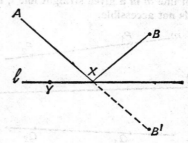

Fig. 31.26

Reflect B in l (Construction 22) to obtain B'. Join A to B'. If AB' meets l in X, the route AXB is the route required.

Proof: Take any other point Y in l. Then $AY + YB' > AB'$, so that $AY + YB > AX + XB$.

Construction 27

To enlarge a triangle by a factor *k*.

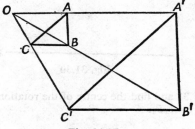

Fig. 31.27

Choose any point *O* not near any of the straight lines *AB*, *BC*, *CA* when produced. Join *OA*, *OB*, *OC*. Produce *OA* and construct the point *A'* in *OA* produced so that $OA' = k\,OA$ (Construction 8 needed if *k* is not an integer). Construct *B'*, *C'* similarly. Then *A'B'C'* is the required triangle.

Proof: by similar triangles, theorem 11, page 285 or by vectors.

Exercise 31.2

Tracing paper required for Q. 1–7.

1. Trace the Fig. 31.28, and construct the image of *P* when reflected in *l*.

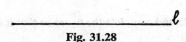

Fig. 31.28

2. Trace Fig. 31.29, and construct the image of *m* when reflected in *l*.

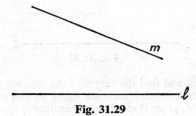

Fig. 31.29

3. Trace Fig. 31.30 and construct the image of *m* when reflected in *l*.

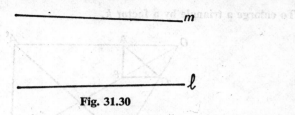

Fig. 31.30

4. Trace Fig. 31.31 and find the centre of the rotation under which *AB* has been rotated to A_1B_1.

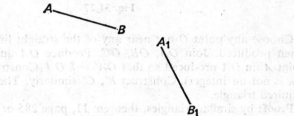

Fig. 31.31

5. Trace Fig. 31.32 and complete the figure so that it is symmetrical about *l*.

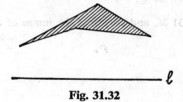

Fig. 31.32

6. Trace Fig. 31.33 and complete the figure so that it is symmetrical about *l*.

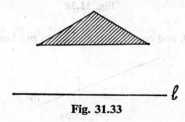

Fig. 31.33

7. Trace Fig. 31.34 and find the centre of the rotation under which *AB* has been rotated to A_1B_1. Complete the figure so that it has rotational symmetry of order 4, given that *AB* is perpendicular to A_1B_1.

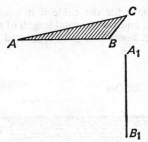

Fig. 31.34

8. A cowboy C is 5 km due East of a long straight river (Fig. 31.35, not to scale). He wishes to water his horse at the river and then return to his log cabin L. At present he is 6 km North and 2 km West of L. Make a scale drawing showing the shortest route he can take, and find the length of this shortest route.

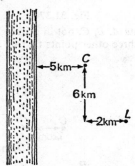

Fig. 31.35

9. Houses P, Q are 80 m and 100 m respectively from a long straight road (Fig. 31.36). A watermain is to be laid to each house from the main water supply, which runs along the road. The water supply is to be breached only at one point X. Make a scale drawing showing the position of X if the sum of the lengths PX and QX is to be as small as possible.

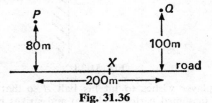

Fig. 31.36

10. Electricity substations A and B are to be connected by cable, which must cross a straight canalized river 100 m wide, as in Fig. 31.37. A is 300 m from the river, B 500m; A is 900 m N, 600 m W of B. Make a scale drawing

showing the best path for the cable if it must cross the river at right angles to the banks, and yet the total length of cable from A to B is to be as short as possible. Show that your suggested route is the shortest.

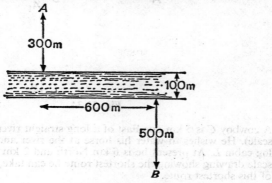

Fig. 31.37

11. Mark any three points A, B, C, not in a straight line. Use the symmetry of the circle to find three other points that lie on the circle, without first drawing the circle.

12.

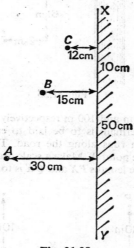

Fig. 31.38

A snooker player wishes to hit his ball A so that it bounces off the cushion XY (assumed perfectly elastic) and strikes ball C without first hitting ball B. Make a scale drawing of the data from Fig. 31.38, and mark the point P in XY towards which the snooker player should hit ball A.

32 Symmetry

The symmetry of an object may explain why we find it beautiful; the symmetry in an engineering structure will often reduce the stresses in the structure. In nature, the manner of life of the creature will determine its symmetry, so that an animal that moves will usually have symmetry about a central *plane* whereas animals that are fixed and wait for their food to arrive, like sea-urchins or sea-anemones, usually have symmetry about a central *axis*.

Symmetry about an axis

We first consider shapes in a plane. It is easy to test if they are symmetrical about the axis in their plane, for if folded about that axis, one half should be superimposed on the other. Fig. 32.1 can be traced

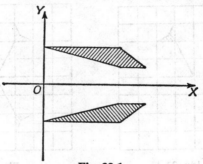

Fig. 32.1

and its symmetry about the x-axis confirmed by folding. If we need a more theoretical proof of its symmetry, we can see that each part has a corresponding part the same distance from the axis of symmetry.

Example 1. Complete Fig. 32.2 so that it is symmetrical about the line ----------------

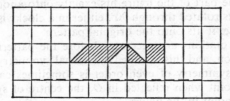

Fig. 32.2

We can fold the figure along the hatched line ----- and trace out the pattern required, or we can complete the figure by eye, so that each part above the line is matched by a corresponding part below the line -------

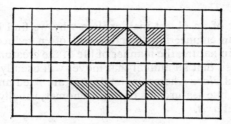

Fig. 32.3

Example 2. Complete Fig. 32.4 so that it is symmetrical about each of the lines ----------

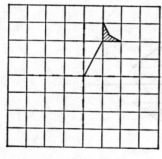 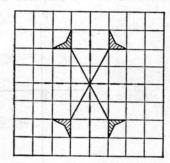

Fig. 32.4   **Fig. 32.5**

Proceeding as in Example 1, we have Fig. 32.5.

Rotational symmetry

Some plane figures are not symmetrical about a straight line in their plane, but are such that if rotated about an axis perpendicular to their plane, then one part of the figure fits exactly onto another (Fig. 32.6). If the pattern is rotated through 90°, either in a clockwise or in an anti-clockwise sense, it fits onto the original pattern.

Four such rotations are necessary before the pattern returns to its original position, and this is called rotational symmetry of order four.

Rotational symmetry of even order is sometimes called **point symmetry**; each point when reflected in *O*, the centre of symmetry, maps onto another point in the figure.

Notice that Fig. 32.6 does not have an axis of symmetry in the plane

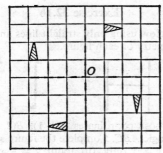

Fig. 32.6

of the figure, whereas Fig. 32.5, which has two perpendicular axes of symmetry, also has rotational symmetry of order two.

Exercise 32.1a

1. Copy the letters below, using straight lines only, and mark on your paper the axes of symmetry in the plane of the paper.

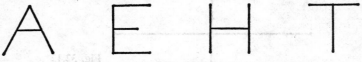

Fig. 32.7

2. Copy and complete Fig. 32.8 so that the figure has
   (i) only the hatched line ---- as an axis of symmetry,
   (ii) both the hatched ---- and the dotted ...... lines as axes of symmetry.

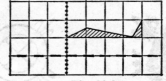
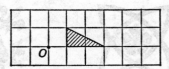

Fig. 32.8   **Fig. 32.9**

3. Plot the points $A(2, 1)$, $B(3, 2)$, $C(4, 5)$, $D(4, 1)$. Join $ABCD$. Complete the figure so that it is symmetrical (i) about Ox only, (ii) about Oy only, (iii) about both Ox and Oy.

4. Copy and complete Fig. 32.9 so that it has rotational symmetry about O of order three.

5. What is the next symbol in the sequence?

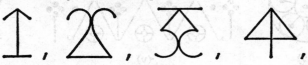

Fig. 32.10

Exercise 32.1b

1. Copy the letters below using only straight lines and say what is the order of rotational symmetry of each figure.

Fig. 32.11

2. Plot the points whose coordinates $A(1, 1)$, $B(2, 1)$, $C(3, 2)$. Shade the interior of the triangle ABC. Complete the figure so that it is symmetrical (i) about the line $y = x$ only, (ii) about $y = -x$ only, (iii) about both $y = x$ and $y = -x$.

3. Draw, side 4 cm, (i) a square, (ii) a regular hexagon, (iii) a regular octagon. Mark the axes of symmetry of each figure. What is the order of rotational symmetry that each figure has?

4. What is the least number of line-segments that must be added to Fig. 32.12 to give it one and only one axis of symmetry? In how many ways can this be done?

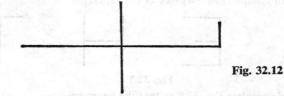

Fig. 32.12

5. The patterns below were among those used by the architect William Butterfield to ornament the Chapel at Rugby School. Describe the symmetry of each.

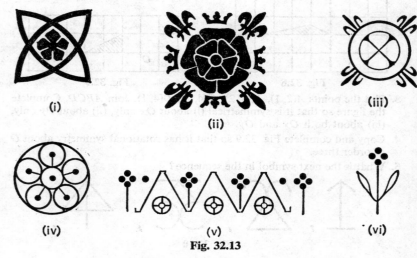

(i)

(ii)

(iii)

(iv)

(v)

(vi)

Fig. 32.13

Use of symmetry in solving geometrical problems

The symmetry of a figure often enables us to solve geometrical problems, for lines which cross an axis of symmetry, unless they have reflections in the other half of the figure, must cross at right angles; line-segments must be bisected by an axis of symmetry, and a line meets its own image on an axis of symmetry.

The isosceles triangle

If the equal sides of an isosceles triangle are AB and AC, then the internal bisector of angle BAC is the axis of symmetry. Thus BC must be bisected by the axis of symmetry, and BC is perpendicular to the axis of symmetry, the internal bisector of the angle BAC is the perpendicular bisector of BC. Since the figure is symmetrical about the axis AD, the 'base angles' at B and C are equal.

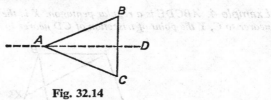

Fig. 32.14

The circle

The circle is particularly interesting because it is symmetrical about *every* diameter, so has an infinite number of axes of symmetry. The diameter through the midpoint of any chord is perpendicular to that chord (Fig. 32.15(a)) and the tangent at any point is perpendicular to the diameter through that point (Fig. 32.15(b)).

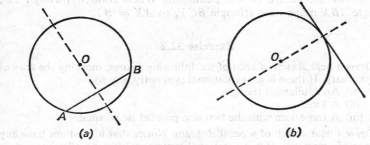

(a)   (b)

Fig. 32.15

Example 3. The tangents from a point P to a circle centre O touch the circle at A and B. A line perpendicular to OP cuts the circle at X and Y. Prove that AX and BY meet on OP.

The figure is symmetrical about the hatched line *l* before the addition of the line *XY*. The line *XY* is perpendicular to *l*, as *XY* is *any* line perpendicular to *l*, and so is symmetrical about *l*, therefore the completed figure is symmetrical about *l*, and *AX* and *BY* meet on *l*.

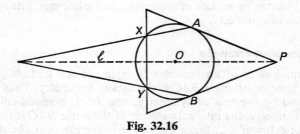

Fig. 32.16

Example 4. *ABCDE is a regular pentagon. X is the point of trisection of BC nearer to C, Y the point of trisection of CD nearer to D. Prove that AX = BY.*

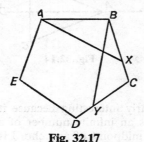

Fig. 32.17

The regular pentagon *ABCDE* has rotational symmetry (of order five) about the centre of the pentagon. When rotated through 72°, triangle *ABX* maps onto triangle *BCY*, so *AX = BY*.

Exercise 32.2

1. Draw a neat sketch of each of the following figures, marking the axes of symmetry. If there is also rotational symmetry, say so.
   (i) An equilateral triangle.
   (ii) A kite.
   (iii) A trapezium with the two non-parallel sides equal.

2. Draw a neat sketch of a parallelogram. Notice that it need not have any axes of symmetry. Has it rotational symmetry? If a parallelogram does have an axis of symmetry, what kind of quadrilateral is it?

3. In Fig. 32.18, *ABY* and *CBX* are straight lines meeting at *B*. *AB = BC* and *BX = BY*. Prove that *AC* is parallel to *XY*.

4. In Fig. 32.19, *ABCD* is a square and *CDE* an isosceles triangle. The

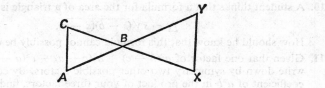

Fig. 32.18

perpendicular bisectors of *DE* and *CE* meet at *X*; the perpendicular bisectors of *BE* and *AE* meet at *Y*. Prove that *EXY* is a straight line.

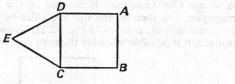

Fig. 32.19

5. In Fig. 32.20, *ABCD* and *DEFG* are squares, *CDE* an equilateral triangle. Prove *AE* = *DF*.

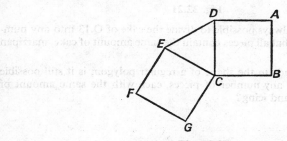

Fig. 32.20

6. In the isosceles triangle *ABC*, *AB* = *AC*. *P* is the point of trisection of *AB* nearer to *A*, *Q* the point of trisection of *AC* nearer to *A*. Prove
$$\widehat{BCP} = \widehat{CBQ}.$$

7. *ABC* is an isosceles triangle. Squares *ABXY*, *CBLM* are drawn outside the triangle *ABC*. Prove that *XL* and *YM* are both parallel to *AC*.

8. *ABCD* is a square. Equilateral triangles *DCE*, *BCF* are drawn outside the square. Prove that *AC*, *DF* and *BE* concur.

9. It can be proved that, with the usual notation,

$$\sin (\tfrac{1}{2}A) = \sqrt{\frac{(s-b)(s-c)}{bc}}$$

By symmetry, write down the corresponding formulae for sin ($\tfrac{1}{2}B$) and sin ($\tfrac{1}{2}C$).

10. A student thinks that a formula for the area of a triangle is

$$\triangle = s\sqrt{(s - b)(s - c)}.$$

How should he know that this formula cannot possibly be correct?

11. Given that one factor of $a^2(b - c) + b^2(c - a) + c^2(a - b)$ is $(b - c)$, write down by symmetry two other possible factors. By considering the coefficient of a^2b in the product of your three factors, find whether they are correct.

12. Given that $(a + b + c)$ is a factor of $a^3 + b^3 + c^3 - 3abc$, show by symmetry that there cannot be any other real linear factors of that expression.

13. A square Christmas cake is covered on the top by a uniform layer of marzipan, and then on the top and sides by uniform layers of icing. A cut is made in the cake, as in Fig. 32.21. Show that wherever this first cut is made, it is possible to divide the cake into four identical pieces.

Fig. 32.21

14. Show that it is always possible to divide the cake of Q.13 into any number of pieces so that all pieces contain the same amount of cake, marzipan and icing.

15. If the cake is made in the shape of a regular polygon, is it still possible to divide it into any number of pieces, each with the same amount of cake, marzipan and icing?

33 Matrices

In the same way that numbers can be used to give information, the price of a garment, the score in a test, etc., so a matrix can be used to give information. Suppose that a garage owner has three garages and sells only the cars of one manufacturer, which come in four types, 1 litre, 2 litre, 3 litre saloons and an estate car. Then he can see clearly the number he has in stock, and where the cars are, if he displays the information in a matrix in this manner:

$$
\begin{array}{c}
\text{High St Garage} \\
\text{Park Rd Garage} \\
\text{Central Garage}
\end{array}
\begin{array}{cccc}
\text{1 litre} & \text{2 litre} & \text{3 litre} & \text{Estate} \\
\begin{pmatrix} 3 & 1 & 0 & 2 \\ 1 & 2 & 3 & 3 \\ 1 & 3 & 0 & 1 \end{pmatrix}
\end{array}
$$

Sales one week might be as displayed by this matrix

$$
\begin{pmatrix} 2 & 0 & 0 & 1 \\ 1 & 1 & 2 & 0 \\ 1 & 2 & 0 & 0 \end{pmatrix}
$$

and to find the stock at the end of that week, if there were no deliveries of new cars, we have to subtract the corresponding entries, and the stock is given by

$$
\begin{pmatrix} 1 & 1 & 0 & 1 \\ 0 & 1 & 1 & 3 \\ 0 & 1 & 0 & 1 \end{pmatrix}
$$

Notice that to subtract (or to add) matrices, the matrices must have the same number of rows and columns. Even though the High St Garage did not sell any 2 litre saloons, we had to have a matrix with a zero in the second column of the first row. Likewise if we want the matrix to give all the information possible we must have a rectangular lay-out.

Definition

A matrix can be defined as a rectangular array of numbers, subject to certain laws of combination. Addition and subtraction we have defined: that we add or subtract corresponding elements. Since addition is associative and commutative over the set of real numbers, matrix addition is both associative and commutative over the set of all matrices whose entries are real numbers.

The order of a matrix

A matrix with m rows and n columns is called an m by n matrix (sometimes written $m \times n$). The matrices we have met so far have been 3 by 4 matrices.

Zero matrix, identity matrix

A matrix with every element zero, e.g. $\begin{pmatrix} 0 & 0 \\ 0 & 0 \end{pmatrix}$ is called a zero matrix, $\mathbf{0}$ (of order two, if it has two rows and two columns). Clearly when added to any other matrix of the same order, that matrix is unaltered,

i.e. $$\mathbf{A} + \mathbf{0} = \mathbf{A} \text{ for all } \mathbf{A}.$$

A matrix which has a 1 in every element of the leading diagonal, and 0 elsewhere is called an identity (unit) matrix $\mathbf{I}$, of appropriate order. All identity matrices are square matrices; the identity matrix of order 2 is

$$\begin{pmatrix} 1 & 0 \\ 0 & 1 \end{pmatrix}$$

Clearly there is a similarity between the zero matrix $\mathbf{0}$ and the number 0, between the identity matrix $\mathbf{I}$ and the number 1.

Example 1. If

$$\mathbf{A} = \begin{pmatrix} 2 & 3 \\ 1 & 2 \end{pmatrix} \text{ and } \mathbf{B} = \begin{pmatrix} 0 & -1 \\ 2 & 3 \end{pmatrix}$$

find (i) $\mathbf{A} + \mathbf{B}$, (ii) $\mathbf{A} - \mathbf{B}$, (iii) $\mathbf{A} + \mathbf{A}$ *(usually written 2A)*, (iv) $5\mathbf{B}$.

(i) Adding corresponding elements,

$$\mathbf{A} + \mathbf{B} = \begin{pmatrix} 2 & 2 \\ 3 & 5 \end{pmatrix}$$

(ii) Subtracting corresponding elements,

$$\mathbf{A} - \mathbf{B} = \begin{pmatrix} 2 & 4 \\ -1 & -1 \end{pmatrix}$$

(iii) Adding corresponding elements,

$$2\mathbf{A} = \mathbf{A} + \mathbf{A} = \begin{pmatrix} 4 & 6 \\ 2 & 4 \end{pmatrix}$$

(iv) Notice that in (iii) we could have multiplied each element by 2. To find $5\mathbf{B}$, we multiply each element by 5.

$$5\mathbf{B} = \begin{pmatrix} 0 & -5 \\ 10 & 15 \end{pmatrix}$$

Exercise 33.1a

1. Given that $A = (2 \quad 3)$, $B = (2 \quad 1 \quad 0)$

$$C = \begin{pmatrix} 2 & 1 \\ 2 & 3 \end{pmatrix} \quad D = \begin{pmatrix} 1 & 0 \\ 2 & 1 \end{pmatrix} \quad E = \begin{pmatrix} 1 & 1 & 1 \\ 2 & 3 & 4 \end{pmatrix}$$

find which pairs of matrices can be added together.

2. Given that

$$A = \begin{pmatrix} 1 & 2 \\ 3 & 0 \end{pmatrix}, \quad B = \begin{pmatrix} 2 & -3 \\ 1 & -1 \end{pmatrix}, \quad C = \begin{pmatrix} 3 & -1 \\ 4 & 3 \end{pmatrix}$$

find   (i) $A + B$,   (ii) $A + C$,   (iii) $A - B$,
       (iv) $B - A$,   (v) $A - C$,   (vi) $3A$,
       (vii) $2A + B$,   (viii) $3B + 2C$,   (ix) $4B - 3C$.

3. A housewife makes the following purchases during one week: Monday, 3 pints of milk, 1 loaf of bread. Tuesday, 3 pints of milk. Wednesday, 2 pints of milk, 1 loaf. Thursday, 2 loaves. Friday, 3 pints of milk. Saturday, 2 pints of milk, 3 loaves. Sunday, 3 pints of milk. Arrange her purchases in a matrix with two rows and seven columns.

4.

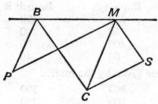

Fig. 33.1

Fig. 33.1 represents the roads in part of a small town, showing those linking the Post Office (P), Bus Station (B), Market Place (M), School (S) and Cinema (C). Copy and complete the matrix below to show whether there are roads linking each pair of these places. Put 0 in each of the gaps in the leading diagonal, to show there is no road from one place back to itself.

$$\begin{array}{c} \\ P \\ B \\ M \\ S \\ C \end{array} \begin{array}{ccccc} P & B & M & S & C \\ \begin{pmatrix} 0 & 1 & 1 & 0 & \\ 1 & 0 & 1 & & \\ 1 & 1 & 0 & & \\ & & & & \\ 0 & & & & \end{pmatrix} \end{array}$$

The entry 1 in the row containing B and the column under P shows that there is a road from B to P. Since all the roads apparently carried two-way traffic, the matrix was symmetrical, as there were as many roads from any one point X to any other point Y as there were from Y to X. This matrix is called a route matrix.

5. With the data of Q.4, the road from M to S is made one-way, so that it is only possible to go from M to S, and the road from B to C is one-way, only allowing passage from B to C. Give the correct route matrix now.

6. Mason's Christmas hampers come in three sizes, Executive, Director, and Cabinet Minister. The executive hamper contains: 1 bottle of port, 1 tin of ham, a box of chocolates, but no whisky; the Director contains: 2 bottles of port, 2 tins of ham and a bottle of whisky, and the Cabinet Minister contains: 2 tins of ham, 1 box of chocolates and 3 bottles of whisky. Display the contents of these hampers in a matrix with four rows and three columns.

7. A firm makes three different breakfast cereals, Krunchie, Sweetie and Bixs. Krunchie contains 3 units of Niacin per 100 g, 0.3 units of Ribo-flavin per 100 g, and 0.2 units of Thiamine per 100 g; Sweetie contains 4 units of Niacin, 0.5 of Riboflavine and 0.1 units of Thiamine, and Bixs contains 2 units of Niacin, 1 of Riboflavine and 0.1 units of Thiamine, all per 100 g. Display this information in a 3×3 matrix, in which the columns give the quantities of any one vitamin in each 100 g of cereal.

8. The stocks of certain items at branches of a supermarket are given by the following matrix:

	Branch A	Branch B	Branch C
Sugar (kg)	350	400	500
Tea (kg)	250	300	400
Butter (kg)	400	80	90
Milk (litres)	60	40	50

The sales one day are given by the matrix

300	300	450
100	250	350
300	70	20
50	30	40

Write down a matrix showing the stock in each branch at the end of that day.

Exercise 33.1b

1. Given that

$$A = \begin{pmatrix} 1 & 2 \\ 2 & 3 \end{pmatrix}, \qquad B = \begin{pmatrix} 1 & 0 & 0 \\ 3 & 1 & 0 \end{pmatrix}, \qquad C = \begin{pmatrix} 1 & 2 & 0 \\ 2 & 3 & 0 \end{pmatrix},$$

$$D = \begin{pmatrix} 1 & 2 & 0 & 0 \\ 2 & 3 & 0 & 0 \end{pmatrix}, \qquad E = \begin{pmatrix} 0 & 1 & 2 \\ 0 & 2 & 3 \end{pmatrix}, \qquad F = \begin{pmatrix} 0 & 0 & 1 & 2 \\ 0 & 0 & 2 & 3 \end{pmatrix}.$$

which of these matrices is it possible to add together?

2. Using the matrices in Q.1, find, where possible,

(i) **B + E,**   (ii) **E − C,**   (iii) **D + F,**
(iv) **4A,**   (v) **3D − E,**   (vi) **F − 2D,**
(vii) **½A,**   (viii) **B − ½C,**   (ix) **B + C + E.**

3. A small country garage sells petrol in four grades, one star *, two star **, three star *** and four star ****. The sales during one week are: Monday, **, 20 litres, ****, 10 litres; Tuesday, *, 30 litres, **, 20 litres, ***, 10 litres, ****, 25 litres; Wednesday, ***, 100 litres; Thursday, **, 20 litres, ***, 80 litres, ****, 60 litres; Friday, *, 30 litres, **, 40 litres, ***, 60 litres, ****, 80 litres; Saturday, ***, 100 litres. Display these sales in a matrix with four rows and six columns.

4. On Friday, October 10th, 1975, the results of the top five teams in the First Division of the Association Football League that season was given by the matrix:

	Played	Won	Drawn	Lost
Q.P.R.	11	5	5	1
Man. Utd	11	6	3	2
West Ham	10	6	3	1
Derby Co.	11	6	2	3
Liverpool	10	5	3	2

The results the following day included: Q.P.R. beat Everton, Man. Utd beat Leeds, West Ham beat Newcastle, Liverpool beat Birmingham and Derby drew with Norwich. Write down a 5 × 4 matrix showing the results for these five teams. By the addition of the two matrices, form a matrix showing the results for these five teams at the end of Saturday, October 11th.

5. Find x, y, z, w if

$$\begin{pmatrix} 2 & 4 \\ 3 & z \end{pmatrix} + \begin{pmatrix} x & y \\ 3 & 4 \end{pmatrix} = \begin{pmatrix} 4 & 4 \\ w & 0 \end{pmatrix}$$

6. The midday dinners one week in a certain school were analysed as below:

	Meat (10 g)	Fish (10 g)	Potatoes (10 g)	Other Vegetables (10 g)
Monday	5	0	10	10
Tuesday	0	6	12	8
Wednesday	6	0	10	10
Thursday	4	0	15	12
Friday	0	7	10	10

On Monday 90 pupils stayed to dinner, on Tuesday 80, on Wednesday 100, on Thursday 90, and on Friday 80. Write down a 5 × 4 matrix showing the quantity of each food eaten on each day of that week.

Multiplication of matrices

We have seen that to multiply a matrix by a number we multiply every element of the matrix by that number. Can we find a way of multiplying a matrix by another matrix, in such a manner that the product conveys some useful information?

Suppose that the price of each type of car in the garages (page 345) is: 1 litre saloon, £2000, 2 litre saloon, £3000, 3 litre saloon, £4000, estate, £3500. Then the value of the cars sold that week by the Park Rd

Garage was £$(1 \times 2000 + 1 \times 3000 + 2 \times 4000 + 1 \times 3500)$. Similarly for the other two garages. If we write the prices in a matrix

$$\begin{pmatrix} 2000 \\ 3000 \\ 4000 \\ 3500 \end{pmatrix}$$

we can define matrix multiplication so that we obtain the value of the cars sold that week by each garage, and the value of the stock at the beginning and end of that week. To multiply two matrices, we make each row of the first matrix 'dive down' each column of the second matrix, and we sum the products of the corresponding elements, so that to find the value of the stock at the beginning of that week we need

	first row	second row	third row

$$\begin{pmatrix} 3 & 1 & 0 & 2 \\ 1 & 2 & 3 & 3 \\ 1 & 3 & 0 & 1 \end{pmatrix} \begin{pmatrix} 2000 \\ 3000 \\ 4000 \\ 3500 \end{pmatrix} = \begin{pmatrix} \begin{array}{c|c} \begin{matrix} 2000 \\ 3000 \\ 4000 \\ 3500 \end{matrix} \begin{matrix} 3 \\ 1 \\ 0 \\ 2 \end{matrix} & \begin{matrix} 2000 \\ 3000 \\ 4000 \\ 3500 \end{matrix} \begin{matrix} 1 \\ 2 \\ 3 \\ 3 \end{matrix} \end{array} & \begin{matrix} 2000 \\ 3000 \\ 4000 \\ 3500 \end{matrix} \begin{matrix} 1 \\ 3 \\ 0 \\ 1 \end{matrix} \end{pmatrix}$$

$$= \begin{pmatrix} 6000 + 3000 + 0 + 7000 \\ 2000 + 6000 + 12000 + 10500 \\ 2000 + 9000 + 0 + 3500 \end{pmatrix} \begin{matrix} \text{first row} \\ \text{second row} \\ \text{third row} \end{matrix}$$

$$= \begin{pmatrix} 16\,000 \\ 30\,500 \\ 14\,500 \end{pmatrix}.$$

Fig. 33.2

Notice that the sum of the products when the first row 'dives down' we put in the first row of the product matrix. Similarly for the other rows.

If we want to find how many of each type of car we possess, we can do that using matrix multiplication. For when we multiply by $(1 \quad 1 \quad 1)$ we add the entries in each column, thus

$$(1 \quad 1 \quad 1) \begin{pmatrix} 3 & 1 & 0 & 2 \\ 1 & 2 & 3 & 3 \\ 1 & 3 & 0 & 1 \end{pmatrix} = (5 \quad 6 \quad 3 \quad 6)$$

We notice that the product of a 3 by 4 matrix with a 4 by 1 matrix is a 3 by 1 matrix, and the product of a 1 by 3 matrix and a 3 by 4 matrix is a 1 by 4 matrix. We must, of course, have as many columns in the first matrix as we have rows in the second matrix.

Example 2. *Which matrix products can be formed from pairs of the following matrices?*

$$A = \begin{pmatrix} 2 & 3 \\ 1 & 4 \end{pmatrix}, \quad B = \begin{pmatrix} 4 & 1 \\ 3 & -2 \end{pmatrix}, \quad C = \begin{pmatrix} 4 & 1 & 0 \\ 5 & 2 & 7 \end{pmatrix},$$

$$D = (3 \quad 1 \quad 4), \quad\quad\quad E = \begin{pmatrix} 5 \\ 4 \\ 1 \end{pmatrix}.$$

Since **A** has the same number of columns as **B** has rows, we can find the product **A.B**.

$$\mathbf{A.B} = \begin{pmatrix} 2 & 3 \\ 1 & 4 \end{pmatrix}\begin{pmatrix} 4 & 1 \\ 3 & -2 \end{pmatrix} = \begin{pmatrix} \boxed{2 \quad 3} \\ 1 \quad 4 \end{pmatrix}\begin{pmatrix} 4 & \boxed{2} \\ 3 & \boxed{3} & -2 \end{pmatrix}$$

$$= \begin{pmatrix} 2 \times 4 + 3 \times 3 & 2 \times 1 + 3 \times (-2) \\ 1 \times 4 + 4 \times 3 & 1 \times 1 + 4 \times (-2) \end{pmatrix}$$

$$= \begin{pmatrix} 17 & -4 \\ 16 & -7 \end{pmatrix}.$$

But **B** also has the same number of columns as **A** has rows, so that we can also find the product **B.A**.

$$\mathbf{B.A} = \begin{pmatrix} 4 & 1 \\ 3 & -2 \end{pmatrix}\begin{pmatrix} 2 & 3 \\ 1 & 4 \end{pmatrix} = \begin{pmatrix} 4 & 1 \\ 3 & -2 \end{pmatrix}\begin{pmatrix} 2 & \boxed{4} & 3 \\ 1 & \boxed{1} & 4 \end{pmatrix}$$

$$= \begin{pmatrix} 4 \times 2 + 1 \times 1 & 4 \times 3 + 1 \times 4 \\ 3 \times 2 + (-2) \times 1 & 3 \times 3 + (-2) \times 4 \end{pmatrix}$$

$$= \begin{pmatrix} 9 & 16 \\ 4 & 1 \end{pmatrix}.$$

We see that **A.B**. does not equal **B.A**, that is, matrix multiplication is not commutative. To emphasise the difference, we sometimes say **A** pre-multiplies **B** when we want the product **A.B**.

Looking at other possible products, **A** and **B** have two columns, so they can pre-multiply any matrix which has two rows so that the products **A.C** and **B.C** exist. Similarly the products **E.D**, **D.E** can be formed, and also the product **C.E**. In the last case,

$$\mathbf{C.E} = \begin{pmatrix} 4 & 1 & 0 \\ 5 & 2 & 7 \end{pmatrix}\begin{pmatrix} 5 \\ 4 \\ 1 \end{pmatrix}$$

$$= \begin{pmatrix} 24 \\ 40 \end{pmatrix}, \text{ a } 2 \times 1 \text{ matrix.}$$

We cannot find any other pairs of matrices from which products can be formed, so that the only products of two matrices that can be formed are **A.B**, **B.A**, **A.C**, **B.C**, **E.D**, **D.E** and **C.E**.

Square matrices

Since both **A** and **B** have the same number of rows as columns, they are called square matrices, and we can find the products **A.A** and **B.B**, which we write as A^2 and B^2. Thus

$$A^2 = A.A = \begin{pmatrix} 2 & 3 \\ 1 & 4 \end{pmatrix} \begin{pmatrix} 2 & 3 \\ 1 & 4 \end{pmatrix} = \begin{pmatrix} 7 & 18 \\ 6 & 19 \end{pmatrix}$$

and $\qquad B^2 = B.B = \begin{pmatrix} 4 & 1 \\ 3 & -2 \end{pmatrix} \begin{pmatrix} 4 & 1 \\ 3 & -2 \end{pmatrix} = \begin{pmatrix} 19 & 2 \\ 6 & 7 \end{pmatrix}$

Similarly $\quad A^3 = A.A^2$ (which is equal to $A^2.A$.)

$$= \begin{pmatrix} 2 & 3 \\ 1 & 4 \end{pmatrix} \begin{pmatrix} 7 & 18 \\ 6 & 19 \end{pmatrix} = \begin{pmatrix} 32 & 93 \\ 31 & 94 \end{pmatrix}$$

Higher powers of square matrices are formed in the same way.

Associative Law

It has been shown that matrix multiplication is not necessarily commutative, since we have found one product for thich $A.B \neq B.A$. It can however be shown that matrix multiplication is associative, and we need to use this property as we develop an algebra of matrices.

Exercise 33.2.

1. Which matrix products can be formed from pairs of the following matrices?

$A = \begin{pmatrix} 2 \\ 3 \end{pmatrix}, \qquad B = (3 \quad 4), \qquad C = \begin{pmatrix} 2 & 1 \\ 5 & 2 \end{pmatrix},$

$D = \begin{pmatrix} 2 & 3 & 0 \\ 1 & 0 & 0 \end{pmatrix}, \qquad E = (2 \quad 3 \quad 0).$

2. Given that

$A = \begin{pmatrix} 1 & 0 \\ 2 & 1 \end{pmatrix}, \qquad B = \begin{pmatrix} 1 & 2 \\ 3 & 4 \end{pmatrix}$ and $C = \begin{pmatrix} 4 & -1 \\ 0 & 2 \end{pmatrix}$

   find the products **A.B, B.C, A.C, C.A**.
3. Using the matrices in Q.2 find A^2, B^2, C^2 and A^3.
4. Given that $A = (1 \quad 2 \quad 3)$, $B = (2 \quad 3)$,

$C = \begin{pmatrix} 1 \\ 0 \\ -1 \end{pmatrix}, \qquad D = \begin{pmatrix} 3 \\ 4 \end{pmatrix}$ and $E = \begin{pmatrix} 1 & 0 \\ 4 & 2 \end{pmatrix}$

   form the products **A.C, B.D, E.D, B.E**.
5. With the data in Ex. 33.1a, Q.3, if milk costs 8p a pint and bread 15p a loaf, form a 1 × 2 cost matrix **C**. Premultiply a certain matrix by **C** to find the cost of each day's purchases.
   Form a 7 × 1 matrix **D**, so that when **D** is premultiplied by a certain matrix it gives a matrix (consisting only in one element) which shows the total cost of the week's purchases of milk and bread.

6. Using the data of Ex. 33.1a, Q.6, if port costs £2 a bottle, ham £1 a tin, chocolates £1.50 a box and whisky £4 a bottle, use a 1 by 4 cost matrix to find the cost of each of Mason's Christmas hampers.

A firm decides to send as presents to its largest customers a total of 5 Executive hampers, 8 Director hampers and 3 Cabinet Minister hampers. Display this information in a 3 by 1 matrix and so find the total cost to the firm of sending Christmas hampers.

7. Use the information in Ex. 33.1a, Q.8. If sugar is worth 12p a kg, tea 80p a kg, butter 70p a kg, and milk 20p a litre, find a 1 by 4 matrix which when multiplying the given matrix shows the value in £ of goods in each branch of the supermarket.

8. Use the information in Ex. 33.1b, Q.3. If * petrol sells at 20p a litre, ** at 21p, *** at 22p and **** at 23p, write down a 1 by 4 matrix displaying these prices. By matrix multiplication form a 1 by 6 matrix showing the total receipts in £ at the garage each day.

9. Use the information in Ex. 33.1b, Q.4. If each team gains 2 points for a win, 1 point for a draw and no points when they lose, write down a 4 by 1 matrix which when multiplied by the given matrix gives a 5 by 1 matrix showing how many points each team had on the Friday. By matrix multiplication, find how many points they had at the end of Saturday.

10. Use the data in Ex. 33.1b, Q.6. If meat costs 2p for 10 g, fish 3p, potatoes $\frac{1}{2}$p and other vegetables 1p, write down a 4 by 1 matrix which when multiplied by the given matrix shows the cost in pence of one dinner on each day of the week.

Transpose of a matrix

If the rows and columns of a matrix **A** are interchanged, the resulting matrix **A'** (sometimes **A**^{T}) is called the transpose of **A**. Thus the transpose of $\begin{pmatrix} 2 & 4 \\ 1 & 5 \end{pmatrix}$ is $\begin{pmatrix} 2 & 1 \\ 4 & 5 \end{pmatrix}$

and the transpose of $\begin{pmatrix} 2 & 4 & 3 \\ 1 & 5 & 0 \end{pmatrix}$ is $\begin{pmatrix} 2 & 1 \\ 4 & 5 \\ 3 & 0 \end{pmatrix}$

Example 3. If $\mathbf{A} = \begin{pmatrix} 1 & 3 \\ 4 & 2 \end{pmatrix}$ *and* $\mathbf{B} = \begin{pmatrix} 5 & 1 \\ -1 & 0 \end{pmatrix}$ *find* **A', B', A.B, A'.B'** *and* **B.A.**

From the definitions of the transpose, $\mathbf{A'} = \begin{pmatrix} 1 & 4 \\ 3 & 2 \end{pmatrix}$, $\mathbf{B'} = \begin{pmatrix} 5 & -1 \\ 1 & 0 \end{pmatrix}$

Also $\qquad \mathbf{A.B} = \begin{pmatrix} 1 & 3 \\ 4 & 2 \end{pmatrix}\begin{pmatrix} 5 & 1 \\ -1 & 0 \end{pmatrix} = \begin{pmatrix} 2 & 1 \\ 18 & 4 \end{pmatrix}$

and $\qquad \mathbf{A'.B'} = \begin{pmatrix} 1 & 4 \\ 3 & 2 \end{pmatrix}\begin{pmatrix} 5 & -1 \\ 1 & 0 \end{pmatrix} = \begin{pmatrix} 9 & -1 \\ 17 & -3 \end{pmatrix}$

and $\qquad \mathbf{B.A} = \begin{pmatrix} 5 & 1 \\ -1 & 0 \end{pmatrix}\begin{pmatrix} 1 & 3 \\ 4 & 2 \end{pmatrix} = \begin{pmatrix} 9 & 17 \\ -1 & -3 \end{pmatrix}$

Notice that $(\mathbf{B}.\mathbf{A})' = \mathbf{A}'.\mathbf{B}'$. Similarly we can verify that $(\mathbf{A}.\mathbf{B})' = \mathbf{B}'.\mathbf{A}'$. We have proved these results for only one pair of matrices, but they can be shown to be true for all matrices for which the appropriate products exist.

The inverse matrix

To solve the equation $3x = 2$, we multiply both sides by $\frac{1}{3}$, since $\frac{1}{3} \times 3 = 1$. To solve the matrix equation $\mathbf{A}.\mathbf{x} = \mathbf{B}$, we need to find the matrix, written $\mathbf{A}^{-1}$, such that $\mathbf{A}^{-1}.\mathbf{A} = \mathbf{I}$, $\mathbf{I}$ being the appropriate identity matrix, so that

$$\mathbf{A}.\mathbf{x} = \mathbf{B}$$
$$\Rightarrow \quad \mathbf{A}^{-1}.(\mathbf{A}.\mathbf{x}) = \mathbf{A}^{-1}.\mathbf{B}$$
i.e. $$\mathbf{x} = \mathbf{A}^{-1}.\mathbf{B},$$

using the associative property of matrix multiplication. We shall see that there are many other uses for the inverse matrix.

If the 2 by 2 matrix is $\begin{pmatrix} 2 & 1 \\ 5 & 3 \end{pmatrix}$, then the inverse matrix will be such

that $\begin{pmatrix} x & y \\ z & w \end{pmatrix} \begin{pmatrix} 2 & 1 \\ 5 & 3 \end{pmatrix} = \begin{pmatrix} 1 & 0 \\ 0 & 1 \end{pmatrix}$

i.e. $\begin{pmatrix} 2x + 5y & x + 3y \\ 2z + 5w & z + 3w \end{pmatrix} = \begin{pmatrix} 1 & 0 \\ 0 & 1 \end{pmatrix}$

Equating corresponding elements we have, $2x + 5y = 1$ and $x + 3y = 0$, giving $x = 3$ and $y = -1$, and $2z + 5w = 0$ and $z + 3w = 1$,

giving $z = -5$ and $w = 2$. Thus the inverse of $\begin{pmatrix} 2 & 1 \\ 5 & 3 \end{pmatrix}$ is $\begin{pmatrix} 3 & -1 \\ -5 & 2 \end{pmatrix}$

We notice also that the order in which we multiply a matrix and its inverse does not matter, for

$\begin{pmatrix} 2 & 1 \\ 5 & 3 \end{pmatrix} \begin{pmatrix} 3 & -1 \\ -5 & 2 \end{pmatrix}$ is also equal to $\begin{pmatrix} 1 & 0 \\ 0 & 1 \end{pmatrix}$.

The method of finding the inverse of a 2×2 matrix then *may* be to interchange the terms in the leading diagonal (here 2 and 3) and to change the signs of the other terms. But if we apply that method to

the matrix $\begin{pmatrix} 4 & 3 \\ 1 & 2 \end{pmatrix}$, we see that

$$\begin{pmatrix} 2 & -3 \\ -1 & 4 \end{pmatrix} \begin{pmatrix} 4 & 3 \\ 1 & 2 \end{pmatrix} = \begin{pmatrix} 5 & 0 \\ 0 & 5 \end{pmatrix}$$

so that some modification of the method is necessary.

To find the inverse of a general matrix $\mathbf{A}$, $\begin{pmatrix} a & b \\ c & d \end{pmatrix}$, let the inverse be $\begin{pmatrix} x & y \\ z & w \end{pmatrix}$ as before. Then

$$\begin{pmatrix} x & y \\ z & w \end{pmatrix} \begin{pmatrix} a & b \\ c & d \end{pmatrix} = \begin{pmatrix} 1 & 0 \\ 0 & 1 \end{pmatrix}$$

so that

$$\begin{pmatrix} xa + yc & xb + yd \\ za + wc & zb + wd \end{pmatrix} = \begin{pmatrix} 1 & 0 \\ 0 & 1 \end{pmatrix}$$

i.e.
$$xa + yc = 1 \qquad xb + yd = 0$$
$$za + wc = 0 \quad \text{and} \quad zb + wd = 1.$$

Solving these equations in pairs for x, y, z and w, we have

$$x = \frac{d}{ad - bc}, \ y = \frac{-b}{ad - bc}, \ z = \frac{-c}{ad - bc} \text{ and } w = \frac{a}{ad - bc}$$

so that the inverse of the matrix **A** is

$$\begin{pmatrix} \dfrac{d}{ad - bc} & \dfrac{-b}{ad - bc} \\ \dfrac{-c}{ad - bc} & \dfrac{a}{ad - bc} \end{pmatrix}$$

which we can write as

$$\mathbf{A}^{-1} = \frac{1}{(ad - bc)} \begin{pmatrix} d & -b \\ -c & a \end{pmatrix}$$

Determinant of a matrix

The expression $ad - bc$ occurs in several contexts, and is called the **determinant** of the 2×2 matrix **A**. Only square matrices have determinants, and we shall only be concerned with the determinants of 2×2 matrices; some of their geometrical applications are given in Chapter 34.

Singular matrix

In solving the simultaneous equations we have assumed that $ad - bc \neq 0$. If $ad = bc$, then we cannot find an inverse matrix, and the matrix is called a singular matrix. In geometry that means that there is no inverse transformation, and we shall see the significance of that later.

Example 4. *Find the inverse of the matrix $A \begin{pmatrix} 4 & 2 \\ 5 & 1 \end{pmatrix}$ and show that $A^{-1}A = I$* and $AA^{-1} = I$.

Using the method above, $ad - bc = 4 \times 1 - 2 \times 5 = -6$,

so
$$A^{-1} = -\frac{1}{6} \begin{pmatrix} 1 & -2 \\ -5 & 4 \end{pmatrix}$$

We see that $-\dfrac{1}{6}\begin{pmatrix} 1 & -2 \\ -5 & 4 \end{pmatrix}\begin{pmatrix} 4 & 2 \\ 5 & 1 \end{pmatrix} = \begin{pmatrix} 1 & 0 \\ 0 & 1 \end{pmatrix}$

and $\quad -\dfrac{1}{6}\begin{pmatrix} 4 & 2 \\ 5 & 1 \end{pmatrix}\begin{pmatrix} 1 & -2 \\ -5 & 4 \end{pmatrix} = \begin{pmatrix} 1 & 0 \\ 0 & 1 \end{pmatrix}$

Matrix method for solving simultaneous equations

Consider the equations $\qquad 4x + 5y = 2$
$$2x + 3y = 0$$

These can be written $\begin{pmatrix} 4 & 5 \\ 2 & 3 \end{pmatrix}\begin{pmatrix} x \\ y \end{pmatrix} = \begin{pmatrix} 2 \\ 0 \end{pmatrix}$

Multiplying by the inverse,

$$\dfrac{1}{2}\begin{pmatrix} 3 & -5 \\ -2 & 4 \end{pmatrix}\begin{pmatrix} 4 & 5 \\ 2 & 3 \end{pmatrix}\begin{pmatrix} x \\ y \end{pmatrix} = \dfrac{1}{2}\begin{pmatrix} 3 & -5 \\ -2 & 4 \end{pmatrix}\begin{pmatrix} 2 \\ 0 \end{pmatrix}$$

i.e. $\qquad\qquad\qquad \begin{pmatrix} x \\ y \end{pmatrix} = \begin{pmatrix} 3 \\ -2 \end{pmatrix}$

so the solution of the equations is $x = 3$, $y = -2$.

Unless we have many pairs of simultaneous equations, the left-hand sides of which can be represented by the same matrix **A**, it is difficult to pretend that the solution of two simultaneous equations is easier by this method than by either of the methods in Chapter 11.

Example 5. Solve the following pairs of simultaneous equations:
  (i) $4x + 2y = 1$, $x + 3y = 0$
  (ii) $4x + 2y = 2$, $x + 3y = -1$
  (iii) $4x + 2y = 3$, $x + 3y = -2$

The coefficients of x and y are given by the matrix $\begin{pmatrix} 4 & 2 \\ 1 & 3 \end{pmatrix}$ whose inverse

is $\dfrac{1}{10}\begin{pmatrix} 3 & -2 \\ -1 & 4 \end{pmatrix}$

The solution of (i) may be written

$$\begin{pmatrix} 4 & 2 \\ 1 & 3 \end{pmatrix}\begin{pmatrix} x \\ y \end{pmatrix} = \begin{pmatrix} 1 \\ 0 \end{pmatrix}$$

$$\Rightarrow \dfrac{1}{10}\begin{pmatrix} 3 & -2 \\ -1 & 4 \end{pmatrix}\begin{pmatrix} 4 & 2 \\ 1 & 3 \end{pmatrix}\begin{pmatrix} x \\ y \end{pmatrix} = \dfrac{1}{10}\begin{pmatrix} 3 & -2 \\ -1 & 4 \end{pmatrix}\begin{pmatrix} 1 \\ 0 \end{pmatrix}$$

$$\Rightarrow \qquad\qquad\qquad \begin{pmatrix} x \\ y \end{pmatrix} = \begin{pmatrix} 0.3 \\ -0.1 \end{pmatrix}$$

so that $x = 0.3$ and $y = -0.1$ is the solution of equations (i). Similarly the
solutions of (ii) are given by $\begin{pmatrix} x \\ y \end{pmatrix} = \dfrac{1}{10}\begin{pmatrix} -3 & -2 \\ -1 & 4 \end{pmatrix}\begin{pmatrix} 2 \\ -1 \end{pmatrix}$

i.e.
$$x = 0.8 \quad \text{and} \quad y = -0.6$$

and the solutions of (iii) by $\begin{pmatrix} x \\ y \end{pmatrix} = \frac{1}{10} \begin{pmatrix} 3 & -2 \\ -1 & 4 \end{pmatrix} \begin{pmatrix} 3 \\ -2 \end{pmatrix}$

i.e.
$$x = 1.3 \quad \text{and} \quad y = -1.1.$$

Exercise 33.3

1. Find the inverse of each of the following matrices:

$$\begin{pmatrix} 3 & 1 \\ 2 & 1 \end{pmatrix}, \quad \begin{pmatrix} 1 & 3 \\ 1 & 5 \end{pmatrix}, \quad \begin{pmatrix} 4 & -2 \\ 1 & 3 \end{pmatrix}, \quad \begin{pmatrix} 1 & 0 \\ 0 & 1 \end{pmatrix}.$$

2. Which of the following matrices has (have) not an inverse?

(i) $\begin{pmatrix} 2 & 1 \\ 1 & 2 \end{pmatrix}$ (ii) $\begin{pmatrix} 2 & 1 \\ 4 & 2 \end{pmatrix}$ (iii) $\begin{pmatrix} 1 & 0 \\ 0 & 0 \end{pmatrix}$ (iv) $\begin{pmatrix} 1 & 1 \\ 0 & 0 \end{pmatrix}$ (v) $\begin{pmatrix} 2 & -2 \\ 4 & -4 \end{pmatrix}$.

3. Find the transpose of each of the following matrices:

(i) $\begin{pmatrix} 3 \\ 1 \end{pmatrix}$ (ii) $\begin{pmatrix} 3 & 2 \\ 1 & 4 \end{pmatrix}$ (iii) $(3 \quad 2)$ (iv) $\begin{pmatrix} 3 & 2 & 1 \\ 4 & 5 & 6 \end{pmatrix}$.

4. Which of the following matrices is (are) singular?

(i) $\begin{pmatrix} 3 & 2 \\ 2 & 1 \end{pmatrix}$ (ii) $\begin{pmatrix} 2 & 3 \\ 4 & -6 \end{pmatrix}$ (iii) $\begin{pmatrix} 3 & 2 \\ 6 & 4 \end{pmatrix}$ (iv) $\begin{pmatrix} 2 & 3 \\ 0 & 0 \end{pmatrix}$.

5. Use a matrix method to solve the following pairs of simultaneous equations:

(i) $3x + y = 1, \quad x + 2y = 2,$
(ii) $3x + y = 0, \quad x + 2y = 1,$
(iii) $3x + y = -1, \quad x + 2y = -1.$

6. Try to solve the following simultaneous equations
(i) using matrices,
(ii) without using matrices.

$$x + 2y = 1, \quad x + 2y = 2.$$

Exercise 33.4: Miscellaneous

1. Solve for x and y:

(i) $\begin{pmatrix} x + 2 \\ y - 4 \end{pmatrix} = \begin{pmatrix} 3 \\ -1 \end{pmatrix}$ (ii) $\begin{pmatrix} 2x \\ 3y \end{pmatrix} = \begin{pmatrix} 10 \\ 12 \end{pmatrix}$

(iii) $\begin{pmatrix} x & 3x \\ y & 4y \end{pmatrix} \begin{pmatrix} 2 \\ 1 \end{pmatrix} = \begin{pmatrix} 5 \\ 12 \end{pmatrix}$ (iv) $\begin{pmatrix} 3 & 1 \\ 1 & 2 \end{pmatrix} \begin{pmatrix} 2 \\ 1 \end{pmatrix} = \begin{pmatrix} x \\ y \end{pmatrix}$

(v) $\begin{pmatrix} 3 & 4 \\ 5 & 6 \end{pmatrix} \begin{pmatrix} 2 \\ -1 \end{pmatrix} = \begin{pmatrix} x \\ y \end{pmatrix}$ (vi) $\begin{pmatrix} 3 & 4 \\ 5 & 6 \end{pmatrix} \begin{pmatrix} x \\ y \end{pmatrix} = \begin{pmatrix} -5 \\ -7 \end{pmatrix}$

(vii) $\begin{pmatrix} 0 & 3 \\ 4 & 2 \end{pmatrix} \begin{pmatrix} x \\ y \end{pmatrix} = \begin{pmatrix} 6 \\ 2 \end{pmatrix}$ (viii) $\begin{pmatrix} 1 & 1 \\ 3 & y \end{pmatrix} \begin{pmatrix} x \\ 1 \end{pmatrix} = \begin{pmatrix} 4 \\ 1 \end{pmatrix}$

2. Given that $A = \begin{pmatrix} 3 & 0 \\ 0 & 2 \end{pmatrix}$, find A^2, A^3, A^{-1}.

Suggest a likely matrix for A^{10}.

3. Given that $A = \begin{pmatrix} 6 & 4 \\ 3 & 10 \end{pmatrix}$ and $B = \begin{pmatrix} 2 & -2 \\ 3 & 1 \end{pmatrix}$

find $A.B$, B^{-1}, $B^{-1}.A.B$ and $(B^{-1}.A.B)^2$.

Use $(B^{-1}.A.B)^2 = (B^{-1}.A.B)(B^{-1}.A.B)$

$\qquad\qquad\qquad\quad = B^{-1}.A^2.B$

to find A^2. Find $(B^{-1}.A.B)^4$ and hence find A^4.

4. Given that $F = \begin{pmatrix} 1 & 1 \\ 1 & 0 \end{pmatrix}$, find f_1 and f_2 when $F\begin{pmatrix} 1 \\ 0 \end{pmatrix} = \begin{pmatrix} f_2 \\ f_1 \end{pmatrix}$.

If $F\begin{pmatrix} f_2 \\ f_1 \end{pmatrix} = \begin{pmatrix} f_3 \\ f_2 \end{pmatrix}$, find f_3.

If $F\begin{pmatrix} f_3 \\ f_2 \end{pmatrix} = \begin{pmatrix} f_4 \\ f_3 \end{pmatrix}$ and $F\begin{pmatrix} f_4 \\ f_3 \end{pmatrix} = \begin{pmatrix} f_5 \\ f_4 \end{pmatrix}$, find f_4, f_5.

f_1, f_2, f_3, f_4, and f_5 are the first five terms in a well-known series. Find the

relation between f_{r+1}, f_r and f_{r-1} if $F\begin{pmatrix} f_r \\ f_{r-1} \end{pmatrix} = \begin{pmatrix} f_{r+1} \\ f_r \end{pmatrix}$ to show that the

terms in $F^n\begin{pmatrix} 1 \\ 0 \end{pmatrix}$ will always be terms in this series, for all integral values
of n.

5. In a certain colony of hamsters, x are in their first year of life, y in their second year and z in their third year. The life-cycle of these hamsters can be described by the matrix A, where

$$A = \begin{pmatrix} 0 & 3 & 2 \\ \tfrac{1}{2} & 0 & 0 \\ 0 & \tfrac{1}{4} & 0 \end{pmatrix}, \text{ for if } X = \begin{pmatrix} x \\ y \\ z \end{pmatrix}$$

then $A.X$ gives the number of each age group the following year, $A^2.X$ the number the year after that, and so on. The non-zero entries in the matrix A vary with different diets and differing predators. If the colony has initially 128 first year hamsters, 64 second year hamsters and 16 third year hamsters, use the product

$$\begin{pmatrix} 0 & 3 & 2 \\ \tfrac{1}{2} & 0 & 0 \\ 0 & \tfrac{1}{4} & 0 \end{pmatrix}\begin{pmatrix} 128 \\ 64 \\ 16 \end{pmatrix}$$

and so find how many of each age group there are after
(i) one year, (ii) two years, (iii) three years, (iv) four years.
Under different conditions the life cycle is described by

$$B = \begin{pmatrix} 0 & 3 & 2 \\ \tfrac{1}{4} & 0 & 0 \\ 0 & \tfrac{1}{2} & 0 \end{pmatrix}$$

Form the products $B.X$, $B^2.X$, $B^3.X$ and $B^4.X$ and compare the number of each age group each year with those in the other colony. What do these figures indicate will eventually happen in each colony?

6. Make a wild guess at the square root of 2. If your guess is 23, i.e. 23/1,

write it as $S = \begin{pmatrix} 23 \\ 1 \end{pmatrix}$.

If $A = \begin{pmatrix} 1 & 2 \\ 1 & 1 \end{pmatrix}$, $A.S = \begin{pmatrix} 25 \\ 24 \end{pmatrix}$ and $25/24 = 1.041 \ldots$

Form the products $A^2.S$, $A^3.S$, etc., and find the ratio of the two terms in each product, as above. Hence find $\sqrt{2}$, correct to as many decimal places as convenient. Note that it does not matter what number you 'guess', even zero will do. It merely makes the arithmetic longer.

Can you find a matrix which will enable you to calculate $\sqrt{3}$ to as many decimal places as convenient? Investigate this, to see if it appears to give $\sqrt{3}$. Can you find matrices for $\sqrt{4}$, $\sqrt{5}$?

7. Suppose that owing to 'flu' the inter-house hockey competition had to be cancelled when only seven matches in one pool had been played. Can we find an order of merit among these houses, if Cotton has beaten Whitelaw and Michell, Whitelaw has beaten School Field and Michell, School Field has beaten Cotton and School House, and Michell has beaten School House?

Since Cotton has beaten Whitelaw, who in turn have beaten School Field, Cotton is said to have 'two-stage dominance' over School Field, even though they lost to School Field. Clearly 'two-stage dominance' is not as strong as an outright win. 'Three-stage dominance' is defined in a similar way. The matrix **M** has 1's in the first row to record victories

$$
\mathbf{M} = \begin{array}{c} \\ C \\ W \\ SF \\ M \\ SH \end{array}
\begin{array}{c} \begin{array}{ccccc} C & W & SF & M & SH \end{array} \\
\begin{pmatrix} 0 & 1 & 0 & 1 & 0 \\ 0 & 0 & 1 & 1 & 0 \\ 1 & 0 & 0 & 0 & 1 \\ 0 & 0 & 0 & 0 & 1 \\ 0 & 0 & 0 & 0 & 0 \end{pmatrix} \end{array}
\qquad
\mathbf{N} = \begin{pmatrix} 2 \\ 2 \\ 2 \\ 1 \\ 0 \end{pmatrix}
$$

of C over W and M, etc., and the matrix **N** records the number of wins by each house. The product **M.N** gives the number of two-stage dominances, **M²·N** the number of three-stage dominances, etc. Form the products **M.N**, **M²·N**, **M³·N** and so find an order of merit for the houses.

If an additional match is played in which School House beats Cotton, write down new matrices **M**, **N** to display the results of the eight matches. Form products **M.N**, **M²·N**, etc. to establish an order of merit. Compare it with the previous order of merit.

8.

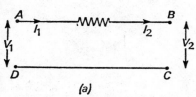

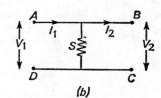

(a)   *(b)*

Fig. 33.3

Fig. 33.3 shows part of an electrical circuit, with four terminals A, B, C, D. The p.d. across D to A is V_1, from C to B is V_2. The current leaving A is I_1; the current arriving at B is I_2. In Fig. 33.3(a) there is a wire resistance R ohms between A and B; clearly $I_1 = I_2$ and Ohm's law gives $V_1 = V_2 + RI_2$. In Fig. 33.3(b), if there is a wire resistance S ohms linking AB and CD, then $V_1 = V_2$ and $V_2 = S(I_1 - I_2)$ by Ohm's law. Thus for Fig. 33.3(a) we can write

$$\begin{pmatrix} V_1 \\ I_1 \end{pmatrix} = \begin{pmatrix} 1 & R \\ 0 & 1 \end{pmatrix} \begin{pmatrix} V_2 \\ I_2 \end{pmatrix}$$

and for Fig. 33.3(b) $\quad \begin{pmatrix} V_1 \\ I_1 \end{pmatrix} = \begin{pmatrix} 1 & 0 \\ 1/S & 1 \end{pmatrix} \begin{pmatrix} V_2 \\ I_2 \end{pmatrix}$

Use matrices to find V_1 and I_1 in terms of V_2 and I_2 for the following:

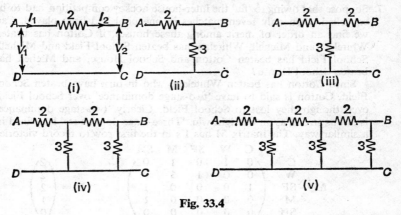

Fig. 33.4

Use inverse matrices to find V_2 and I_2 in terms of V_1 and I_1 in each case.

34 Geometrical Applications of Matrices

A matrix is a rectangular array of numbers which can be used to give information. The coordinates in two dimensions of a point can be represented by a matrix, so that the coordinates (2, 3) of a point can be represented by a matrix $(2 \quad 3)$ or by a matrix $\begin{pmatrix} 2 \\ 3 \end{pmatrix}$. The components of a vector can be represented by a matrix (see page 371). We can even write the coordinates of several points in a single matrix, so that the points coordinates $O(0,0)$, $P(1,0)$, $Q(1,1)$ and $R(0,1)$, the vertices of the unit square, can be displayed in the matrix $\begin{pmatrix} 0 & 1 & 1 & 0 \\ 0 & 0 & 1 & 1 \end{pmatrix}$ and the points $O(0,0)$, $P(2,0)$, $Q(2,1)$ and $R(0,1)$ can be displayed in the matrix $\begin{pmatrix} 0 & 2 & 2 & 0 \\ 0 & 0 & 1 & 1 \end{pmatrix}$. Since these points are the vertices of a rectangle, call this matrix **R**.

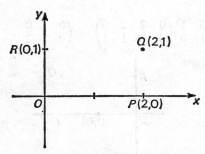

Fig. 34.1

When the matrix **R** is multiplied by the 2×2 unit matrix, it is unaltered,

$$\begin{pmatrix} 1 & 0 \\ 0 & 1 \end{pmatrix}\begin{pmatrix} 0 & 2 & 2 & 0 \\ 0 & 0 & 1 & 1 \end{pmatrix} = \begin{pmatrix} 0 & 2 & 2 & 0 \\ 0 & 0 & 1 & 1 \end{pmatrix}$$

Reflection

When **R** is multiplied by $\begin{pmatrix} 1 & 0 \\ 0 & -1 \end{pmatrix}$,

$$\begin{pmatrix} 1 & 0 \\ 0 & -1 \end{pmatrix}\begin{pmatrix} 0 & 2 & 2 & 0 \\ 0 & 0 & 1 & 1 \end{pmatrix} = \begin{pmatrix} 0 & 2 & 2 & 0 \\ 0 & 0 & -1 & -1 \end{pmatrix}$$

the rectangle $OPQR$ has been reflected in the x-axis. The rectangle

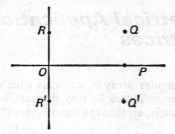

Fig. 34.2

$OPQ'R'$ is called the **image** of $OPQR$ under the transformation described by the **matrix** $\begin{pmatrix} 1 & 0 \\ 0 & -1 \end{pmatrix}$, or more simply, the image under the transformation $\begin{pmatrix} 1 & 0 \\ 0 & -1 \end{pmatrix}$.

Rotation

Similarly $\begin{pmatrix} 0 & 1 \\ -1 & 0 \end{pmatrix}\begin{pmatrix} 0 & 2 & 2 & 0 \\ 0 & 0 & 1 & 1 \end{pmatrix} = \begin{pmatrix} 0 & 0 & 1 & 1 \\ 0 & -2 & -2 & 0 \end{pmatrix}$ showing

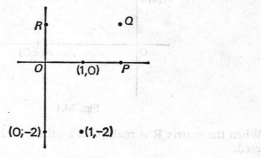

Fig. 34.3

that $OPQR$ has been rotated about the origin through 90° in clockwise sense, and we see that the matrix $\begin{pmatrix} 0 & 1 \\ -1 & 0 \end{pmatrix}$ describes a rotation through 90° in a clockwise sense. The matrix $\begin{pmatrix} 0 & -1 \\ 1 & 0 \end{pmatrix}$ describes a rotation about the origin through 90° in an anticlockwise sense.

Enlargement

If we consider the effect of the matrix $\begin{pmatrix} 3 & 0 \\ 0 & 3 \end{pmatrix}$ on the unit square,

$$\begin{pmatrix} 3 & 0 \\ 0 & 3 \end{pmatrix}\begin{pmatrix} 0 & 1 & 1 & 0 \\ 0 & 0 & 1 & 1 \end{pmatrix} = \begin{pmatrix} 0 & 3 & 3 & 0 \\ 0 & 0 & 3 & 3 \end{pmatrix}$$

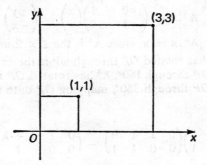

Fig. 34.4

we see that each side has been enlarged by a factor 3, and the area has been increased by 9, the figure remaining a square. Thus we see that some of the simplest geometrical operations, reflection, rotation and enlargement can be described by matrices. We should, of course, show that these matrices will make the appropriate transformations on ANY figure, not just on the unit square and a rectangle (see Exercise 34). Other transformations are shown in the following examples.

Example 1. Given that $A = \begin{pmatrix} 0 & -1 \\ 1 & 0 \end{pmatrix}$, *find* A^2, A^3 *and* A^4. *Investigate the transformations described by these matrices.*

$$A^2 = \begin{pmatrix} 0 & -1 \\ 1 & 0 \end{pmatrix}\begin{pmatrix} 0 & -1 \\ 1 & 0 \end{pmatrix} = \begin{pmatrix} -1 & 0 \\ 0 & -1 \end{pmatrix}$$

$$A^3 = A.A^2 = \begin{pmatrix} 0 & -1 \\ 1 & 0 \end{pmatrix}\begin{pmatrix} -1 & 0 \\ 0 & -1 \end{pmatrix}$$

$$= \begin{pmatrix} 0 & 1 \\ -1 & 0 \end{pmatrix}$$

and
$$A^4 = A.A^3 = \begin{pmatrix} 0 & -1 \\ 1 & 0 \end{pmatrix}\begin{pmatrix} 0 & 1 \\ -1 & 0 \end{pmatrix}$$

$$= \begin{pmatrix} 1 & 0 \\ 0 & 1 \end{pmatrix}$$

The coordinates of any point P (x,y) can be written as a matrix $\begin{pmatrix} x \\ y \end{pmatrix}$, denoted by **x** and we see that

$$\mathbf{A} . \mathbf{x} = \begin{pmatrix} 0 & -1 \\ 1 & 0 \end{pmatrix}\begin{pmatrix} x \\ y \end{pmatrix} = \begin{pmatrix} -y \\ x \end{pmatrix}$$

$$\mathbf{A}^2 . \mathbf{x} = \begin{pmatrix} -1 & 0 \\ 0 & -1 \end{pmatrix}\begin{pmatrix} x \\ y \end{pmatrix} = \begin{pmatrix} -x \\ -y \end{pmatrix}$$

$$\mathbf{A}^3 . \mathbf{x} = \begin{pmatrix} 0 & 1 \\ -1 & 0 \end{pmatrix}\begin{pmatrix} x \\ y \end{pmatrix} = \begin{pmatrix} y \\ -x \end{pmatrix}$$

and $\mathbf{A}^4 . \mathbf{x} = \mathbf{x}$, since $\mathbf{A}^4$ is the 2×2 unit matrix.

The matrix **A** has rotated OP through 90° in the anticlockwise sense, $\mathbf{A}^2$ has rotated OP through 180°, $\mathbf{A}^3$ has rotated OP through 270° and $\mathbf{A}^4$ has rotated OP through 360°, mapping OP onto itself.

Shear

Since $$\begin{pmatrix} 1 & k \\ 0 & 1 \end{pmatrix}\begin{pmatrix} 0 & 1 & 1 & 0 \\ 0 & 0 & 1 & 1 \end{pmatrix} = \begin{pmatrix} 0 & 1 & 1+k & k \\ 0 & 0 & 1 & 1 \end{pmatrix}$$

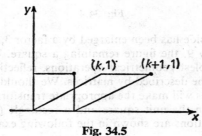

Fig. 34.5

the matrix $\begin{pmatrix} 1 & k \\ 0 & 1 \end{pmatrix}$ has mapped the unit square into a parallelogram as in Fig. 34.5. Such a transformation is called a shear parallel to the x-axis.

Finding the matrix to describe a transformation

So far we have considered the geometrical transformation produced by a given matrix. How do we proceed to find the matrix, if we know the geometrical transformation?

Supposing we wish to find the matrix which reflects in the line $y = x$. Then every point (x,y) must be transformed into the corresponding point (y,x). If the required matrix is $\begin{pmatrix} a & b \\ c & d \end{pmatrix}$ then

$$\begin{pmatrix} a & b \\ c & d \end{pmatrix}\begin{pmatrix} x \\ y \end{pmatrix} = \begin{pmatrix} y \\ x \end{pmatrix}$$

for all x,y

i.e.
$$ax + by \equiv y$$
$$cx + dy \equiv x.$$

Since this is an identity, the solution is $a = 0$, $b = 1$, $c = 1$ and $d = 0$, so the matrix which reflects in the line $y = x$ is

$$\begin{pmatrix} 0 & 1 \\ 1 & 0 \end{pmatrix}$$

Example 2. *Find the matrix which maps the unit square into the parallelogram*

$$P \begin{pmatrix} 0 & 3 & 4 & 1 \\ 0 & 1 & 3 & 2 \end{pmatrix}$$ *and the matrix which maps P back into the unit square.*

Let the required matrix be
$$\begin{pmatrix} a & b \\ c & d \end{pmatrix}$$

$$\begin{pmatrix} a & b \\ c & d \end{pmatrix}\begin{pmatrix} 0 & 1 & 1 & 0 \\ 0 & 0 & 1 & 1 \end{pmatrix} = \begin{pmatrix} 0 & 3 & 4 & 1 \\ 0 & 1 & 3 & 2 \end{pmatrix}$$

so that
$$\begin{pmatrix} 0 & a & a+b & b \\ 0 & c & c+d & d \end{pmatrix} = \begin{pmatrix} 0 & 3 & 4 & 1 \\ 0 & 1 & 3 & 2 \end{pmatrix}$$

Two matrices are equal only if corresponding elements are equal, so that $a = 3$, $b = 1$, $c = 1$ and $d = 2$. Notice that we had six equations for four unknowns; we could only find solutions because the element 4 was the sum of the 3 and the 1 in the first row, and the 3 in the second row was the sum of the 2 and 1 on either side of it.

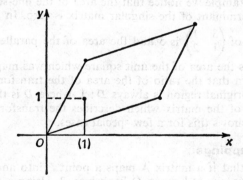

Fig. 34.6

The mapping of P back into the unit square is the inverse mapping, so will be described by the inverse matrix, i.e. $\dfrac{1}{5}\begin{pmatrix} 2 & -1 \\ -1 & 3 \end{pmatrix}$.

Verifying this, $\dfrac{1}{5}\begin{pmatrix} 2 & -1 \\ -1 & 3 \end{pmatrix}\begin{pmatrix} 0 & 3 & 4 & 1 \\ 0 & 1 & 3 & 2 \end{pmatrix} = \begin{pmatrix} 0 & 1 & 1 & 0 \\ 0 & 0 & 1 & 1 \end{pmatrix}$.

Mapping described by a singular matrix

Since
$$\begin{pmatrix} a & b \\ ka & kb \end{pmatrix}\begin{pmatrix} 0 & 1 & 1 & 0 \\ 0 & 0 & 1 & 1 \end{pmatrix} = \begin{pmatrix} 0 & a & a+b & b \\ 0 & ka & ka+kb & kb \end{pmatrix}$$

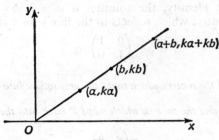

Fig. 34.7

the singular matrix $\begin{pmatrix} a & b \\ ka & kb \end{pmatrix}$ maps the vertices of the unit square into

the points (a,ka), (b,kb), $(a+b, ka+kb)$. These points all lie on the straight line $y = kx$, so that the unit square itself is mapped onto that segment of the line $y = kx$ that lies between $(0,0)$ and $(a+b, ka+kb)$, if a,b and k are all positive. What happens when some of these are negative?

Change of area

In the last example we notice that the area of the line-segment is zero, and the determinant of the singular matrix is zero. In Example 2 the

determinant of $\begin{pmatrix} 3 & 1 \\ 1 & 2 \end{pmatrix}$ is 5 and the area of the parallelogram P is 5,

i.e. five times the area of the unit square which was mapped into P. It can be shown that the ratio of the area of the transformed region to that of the original region is always $D:1$, where D is the value of the determinant of the matrix which describes the transformation. Q.5 in Exercise 34 shows this for a few special cases.

Inverse mappings

We can see that if a matrix A maps a point P into another point Q, the inverse matrix A^{-1} maps Q back into P. Using a column matrix

$\begin{pmatrix} x \\ y \end{pmatrix}$ to give the coordinates of P, $\begin{pmatrix} X \\ Y \end{pmatrix}$ to give the coordinates of Q and

denoting A by $\begin{pmatrix} a & b \\ c & d \end{pmatrix}$, we can write

$$\begin{pmatrix} a & b \\ c & d \end{pmatrix}\begin{pmatrix} x \\ y \end{pmatrix} = \begin{pmatrix} X \\ Y \end{pmatrix}$$

For example, $\begin{pmatrix} 4 & 2 \\ 1 & 3 \end{pmatrix}$ maps $P \begin{pmatrix} 1 \\ 1 \end{pmatrix}$ into $Q \begin{pmatrix} 6 \\ 4 \end{pmatrix}$ (Fig. 34.8)

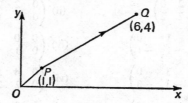

Fig. 34.8

and we can find the image of any point P under $\mathbf{A}$; using the inverse matrix we can find the point L which has been mapped into any given point M.

Suppose that $\mathbf{M}$ is given by the matrix $\begin{pmatrix} 6 \\ -1 \end{pmatrix}$. Then we wish to find

$L \begin{pmatrix} x \\ y \end{pmatrix}$ such that $\begin{pmatrix} 4 & 2 \\ 1 & 3 \end{pmatrix}\begin{pmatrix} x \\ y \end{pmatrix} = \begin{pmatrix} 6 \\ -1 \end{pmatrix}$.

Multiplying by the inverse of $\mathbf{A}$,

$$\frac{1}{10}\begin{pmatrix} 3 & -2 \\ -1 & 4 \end{pmatrix}\begin{pmatrix} 4 & 2 \\ 1 & 3 \end{pmatrix}\begin{pmatrix} x \\ y \end{pmatrix} = \frac{1}{10}\begin{pmatrix} 3 & -2 \\ -1 & 4 \end{pmatrix}\begin{pmatrix} 6 \\ -1 \end{pmatrix}$$

i.e. $\begin{pmatrix} 1 & 0 \\ 0 & 1 \end{pmatrix}\begin{pmatrix} x \\ y \end{pmatrix} = \frac{1}{10}\begin{pmatrix} 20 \\ -10 \end{pmatrix}, \begin{pmatrix} x \\ y \end{pmatrix} = \begin{pmatrix} 2 \\ -1 \end{pmatrix}$

the point whose coordinates are $(2,-1)$ was mapped by $\mathbf{A}$ into the point $(6,-1)$.

Exercise 34

1. Describe the transformation of each of the following matrices on the rectangle $\begin{pmatrix} 0 & 2 & 2 & 0 \\ 0 & 0 & 1 & 1 \end{pmatrix}$. Illustrate each by a sketch.

(i) $\begin{pmatrix} 1 & 0 \\ 0 & -1 \end{pmatrix}$　　(ii) $\begin{pmatrix} 2 & 0 \\ 0 & 2 \end{pmatrix}$　　(iii) $\begin{pmatrix} \frac{1}{2} & 0 \\ 0 & \frac{1}{2} \end{pmatrix}$

(iv) $\begin{pmatrix} 2 & 0 \\ 0 & -2 \end{pmatrix}$　　(v) $\begin{pmatrix} 0 & -1 \\ -1 & 0 \end{pmatrix}$　　(vi) $\begin{pmatrix} 0 & 1 \\ -1 & 0 \end{pmatrix}$

(vii) $\begin{pmatrix} 0 & 3 \\ 3 & 0 \end{pmatrix}$　　(viii) $\begin{pmatrix} 1 & 2 \\ 0 & 1 \end{pmatrix}$　　(ix) $\begin{pmatrix} 1 & \frac{1}{2} \\ 0 & 1 \end{pmatrix}$

(x) $\begin{pmatrix} 1 & 0 \\ 2 & 1 \end{pmatrix}$　　(xi) $\begin{pmatrix} 1 & -2 \\ 0 & 1 \end{pmatrix}$　　(xii) $\begin{pmatrix} 1 & 0 \\ -2 & 1 \end{pmatrix}$

2. Find the matrix which maps the unit square into each of the following and illustrate each mapping by a sketch:

(i) $\begin{pmatrix} 0 & 2 & 2 & 0 \\ 0 & 0 & 2 & 2 \end{pmatrix}$

(ii) $\begin{pmatrix} 0 & 2 & 2 & 0 \\ 0 & 0 & 1 & 1 \end{pmatrix}$

(iii) $\begin{pmatrix} 0 & 1 & 1 & 0 \\ 0 & 0 & -1 & -1 \end{pmatrix}$

(iv) $\begin{pmatrix} 0 & 0 & -1 & -1 \\ 0 & 1 & 1 & 0 \end{pmatrix}$

(v) $\begin{pmatrix} 0 & 3 & 3 & 0 \\ 0 & 0 & -3 & -3 \end{pmatrix}$

(vi) $\begin{pmatrix} 0 & 4 & 5 & 1 \\ 0 & 1 & 2 & 1 \end{pmatrix}$

(vii) $\begin{pmatrix} 0 & 1 & -1 & -2 \\ 0 & 0 & 1 & 1 \end{pmatrix}$

(viii) $\begin{pmatrix} 0 & 0 & 1 & 1 \\ 0 & \frac{1}{2} & 1\frac{1}{2} & 1 \end{pmatrix}$

3. Find the image of the unit square under each of the following mappings, and the matrix which describes the inverse mapping:

(i) $\begin{pmatrix} 4 & 0 \\ 0 & 4 \end{pmatrix}$

(ii) $\begin{pmatrix} \frac{1}{8} & 0 \\ 0 & \frac{1}{8} \end{pmatrix}$

(iii) $\begin{pmatrix} 0 & 1 \\ -1 & 0 \end{pmatrix}$

(iv) $\begin{pmatrix} -1 & 0 \\ 0 & -1 \end{pmatrix}$

(v) $\begin{pmatrix} 1 & 0 \\ 0 & -1 \end{pmatrix}$

(vi) $\begin{pmatrix} 0 & -1 \\ -1 & 0 \end{pmatrix}$

(vii) $\begin{pmatrix} 2 & 1 \\ -1 & -3 \end{pmatrix}$

(viii) $\begin{pmatrix} 3 & 0 \\ 0 & 2 \end{pmatrix}$

(ix) $\begin{pmatrix} 4 & 2 \\ 1 & 2 \end{pmatrix}$

(x) $\begin{pmatrix} 3 & -2 \\ 1 & 1 \end{pmatrix}$

(xi) $\begin{pmatrix} 2 & \frac{1}{2} \\ 4 & -1 \end{pmatrix}$

(xii) $\begin{pmatrix} -1 & -1 \\ 1 & -1 \end{pmatrix}$

4. Find the image of the unit square under each of the mappings described by the following matrices and illustrate each by a sketch.

(i) $\begin{pmatrix} 1 & 2 \\ 2 & 4 \end{pmatrix}$

(ii) $\begin{pmatrix} -1 & 2 \\ -2 & 4 \end{pmatrix}$

(iii) $\begin{pmatrix} 1 & -2 \\ 2 & -4 \end{pmatrix}$

(iv) $\begin{pmatrix} 1 & \frac{1}{2} \\ 2 & 1 \end{pmatrix}$

(v) $\begin{pmatrix} 1 & 2 \\ -1 & -2 \end{pmatrix}$

(vi) $\begin{pmatrix} 1 & 1 \\ 0 & 0 \end{pmatrix}$

5. Find the area of the region into which the interior of the unit square is mapped by each of the following matrices:

(i) $\begin{pmatrix} 4 & 2 \\ 5 & 3 \end{pmatrix}$

(ii) $\begin{pmatrix} 1 & 1 \\ 1 & 1 \end{pmatrix}$

(iii) $\begin{pmatrix} 3 & 5 \\ 1 & 2 \end{pmatrix}$

(iv) $\begin{pmatrix} 3 & -2 \\ 1 & 2 \end{pmatrix}$

(v) $\begin{pmatrix} 3 & 4 \\ 2 & 3 \end{pmatrix}$

(vi) $\begin{pmatrix} 6 & 2 \\ 2 & 1 \end{pmatrix}$

6. Find the image of the point $P\begin{pmatrix} x \\ y \end{pmatrix}$ under the transformation described by

the matrix $\begin{pmatrix} 2 & -1 \\ 1 & 2 \end{pmatrix}$

(i) when $\begin{pmatrix} x \\ y \end{pmatrix} = \begin{pmatrix} 1 \\ -1 \end{pmatrix}$ (ii) when $\begin{pmatrix} x \\ y \end{pmatrix} = \begin{pmatrix} 0 \\ 1 \end{pmatrix}$

and find which point P has been mapped into the point Q by the same transformation when Q is given by (iii) $\begin{pmatrix} 6 \\ 3 \end{pmatrix}$, (iv) $\begin{pmatrix} -3 \\ -4 \end{pmatrix}$.

7. Show the matrix $\begin{pmatrix} 3 & 2 \\ 1 & 1 \end{pmatrix}$ maps the unit square $\begin{pmatrix} 0 & 1 & 1 & 0 \\ 0 & 0 & 1 & 1 \end{pmatrix}$ into the

parallelogram $\begin{pmatrix} 0 & 3 & 5 & 2 \\ 0 & 1 & 2 & 1 \end{pmatrix}$. Find the figure which this matrix maps

into $\begin{pmatrix} 0 & 6 & 8 & 2 \\ 0 & 2 & 3 & 1 \end{pmatrix}$.

8. Find the matrix which maps the unit square $\begin{pmatrix} 0 & 1 & 1 & 0 \\ 0 & 0 & 1 & 1 \end{pmatrix}$ into the

parallelogram $\begin{pmatrix} 0 & 3 & 4 & 1 \\ 0 & 0 & 1 & 1 \end{pmatrix}$ and find the matrix which maps this

parallelogram back into the unit square.

9. Find the image of the unit square under the transformation described by

the matrix $A \begin{pmatrix} 3 & 1 \\ 0 & 3 \end{pmatrix}$ and the image of the unit square under the matrix

A^{-1}. Describe each transformation geometrically.

10. The vertices of a triangle T_1 are given by $\begin{pmatrix} 1 & 2 & 3 \\ 0 & 1 & 1 \end{pmatrix}$. The matrix

$R = \begin{pmatrix} 0 & 1 \\ -1 & 0 \end{pmatrix}$ maps T_1 into T_2, and the matrix $S = \begin{pmatrix} 1 & 3 \\ 0 & 1 \end{pmatrix}$ maps T_2

into T_3. Find the matrices that give the vertices of T_2 and T_3. Show that
the matrix $(S.R)$ maps T_1 into T_3 and the matrix $(R^{-1}.S^{-1})$ maps T_3
into T_1.

35 Vectors

If we walk 3 m in a straight line, then 4 m, again in a straight line, how far are we from our starting point? The answer of course depends on the direction in which we were walking in each of the two stages. To fix our displacements, we had to know each magnitude and direction.

Vector; definition

A vector can be defined as a physical quantity having magnitude and direction,* and combined according to certain laws. By contrast, a scalar quantity, like time, mass, temperature, needs only one value to determine it. Vector quantities we may have met in Physics include displacement, velocity, acceleration, force and momentum.

Displacement vectors

The simplest applications of vectors are to displacements, distances in particular directions. If we travel 3 m due East, then 4 m due North, we are 5 m from our starting point; if, instead, we travel 3 m due East then 4 m due West, we are only 1 m from our starting point, and in a different direction to that in the first example.

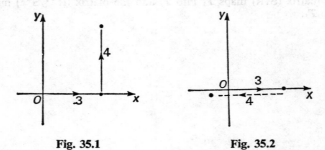

Fig. 35.1   Fig. 35.2

We shall usually work in two dimensions at this stage, but the great advantage of vector algebra is that it can be used easily in three or more dimensions.

To determine the position of a point relative to a fixed origin, it is sufficient if we know the displacements parallel to two perpendicular

*Some writers say 'magnitude, sense and direction', but surely e.g. NE is not the same direction as SW, so direction includes sense.

axes, Ox, Oy, and they are the **components** in these directions of the vector. Thus if P is the point $(4,3)$, the position vector of P relative to the

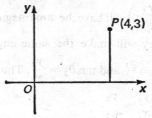

Fig. 35.3

origin O has component 4 units parallel to Ox, 3 units parallel to Oy. We can write the vector which represents the displacement from O to P as $\overrightarrow{OP}$; the components we can write as the entries in a matrix with 2 rows and 1 column, or as a matrix with 1 row and 2 columns, or as multiples of unit vectors $\mathbf{i}$ and $\mathbf{j}$ along Ox and Oy. Thus

$$\overrightarrow{OP} = \begin{pmatrix} 4 \\ 3 \end{pmatrix} = (4\mathbf{i} + 3\mathbf{j}) = (4 \quad 3).$$

We shall not use the last form, as it is easily confused with the co-ordinates of points.

Magnitude and direction of a vector

The distance of the point P above from O is $\sqrt{(4^2 + 3^2)}$, and we define the magnitude of the vector $\begin{pmatrix} x \\ y \end{pmatrix}$ to be $\sqrt{(x^2 + y^2)}$. The direction of a vector in two dimensions is determined by the angle it makes with a fixed line, usually the x-axis. If θ is the angle made by the vector $\overrightarrow{OP} = \begin{pmatrix} x \\ y \end{pmatrix}$ with the x-axis, then $\tan \theta = y/x$, if both x and y are positive. If either or both is negative, we can draw a suitable diagram to enable us to find an acute angle, and then deduce the angle made with

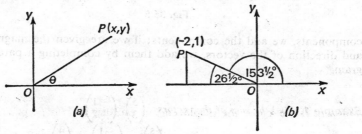

Fig. 35.4

the positive x-axis. (Fig. 35.4(b), where the angle made by OP with the positive x-axis is about $153\frac{1}{2}°$, since $\tan 26\frac{1}{2}° \simeq \frac{1}{2}$.)

Two vectors $\begin{pmatrix} x_1 \\ y_1 \end{pmatrix}$, $\begin{pmatrix} x_2 \\ y_2 \end{pmatrix}$ will have the same magnitude if $\sqrt{(x_1{}^2 + y_1{}^2)}$ $= \sqrt{(x_2{}^2 + y_2{}^2)}$; they will make the same angle with the x-axis if $\theta_1 = \theta_2$, where $\tan \theta_1 = \dfrac{y_1}{x_1}$ and $\tan \theta_2 = \dfrac{y_2}{x_2}$. Thus $\begin{pmatrix} x_1 \\ y_1 \end{pmatrix}$, $\begin{pmatrix} x_2 \\ y_2 \end{pmatrix}$ will be parallel if and only if $\dfrac{y_1}{x_1} = \dfrac{y_2}{x_2}$.

Addition of vectors

If we give a point initially at the origin O a displacement $\begin{pmatrix} 3 \\ 1 \end{pmatrix}$ it moves to the point (3,1); if it then receives a displacement $\begin{pmatrix} 1 \\ 2 \end{pmatrix}$ it moves to the point (4,3), adding the x components and the y components. Thus it is reasonable to define vector addition as the addition of the corresponding components, i.e. the corresponding entries in the matrix representing each vector. This can be shown to be equivalent to completing the parallelogram in Fig. 35.5. Thus if we are given a vector in terms of its

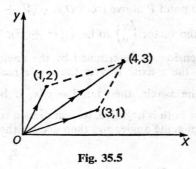

Fig. 35.5

components, we add the components; if we are given the magnitude and direction of the vectors, we add them by completing a parallelogram.

Example 1. Find the sum of the vectors $\begin{pmatrix} 5 \\ 3 \end{pmatrix}$, $\begin{pmatrix} -1 \\ 2 \end{pmatrix}$.

Adding the corresponding entries $\begin{pmatrix} 5 \\ 3 \end{pmatrix} + \begin{pmatrix} -1 \\ 2 \end{pmatrix} = \begin{pmatrix} 4 \\ 5 \end{pmatrix}$.

Example 2. *Find the sum of two vectors, magnitudes 2 and 3 inclined to each other at 40°.*

Solution by drawing recommended.

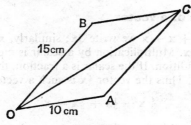

Fig. 35.6

Take a suitable scale, say 5 cm to 1 unit.
Draw straight lines, 10 cm and 15 cm long, containing an angle of 40°. Complete the parallelogram *OACB*. Measure *OC*, 23.5 cm and the angle it makes with *OA*. Interpreting our scale, the sum of the two vectors is a vector magnitude 4.7 at an angle 24° with the first vector.

To solve by calculation, it is necessary to use the cosine formula to find the magnitude of *OC*, and then the sine formula to find the angle *AOC*. In some cases, where there are right angled or isosceles triangles, it is possible to use the properties of these triangles to shorten the calculations.

Subtraction of vectors

Since
$$\begin{pmatrix} a \\ b \end{pmatrix} + \begin{pmatrix} c \\ d \end{pmatrix} = \begin{pmatrix} a + c \\ b + d \end{pmatrix},$$
$$\begin{pmatrix} a + c \\ b + d \end{pmatrix} - \begin{pmatrix} a \\ b \end{pmatrix} = \begin{pmatrix} c \\ d \end{pmatrix};$$

thus to subtract vectors, we subtract the corresponding entries in the matrices that represent the vectors.

If vectors are not given in matrix form, then we have to use a parallelogram to subtract vectors. If **a** and **b** are two vectors, **a** − **b** = **a** + (−**b**), so that a vector equal in magnitude but opposite in direction

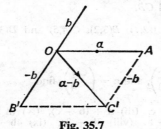

Fig. 35.7

to **b** is drawn as in Fig. 35.7, and the parallelogram $OAC'B'$ is completed.

Scalar multiples of a vector

If **x** is a vector, $\mathbf{x} + \mathbf{x} + \mathbf{x}$ we write $3\mathbf{x}$; similarly, $\mathbf{x} + \mathbf{x} + \mathbf{x}....$, to n terms, we write $n\mathbf{x}$. Multiplication by a scalar is merely an abbreviation for repeated addition. If the scalar is a fraction, then multiplication is defined similarly. Thus the vector $\frac{1}{3}\mathbf{x}$ is such a vector that

$$\tfrac{1}{3}\mathbf{x} + \tfrac{1}{3}\mathbf{x} + \tfrac{1}{3}\mathbf{x} = \mathbf{x}.$$

Example 3. If $\mathbf{a} = \begin{pmatrix} 4 \\ 6 \end{pmatrix}$, find $3\mathbf{a}$, $\frac{1}{2}\mathbf{a}$.

Using the above, $3\mathbf{a} = \begin{pmatrix} 12 \\ 18 \end{pmatrix}$, $\frac{1}{2}\mathbf{a} = \begin{pmatrix} 2 \\ 3 \end{pmatrix}$.

Example 4. If the vector **b** has magnitude 4, the vector $3\mathbf{b}$ has magnitude 12, the vector $\frac{1}{2}\mathbf{b}$ has magnitude 2. The directions of $3\mathbf{b}$ and $\frac{1}{2}\mathbf{b}$ are the same as the directions of **b**.

Exercise 35.1a

1. The points A, B, C and D have the following coordinates: $A(3,1)$, $B(1,-2)$, $C(-1,3)$, $D(-1,-2)$. Plot these points, and write down the position vectors relative to the origin O of A, B, C and D.
   Use your diagram to write down
   (i) position vectors of A relative to B, C and D.
   (ii) the position vectors of B relative to A, C and D.

2. Plot the points $A(2,0)$, $B(0,3)$, $C(-1,1)$ and $D(-2,-1)$. The position vector of A relative to B is written $\overrightarrow{BA}$. Write down the following position vectors: $\overrightarrow{BA}$, $\overrightarrow{BC}$, $\overrightarrow{BD}$, $\overrightarrow{CA}$, $\overrightarrow{CB}$. Find a relation (i) between $\overrightarrow{BA}$ and $\overrightarrow{AB}$; (ii) between $\overrightarrow{BC}$ and $\overrightarrow{CB}$.

3. Plot the positions $A(2,1)$, $B(3,2)$, $C(4,5)$ and $D(3,4)$. Find the vectors $\overrightarrow{AB}$, $\overrightarrow{CD}$, $\overrightarrow{AD}$, $\overrightarrow{BC}$.

4. If $\mathbf{a} = \begin{pmatrix} 3 \\ 2 \end{pmatrix}$, $\mathbf{b} = \begin{pmatrix} 2 \\ 3 \end{pmatrix}$, and $\mathbf{c} = \begin{pmatrix} -1 \\ 4 \end{pmatrix}$, find

   (i) $\mathbf{a} + \mathbf{b}$, (ii) $\mathbf{b} + \mathbf{c}$, (iii) $\mathbf{a} + (\mathbf{b} + \mathbf{c})$, (iv) $(\mathbf{a} + \mathbf{b}) + \mathbf{c}$, (v) $\mathbf{a} - \mathbf{b}$, (vi) $2\mathbf{a}$, (vii) $\frac{1}{2}\mathbf{a} - \mathbf{b}$, (viii) $3\mathbf{b} + 2\mathbf{c}$, (ix) $4\mathbf{b} - \mathbf{c}$, (x) $\mathbf{a} - \mathbf{b} - 2\mathbf{c}$, (xi) $3\mathbf{a} - 2\mathbf{b} - \mathbf{c}$, (xii) $\frac{1}{2}\mathbf{a} - \mathbf{c}$.

5. Find the magnitude of each of the following vectors, and the angle, correct if necessary to the nearest degree, each makes with the vector $\begin{pmatrix} 1 \\ 0 \end{pmatrix}$:

(i) $\begin{pmatrix} 1 \\ 1 \end{pmatrix}$ (ii) $\begin{pmatrix} -2 \\ 0 \end{pmatrix}$ (iii) $\begin{pmatrix} 0 \\ 2 \end{pmatrix}$

(iv) $\begin{pmatrix} -1 \\ 1 \end{pmatrix}$ (v) $\begin{pmatrix} -2 \\ -1 \end{pmatrix}$ (vi) $\begin{pmatrix} -1 \\ 2 \end{pmatrix}$

(vii) $\begin{pmatrix} 2 \\ 1 \end{pmatrix}$ (viii) $\begin{pmatrix} 3 \\ -1 \end{pmatrix}$ (ix) $\begin{pmatrix} 1 \\ 3 \end{pmatrix}$

6. If $a = \begin{pmatrix} 2 \\ 1 \end{pmatrix}$, $b = \begin{pmatrix} -1 \\ 2 \end{pmatrix}$ and $c = \begin{pmatrix} -1 \\ 7 \end{pmatrix}$, find a scalar λ such that $a + \lambda b = c$.

7. If $a = \begin{pmatrix} 2 \\ 0 \end{pmatrix}$, $b = \begin{pmatrix} 0 \\ 3 \end{pmatrix}$ and $c = \begin{pmatrix} 6 \\ 6 \end{pmatrix}$, find scalars λ and μ such that $\lambda a + \mu b = c$.

8. If $a = \begin{pmatrix} 2 \\ 1 \end{pmatrix}$, $b = \begin{pmatrix} -1 \\ 3 \end{pmatrix}$ and $c = \begin{pmatrix} 1 \\ 11 \end{pmatrix}$, find scalars λ and μ such that $\lambda a + \mu b = c$.

9. If $a = \begin{pmatrix} 7 \\ 24 \end{pmatrix}$ and $b = \begin{pmatrix} 15 \\ -20 \end{pmatrix}$, show that a and b have the same magnitude.

10. If the position vector of the point A relative to the origin is $\begin{pmatrix} 4 \\ 2 \end{pmatrix}$, find the position vector relative to the origin of M, the midpoint of OA.

Exercise 35.1b

1. The points A, B, C, D have the following coordinates; $A(2,-1)$, $B(1,-3)$, $C(-1,-3)$, $D(2,2)$. Plot these points and write down the position vectors relative to the origin O of A, B, C and D.

   Use your diagram to write down the position vectors relative to D of A, B and C.

2. Plot the points $A(3,0)$, $B(0,-3)$, $C(-1,-2)$ and $D(-2,-3)$. Write down the following vectors: $\overrightarrow{AB}$, $\overrightarrow{BA}$, $\overrightarrow{CD}$, $\overrightarrow{CB}$. Find a relation between $\overrightarrow{AB}$ and $\overrightarrow{BA}$, between $\overrightarrow{AB}$ and $\overrightarrow{CD}$.

3. Plot the points $A(2,-1)$, $B(-2,0)$, $C(0,-3)$, $D(-4,0)$. Write down the vectors $\overrightarrow{AB}$, $\overrightarrow{CD}$, $\overrightarrow{AD}$, $\overrightarrow{BC}$.

4. If $a = \begin{pmatrix} 2 \\ 3 \end{pmatrix}$, $b = \begin{pmatrix} -1 \\ -2 \end{pmatrix}$, and $c = \begin{pmatrix} -2 \\ 0 \end{pmatrix}$, find

   (i) $a - b$, (ii) $2b - c$, (iii) $\frac{1}{2}(a + c)$, (iv) $3c$, (v) $-2b$, (vi) $\frac{1}{3}c$, (vii) $b - a$, (viii) $c - 2b$, (ix) $\frac{1}{2}(b - c)$, (x) $\frac{1}{3}(a + 2b)$, (xi) $a - b - c$, (xii) $c - b - a$.

5. Find the magnitude of each of the following vectors, and the angle, correct if necessary to the nearest degree, which each vector makes with the x-axis.

(i) $\begin{pmatrix} 1 \\ -1 \end{pmatrix}$ (ii) $\begin{pmatrix} -2 \\ -2 \end{pmatrix}$ (iii) $\begin{pmatrix} -1 \\ 0 \end{pmatrix}$

(iv) $\begin{pmatrix} 0 \\ -1 \end{pmatrix}$ (v) $\begin{pmatrix} 2 \\ -1 \end{pmatrix}$ (vi) $\begin{pmatrix} 4 \\ -2 \end{pmatrix}$

(vii) $\begin{pmatrix} 2 \\ -2 \end{pmatrix}$ (viii) $\begin{pmatrix} \frac{1}{2} \\ \frac{1}{2} \end{pmatrix}$ (ix) $\begin{pmatrix} 1 \\ \frac{1}{2} \end{pmatrix}$

Which vectors are parallel?

6. If $a = \begin{pmatrix} 2 \\ 1 \end{pmatrix}$, $b = \begin{pmatrix} -1 \\ 1 \end{pmatrix}$ and $c = \begin{pmatrix} 0 \\ 3 \end{pmatrix}$, find a scalar λ such that $a + \lambda b = c$.

7. If $a = \begin{pmatrix} 1 \\ 1 \end{pmatrix}$, $b = \begin{pmatrix} 2 \\ 3 \end{pmatrix}$ and $c = \begin{pmatrix} 0 \\ -1 \end{pmatrix}$, find scalars λ and μ such that $\lambda a + \mu b = c$.

8. If $a = \begin{pmatrix} 1 \\ 1 \end{pmatrix}$, $b = \begin{pmatrix} 2 \\ 2 \end{pmatrix}$ and $c = \begin{pmatrix} 3 \\ 1 \end{pmatrix}$, what happens when we try to find scalars λ, μ such that $\lambda a + \mu b = c$?

9. If $a = \begin{pmatrix} 3 \\ 4 \end{pmatrix}$ and $b = \begin{pmatrix} -4 \\ 3 \end{pmatrix}$, plot the points A, B whose position vectors relative to an origin O are a and b respectively. Find the size of angle AOB.

10. If the position vectors of points P_1, P_2 relative to an origin O are $\begin{pmatrix} x_1 \\ y_1 \end{pmatrix}$, $\begin{pmatrix} x_2 \\ y_2 \end{pmatrix}$ respectively, show that $\overrightarrow{OP_1}$ and $\overrightarrow{OP_2}$ are perpendicular if and only if $x_1 x_2 + y_1 y_2 = 0$.

Transformation of a vector by a matrix

We have seen that some mathematical operations, reflections, rotations, shears, can be described by matrices. If we write a vector in matrix form, e.g. $\begin{pmatrix} x \\ y \end{pmatrix}$, then we can consider the transformations made on this vector by certain matrices, as we considered transformations made on squares and rectangles in Chapter 34.

Example 5. Find the image of the vector $\begin{pmatrix} 2 \\ 1 \end{pmatrix}$ under the transformations

(i) $\begin{pmatrix} -1 & 0 \\ 0 & 1 \end{pmatrix}$ (ii) $\begin{pmatrix} 1 & -1 \\ 1 & 1 \end{pmatrix}$.

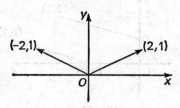

Fig. 35.8

(i) $\begin{pmatrix} -1 & 0 \\ 0 & 1 \end{pmatrix} \begin{pmatrix} 2 \\ 1 \end{pmatrix} = \begin{pmatrix} -2 \\ 1 \end{pmatrix}$ so $\begin{pmatrix} 2 \\ 1 \end{pmatrix} \mapsto \begin{pmatrix} -2 \\ 1 \end{pmatrix}$,

that is reflection in the y-axis.

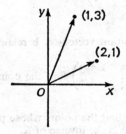

Fig. 35.9

(ii) $\begin{pmatrix} 1 & -1 \\ 1 & 1 \end{pmatrix} \begin{pmatrix} 2 \\ 1 \end{pmatrix} = \begin{pmatrix} 1 \\ 3 \end{pmatrix}$ so $\begin{pmatrix} 2 \\ 1 \end{pmatrix} \mapsto \begin{pmatrix} 1 \\ 3 \end{pmatrix}$.

This shows the magnitude of the vector is increased from $\sqrt{5}$ to $\sqrt{10}$, i.e. by a factor $\sqrt{2}$. It can be shown by trigonometry that the vector is rotated through $45°$ in an anticlockwise sense.

Translation

A translation can be described by adding a vector, $\begin{pmatrix} p \\ q \end{pmatrix}$ being a translation p units along the x-axis and q units parallel to the y-axis, since it adds p to every x coordinate and q to every y coordinate.

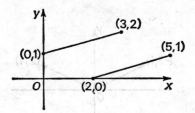

Fig. 35.10

Translation described by $+\begin{pmatrix} 3 \\ 1 \end{pmatrix}$.

Exercise 35.2

1. Points A, B have position vectors **a**, **b** relative to O, where $\mathbf{a} = \begin{pmatrix} 2 \\ 1 \end{pmatrix}$, $\mathbf{b} = \begin{pmatrix} 3 \\ -2 \end{pmatrix}$. If $\mathbf{R} = \begin{pmatrix} 1 & -1 \\ 1 & 1 \end{pmatrix}$ find the points whose position vectors are $\mathbf{R}.\mathbf{a}$ and $\mathbf{R}.\mathbf{b}$.

2. Points A, B have position vectors **a**, **b** relative to O, where $\mathbf{a} = \begin{pmatrix} 7 \\ 1 \end{pmatrix}$, $\mathbf{b} = \begin{pmatrix} 6 \\ 8 \end{pmatrix}$. If $\mathbf{R} = \begin{pmatrix} 4/5 & -3/5 \\ 3/5 & 4/5 \end{pmatrix}$ find the points whose position vectors are $\mathbf{R}.\mathbf{a}$ and $\mathbf{R}.\mathbf{b}$.

3. With the data of Q.1, find the points whose position vectors are $\mathbf{R}^{-1}.\mathbf{b}$ and $\mathbf{R}^{-1}.\mathbf{b}$, where $\mathbf{R}^{-1}$ is the inverse of $\mathbf{R}$.

4. With the data of Q.2, find the points whose position vectors are $\mathbf{R}^{-1}.\mathbf{a}$ and $\mathbf{R}^{-1}.\mathbf{b}$.

5. With the data of Q.2, find the points whose position vectors are $\mathbf{R}^2.\mathbf{a}$ and $\mathbf{R}^2.\mathbf{b}$.

Exercise 35.3: Miscellaneous

1. If **u**, **v** are unit vectors inclined at 45°, find by drawing the magnitude of each of the following vectors: (i) $\mathbf{u} + \mathbf{v}$, (ii) $\mathbf{u} - \mathbf{v}$, (iii) $\mathbf{u} + 2\mathbf{v}$, (iv) $3\mathbf{u} + \mathbf{v}$.

2. If **u** and **v** are unit vectors inclined at 30°, find by drawing the magnitude of each of the following vectors: (i) $\mathbf{u} + \mathbf{v}$, (ii) $2\mathbf{u} - \mathbf{v}$, (iii) $\mathbf{u} - 2\mathbf{v}$, (iv) $3\mathbf{u} + 2\mathbf{v}$.

3. If **u** and **v** are unit vectors, find the angle between **u** and **v** if the magnitude of $\mathbf{u} - \mathbf{v}$ is 1.

4. If **u** and **v** are unit vectors, find the angle between them if the magnitude of $\mathbf{u} + \mathbf{v}$ is 1.

5. If **u** and **v** are unit vectors inclined at 40°, find by calculation the magnitude of (i) **u** + **v**, (ii) 2**u** + 2**v**, (ii) **u** − **v**.

6. The position vectors of points A, B relative to O are $\begin{pmatrix} 3 \\ 6 \end{pmatrix}$ and $\begin{pmatrix} 9 \\ -3 \end{pmatrix}$ respectively. Points L, M divide OA, OB in the ratio 2 : 1. Find (i) the position vectors of L and M, (ii) the vector AB, (iii) the vector LM. Show that AB is parallel to LM, and find the ratio of the lengths $LM : AB$.

7. A man starts from a point O, and walks with a velocity given in m/s by the vector $\begin{pmatrix} 1.2 \\ 1.6 \end{pmatrix}$. Find his position vector relative to O after 1 s, after 2 s, after 3 s and after 5 s. After how many seconds is he 50 m from O?

8. A boat sails from a buoy O with velocity given in m/s by the vector $\begin{pmatrix} 2 \\ 1 \end{pmatrix}$. Find its position vector relative to O after 1 s, after 2 s and after 5 s. After how many seconds is it 50 m from O?

9. An aircraft flies from an airport A, position vector relative to an origin O $\begin{pmatrix} 100 \\ 50 \end{pmatrix}$ km, the velocity of the aircraft being $\begin{pmatrix} 10 \\ 12 \end{pmatrix}$ km/min. Find the position vector of the aircraft relative to O after 10 minutes.

10. A man starts to swim from a point O with velocity relative to the water $\begin{pmatrix} 0.5 \\ 0.4 \end{pmatrix}$ m/s. Unknown to him, there is a current $\begin{pmatrix} 2 \\ -1 \end{pmatrix}$ m/s. Find his position vector relative to O after 10 s.

11. A boy wishes to swim directly across a river, in which there is a current of 0.2 m/s. The boy can swim at 0.5 m/s. At what angle to the bank should he head in order to travel directly across the river?

12. If the river in Q.11 is 50 m wide, with straight parallel banks, how should the boy head if he wishes to land 10 m below the point at which he enters the water? How long will it take him to cross along this path?

36 Vector Geometry

Parallel vectors

We saw that the tangent of the angle θ made with the x-axis by the vector $\mathbf{a} = \begin{pmatrix} x \\ y \end{pmatrix}$ is such that $\tan \theta = y/x$. Thus the angle made by the vector $2\mathbf{a}$ with the x-axis is also θ, so that $\mathbf{a}$ and $2\mathbf{a}$ are parallel. Similarly we can show that, for any non-zero scalar k, $k\mathbf{a}$ is parallel to $\mathbf{a}$.

Position vector of a point A relative to a point B

In many of the exercises in the last chapter, we saw that in numerical

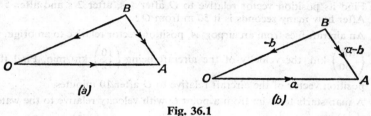

(a)   (b)

Fig. 36.1

examples, to find the position vector of a point A relative to a point B, we subtracted OB from OA, thus

$$\overrightarrow{BA} = \overrightarrow{OA} - \overrightarrow{OB},$$

or $\overrightarrow{BA} = \mathbf{a} - \mathbf{b}$, if $OA = \mathbf{a}$, $OB = \mathbf{b}$.

Midpoint theorem

If the position vectors of points A, B relative to an origin O are $\mathbf{a}$, $\mathbf{b}$ respectively, the position vectors of the midpoints X of OA and Y of OB are $\frac{1}{2}\mathbf{a}$ and $\frac{1}{2}\mathbf{b}$. The vector describing $\overrightarrow{AB}$ is $\mathbf{b} - \mathbf{a}$; that describing $\overrightarrow{XY}$ is $\frac{1}{2}(\mathbf{b} - \mathbf{a})$, so $\overrightarrow{XY}$ is parallel to $\overrightarrow{AB}$ and the length $\overrightarrow{XY}$ is half the length $\overrightarrow{AB}$.

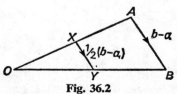

Fig. 36.2

The diagonals of a parallelogram bisect each other

Let points A, B have position vectors relative to O, **a** and **b** respectively.

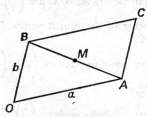

Fig. 36.3

Then the position vector of C, the fourth vertex of the parallelogram $OACB$ is $(\mathbf{a} + \mathbf{b})$, and the position vector of the midpoint of the diagonal OC is $\frac{1}{2}(\mathbf{a} + \mathbf{b})$. The vector describing the displacement $\overrightarrow{AB}$ is $(\mathbf{b} - \mathbf{a})$, so the vector describing the displacement from A to the midpoint M of AB is $\frac{1}{2}(\mathbf{b} - \mathbf{a})$. But $\overrightarrow{OM} = \overrightarrow{OA} + \overrightarrow{AM}$,

$$\therefore \overrightarrow{OM} = \mathbf{a} + \tfrac{1}{2}(\mathbf{b} - \mathbf{a})$$
$$= \tfrac{1}{2}(\mathbf{a} + \mathbf{b}),$$

so that the midpoint of AB has the same position vector as the midpoint of OC, i.e. AB and OC have the same midpoint, showing that the diagonals bisect each other.

Example 1. *ABCD is a quadrilateral. The midpoints of AB, BC, CD, DA are X, Y, Z, W respectively. Show that XYZW is a parallelogram.*

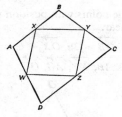

Fig. 36.4

Let the position vectors of A, B, C, D relative to O be **a**, **b**, **c** and **d** respectively. Then, using the result just proved, the position vectors relative to O of X, Y, Z, W are given by

$$\overrightarrow{OX} = \tfrac{1}{2}(\mathbf{a} + \mathbf{b}), \ \overrightarrow{OY} = \tfrac{1}{2}(\mathbf{b} + \mathbf{c})$$
$$\overrightarrow{OZ} = \tfrac{1}{2}(\mathbf{c} + \mathbf{d}) \text{ and } \overrightarrow{OW} = \tfrac{1}{2}(\mathbf{d} + \mathbf{a}).$$

The vector $\quad \overrightarrow{XY} = \tfrac{1}{2}(\mathbf{b} + \mathbf{c}) - \tfrac{1}{2}(\mathbf{a} + \mathbf{b}) = \tfrac{1}{2}(\mathbf{c} - \mathbf{a}),$

and $\quad \overrightarrow{WZ} = \tfrac{1}{2}(\mathbf{c} + \mathbf{d}) - \tfrac{1}{2}(\mathbf{d} + \mathbf{a}) = \tfrac{1}{2}(\mathbf{c} - \mathbf{a}).$

Thus XY and WZ are equal and parallel, so $XYZW$ is a parallelogram.

Exercise 36

1. The position vectors of points A, B, C are $\begin{pmatrix} 1 \\ 1 \end{pmatrix}$, $\begin{pmatrix} 3 \\ 1 \end{pmatrix}$, $\begin{pmatrix} 2 \\ 4 \end{pmatrix}$ respectively.

   Find the position vector of the fourth vertex D of the parallelograms (i) $ABCD$, (ii) $ABDC$.

2. Show that the points position vectors $\begin{pmatrix} 1 \\ 0 \end{pmatrix}$, $\begin{pmatrix} 7 \\ 3 \end{pmatrix}$, $\begin{pmatrix} 6 \\ 5 \end{pmatrix}$, $\begin{pmatrix} 0 \\ 2 \end{pmatrix}$ are the vertices of a rectangle.

3. Show that the points position vectors $\begin{pmatrix} 1 \\ 2 \end{pmatrix}$, $\begin{pmatrix} 4 \\ 1 \end{pmatrix}$, $\begin{pmatrix} 5 \\ 4 \end{pmatrix}$, $\begin{pmatrix} 2 \\ 5 \end{pmatrix}$ are the vertices of a rhombus.

4. Show that the points with position vectors $\begin{pmatrix} 0 \\ -1 \end{pmatrix}$, $\begin{pmatrix} 1 \\ 0 \end{pmatrix}$, $\begin{pmatrix} 0 \\ 1 \end{pmatrix}$, $\begin{pmatrix} 2 \\ 3 \end{pmatrix}$ are the vertices of a trapezium in which one of the parallel sides is twice the length of the other.

5. In this question, points A, B, C, D have position vectors $\mathbf{a}$, $\mathbf{b}$, $\mathbf{c}$, $\mathbf{d}$ respectively.
   (i) What is the condition that AB is parallel to CD?
   (ii) What is the condition that AB is equal and parallel to CD?
   (iii) What is the condition that $ABCD$ is a trapezium with AD parallel to BC?
   (iv) What is the condition that AX is equal and parallel to BC, where X is the midpoint of CD?

6. Show that the points with position vectors $\mathbf{a}$, $2\mathbf{a} + 3\mathbf{b}$, $3\mathbf{a} + 6\mathbf{b}$ are colinear.

7. Find which three of the points with position vectors $\mathbf{a}$, $2\mathbf{a} + \mathbf{b}$, $4\mathbf{a} + 2\mathbf{b}$, and $5\mathbf{a} + 4\mathbf{b}$ are colinear.

8. In Fig. 36.5, $\overrightarrow{OA} = 3\mathbf{a}$, $\overrightarrow{OB} = 3\mathbf{b}$, $\overrightarrow{OX} = 6\mathbf{a}$ and $\overrightarrow{OY} = 6\mathbf{b}$. Find, in terms of $\mathbf{a}$ and $\mathbf{b}$, the vectors $\overrightarrow{AB}$, $\overrightarrow{XY}$, $\overrightarrow{AG}$, $\overrightarrow{OG}$.

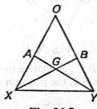

Fig. 36.5

9. In the triangle OAB, X is the point in OA such that $OX : XA = 3 : 1$, Y is the midpoint of OB and Z is the midpoint of XB. If $OA = \mathbf{a}$, $OB = \mathbf{b}$, find the vectors OX, OY, YZ in terms of $\mathbf{a}$ and $\mathbf{b}$. Hence show that YZ is parallel to OA, and find the ratio of the lengths $YZ : OA$.

10.

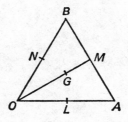

Fig. 36.6

In Fig. 36.6, *L*, *M*, *N* are the midpoints of *OA*, *AB* and *OB* respectively. *G* is the point on *OM* such that $OG = \frac{2}{3}OM$. If $\overrightarrow{OA} = \mathbf{a}$, $\overrightarrow{OB} = \mathbf{b}$, find $\overrightarrow{OM}$ and $\overrightarrow{OG}$. Find *BG* and *GL*, show that *BGL* is a straight line and that $BG : GL = 2 : 1$. Similarly, show that *AGN* is a straight line and that $AG : GN = 2 : 1$. Deduce that the lines joining each vertex of a triangle to the midpoint of the opposite sides meet at a point.

Fig. 16.5

In Fig. 16.5, L, M, N are the midpoints of OA, AB and OB respectively. C is the point on OM such that OC = ⅔OM. If OA = **a**, OB = **b**, find OM and OC. Find BC and CL. Show that BCL is a straight line and that BC = ⅔CL. Similarly, show that ACN is a straight line and that AC = ⅔CN. Deduce that the lines joining each vertex of a triangle to the midpoint of the opposite sides meet at a point.

TRIGONOMETRY

TRIGONOMETRY

The ratios

$\sin$ is $\dfrac{\text{opposite}}{\text{hypotenuse}}$ or the projection onto Oy

$\cos$ is $\dfrac{\text{adjacent}}{\text{hypotenuse}}$ or the projection onto Ox

$\tan$ is $\dfrac{\sin}{\cos}$

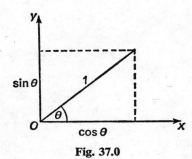

Fig. 37.0

$\cot$ is $\dfrac{1}{\tan}$; $\sec$ is $\dfrac{1}{\cos}$, cosec is $\dfrac{1}{\sin}$.

Some useful results

$$\sin \ 0° = \cos 90° = 0$$
$$\sin 30° = \cos 60° = \tfrac{1}{2}$$
$$\sin 60° = \cos 30° = \tfrac{1}{2}\sqrt{3}$$
$$\sin 90° = \cos \ 0° = 1$$
$$\sin^2 A + \cos^2 A = 1$$

Sine formula

$$\frac{a}{\sin A} = \frac{b}{\sin B} = \frac{c}{\sin C}$$

Cosine formula

$$c^2 = a^2 + b^2 - 2ab \cos C$$
$$\text{i.e. } \cos C = \frac{a^2 + b^2 - c^2}{2ab}$$

Area of triangle

$$\Delta = \tfrac{1}{2}bc \sin A = \sqrt{s(s-a)(s-b)(s-c)}.$$

Arc and sector

Length of arc $= \dfrac{\theta}{360}(2\pi r)$

area of sector $= \dfrac{\theta}{360}(\pi r^2).$

37 Ratios of an Acute Angle

Sine

Draw a triangle ABC, right-angled at B and with the angle A equal say, to 40°. Take points C' on AC and B' on AB so that $C'B'$ is parallel to CB. By similar triangles, $\dfrac{CB}{AC} = \dfrac{C'B'}{AC'}$, and the ratio $\dfrac{C'B'}{AC'}$ is the same wherever $C'B'$ is drawn, providing it is parallel to CB. This ratio depends only on the angle A, which in this case we made 40°. This is called the sine of angle A.

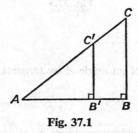

Fig. 37.1

If AC is 1 unit in length, sine $A = \dfrac{BC}{1}$; alternatively we can define sine A as the projection of a line 1 unit in length on the line through C perpendicular to AB (Fig. 37.2).

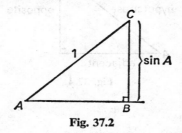

Fig. 37.2

Cosine

In Fig. 37.1, $\dfrac{AB}{AC} = \dfrac{AB'}{AC'}$ for all positions of $B'C'$ parallel to BC. This ratio is called the cosine of angle A.

Again, if AC is 1 unit in length, cosine $A = \dfrac{AB}{1}$; an alternative definition of cosine A is the projection of a line 1 unit in length onto the other arm bounding the angle (Fig. 37.3).

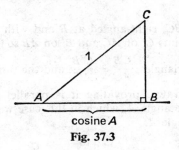

cosine A

Fig. 37.3

Tangent

We define the tangent of an angle A by tangent $A = \dfrac{\text{sine } A}{\text{cosine } A}$.

The three ratios

These are the commonest trigonometric ratios. If A is an angle in a right-angled triangle, they are usually abbreviated and remembered

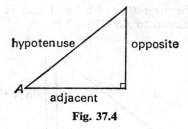

Fig. 37.4

thus:

$$\text{sine } A = \sin A = \frac{\text{opposite}}{\text{hypotenuse}}$$

$$\text{cosine } A = \cos A = \frac{\text{adjacent}}{\text{hypotenuse}}$$

$$\text{tangent } A = \tan A = \frac{\text{opposite}}{\text{adjacent}}$$

Three other ratios

Sometimes the reciprocals of these ratios are required, and it is useful to have three other ratios defined:

$$\text{secant } A = \sec A = \frac{1}{\cos A} = \frac{\text{hypotenuse}}{\text{adjacent}}$$

$$\text{cosecant } A = \text{cosec } A = \frac{1}{\sin A} = \frac{\text{hypotenuse}}{\text{opposite}}$$

$$\text{cotangent } A = \cot A = \frac{1}{\tan A} = \frac{\text{adjacent}}{\text{opposite}}$$

Exercise 37.1a

1. From Fig. 37.5, write down the values of sin A, cos C, tan A, cot C, sec A, cosec C.

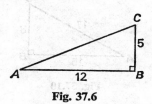

Fig. 37.5

2. From Fig. 37.6, write down the values of tan A, sin C, cot A, cosec C.

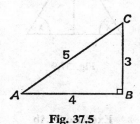

Fig. 37.6

3. From Fig. 37.7, write down the values of sin x and cot y.

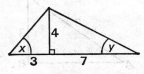

Fig. 37.7

4. From Fig. 37.8, write down the values of tan A and cot C.

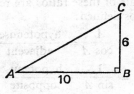

Fig. 37.8

5. From Fig. 37.9, write down the value of cos x and calculate the value of tan x.

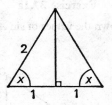

Fig. 37.9

Exercise 37.1b

1. From Fig. 37.10, write down the values of sin A, cos C, tan A, cot C.

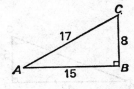

Fig. 37.10

2. From Fig. 37.11, write down the values of sin A and sin C.

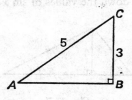

Fig. 37.11

3. From Fig. 37.12, write down the values of cos *x* and tan *y*.

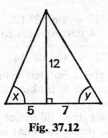

Fig. 37.12

4. From Fig. 37.13, write down the values of sec *A* and cosec *C*.

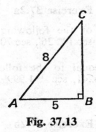

Fig. 37.13

5. From Fig. 37.14 ,write down the values of tan *x* and calculate the value of sin *x*.

Fig. 37.14

Use of tables

Since the value of each ratio of any given angle is known, we can buy books of tables giving these ratios. They are also marked on some slide rules and obtainable from some calculators. Slide rules and calculators have their own booklets of instructions, as the procedures vary for different models of each.

To find, say, sin 40°, we merely look in the appropriate table. Conversely, we can also find which angle has sine equal to a given value, if we know that the angle is acute. Supposing that sin *A* = 0.6721. Look in the body of the tables for the angle next below 0.6721 (or 0.672, if we have three-figure tables). This is 0.6717, to four places of decimals, which is 4 short of 0.6721. The difference 4 corresponds to 2'. We have

$$0.6717 = \sin 42° 12'$$
$$\underline{4 \text{ corresponds to } 2'}$$
$$0.6721 = \sin 42° 14'$$

Most books of tables also give tabulated values of log sin, log cos, etc. They save the double operation of looking up first the trig ratio, then the logarithm of that ratio.

N.B. Remember to **subtract** the difference in the tables for those ratios which begin with the prefix **co**. That is for cosine, cosecant and cotangent.

Exercise 37.2a

1. Write down the values of the following: sin 20°, cos 70°, cot 40°, tan 13° 12', tan 63° 27', cosec 14° 26', sec 70° 33', cot 14° 3', sin 24° 15', cos 32° 16'.

2. Find the acute angles equal to the following: $\sin^{-1} 0.5$, $\cos^{-1} 0.5$, $\tan^{-1} 0.8341$, $\mathrm{cosec}^{-1} 1.6241$, $\sec^{-1} 1.9999$, $\cot^{-1} 2$, $\sin^{-1} \frac{1}{3}$, $\cos^{-1} \frac{2}{3}$, $\tan^{-1} 1$, $\tan^{-1} 2$.

Exercise 37.2b

1. Write down the values of the following: sin 25°, cosec 42°, tan 18°, sin 68° 13', cos 72° 11', cot 11° 22', sec 75° 32', cosec 14° 28', cosec 22° 3', cosec 60°.

2. Find the acute angles equal to the following: $\sin^{-1} 0.866$, $\cos^{-1} 0.866$, $\tan^{-1} 3.0$, $\mathrm{cosec}^{-1} 2$, $\sec^{-1} 2$, $\mathrm{cosec}^{-1} 2.413$, $\sin^{-1} \frac{1}{3}$, $\tan^{-1} \frac{4}{3}$, $\cot^{-1} \frac{4}{3}$, $\cot^{-1} 1$.

Given one ratio, to find the others

Suppose that we are given one value of sin A and need to find the other ratios of A without using tables. Sketch a right-angled triangle, and mark the appropriate sides as in Example 1. Calculate the third side of the triangle by Pythagoras' Theorem, and write down the value of the required ratio.

Example 1. Given that A is an acute angle and that $\tan A = \frac{4}{3}$, find the value of cosec A.

Sketch a right-angled triangle so that the lengths of the sides containing the right angle are 4 cm and 3 cm (*see* Fig. 37.15). The angle opposite the side of length 4 cm is equal to *A*. The other side of the triangle is given by

$$AC^2 = 3^2 + 4^2, \quad \text{and therefore} \quad AC = 5 \text{ cm}.$$
$$\therefore \mathrm{cosec}\, A = \tfrac{5}{4}.$$

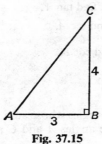

Fig. 37.15

Example 2. *Given that x is an acute angle and that $\sin x = \dfrac{m}{n}$, find $\cot x$.*

The triangle shown has $\sin x = \dfrac{m}{n}$. By Pythagoras, the length of the third

side is $\sqrt{n^2 - m^2}$.

$$\therefore \cot x = \frac{\sqrt{n^2 - m^2}}{m}.$$

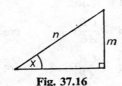

Fig. 37.16

N.B. Note that, since the hypotenuse must be the longest side of a triangle, the sine and cosine of an angle are always less than 1; for the same reason, the secant and cosecant of an angle are always greater than 1. There is no restriction on the value of the tangent or the cotangent of an angle.

Exercise 37.3a

1. Given that $\sin A = 0.6$, find $\cos A$.
2. Given that $\tan A = \frac{5}{12}$, find $\operatorname{cosec} A$.
3. Given that $\sec X = \frac{17}{15}$, find $\sin X$.
4. Given that $\cos Y = \frac{21}{29}$, find $\tan Y$.
5. Given that $\tan B = \dfrac{m}{n}$, find $\operatorname{cosec} B$.

Exercise 37.3b

1. Given that $\tan A = 0.75$, find $\sin A$.
2. Given that $\sin X = \frac{5}{13}$, find $\cot X$.

3. Given that cosec $Y = \frac{17}{8}$, find tan Y.

4. Given that sin $A = \frac{20}{29}$, find sec A.

5. Given that sec $B = \dfrac{p}{q}$, find cot B.

Complementary angles

Complementary angles are angles whose sum is 90° and so in a triangle ABC, right-angled at B, the angles A and C are complementary.

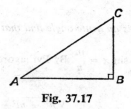

Fig. 37.17

From the definitions of the ratios,

$$\sin A = \frac{CB}{AC}; \qquad \cos C = \frac{CB}{AC};$$

$$\tan A = \frac{CB}{AB}; \qquad \cot C = \frac{CB}{AB};$$

$$\sec A = \frac{AC}{AB}; \qquad \operatorname{cosec} C = \frac{AC}{AB}.$$

So
$$\sin A = \cos C;$$
$$\tan A = \cot C;$$

and
$$\sec A = \operatorname{cosec} C.$$

Since $C = 90° - A$, these equations may be written

$$\sin A = \cos (90° - A);$$
$$\tan A = \cot (90° - A);$$
$$\sec A = \operatorname{cosec} (90° - A).$$

Notice that three of the names (co-sine, co-tangent and co-secant) are formed by the addition of the prefix *co* to the other names. This prefix is short for 'complementary' and means that the ratio of any angle is equal to the co-ratio of the complementary angle.

To give a few examples:

$$\sin 20° = \cos 70°; \qquad \cos 20° = \sin 70°;$$
$$\sec 42° = \operatorname{cosec} 48°; \qquad \operatorname{cosec} 42° = \sec 48°;$$
$$\tan 37° = \cot 53°; \qquad \cot 37° = \tan 53°.$$

The isosceles triangle

Suppose that ABC is an isosceles triangle with AB equal to AC. Since the triangle is not right-angled, we cannot write down immediately an expression for sin B, for example. However, by drawing the perpendicular AD from A to BC, the ratios of the angle B may be calculated easily.

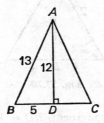

Fig. 37.18

Let $AB = AC = 13$ cm and $BC = 10$ cm.
The triangles ABD and ACD are congruent and therefore $DB = 5$ cm.
By Pythagoras, $AD^2 = AB^2 - BD^2 = 13^2 - 5^2$.
$$\therefore AD = 12 \text{ cm.}$$
From the right-angled triangle ABD,
$$\sin B = \tfrac{12}{13}, \cos B = \tfrac{5}{13}, \text{ and } \tan B = \tfrac{12}{5}.$$

The ratios of 45°

Draw an isosceles right-angled triangle XYZ. Then angle $X =$ angle $Z = 45°$. The size of the triangle will not affect the ratios of 45° so let $XY = YZ = 1$ cm and then by Pythagoras $XZ = \sqrt{2}$ cm.

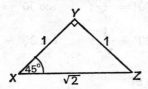

Fig. 37.19

From the definitions:

$$\sin 45° = \frac{1}{\sqrt{2}}; \quad \operatorname{cosec} 45° = \sqrt{2};$$

$$\cos 45° = \frac{1}{\sqrt{2}}; \quad \sec 45° = \sqrt{2};$$

$$\tan 45° = 1; \quad \cot 45° = 1.$$

The ratios of 30° and 60°

Draw an equilateral triangle. For convenience, choose each side to be 2 cm. Draw the perpendicular AD from A to BC.

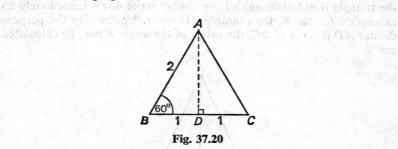

Fig. 37.20

Since ABD and ACD are congruent, $BD = 1$ cm.

∴ $AD^2 = 2^2 - 1^2$ and $AD = \sqrt{3}$ cm.

ABD is a right-angled triangle with B equal to 60° and angle BAD equal to 30°.

From the definitions:

$$\sin 60° = \frac{\sqrt{3}}{2}; \qquad \operatorname{cosec} 60° = \frac{2}{\sqrt{3}};$$

$$\cos 60° = \tfrac{1}{2}; \qquad \sec 60° = 2;$$

$$\tan 60° = \sqrt{3}; \qquad \cot 60° = \frac{1}{\sqrt{3}}.$$

Also

$$\sin 30° = \tfrac{1}{2}; \qquad \operatorname{cosec} 30° = 2;$$

$$\cos 30° = \frac{\sqrt{3}}{2}; \qquad \sec 30° = \frac{2}{\sqrt{3}};$$

$$\tan 30° = \frac{1}{\sqrt{3}}; \qquad \cot 30° = \sqrt{3}.$$

The ratios of 0° and 90°

Draw a triangle LMN in which M is a right angle and in which L is a very small angle. The angle N will consequently be very nearly 90°.

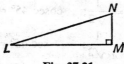

Fig. 37.21

As the angle L gets smaller, LM will get more nearly equal to LN. The ratio $\dfrac{LM}{LN}$ will tend to become 1; the ratios $\dfrac{MN}{ML}$ and $\dfrac{MN}{NL}$ will tend to become zero.

The ratios $\dfrac{ML}{MN}$ and $\dfrac{NL}{MN}$, on the other hand, become infinitely large as the denominators become zero.

Hence we are able to write down the ratios of 0° and 90°.

$$\sin 90° = 1; \qquad \operatorname{cosec} 90° = 1;$$
$$\cos 90° = 0; \qquad \sec 90° = \infty;$$
$$\tan 90° = \infty; \qquad \cot 90° = 0.$$

Also

$$\sin 0° = 0; \qquad \operatorname{cosec} 0° = \infty;$$
$$\cos 0° = 1; \qquad \sec 0° = 1;$$
$$\tan 0° = 0; \qquad \cot 0° = \infty.$$

Exercise 37.4a

Without using tables, write down the values of the following:

1. $\dfrac{\sin 20°}{\cos 70°}.$
2. $\dfrac{\sec 40°}{\operatorname{cosec} 50°}.$
3. $\dfrac{\cot 25°}{\tan 65°}.$

4. $\tan^2 60° + \tan^2 45°.$
5. $\cos^2 30° + \sin^2 30°.$

6. $\sec^2 60° - \tan^2 60°.$
7. $\operatorname{cosec}^2 45° - \cot^2 45°.$

8. $1 + \tan^2 30°.$
9. $1 - \sin 45°.$

10. $\operatorname{cosec}^2 60° - 1.$

Exercise 37.4b

Without using tables, write down the values of the following.

1. $\dfrac{\cos 10°}{\sin 80°}.$
2. $\dfrac{\operatorname{cosec} 15°}{\sec 75°}.$
3. $\dfrac{\tan 72°}{\cot 18°}.$

4. $\sin^2 60° + \tan^2 45°.$
5. $\cos^2 60° + \sin^2 60°.$

6. $\sec^2 45° - \tan^2 45°.$
7. $\operatorname{cosec}^2 30° - \cot^2 30°.$

8. $1 + \tan^2 60°.$
9. $1 - \sin^2 30°.$

10. $1 - \cos^2 60°.$

Exercise 37.5. Miscellaneous

1. Find by drawing an approximate value for $\tan 37°$.
2. Find by drawing and measurement the acute angle whose sine is 0.7.

3. In Fig. 37.22, B is a right angle and X is the middle point of BC. Calculate the ratio $\dfrac{\tan AXB}{\tan ACB}$.

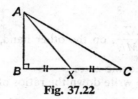

Fig. 37.22

4. PQR is an isosceles triangle in which $PQ = PR = 5$ cm and $QR = 8$ cm. Calculate sin Q.

5. In Fig. 37.23, prove that $\theta + \phi = 90°$. Prove also that there are two

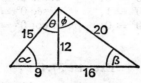

Fig. 37.23

other right angles in the figure. Write down the values of (i) sin α; (ii) cos β; (iii) sec ϕ; (iv) cosec θ.

6. Evaluate without using tables: (i) tan 20° cot 20°, (ii) cosec 25° cos 65°, (iii) sin² 30° + cos² 30°.

7. Given that x is an acute angle such that sin $x = p/q$, find the value of tan x.

8. Given that x is an acute angle and that cot $x = \frac{8}{15}$, find, without using tables, the values of (i) cos x, (ii) sin x, (iii) cos² x + sin² x.

9. Find values of x from the following:
(i) sin $x° = \cos 24°$, (ii) tan $x° = \cot 18°$, (iii) cosec $x° = \sec 51°$.

10. In Fig. 37.24, D is a right angle. Write down, in terms of a and b, (i) sin B, (ii) cot $\widehat{BAD}$.

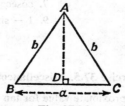

Fig. 37.24

11. In Fig. 37.25, *BD* is an altitude. Write down, in terms of *a*, *c* and *h*, (i) sin *A*, (ii) tan $\widehat{ABD}$, (iii) sec *C*.

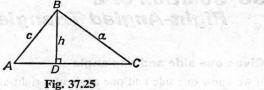

Fig. 37.25

12. Which of the following equations do not have solutions? (i) sin *x*° = 1.2, (ii) cos *y*° = 0.8, (iii) tan *y*° = 2.4, (iv) cosec α° = 0.7, (v) sec α° = ½.

38 Solution of a Right-Angled Triangle

Given one side and one angle

If we know one side and one angle of a right-angled triangle, the third angle is of course the complement of the known angle. We can find the lengths of the other two sides by trigonometry.

Example 1. The hypotenuse of a right-angled triangle is 18.1 cm and one of the angles is 32°. Find the lengths of the other two sides.

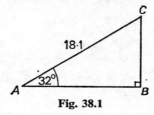

Fig. 38.1

From the figure,

$$\frac{BC}{18.1} = \sin 32°$$

$$\therefore BC = 18.1 \sin 32° = 9.59 \text{ cm.}$$

Since $C = 90° - 32° = 58°,$

$$\frac{AB}{AC} = \sin 58°$$

$$\therefore AB = 18.1 \sin 58° = 15.3 \text{ cm.}$$

We could have used $\frac{AB}{AC} = \cos 32°$, but it is often easier to use sine rather than cosine. The sine scale is easier to use on a slide rule, some calculators do not have cos functions, and if using tables we have to remember to subtract the difference in the cosine tables.

Example 2. In the triangle ABC, a = 8 cm, A = 28°. Find the lengths of the other two sides.

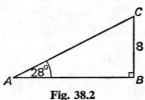

Fig. 38.2

From the figure, $\qquad C = 62°.$

$$\tan 62° = \frac{c}{8},$$

$$\therefore c = 15.0 \text{ cm}.$$

Also $\qquad\qquad \dfrac{8}{b} = \sin 28°,$

$$\therefore b = \frac{8}{\sin\ 28°} = 17.0 \text{ cm}.$$

There are many other ratios we could have used, and our choice of ratio will be guided by the calculating aids we have. In general, use sin instead of cos, and if possible, choose the ratio to be used so that we can multiply rather than divide.

Exercise 38.1a

Find the other sides of the triangles given that angle $B = 90°$.
1. $a = 13.6$ cm, $A = 72°$.   2. $b = 4.7$ cm, $A = 62°$.
3. $c = 8.3$ cm, $C = 20°\ 40'$.   4. $b = 12.0$ m, $C = 28°\ 32'$.
5. $c = 22.6$ cm, $A = 30°$.

Exercise 38.1b

Find the other sides of the triangles given that angle $B = 90°$.
1. $a = 16.7$ cm, $A = 22°$.   2. $a = 23.4$ cm, $C = 72°$.
3. $b = 16.4$ cm, $A = 21°\ 30'$.   4. $b = 6.2$ m, $C = 40°\ 20'$.
5. $c = 7.6$ cm, $A = 42°\ 10'$.

Given two sides

If we know two sides of a triangle, we can find the third side by Pythagoras' Theorem. To find one of the angles, use the two known sides, just in case we have made an error in calculating the length of the third side.

Example 3. *Given that b = 18 cm and c = 15 cm in a right-angled triangle, find a and angle A.*

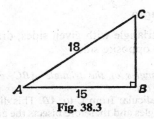

Fig. 38.3

By Pythagoras' Theorem
$$18^2 = 15^2 + a^2,$$
$$a^2 = 324 - 225,$$
$$a = 9.95 \text{ cm.}$$

$$\text{Sin } C = \frac{15}{18}$$

$$\therefore C = 56° \ 26'.$$
$$A = 90° - 56° \ 26' = 33° \ 34'.$$

Example 4. *Given that* $B = 90°$, $a = 8.1$ *cm and* $c = 7.5$ *cm, find angle* A.

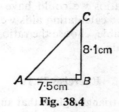

8·1cm

A 7·5cm B

Fig. 38.4

$$\text{Tan } A = \frac{8.1}{7.5}$$

$$\therefore A = 47° \ 12'.$$

Exercise 38.2a

Find the angles of the following triangles, given that angle $B = 90°$.

1. $a = 12.6$ cm, $b = 18.0$ cm.   **2.** $a = 2.4$ cm, $c = 3.2$ cm.
3. $b = 7.2$ cm, $c = 2.43$ cm.   **4.** $a = 8.2$ cm, $b = 9.1$ cm.
5. $c = 24.6$ cm, $b = 28.3$ cm.

Exercise 38.2b

Find the angles of the following triangles, given that angle $B = 90°$.

1. $a = 2$ cm, $b = 3$ cm.   **2.** $a = 7.6$ cm, $c = 4.0$ cm.
3. $b = 20.4$ cm, $c = 7.62$ cm.   **4.** $a = 3.41$ cm, $b = 7.16$ cm.
5. $c = 12.2$ cm, $b = 73.4$ cm.

The isosceles triangle

To solve an isosceles triangle with given sides, draw the perpendicular from the vertex to the opposite side.

Example 5. *Find the angles of the triangle* ABC, *given that* $a = 8.2$ *cm,* $b = 8.2$ *cm and* $c = 9.6$ *cm.*

Draw CD, the perpendicular from C to AB. This divides the triangle ABC into two congruent triangles and therefore bisects the angle C and the side AB.

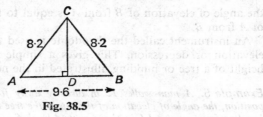

Fig. 38.5

$$\therefore AD = 4.8 \text{ cm,}$$

and in the right-angled triangle ADC,

$$\frac{4.8}{8.2} = \cos A.$$

$$\therefore \cos A = \frac{24}{41} = 0.5854$$

and
$$A = 54° 10'.$$

Since the triangle ABC is isosceles,
$$B = 54° 10' \quad \text{and} \quad C = 180° - 2(54° 10') = 71° 40'.$$

Angles of elevation and depression

The angle of elevation of an object B from an observer at A who is below the level of B is the angle which the line BA makes with the horizontal.

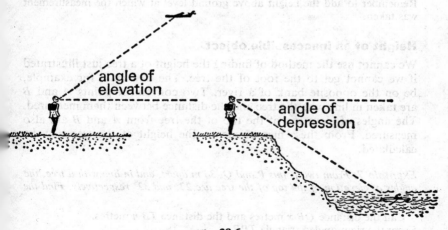

Fig. 38.6

If C is below A, **the angle of depression** of C from A is the angle which AC makes with the horizontal.

N.B. Notice that both angles are measured with the horizontal and that the angle of elevation of B from A is equal to the angle of depression of A from B.

An instrument called the theodolite is used for measuring angles of elevation or depression. This gives a simple method of finding the height of a tree or building, illustrated in the next example.

Example 6. A man walks 12 m directly away from a tree and from this position, the angle of elevation of the top of the tree is 24°. If the measurement is taken from a point 1.5 m above ground level, find the height of the tree.

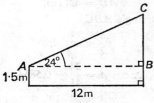

Fig. 38.7

From the figure,

$$\frac{BC}{12} = \tan 24°.$$

$$\therefore \frac{BC}{12} = \tan 24° = 0.4452,$$

and

$$BC = 5.34 \text{ m (to 3 s.f.)}.$$

The height of the tree is therefore 6.84 m (to 3 s.f.).

Remember to add the height above ground level at which the measurement was taken.

Height of an inaccessible object

We cannot use the method of finding the height of a tree just illustrated if we cannot get to the foot of the tree. The tree may, for example, be on the opposite bank of a river. Two convenient points A and B are taken in line with the tree and the distance between them measured. The angles of elevation of the top of the tree from A and B are also measured. From these measurements, the height of the tree may be calculated.

Example 7. From two points P and Q, 30 m apart, and in line with a tree, the angles of elevation of the top of the tree are 22° and 32° respectively. Find the height of the tree.

Call the distance QB x metres and the distance TB h metres.
From the right-angled triangle TPB,

$$\frac{PB}{TB} = \cot 22°.$$

$$\therefore \frac{x + 30}{h} = \cot 22°$$

or $x + 30 = h \cot 22° = 2.4751h$ (i)

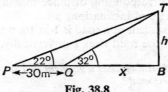

Fig. 38.8

From the right-angled triangle TQB,

$$\frac{QB}{h} = \cot 32° \quad \text{or} \quad x = h \cot 32° = 1.6003h \qquad \text{(ii)}$$

Subtracting (ii) from (i),
$$30 = 2.4751h - 1.6003h$$
$$= 0.8748h.$$

$$\therefore h = \frac{30}{0.8748} = 30(1.143) = 34.3 \text{ m (to 3 s.f.)}.$$

Example 8. *Find a formula for the length AD of the perpendicular from A to BC in the triangle ABC in terms of A, B and C.*

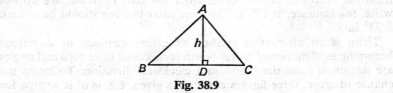

Fig. 38.9

Let $AD = h$. In the right-angled triangle ABD,

$$\frac{BD}{h} = \cot B \quad \therefore BD = h \cot B.$$

In the right-angled triangle ADC,

$$\frac{DC}{h} = \cot C \quad \therefore DC = h \cot C.$$

By addition, $BD + DC = h (\cot B + \cot C)$
$$\therefore a = h (\cot B + \cot C)$$

and $$h = \frac{a}{\cot B + \cot C}.$$

Gradient

It is unfortunate that the word **gradient** has two different meanings. It is used in graphical work to denote the ratio of the vertical distance moved to the horizontal distance. When used in connection with roads or railways, however, the gradient means the ratio of the vertical distance moved to the corresponding distance measured along the slope, i.e. the sine of the angle of inclination.

If we say that the gradient of a road is 1 in 16, we mean that for every 16 m along the slope, the road rises 1 m vertically.

Example 9. A mountain railway has a gradient of 1 in 5. What angle does the track make with the horizontal?

If θ is the required angle,

$$\sin \theta = \tfrac{1}{5} = 0.2.$$
$$\therefore \theta = 11° \ 32'.$$

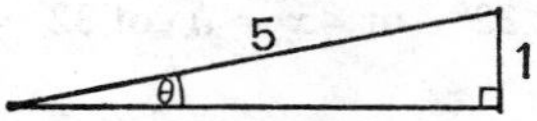

Fig. 38.10

Bearings

The four cardinal directions are North, South, East and West. A bearing of N 27° E means a turn of 27° from the North towards the East; S 31° E means a turn of 31° from the South towards the East.

When bearings are measured in this way, they are always measured from the North or South, never from the East or West. We do not write, for instance, E 17° S. This particular bearing should be written S 73° E.

There is an alternative method of giving bearings or directions becoming used more and more. North is reckoned to be zero and angles are measured from the North in a clockwise direction. To lessen the chance of error, three figures are always given, e.g. 006° is written for 6°, 034° is written for 34°. In this system East is written 090°, South 180° and West 270°. The bearing considered first in this paragraph (N 27° E) may also be written 027°; N 27° W is the same bearing as 333°.

Example 10. A ship sails 3 km due E and then 4 km due N. Find its bearing from the original position.

From the figure,

$$\tan \theta = \tfrac{4}{3} = 1.3333.$$
$$\therefore \theta = 53° \ 8'.$$

and the bearing to the nearest degree is 037°.

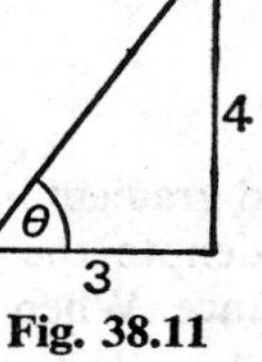

Fig. 38.11

Projections

If AA', BB' are perpendiculars drawn from two points A, B to a given line l, then $A'B'$ is said to be the projection of AB on the line l. We saw

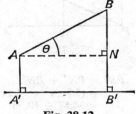

Fig. 38.12

(page 389) that the trig ratios can be defined in terms of projections so that if AB is 1 unit in length, $A'B' = \cos \theta$. If AB is enlarged by a factor k, $A'B'$ is now $k \cos \theta$. Projections are very useful when solving problems on courses and bearings, and when solving quadrilaterals.

Example 11. *A ship steams 3 km from a port P on a bearing of 080° and then 4 km on a bearing of 047°. Find its distance and bearing from P.*

Suppose PQ and QR represent the two courses. Take the East and North directions as axes of reference.

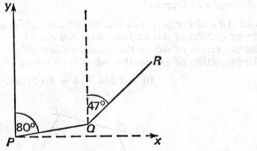

Fig. 38.13

PQ makes an angle of $10°$ with the East; QR makes an angle of $43°$ with the East.
The projection of PQ on the x-axis is $3 \cos 10°$.
The projection of QR on the x-axis is $4 \cos 43°$.
The distance the ship is East of P is

$$3 \cos 10° + 4 \cos 43° = 2.954 + 2.926 = 5.880 \text{ km.}$$

PQ makes an angle of $80°$ with the North;
QR makes an angle of $47°$ with the North.
The distance the ship is North of P is

$$3 \cos 80° + 4 \cos 47° = 3.249 \text{ km.}$$

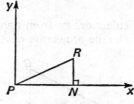

Fig. 38.14

$$PR^2 = PN^2 + RN^2$$
$$= (5.880)^2 + (3.249)^2$$
$$= 34.57 + 10.56$$
$$= 45.13.$$
$$\therefore PR = 6.72 \text{ km (to 3 s.f.).}$$

$$\tan N\widehat{P}R = \frac{RN}{PN} = \frac{3.249}{5.880}$$

and
$$N\widehat{P}R = 28°\ 55'.$$

The distance and bearing of R from P are 6.72 km and 061° to the nearest degree.

Example 12. ABCD is a quadrilateral such that AB = 10 cm, BC = 8 cm, AD = 6 cm, the angle A = 72° and the angle B = 60°. Find the length of CD and the angles at C and D.

Take AB and the perpendicular to AB through A as axes of reference.
The projection of AD on the x-axis is 6 cos 72°.
The projection of BC on the x-axis is 8 cos 60°.
The projection of DC on the x-axis is therefore

$$10 - 1.854 - 4 = 4.146 \text{ cm.}$$

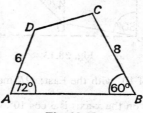

Fig. 38.15

The projection of AD on the y-axis $= 6 \cos 18°$ (or 6 sin 72°).
The projection of BC on the y-axis $= 8 \cos 30°$ (or 8 sin 60°).
The projection of DC on the y-axis $= 8 \cos 30° - 6 \cos 18°$.
$$= 6.928 - 5.707$$
$$= 1.221 \text{ cm.}$$

If CN and DN are drawn parallel to the y-axis and the x-axis respectively:

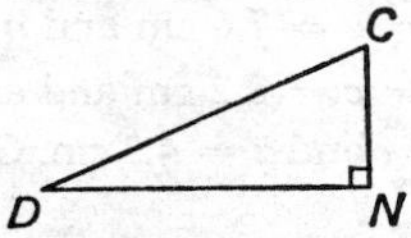

Fig. 38.16

$$CN = 1.221 \quad \text{and} \quad DN = 4.146 \text{ cm.}$$

$$\therefore CD^2 = CN^2 + DN^2$$
$$= (1.221)^2 + (4.146)^2$$
$$= 1.49 + 17.19$$
$$= 18.68.$$
$$\therefore CD = 4.32 \text{ cm (to 3 s.f.).}$$

Also $\qquad \tan N\widehat{D}C = \dfrac{CN}{DN} = \dfrac{1.221}{4.146}$ and $N\widehat{D}C = 16° 24'.$

Angle $D = 108° + N\widehat{D}C = 124° 24'.$

Angle $C = 120° - N\widehat{D}C = 103° 36'.$

Exercise 38.3a

1. In the triangle ABC, $b = c = 12.2$ cm and $a = 8.4$ cm. Find the angle B.
2. In the triangle ABC, $b = c = 6.4$ cm and angle $A = 80°$. Find a.
3. In the triangle ABC, $b = c$ and $a = 7.2$ cm. Given also that $A = 72°$, find b.
4. A man 1.8 m tall is 12 m away from a tree 7.2 m high. What is the angle of elevation of the top of the tree from his eyes?
5. A man 1.8 m tall observes that the angle of elevation of the top of a tree 12m distant is 32°. What is the height of the tree?
6. A man on the top of a cliff 80 m high observes that the angle of depression of a buoy at sea is 12°. How far is the buoy from the cliff?
7. A road has a gradient of 1 in 7.2. What angle does the road make with the horizontal?
8. A mountain railway track is inclined at 14° to the horizontal. Express its gradient in the form $1 : n$.
9. A man on the top of a cliff 100 m high is in line with two buoys whose angles of depression are 17° and 19°. Find the distance between the buoys.
10. A and B are two points in line with a tree such that $AB = 20$ m. The angles of elevation of the top of a tree from A and B are 17° 30′ and 19° respectively. Find the height of the tree.
11. A man walks 1 km due E and then 2 km due S. Find his bearing from his original position.
12. A ship steams 2 km due N and then 3 km on a bearing of 060°. Find its distance and bearing from the original position.

Exercise 38.3b

1. In the triangle ABC, $b = c = 7.6$ cm and $a = 4.2$ cm. Find the angle B.
2. In the triangle ABC, $b = c = 8.2$ cm and angle $A = 80°$. Calculate a.
3. In the triangle ABC, $b = c$ and $a = 4.3$ cm. Given also that angle $A = 64°$, calculate b.
4. A man 1.5 m tall is 15 m away from a building, 24 m high. What is the angle of elevation of the top of the building?
5. A man 1.5 m tall observes that the angle of elevation of the top of a building 24 m away is 41°. What is the height of the building?
6. A man on the top of a cliff 100 m high observes that the angle of depression of a boat at sea is 18°. How far is the boat from the cliff?
7. A road has a gradient of 1 in 6.4. What angle does it make with the horizontal?
8. A funicular railway track is inclined at 17° to the horizontal. If its gradient is 1 in n, find n.
9. A man on the top of a cliff 80 m high is in line with two buoys whose angles of depression are 15° 20′ and 12° 30′. What is the distance between the buoys?
10. P and Q are two points in line with a tree. If PQ is 30 m and the angles of elevation of the top of the tree from P and Q are 12° and 15° respectively, find the height of the tree.
11. A boat sails 4 km due W and then 8 km due S. Find its bearing from its original position to the nearest degree.
12. $AB = 2$ km and $BC = 3$ km. If the bearing of B from A is 024° and the bearing of C from B is 282°, find the bearing of C from A.

Exercise 38.4. Miscellaneous

1. A ladder leaning against a vertical wall makes an angle of 24° with the wall. The foot of the ladder is 5 m from the wall. Find the length of the ladder.
2. The semi-vertical angle of a cone is 38° and the diameter of its base is 4 cm. Find its height.
3. A vertical stick is 8 m high and the length of its shadow is 6 m. What is the angle of elevation of the sun?
4. A bell tent has a radius of 8 m and its pole is 7 m high. What angle does a slant side make with the ground?
5. One angle of a rhombus is 72°. The shorter diagonal is 8 cm long. Find the length of the other.
6. A chord of a circle is 18 cm long and subtends an angle of 110° at the centre. Find the radius of the circle.
7. The greatest and least heights of a bicycle shed are 3 m and 2.5 m. The shed is 3.5 m wide. Find the angle of slope of the roof.
8. An equilateral triangle is inscribed in a circle of radius 5 cm. Find its side.

9. Two men are on opposite sides of a tower. They measure the angles of elevation of the top of the tower as 22° and 18° respectively. If the height of the tower is 100 m, find the distance between the men.

10. Find the angle between the diagonals of a rectangle whose sides are 3 cm and 4 cm.

11. A man standing 60 m away from a tower notices that the angles of elevation of the top and bottom of a flagstaff on top of the tower are 64° and 62° respectively. Find the length of the flagstaff.

12. From a man, the angle of elevation of the top of a tree is 28°. What is the angle of elevation from the man of a bird perched halfway up the tree?

13. A man walks 2 km up a hill whose slope is 1 in 12 and 3 km up a hill whose slope is 1 in 15. How much higher is he than when he started?

14. A stick 4 m long casts a shadow 3 m long when the sun is vertically overhead. What is the inclination of the stick?

15. The legs of a pair of compasses are each 6 cm long and they are used to draw a circle of radius 4 cm. Find the angle between the legs.

16. A cliff 1 km long is represented on a map of scale 2 cm to 1 km by a line of length 0.42 cm. What is the average inclination of the cliff to the horizontal?

17. A man walks 100 m up a slope of 14° and then 50 m up a slope of 12°. How much higher is he than when he started?

18. The bob of a pendulum 4 m long is 25 cm higher at the top of its swing that it is at the bottom. Find the angle of swing on each side of the vertical.

19. A man starts at *A* and walks 2 km on a bearing of 017°. He then walks 3 km on a bearing of 107° to *C*. What is the bearing of *C* from *A*?

20. A slab of stone 2 m by 5 m rests as shown in Fig. 38.17. What is the height of *D* above the ground?

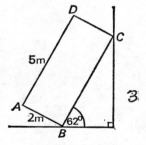

Fig. 38.17

21. A kite flying at a height of 67.2 m is attached to a string inclined at 55° to the horizontal. What is the length of string?

22. A boat 4 km South of the Needles is sailing on a course of 084° at 12 km/h. What is the bearing of the boat from the Needles after half an hour?

23. A man 1.8 m tall can just see the sun over a fence 3 m high, which is 4 m away from him. What is the angle of elevation of the sun?

24. In a circle of radius 6 cm, two radii are drawn making an angle of 40° with each other. Calculate the length of the chord joining the ends of the radii and the length of the perpendicular from the centre to the chord.

25. The vertical angle of a cone of height 6.2 cm is 44°. Find the area of the curved surface.

39 The General Angle

We have considered so far the ratios of acute angles only. Although the definitions in terms of the sides of a right-angled triangle are simple to use, they only apply to right-angled triangles, and when we wish to consider the ratios of angles greater than 90°, or of negative angles, we have to use the definitions in terms of projections (pages 389 and 409) which we saw were consistent with the other definitions.

Projections

If OP is a line length 1 unit, inclined to a base line (taken as Ox) at an

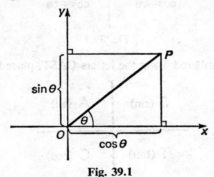

Fig. 39.1

angle θ, then the projection on the x-axis is $\cos \theta$, and the projection on the y-axis is $\sin \theta$. As before, $\tan \theta$ is defined as $\dfrac{\sin \theta}{\cos \theta}$. Cot, sec and cosec are similarly defined (page 391).

Obtuse angles

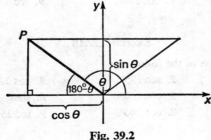

Fig. 39.2

From Fig. 39.2, if $90° < \theta < 180°$, then the projection of OP on the x-axis is negative, and equal in magnitude to cos $(180° -\theta)$. Sin θ, we see, is positive, and equal to sin $(180° - \theta)$.

Signs of the ratios

Whether any one trig ratio is positive or negative will depend on whether the projection is on the positive or negative part of the appropriate coordinate axes, illustrated in Fig. 39.3. The signs of sin, cos and

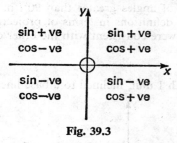

Fig. 39.3

tan can be remembered using the letters CAST, placed as in Fig. 39.4.

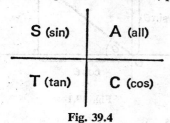

Fig. 39.4

Exercise 39.1a

Write down the signs of the following:

1. tan 170°.
2. sec 320°.
3. cos 160°.
4. sin 220°.
5. cosec 320°.
6. sin (−40°).
7. cos (−100°).
8. cos 310°.
9. sec 170°.
10. sec 190°.

Exercise 39.1b

Write down the signs of the following:

1. tan 210°.
2. sec 195°.
3. cos 125°.
4. sin 236°.
5. cosec 400°.
6. sin (−20°).
7. cos (−80°).
8. cot 240°.
9. sec 250°.
10. cosec 250°.

Magnitudes of the ratios

We have seen that if $90° < \theta < 180°$, then $\sin \theta = \sin (180° - \theta)$, and $\cos \theta = -\cos (180° - \theta)$. Fig. 39.5 shows that if θ is an angle in the third quadrant, $\sin \theta = -\sin (\theta - 180°)$, $\cos \theta = -\cos (\theta - 180°)$, and if θ is in the fourth quadrant, $\sin \theta = -\sin (360° - \theta)$, $\cos \theta = \cos (360° - \theta)$. This can be extended for positive and negative angles of any size, by drawing a figure and marking the appropriate projections.

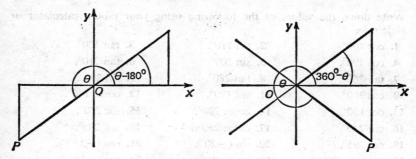

Fig. 39.5

The ratios of any angle are therefore equal numerically to the same ratios of the **acute** angle which the radius defining the angle makes with the **x-axis**.

Example 1. *Find the value of tan 220°.*

First, find its sign.

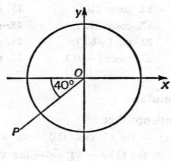

Fig. 39.6

220° is in the third quadrant; its tangent is positive.
The acute angle OP makes with the x-axis is 40°.

$$\therefore \ \tan 220° = + \tan 40° = 0.8391.$$

Example 2. Find the value of cos (−120°).

−120° is measured clockwise from Ox and so is in the third quadrant. Its cosine is therefore negative.

The acute angle OP makes with the x-axis is 60°.

$$\therefore \cos(-120°) = -\cos 60° = -0.5.$$

Exercise 39.2a

Write down the values of the following using your tables, calculator or slide rule.

1. cos 100°.	**2.** sin 110°.	**3.** tan 120°.
4. cos 190°.	**5.** sin 200°.	**6.** tan 210°.
7. tan 260°.	**8.** tan 280°.	**9.** sin 290°.
10. cos 300°.	**11.** sec 120°.	**12.** cosec 140°.
13. cot 150°.	**14.** cosec 200°.	**15.** sec 210°.
16. cot 220°.	**17.** cosec 280°.	**18.** sec 285°.
19. cot 295°.	**20.** sin (−20°).	**21.** cos (−130°).
22. cosec (−210°).	**23.** sec (−190°).	**24.** cot (−185°).

Exercise 39.2b

Write down the values of the following:

1. cos 110°.	**2.** sin 120°.	**3.** tan 130°.
4. cos 200°.	**5.** sin 220°.	**6.** tan 220°.
7. tan 290°.	**8.** sin 180°.	**9.** sin 300°.
10. cos 310°.	**11.** sec 130°.	**12.** cosec 150°.
13. cot 160°.	**14.** cosec 210°.	**15.** sec 220°.
16. cot 230°.	**17.** cosec 290°.	**18.** sec 300°.
19. cot 310°.	**20.** tan (−80°).	**21.** cos (−170°).
22. sec (−170°).	**23.** cosec (−80°).	**24.** cot (−160°).

General angle formulae

We have already mentioned that

$$\sin(180 - x)° = \sin x°; \quad \cos(180 - x)° = -\cos x°$$

and

$$\tan(180 - x)° = -\tan x°.$$

Other similar formulae may be deduced as follows:

Example 3. Simplify cos (270 + x)°.

For convenience consider $x°$ to be acute and then $(270 + x)°$ is in the fourth quadrant. The cosine of an angle in the fourth quadrant is positive.

The arm making $(270 + x)°$ will make an acute angle of $(90 - x)°$ with the x-axis.

$$\therefore \cos (270 + x)° = + \cos (90 - x)°$$
$$= + \sin x°.$$

Example 4. *Simplify cos (360 − x)°.*

If x is considered as acute, $(360 - x)°$ is in the fourth quadrant. $\cos (360 - x)°$ is therefore positive.

The arm making $(360 - x)°$ makes an acute angle $x°$ with the x-axis.

$$\therefore \cos (360 - x)° = + \cos x°.$$

N.B. If x is not acute, the formulae proved in these two examples will still hold.

Exercise 39.3a

Simplify:

1. $\sin (180 + x)°$.
2. $\cos (90 + x)°$.
3. $\tan (270 - x)°$.
4. $\cot (360 - x)°$.
5. $\sec (-x)°$.
6. $\sec (x - 90)°$.
7. $\csc (x - 180)°$.
8. $\tan (270 + x)°$.
9. $\tan (360 - x)°$.
10. $\sec (180 + 2x)°$.

Exercise 39.3b

Simplify:

1. $\cos (180 + x)°$.
2. $\sin (90 + x)°$.
3. $\cos (270 - x)°$.
4. $\tan (360 - x)°$.
5. $\sin (-x)°$.
6. $\csc (x - 90)°$.
7. $\sec (x - 180)°$.
8. $\cot (360 - x)°$.
9. $\sec (360 - x)°$.
10. $\cot (180 + 2x)°$.

Ratios of 0°, 90°, 180°, 270°

The coordinates of A are $(1, 0)$.

$$\therefore \cos 0° = 1; \sin 0° = 0.$$

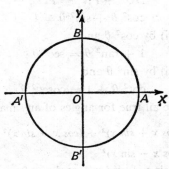

Fig. 39.7

The coordinates of B are $(0, 1)$.

$$\therefore \cos 90° = 0; \sin 90° = 1.$$

The coordinates of A' are $(-1, 0)$.

$$\therefore \cos 180° = -1; \sin 180° = 0.$$

The coordinates of B' are $(0, -1)$.

$$\therefore \cos 270° = 0; \sin 270° = -1.$$

Identities

There are four important identities which exist between the ratios. The first of these follows from the definition of tangent,

$$\tan \theta = \frac{\sin \theta}{\cos \theta} \qquad\qquad \text{(i)}$$

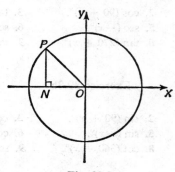

Fig. 39.8

If P is the point on the circle of unit radius such that $x\widehat{O}P = \theta$, then the coordinates of P are $(\cos \theta, \sin \theta)$.

But

$$NO^2 + PN^2 = PO^2,$$

or

$$\cos^2 \theta + \sin^2 \theta = 1 \qquad\qquad \text{(ii)}$$

Divide equation (ii) by $\cos^2 \theta$ and

$$1 + \tan^2 \theta = \sec^2 \theta \qquad\qquad \text{(iii)}$$

Divide equation (ii) by $\sin^2 \theta$ and

$$\cot^2 \theta + 1 = \operatorname{cosec}^2 \theta \qquad\qquad \text{(iv)}$$

These formulae are all true for angles of any magnitude.

Example 5. *Prove* $(\cos x + \sin x)^2 + (\cos x - \sin x)^2 = 2.$

$(\cos x + \sin x)^2 + (\cos x - \sin x)^2$
$$= \cos^2 x + 2 \cos x \sin x + \sin^2 x + \cos^2 x - 2 \cos x \sin x + \sin^2 x$$
$$= 2(\cos^2 x + \sin^2 x) = 2.$$

Example 6. *Prove that cot x + tan x = sec x cosec x.*

(In proving identities, it is often a good plan to express all quantities in terms of cos and sin.)

$$\cot x + \tan x = \frac{\cos x}{\sin x} + \frac{\sin x}{\cos x} = \frac{\cos^2 x + \sin^2 x}{\cos x \sin x}$$

$$= \frac{1}{\cos x \sin x} = \sec x \ \text{cosec} \ x.$$

Example 7. *Find all values of x between 0° and 360° which satisfy the equation*
$$5 \cos x = 2 \sin x.$$

Divide both sides of the equation by cos x
$$5 = 2 \tan x.$$
$$\therefore \ \tan x = 2.5.$$

$x = 68°\ 12'$ is one solution, tan x is positive in the first and third quadrants and so the only other possible solution is the angle in the third quadrant which makes 68° 12′ with the x-axis, i.e.
$$180° + 68°\ 12' = 248°\ 12'.$$
$$x = 68°\ 12' \quad \text{or} \quad 248°\ 12'.$$

Exercise 39.4. Miscellaneous

1. What can you say about x if sin x is positive and cos x negative?

2. By drawing and measurement, find values for cos 80°, sin 100°, cos 220°, sin 340°.

3. If cos $x°$ = − cos 200° and x is acute, find x.

4. If sin $x°$ = sin 160° and x is acute, find x.

5. What values between 0 and 360 satisfy the equation
$$\cos x° = -0.4?$$

6. What values between 0 and 360 satisfy the equation
$$\sin x° = -0.4?$$

7. If sin $x° > \frac{1}{2}$ and x lies between 0 and 360, what further limits can you impose on x?

8. If cos $x° < \frac{1}{2}$, what do you know about x?

Find two values of x between 0° and 360° to satisfy the following equations:

9. sin $x = 0.7$. 10. cos $x = -0.6$. 11. sin $x = 2 \cos x$.

12. cot $x = -2.1$. 13. tan $x = -0.4$. 14. sec $x = -2$.

Simplify the following expressions:

15. cosec $(90° + x)$. 16. sec $(270° - x)$. 17. sec $(-x)$.

18. cos $(180° + x)$. 19. cot $(270° + x)$. 20. tan $(360° - x)$.

21. Find x if sin $x = \frac{4}{5}$ and cos $x = -\frac{3}{5}$.

22. Find x if $\sin x = -\frac{4}{5}$ and $\cos x = -\frac{3}{5}$.

23. Find x if $\sin x = -\frac{4}{5}$ and $\cos x = \frac{3}{5}$.

24. Find x if $\sin x = \frac{4}{5}$ and $\tan x = -\frac{4}{3}$.

25. Find x if $\sin x = -\frac{4}{5}$ and $\tan x = -\frac{4}{3}$.

26. Find x if $\sin x = -\frac{4}{5}$ and $\tan x = +\frac{4}{3}$.

27. If A, B, C are the angles of a triangle, express in terms of A:
    (i) $\sin (B + C)$, (ii) $\cos (B + C)$, (iii) $\tan (B + C)$, (iv) $\sin \frac{1}{2}(B + C)$,
    (v) $\cos 2(B + C)$.

28. What do you know about θ if $\tan \theta$ is less than 1 and $\sin \theta$ is negative?
    ($0° < \theta < 360°$.)

29. Find the values of $\theta(0° < \theta < 360°)$ from the equations (i) $\sin \theta = -0.3$,
    (ii) $\cos \theta = -0.2$, (iii) $\tan \theta = 1.2$.

30. The sine of an obtuse angle is $\frac{3}{4}$. Find the cosine.

40 Graphs

The graphs of sine, cosine and tangent are very important and we should be familiar with the general shape of these three curves. The sine and cosine curves illustrate many types of oscillatory motion, currents in some electrical circuits, the rise and fall of the tides, the motion of bodies on springs, etc.

The sine curve

The sine curve is a wave which lies between the values $+1$ and -1; it is zero at multiples of 180.

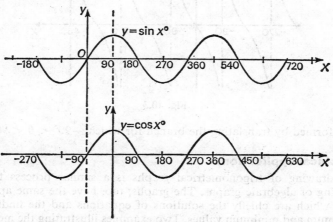

Fig. 40.1

The cosine curve

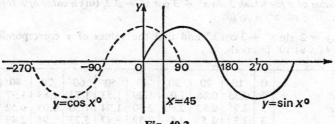

Fig. 40.2

Since we can show that $\cos \theta° = \sin (\theta + 90)°$, the cosine curve is the sine curve translated 90 units to the left (Fig. 40.1). The value of the cosine is zero at odd multiples of 90°. Moreover, since $\cos \theta° = \sin (90-\theta)°$, the cosine curve is the reflection of the sine curve in the line $x = 45°$ (Fig. 40.2).

The tangent curve

The tangent curve is quite different. Since $\tan \theta° = \dfrac{\sin \theta°}{\cos \theta°}$ and $\cos \theta° = 0$ when θ is an odd multiple of 90, $\tan \theta°$ is infinite when θ is an odd multiple of 90. The curve consists of an infinite number of branches,

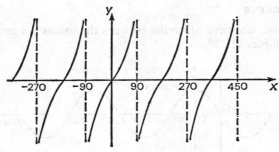

Fig. 40.3

each formed by translating the branch for which $-90° < \theta < 90°$.

Graphical applications

The drawing of trigonometrical graphs is a similar process to the drawing of algebraic graphs. The graphs, too, have the same applications which are chiefly the solutions of equations and the finding of maximum and minimum values. Two examples illustrating the methods used are given.

Example 1. Draw the graph of $2 \sin x° + 3 \cos x°$ between $x = 0$ and $x = 90$. Find from your graph (i) the maximum value of $2 \sin x° + 3 \cos x°$, (ii) a value of x for which $2 \sin x° + 3 \cos x° = 3.2$, (iii) a value of x for which $2 \sin x° + 3 \cos x° = x/10$.

Let $y = 2 \sin x° + 3 \cos x°$ and find the values of y corresponding to values of x at 10° intervals.

x	0	10	20	30	40	50	60	70	80	90
$2 \sin x°$	0	0.35	0.68	1.0	1.29	1.53	1.73	1.88	1.97	2.0
$3 \cos x°$	3	2.95	2.82	2.60	2.30	1.94	1.50	1.03	0.52	0
y	3	3.3	3.5	3.6	3.59	3.47	3.23	2.91	2.49	2.0

(i) The maximum value of 2 sin $x°$ + 3 cos $x°$ is slightly larger than 3.6 and occurs when $x = 34$.

(ii) If $y = 3.2$, from the graph $x = 8$ or 60.5.

(iii) Draw, using the same scales and axes, the graph of $y = x/10$. When $x = 20$, $y = 2$; when $x = 40$, $y = 4$. The graph is the straight line joining these two points.

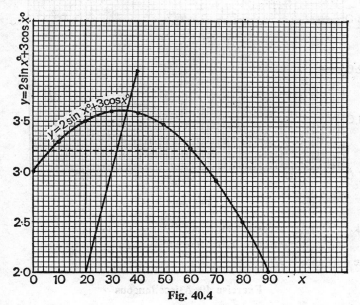

Fig. 40.4

The value of x, 36, at the point of intersection of the straight line with the curve is a value of x for which 2 sin $x°$ + 3 cos $x°$ = $x/10$.

Example 2. Draw, using the same scales and axes, the graphs of $y = tan 2x°$ and $y = 4 cos x° - 1$, for values of x from 10 to 35. Deduce a value of x for which $4 cos x° - tan 2x° = 1$.

First prepare two tables of values.

x	10	15	20	25	30	35
tan 2$x°$	0.36	0.58	0.84	1.19	1.73	2.75

x	10	15	20	25	30	35
4 cos $x°$	3.94	3.86	3.76	3.63	3.46	3.28
4 cos $x°$ − 1	2.94	2.86	2.76	2.63	2.46	2.28

The value of x where the two curves cross (33.5) is a solution of the equation

$$4 \cos x° - 1 = \tan 2x° \quad \text{or of} \quad 4 \cos x° - \tan 2x° = 1.$$

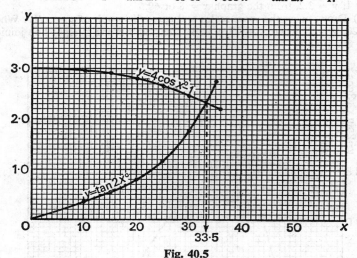

Fig. 40.5

Check. $4 \cos 33° 30' = 4(0.8339) = 3.3356.$

                         $\tan 2(33° 30') = \tan 67° = 2.356.$

The difference is 0.98.

Exercise 40. Miscellaneous

1. Draw the graph of $y = \sin x°$, for values of x from -180 to $+180$. Keep this graph for reference, and compare it with the graphs in Qs. 2 and 3.

2. On a separate sheet of graph paper, draw each of the following graphs for values of x from -180 to $+180$:

    (i) $y = 2 \sin x°$.   (ii) $y = -2 \sin x°$.

    (iii) $y = 1 + \sin x°$.   (iv) $y = \sin x° - 1$.

3. On a separate sheet of graph paper, draw each of the following graphs for the given range of values of x:

    (i) $y = \sin 2x°$, for $-90 < x < 90$.

    (ii) $y = \sin \frac{1}{2}x°$, for $-180 < x < 180$.

    (iii) $y = \sin (x + 45)°$, for $-180 < x < 180$.

4. Draw the graph of $y = \cos x°$, for values of x from -180 to $+180$. Keep this graph for reference, and compare it with the graphs in Qs. 5 and 6.

5. On a separate sheet of graph paper, draw each of the following graphs for values of x from -180 to $+180$.

    (i) $y = 3 \cos x°$.   (ii) $y = -3 \cos x°$.

    (iii) $y = 2 + \cos x°$.   (iv) $y = 2 - \cos x°$.

6. On a separate sheet of graph paper, draw each of the following graphs for the given range of values of x:

   (i) $y = \cos 3x°$, for $-60 < x < 60$.

   (ii) $y = \cos \frac{1}{2}x°$, for $-180 < x < 180$.

   (iii) $y = \cos (30 - x)°$ for $-180 < x < 180$.

7. Draw the graph of $y = \tan x°$ for values of x from -85 to $+85$.

8. Draw the graph of $y = \tan x°$ for values of x from -85 to 85, from 95 to 265, and from 275 to 445.

9. Draw the graph of $y = 4 \cos x° - 1$ for values of x from 0 to 35.

   (i) Read the values of $4 \cos x° - 1$ when $x = 8, 16, 24, 32$.

   (ii) Find the values of x that satisfy $4 \cos x° - 1 = 2.6$,

   (iii) Solve $\cos x° = 0.95$.

10. By adding a suitable straight line graph to that in Q.9, solve the equation $4 \cos x° - 1 = 0.3x$.

11. Draw the graph of $y = \tan 2x°$ for values of x from 0 to 40°. Find the values of x in this range that satisfy (i) $\tan 2x° = 0.4$, (ii) $\tan 2x° = 2.3$.

12. By adding a suitable straight line graph, solve the equation $\tan 2x° = x/15$.

13. Draw the graph of $y = 2 \sin x° + 3 \cos x°$ for values of x from 0 to 90.

   (i) Read the values of $2 \sin x° + 3 \cos x°$ when $x = 8, 25, 72, 77$.

   (ii) Find the values of x in this range that satisfy

$$2 \sin x° + 3 \cos x° = 2.2.$$

14. By adding a suitable straight line graph to that in Q.13, solve the equation $2 \sin x° + 3 \cos x° = x/20$.

15. Find by a graphical method the maximum value of $5 \sin x° + 12 \cos x°$.

41 Problems in Three Dimensions

The plane

A plane is a surface such as the cover of this book in which the line joining any two points of the surface lies entirely in that surface.

A plane is determined by any one of the following:

    (i) two intersecting straight lines,
    (ii) two parallel lines,
    (iii) three points not in the same straight line,
    (iv) a line and a point not on the line.

Two planes which are not parallel intersect in a straight line. An example of this is the intersection of a wall and the floor of a room.

Three planes will in general intersect in a point. An example of this is two intersecting walls and the floor of a room.

Skew lines

Skew lines are lines which are not parallel and which do not meet. They cannot lie in the same plane. The angle between two skew lines is equal to the angle between lines which intersect and which are parallel to the skew lines.

Angle between a line and a plane

If a line PO intersects a given plane at O and PN is the perpendicular from P to the plane, the angle between the line and the plane is defined as the angle PON.

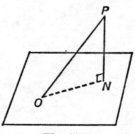

Fig. 41.1

It is the angle between the line and its projection on the plane.

Line of greatest slope

The line of greatest slope in a plane is a line perpendicular to any horizontal line in the plane.

Angle between two planes

Two planes which are not parallel intersect in a straight line. Draw two lines, one in each plane and each perpendicular to the common line of intersection. The angle between these two lines is defined as the angle between the planes.

Problems

When dealing with problems in three dimensions, choose and draw suitable triangles in different planes and find the sides and angles of these triangles as necessary.

Example 1. *The figure shows a room in which AB = 20 m, BC = 16 m and CC' = 12 m. Calculate*

   (i) *the length of the diagonal AC',*
   (ii) *the angle AC' makes with the floor,*
   (iii) *the angle which the plane D'ABC' makes with the floor.*

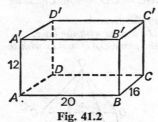

Fig. 41.2

(i) To find AC', consider the triangle ACC'.
From the triangle ACB, which is right angled at B,

$$AC^2 = AB^2 + BC^2 = 20^2 + 16^2 = 656.$$

From the triangle ACC' in Fig. 41.3

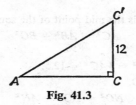

Fig. 41.3

$$AC'^2 = AC^2 + CC'^2$$
$$= 656 + 12^2$$
$$= 800.$$
$$\therefore AC' = 28.3 \text{ m (to 3 s.f.)}$$

(ii) The projection of AC' on the floor is AC.
Therefore the angle between AC' and the floor is CAC'.

From Fig. 41.3, $\sin CAC' = \dfrac{12}{28.3}$.

$$\therefore\ CAC' = 25° 6'.$$

(iii) The intersection of $D'ABC'$ with the floor is AB.
Lines perpendicular to AB, one in each plane, are AD and AD'.

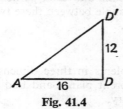

Fig. 41.4

$$\tan DAD' = \tfrac{12}{16} = 0.75.$$
$$\therefore\ DAD' = 36° 52'.$$

Example 2. The figure represents a pyramid with a square base ABCD of side 5 m. OA = OB = OC = OD = 8 m. Calculate

  (i) *the height, ON, of the pyramid,*
 (ii) *the angle which OA makes with the base,*
(iii) *the angle which the plane OAD makes with the base.*

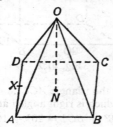

Fig. 41.5

(i) By symmetry, N is the mid point of the square $ABCD$.
$$AC^2 = AB^2 + BC^2$$
$$= 50.$$

Since $AN = \tfrac{1}{2}AC$, $AN^2 = \tfrac{1}{4}AC^2 = 12.5$.

From the triangle ANO,

$$NO^2 = AO^2 - AN^2$$
$$= 64 - 12.5$$
$$= 51.5.$$
$$NO = 7.176 \text{ m} = 7.18 \text{ m (to 3 s.f.)}.$$

(ii) The projection of OA on the base is AN.

Therefore the angle OA makes with the base is $O\widehat{A}N$.
From the triangle OAN,

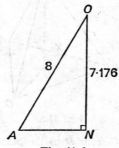

Fig. 41.6

$$\sin O\widehat{A}N = \frac{7.176}{8} = 0.897.$$

$$\therefore\ O\widehat{A}N = 63° 46'.$$

(iii) The intersection of the plane OAD with the base is AD.
If X is the mid point of AD, XN and XO are perpendicular to AD, one lying in each plane.

Fig. 41.7

Therefore the angle between the planes OAD and $ABCD$ is OXN.
From the triangle OXN,

$$\tan O\widehat{X}N = \frac{7.176}{2.5}$$

$$= 2.8704.$$

$$\therefore\ O\widehat{X}N = 70° 48'.$$

Example 3. *ABCD is a tetrahedron (the solid formed by joining four points in space). Given that $AB = BC = CA = 4\,m$ and that $DA = DB = DC = 5\,m$, calculate*

   (i) *the height DG of the tetrahedron,*
   (ii) *the angle BD makes with the plane ABC,*
   (iii) *the angle between the planes DAB and ABC.*

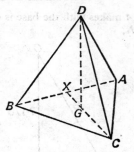

Fig. 41.8

(i) The perpendicular DG from D to the plane ABC meets ABC at the centre of the equilateral triangle ABC.

If X is the mid point of AB, $\widehat{XAG} = 30°$, $AX = 2$ m and
$$\frac{AG}{AX} = \sec 30° = \frac{2}{\sqrt{3}}.$$
$$\therefore AG = \frac{4}{\sqrt{3}} \text{ m.}$$

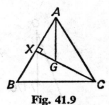

Fig. 41.9

Also $\qquad\qquad XG = 2 \tan 30° = \frac{2}{\sqrt{3}}.$

From the triangle DAG,
$$DG^2 = AD^2 - AG^2$$
$$= 25 - \tfrac{16}{3} = \tfrac{59}{3}$$
$$= 19.67,$$

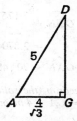

Fig. 41.10

and so $\qquad\qquad DG = 4.435 = 4.44$ m (to 3 s.f.).

(ii) The projection of *BD* on the plane *ABC* is *BG*.
The angle required is therefore *DBG*.

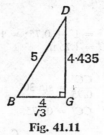

Fig. 41.11

From the triangle *DBG*,

$$\sin \widehat{DBG} = \frac{4.435}{5} = 0.887.$$

$$\therefore \ \widehat{DBG} = 62° \ 30'.$$

(iii) The planes *DAB* and *ABC* intersect in *AB*. *GX* and *XD* are perpendicular to *AB*, one in each plane, and the angle required is *DXG*.

Fig. 41.12

From the triangle *DXG*,

$$\tan \widehat{DXG} = \frac{4.435}{2/\sqrt{3}} = 3.841.$$

$$\therefore \ \widehat{DXG} = 75° \ 24'.$$

Example 4. *ABCD is a desk, 1 m by 0.75 m, which is inclined at 30° to the horizontal. Find the inclination of the diagonal AC to the horizontal.*

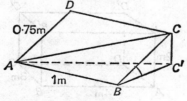

Fig. 41.13

If C' is the foot of the perpendicular from C to the horizontal plane through AB, the projection of AC on this horizontal plane is AC'.

The angle required is CAC'.

From the triangle CAB,

$$AC^2 = 1^2 + 0.75^2 = 1.5625$$
$$\therefore AC = 1.25 \text{ m.}$$

From the triangle BCC',

$$\frac{CC'}{0.75} = \sin 30° = \tfrac{1}{2}$$
$$\therefore CC' = 0.375.$$

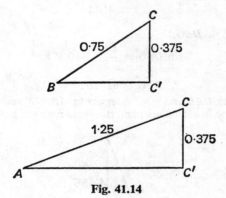

Fig. 41.14

From the triangle CAC',

$$\sin \widehat{CAC'} = \frac{0.375}{1.25} = 0.3$$
$$\therefore \widehat{CAC'} = 17° \; 28'.$$

Exercise 41.1a

The figure shows a box in which $AB = 4$ m, $BC = 3$ m and $CC' = 2$ m. Calculate the following:

1. The angle between the lines AB and AC'.
2. The angle between the lines AC' and AC.

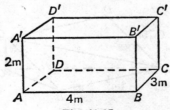

Fig. 41.15

3. The angle between the lines AC' and $A'C$.
4. The angle between the lines AC' and BD'.
5. The angle AC' makes with the plane $ABCD$.
6. The angle AD' makes with the plane $ABCD$.
7. The angle AB' makes with the plane $ABCD$.
8. The angle between the planes $ABCD$ and $ABC'D'$.
9. The angle between the planes $ADC'B'$ and $ABCD$.
10. The angle between the planes $AD'B'$ and $ABCD$.

Exercise 41.1b

The figure shows a box in which $PQ = 3$ m, $QR = 4$ m and $QQ' = 3$ m. Calculate the following:

1. The angle between the lines PQ' and PS.
2. The angle between the lines PQ' and PQ.
3. The angle between the lines PQ and PS'.

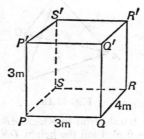

Fig. 41.16

4. The angle between the lines PQ' and PR'.
5. The angle PQ' makes with the plane $PQRS$.
6. The angle PR' makes with the plane $PQRS$.
7. The angle PS' makes with the plane $PQRS$.
8. The angle between the planes $PQR'S'$ and $PQRS$.
9. The angle between the planes $PSR'Q'$ and $PQRS$.
10. The angle between the planes $PQ'S'$ and $PQRS$.

Exercise 41.2. Miscellaneous

1. The edges of a box are 3 cm, 6 cm and 7 cm. Calculate the angles which a diagonal makes with the faces.
2. The face of a desk is a square and slopes at 30° to the horizontal. Find the angle which a diagonal of the square makes with the horizontal.

3. Fig. 41.17 shows a pyramid on a square base, of side 8 m. If $OA = OB = OC = OD = 12$ m, find the height ON of the pyramid.

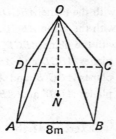

Fig. 41.17

4. In Fig. 41.17, find the angle OA makes with the plane $ABCD$.
5. In Fig. 41.17, find the angle between the planes AOD and $ABCD$.
6. In Fig. 41.17, find the angle between the planes DOA and COB.

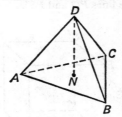

Fig. 41.18

7. Fig. 41.18 shows a tetrahedron in which $AB = AC = CB = 4$ m and $DA = DB = DC = 6$ m. Find the height DN of the tetrahedron.

8. In Fig. 41.18, find the angle DA makes with the plane ABC.

9. In Fig. 41.18, find the angle between the planes ADC and ABC.

10. The face of a clock is inclined at 30° to the vertical. If a hand is horizontal at 15.00h, what is the inclination of a hand to the horizontal at 14.00h?

11. In order to climb a hill of gradient 1 in 6, a cyclist rides so that he makes an angle of 45° with the line of greatest slope. What is the gradient of his route?

12. An isosceles triangle ABC with $AB = AC$ is placed so that BC is horizontal and the plane of the triangle is inclined at 30° to the horizontal. If angle $ABC = 72°$, find the inclination of AB to the horizontal.

13. A ring of radius 10 cm is suspended from a point by four equal strings tied symmetrically to the ring so that it hangs horizontally. If the length of each string is 12 cm, find the inclination of a string to the vertical.

14. Fig. 41.19 shows a pyramid on a rectangular base. If $AB = 2$ m, $BC = 3$ m and $VA = VB = VC = VD = 4$ m, find (i) the height VN of the pyramid, (ii) the angle VA makes with the plane $ABCD$, (iii) the angle between the planes VAB and $ABCD$.

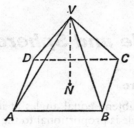

Fig. 41.19

15. *ABCDO* is a pyramid in which the horizontal base *ABCD* is a square of side 8 cm. If $OA = OB = OC = OD = 12$ cm, find (i) the height *ON* of the pyramid, (ii) the angle *OA* makes with the horizontal, (iii) the angle the plane *OAB* makes with the horizontal, (iv) the angle between *OA* and *OC*, (v) the angle between the planes *OAD* and *OBC*.

16. A tetrahedron *ABCD* has its horizontal base *ABC* an equilateral triangle of side 3 m. If $DA = DB = DC = 6$ m, find (i) the height *DN* of the tetrahedron, (ii) the angle *DA* makes with the horizontal, (iii) the angle between the plane *ADC* and the horizontal, (iv) the angle between the planes *DAB* and *DBC*.

17. From a point *A* due south of a tower, the angle of elevation of the top of the tower is 5° 30′. From a point *B* due east of the tower, the angle of elevation of the top of the tower is 8° 20′. If the distance $AB = 200$ m, find the height of the tower.

18. A hillside is a plane which slopes at an angle α to the horizontal. A track on the hillside makes an angle β with the line of greatest slope. Find the inclination of the track to the horizontal.

19. A ring of radius 8 cm is suspended by 5 strings each of length 12 cm attached symmetrically to the ring, which hangs horizontally. Find (i) the angle between adjacent strings, (ii) the angle which each string makes with the horizontal.

20. A square *ABCD* is rotated through an angle of 30° about the side *AB*. Find the angle between the old and new positions of the diagonal *AC*.

42 The Circle and Sphere

Length of circular arc

Equal arcs of a circle subtend equal angles at its centre. Therefore the length of an arc of a circle is proportional to the angle it subtends at the centre. For example, if one arc of a circle is double another, the angle which the first arc subtends at the centre of the circle is double the angle subtended by the second arc.

The circumference of a circle is $2\pi r$ and the angle the whole circum-

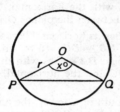

Fig. 42.1

ference subtends at the centre is 360°. Therefore, if an arc PQ of a circle of radius r subtends an angle $x°$ at the centre of the circle,

$$\frac{\text{arc } PQ}{2\pi r} = \frac{x}{360},$$

and the length of the arc PQ is $\dfrac{2\pi rx}{360}$ or $\dfrac{\pi rx}{180}$.

Area of circular sector

Similarly the area of the circular sector POQ is proportional to the angle x. The area of the whole circle is πr^2.

$$\therefore \quad \frac{\text{Area of sector } POQ}{\pi r^2} = \frac{x}{360},$$

and the area of the sector is $\dfrac{\pi r^2 x}{360}$.

Area of segment

The chord PQ divides the area of the circle into two parts, called the major segment and the minor segment. The area of the minor segment is equal to (the area of the sector POQ) — (the area of the triangle POQ).

The area of the triangle POQ is $\frac{1}{2}r^2 \sin x$.

$\therefore$ the area of the minor segment $POQ = \dfrac{\pi r^2 x}{360} - \frac{1}{2}r^2 \sin x$

$$= \frac{1}{2}r^2 \left(\frac{\pi x}{180} - \sin x \right).$$

Radian measure

The formulae for length of arc and area of sector may also be expressed in terms of radian measure.

A **radian** is the angle subtended at the centre of a circle by an arc equal in length to its radius. Since the angle at the centre is proportional to the arc on which it stands, an arc of length πr will subtend an angle of π radians (written π rad) at the centre. But this is the angle subtended at the centre by a semi-circle and therefore

$$\pi \text{ rad} = 180° \quad \text{and} \quad 1 \text{ rad} = 57.3° \text{ approximately.}$$

A table is provided in most books of tables to help in converting degrees to radians and vice-versa; an appropriate key is on most electronic calculators.

If the angle subtended at the centre of a circle of radius r by an arc is x radians, the length of the arc is rx.

In this case, the area of the sector $= \dfrac{x}{2\pi} \times$ (area of circle)

$$= \frac{x}{2\pi}(\pi r^2)$$

$$= \frac{1}{2}xr^2.$$

Example 1. A chord PQ of a circle of radius 5 cm subtends an angle of 70° at the centre. Find (i) the length of the chord PQ, (ii) the length of the arc PQ, (iii) the area of the sector POQ, (iv) the area of the minor segment cut off by PQ.

Let O be the centre of the circle and ON the perpendicular from O to PQ. Then N is the mid point of the chord PQ and NO bisects the angle POQ.

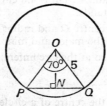

Fig. 42.2

(i) From the triangle PON,

$$\frac{PN}{5} = \sin 35° = 0.5736.$$

$$\therefore PN = 2.868$$

and the length of the chord PQ is 5.74 cm.

(ii)
$$\frac{\text{The arc } PQ}{2\pi(5)} = \frac{70}{360}.$$

$$\therefore \text{ the length of the arc } PQ = \frac{70\pi}{36} = 6.11 \text{ cm.}$$

(iii)
$$\frac{\text{The area of the sector } POQ}{\pi(5)^2} = \frac{70}{360}.$$

$$\therefore \text{ the area of the sector } = \frac{7 \times 25\pi}{36}$$

$$= 15.28 \text{ or } 15.3 \text{ cm (to 3 s.f.).}$$

(iv)
$$\text{The area of the triangle } POQ = \tfrac{1}{2}(5)^2 \sin 70°$$
$$= 11.74 \text{ or } 11.7 \text{ cm}^2$$

$$\therefore \text{ the area of the minor segment } = 15.28 - 11.74$$
$$= 3.54 \text{ cm}^2 \text{ (to 3 s.f.).}$$

Exercise 42.1a

1. A chord PQ of a circle of radius 4 cm subtends an angle of 50° at the centre. Find the length of the arc PQ.

2. A chord XY of a circle of radius 6.2 cm subtends an angle of 38° at the centre of the circle. Find the difference in length between the chord and the minor arc XY.

3. A chord PQ of a circle of radius 5.5 cm subtends an angle of 42° at the centre O. Find the area of the sector POQ.

4. A chord AB of a circle of radius 10.4 cm subtends an angle of 35° at the circumference. Find the area of the minor segment cut off by the chord.

5. A chord of a circle of radius 12 cm is distant 5 cm from the centre. Find the length of the major arc cut off by the chord.

6. A chord of a circle of radius 7 cm is 4.8 cm long. Find the length of the minor arc on which it stands.

7. A chord AB of a circle, centre O and radius 5.6 cm, is 3.0 cm long. Find the ratio of the areas of the major and minor sectors AOB.

8. Find the angles subtended at the circumference of a circle of radius 8 cm by a chord 6 cm long.

9. Convert 117° 10′ to radians.

10. The angle subtended at the centre of a circle of radius 5 cm is 2 radians. Find the area of the sector.

Exercise 42.1b

1. A chord PQ of a circle of radius 5 cm subtends an angle of 40° at the centre. Find the length of the minor arc PQ.

2. A chord XY of a circle of radius 6.8 cm subtends an angle of 42° at the centre of the circle. Find the difference in length between the chord XY and the minor arc XY.

3. A chord PQ of a circle of radius 6.2 cm subtends an angle of 52° at the centre O. Find the area of the minor sector POQ.

4. A chord AB of a circle of radius 8.2 cm subtends an angle of 50° at the circumference. Find the area of the minor segment cut off by AB.

5. A chord of a circle of radius 18 cm is distant 6 cm from the centre. Find the length of the minor arc cut off by the chord.

6. A chord of a circle of radius 8 cm is 4.2 cm long. Find the length of the minor arc on which it stands.

7. A chord AB of a circle, centre O and radius 10 cm, is 4 cm long. Find the ratio of the areas of the major and minor sectors AOB.

8. Find the angles subtended at the circumference of a circle of radius 4.8 cm by a chord 3.0 cm long.

9. Express 1.42 radians in degrees and minutes.

10. The area of a sector of a circle of radius 4 cm is 20 cm². Find the angle of the sector in radians.

Latitude and longitude

In Fig. 42.3, N and S represent the North and South poles of the Earth. The line SN is called the axis of the Earth and O is the centre of the Earth.

The Equator is the line in which the Earth's surface is cut by a plane through the centre perpendicular to NS.

A Great Circle is a section of the Earth's surface by any plane through O. Its radius is equal to the radius of the Earth. The shortest distance along the Earth's surface between any two places is the minor arc of the Great Circle passing through them.

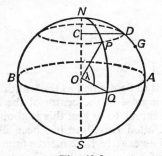

Fig. 42.3

A Small Circle is any other circle lying on the Earth's surface.

A Meridian of Longitude is a great circle which passes through the North and South poles.

The Prime Meridian is that meridian which passes through Greenwich.

A Parallel of Latitude is a section of the Earth's surface by a plane parallel to the equator.

Suppose that any meridian of longitude $NPQS$ cuts the equator at Q and a parallel of latitude at P. The angle which PO makes with the plane of the equator is called the latitude of P. In the figure, it is shown as the angle POQ ($\lambda°$).

All places on the same parallel of latitude have the same latitude. The latitude of a place can vary from 90° N (North Pole) to 90° S (South Pole).

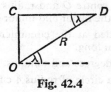

Fig. 42.4

If NAS is the prime meridian, the angle between the planes $NPQS$ and NAS is equal to the longitude of P. It is represented by the angle PCD or by the angle QOA. All places on the same meridian of longitude have the same longitude. The longitude of a place can vary from 180° E to 180° W.

Considering the triangle COD and calling the radius of the Earth R the radius of the parallel of latitude λ is CD.

Since $\dfrac{CD}{R} = \cos \lambda$, the radius of the parallel of latitude λ is $R \cos \lambda$.

If $\theta°$ is the longitude of P, then $PCD = \theta°$. The length of the arc PD is $\dfrac{\theta}{360} \times 2\pi R \cos \lambda$.

The nautical mile

The nautical mile used to be defined as the length of arc of the meridian which subtends one minute at the centre of the Earth. If the Earth is regarded as a sphere radius 6370 km, this is about 1.85 km (6080 feet). But since 1970 the United Kingdom has been using the international nautical mile defined as exactly 1852 metres (about 6076.115 feet).

A **knot** is a speed of 1 nautical mile per hour.

Local time

Local time at any place P depends on the longitude of P. Two places diametrically opposite on the equator differ in time by 12 hours and in longitude by 180°. The difference in time for 1° longitude is therefore $\dfrac{12 \times 60}{180}$ or 4 minutes.

Example 2. Two places P, Q both on the parallel of latitude 26° N differ in longitude by 40°. Find (i) the distance between them along their parallel of latitude, (ii) the shortest distance between them along the Earth's surface.

The radius of the parallel of latitude = 6370 cos 26° km.

(i) The distance between the places along the parallel of latitude
$$= \tfrac{40}{360} \text{ of the circumference of this circle}$$
$$= \tfrac{40}{360} \times 2\pi \times 6370 \cos 26°$$
$$= 3997 \text{ km} = 4000 \text{ km (to 3 s.f.).}$$

(ii) The chord PQ subtends an angle of 40° at the centre of a circle of radius 6370 cos 26°.

$$\therefore \quad \frac{\tfrac{1}{2}PQ}{6370 \cos 26°} = \sin \frac{40°}{2} = \sin 20°$$

and
$$\tfrac{1}{2}PQ = 6370 \cos 26° \sin 20°.$$

Suppose that PQ subtends an angle of $2x°$ at the centre of the Earth.

Then
$$\sin x = \frac{\tfrac{1}{2}PQ}{6370} = \cos 26° \sin 20°$$
$$\log \sin x = \bar{1}.9537 + \bar{1}.5341 = \bar{1}.4878$$
$$\text{and } x = 17° \ 54'.$$

The angle subtended by PQ at the centre of the Earth is $2x$ or 35° 48'. The minor arc of this great circle is the shortest distance between P and Q along the Earth's surface.

$$\text{This distance} = \frac{35° \ 48'}{360°} \times 2\pi \times 6370 = \frac{35.8 \times 2\pi \times 6370}{360}$$
$$= 3980 \text{ km (to 3 s.f.).}$$

Exercise 42.2a

(Take the radius of the Earth to be 6370 km.)

1. Two places on the Equator differ in longitude by 24°. What is the distance between them?

2. Find the radius of the circle of latitude 60° N.

3. Find the distance along the parallel of latitude between two places on the parallel latitude 50° N which differ in longitude by 36°.

4. The Earth rotates on its own axis once in 24 hours. What is the speed of a place whose latitude is 30° N?

5. Find the difference in local time between Paris (48° 50′ N, 2° 20′ E) and Greenwich.

6. A ship sails 100 km due east. If her longitude changes by 4°, find her latitude.

7. What is the distance over the North Pole of two places both in latitude 60° N, if their longitudes differ by 180°?

8. Two places on the same meridian have latitudes 30° N and 20° S. Find the distance between them along their meridian.

9. Compare the distances travelled in an hour by two places of latitudes 30° N and 60° N respectively.

10. Two places P and Q both in latitude 40° N differ in longitude by 30°. Find the distance between them measured along the parallel of latitude.

Exercise 42.2b

(Take the radius of the Earth to be 6370 km.)

1. Two places on the Equator differ in longitude by 32°. Find the distance between them.

2. Find the radius of the circle of latitude 30° S.

3. Find the distance along the parallel of latitude between two places on the parallel of latitude 40° N which differ in longitude by 24°.

4. Find the distance moved in 30 minutes by a place whose latitude is 40° S.

5. Find the difference in local time between two places both in latitude 35° S if their longitudes differ by 26°.

6. A ship sails 100 km due north. Find the change in her latitude.

7. Find the distance over the South Pole between two places both in latitude 70° S if their longitudes differ by 180°.

8. Two places on the same meridian have latitudes 40° N and 50° S. What is their distance apart along their meridian?

9. Compare the distances travelled in an hour by two places of latitudes 30° N and 20° S respectively.

10. Two places P and Q both in latitude 60° N differ in longitude by 28° 20′. Find the distance between them measured along their parallel of latitude.

Exercise 42.3. Miscellaneous

1. The minute hand of a clock is $2\frac{1}{2}$ cm long. Find the distance moved by the tip in 35 minutes.

2. An arc of a circle of radius 5 cm is 4 cm long. Find the length of the chord joining its ends.

3. A piece of wire in the form of an arc of a circle of radius 10 cm and subtending an angle of 50° at the centre is bent into the form of a complete circle. Find its radius.

4. An equilateral triangle is inscribed in a circle of radius 5 cm. Find the area of the minor segment cut off by one of the sides.

5. A piece of wire in the form of a square of side 4 cm is bent into an arc of a circle of radius 10 cm. Find the distance between the ends of the wire.

6. A regular hexagon is inscribed in a circle of radius 5 cm. Find the area of the hexagon.

7. A regular octagon (8-sided figure) is inscribed in a circle of radius 10 cm. Find the area of the octagon.

8. A continuous belt (Fig. 42.5) passes round two circles of radii 3 m and 5 m whose centres are 10 m apart. Find the length of the belt.

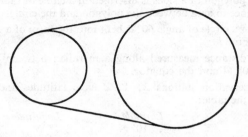

Fig. 42.5

9. A sector of a circle of angle 40° is bent into the form of a cone. Find the semi-vertical angle of the cone.

10. A cone of semi-vertical angle 30° is bent into the form of a sector of a circle. Find the angle of the sector.

11. If P and Q are two points on the parallel of latitude 60° S such that the difference in their longitudes is 90°, find the angle subtended by PQ at the centre of the Earth.

12. A sphere centre C is of radius R. Two points P, Q on its surface are such that the angle PCQ is 20°. Find the length of the chord PQ.

13. Find the distance measured along the surface of the Earth of the North Pole from any place in latitude 60° N.

14. A ship sails 100 km due east and finds that her longitude has altered by 2°. Find her latitude.

15. Find the distance measured along their parallel of latitude between Greenwich (latitude 51° N) and a place of latitude 51° N and longitude 90° E.

16. A chord XY of a circle of radius 10 cm subtends an angle of 80° at the centre O. Find (i) the length of the chord XY; (ii) the area of the sector XOY.

17. Two places P and Q both in latitude 40° N differ in longitude by 15°. Find (i) the distance between P and Q measured along their parallel of latitude, (ii) the length of the straight line joining P to Q.

18. The ropes of a swing are 3.6 m long and the seat when stationary is 60 cm above the ground. If at the highest points of its arc, the seat is 1.8 m above the ground, find the length of arc of the swing.

19. An arc of a circle of radius 5 cm is 4 cm long. What angle does the arc subtend at the centre of the circle?

20. Find the distance travelled in 1 hour due to the rotation of the Earth by a place whose latitude is 48° N.

21. A regular hexagon circumscribes a circle of radius 8 cm. Find the area of the hexagon.

22. What is the area of the minor segment of a circle of radius 4 cm cut off by a chord of length 5 cm?

23. A regular polygon of 8 sides is inscribed in a circle of radius 10 cm. Find the difference in area between the polygon and the circle.

24. A sector of a circle of angle 60° is bent into the form of a cone of vertical angle $\theta°$. Find θ.

25. Find the distance measured along a meridian between the parallel of latitude 30° N and the equator.

26. Three observation stations X, Y, Z have latitudes and longitudes as shown in the table:

	Latitude	Longitude
X	42° 10′ N	1° W
Y	42° 10′ N	1° 30′ W
Z	42° N	1° 30′ W

Find (i) the distance between X, Y measured along their parallel of latitude; (ii) the distance between Y and Z measured along the meridian.

27. Two places both in latitude 35° N differ in longitude by 60°. Calculate the distance between the two places measured along the parallel of latitude.

28. A ship is in latitude 52° N, longitude 24° W. Find how far the ship is from the North Pole measured along the meridian.

29. A regular octagon is circumscribed about a circle of radius 8 cm. Find the area of the octagon.

30. What is the shortest distance over the Earth's surface between P (32° N; 8° W) and Q (40° N; 172° E)?

43 Sine and Cosine Formulae

If a triangle has its angles given in size, its sides can vary in length but will always be in the same ratio (see similar triangles). We do not know, however, how the ratios of the sides depend on the angles. We do know that if two of the angles are equal, the sides opposite the equal angles are also equal (isosceles triangle). But will the side opposite the 60° angle in a 30°, 60°, 90° triangle be twice as long as the side opposite the 30° angle? The answer to this question is no. Sides are not proportional to the opposite angles but to the sines of the opposite angles and this fact is commonly given in the sine formula,

$$\frac{a}{\sin A} = \frac{b}{\sin B} = \frac{c}{\sin C}.$$

THE SINE FORMULA

Proof of the sine formula

This is proved for acute and obtuse-angled triangles.

(i) *When the triangle ABC is acute-angled*
Draw AD perpendicular to BC.

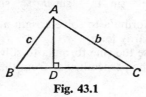

Fig. 43.1

From the triangle ABD,
$$AD = c \sin B;$$
from the triangle ADC,
$$AD = b \sin C.$$
$$\therefore c \sin B = b \sin C$$
i.e.
$$\frac{b}{\sin B} = \frac{c}{\sin C}.$$
Similarly, by drawing the perpendicular from B to AC,
$$\frac{a}{\sin A} = \frac{c}{\sin C}.$$
$$\therefore \frac{a}{\sin A} = \frac{b}{\sin B} = \frac{c}{\sin C}.$$

(ii) *When the angle C is obtuse*

Draw the perpendicular AD from A to BC produced.
From the triangle ACD,

$$AD = b \sin (180° - C)$$
$$= b \sin C.$$

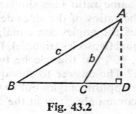

Fig. 43.2

From the triangle ABD,

$$AD = c \sin B.$$

$$\therefore b \sin C = c \sin B \quad \text{or} \quad \frac{b}{\sin B} = \frac{c}{\sin C}$$

As before,

$$\frac{a}{\sin A} = \frac{b}{\sin B} = \frac{c}{\sin C}$$

Never use the sine formula to solve a right-angled triangle; it is not wrong but is unnecessarily long.

Generally do not use the sine formula to solve an isosceles triangle; it is simpler to draw the perpendicular from the vertex to the base.

Given two angles and a side

Given two angles of a triangle and a side, the other sides and angles may be calculated.

Example 1. In a triangle ABC, A = 40°, B = 52° and a = 30.2 cm. Calculate the other sides and angles.

$$C = 180° - 40° - 52° = 88°.$$

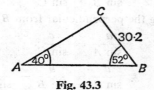

Fig. 43.3

Using the sine formula,

$$\frac{30.2}{\sin 40°} = \frac{b}{\sin 52°} = \frac{c}{\sin 88°}.$$

$$\therefore b = \frac{30.2 \sin 52°}{\sin 40°} \quad \text{and} \quad c = \frac{30.2 \sin 88°}{\sin 40°}.$$

Therefore $b = 37.0$ cm and $c = 46.9$ cm (to 3 s.f.), using calculator, slide rule or logarithms.

The ambiguous case

Suppose that we are given two sides and one angle of a triangle. If the angle is included between the given sides, the shape of the triangle is fixed.

If the angle is not included, there are two possible shapes for the triangle. Suppose that the angle A is given and also the sides c and a.

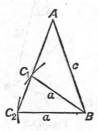

Fig. 43.4

Construct the angle A and mark off AB equal to its given length. With centre B and radius equal to the given value of a draw an arc. This arc will generally cut the third side at two points C_1, C_2, and ABC_1 and ABC_2 are the two triangles satisfying the given conditions.

Since BC_1C_2 is isosceles,

$$\text{angle } BC_1C_2 = \text{angle } BC_2C_1$$

and the angles AC_1B, AC_2B are supplementary.

The ambiguous case arises when, and only when, the arc cuts at two points C_1, C_2 on the same side of A.

For this to occur, the angle A must be acute and a must be smaller than c.

For the arc to cut at all, a must be greater than the perpendicular from B to AC.

$$\therefore a > c \sin A.$$

The two cases arise quite naturally in a trigonometrical solution.

Example 2. Solve the triangle ABC given that a = 7, c = 12 and A = 30°.

Using the sine formula,

$$\frac{7}{\sin 30°} = \frac{12}{\sin C}.$$

$$\therefore \ \sin C = \frac{12 \sin 30°}{7} = 0.8571,$$

and $\qquad\qquad C = 58° \ 59' \quad \text{or} \quad 121° \ 1'.$

(We have found the two supplementary values for the angle C.)

(i) If $C = 58° \ 59'$, $A = 30°$, $B = 180° - 58° \ 59' - 30° = 91° \ 1'$.
Using the sine formula,

$$\frac{7}{\sin 30°} = \frac{b}{\sin 91° \ 1'}$$

and $\qquad\qquad\qquad b = 14.0.$

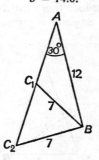

Fig. 43.5

(ii) If $C = 121° \ 1'$ and $A = 30°$, $B = 180° - 151° \ 1' = 28° \ 59'$.
Using the sine formula,

$$\frac{7}{\sin 30°} = \frac{b}{\sin 28° \ 59'}$$

and $\qquad\qquad\qquad b = 6.8.$

Exercise 43.1a

Solve the following triangles:

1. $a = 8.52$, $B = 28°$, $C = 64°$.
2. $b = 7.24$, $B = 86°$, $C = 48°$.
3. $c = 8.4$, $B = 112°$, $C = 42°$.
4. $a = 4.21$, $B = 62°$, $C = 54°$.
5. $b = 28.1$, $A = 81°$, $B = 42° \ 30'$.
6. $A = 42°$, $a = 4.2$, $b = 7.1$.
7. $A = 38°$, $a = 7.1$, $b = 4.2$.
8. $A = 80°$, $a = 12.3$, $b = 12.8$.

9. $A = 26°$, $a = 4.3$, $b = 5.2$.
10. $A = 51°$, $a = 2.7$, $b = 3.1$.

Exercise 43.1b

Solve the following triangles:
1. $a = 21.7$, $B = 42°$, $C = 51°$.
2. $b = 8.34$, $B = 79°$, $C = 24°$.
3. $c = 11.1$, $B = 108°$, $C = 25°$.
4. $a = 14.8$, $B = 62° 30'$, $C = 51°$.
5. $b = 28.7$, $A = 80° 24'$, $B = 61° 27'$.
6. $A = 41° 30'$, $a = 4.8$, $b = 6.0$.
7. $A = 112°$, $a = 7.1$, $b = 4.2$.
8. $A = 80° 27'$, $a = 10.8$, $b = 11.9$.
9. $A = 22° 10'$, $a = 3.7$, $b = 4.4$.
10. $A = 49°$, $a = 22.7$, $b = 31.2$.

THE COSINE FORMULA

The cosine formula is used to solve a triangle given either (i) three sides, or (ii) two sides and the included angle.

It tells us that in any triangle ABC,
$$c^2 = a^2 + b^2 - 2ab \cos C.$$

Angle C acute 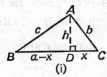   Angle C obtuse

(i)   (ii)

Fig. 43.6

Draw AD perpendicular to BC.

In the triangle ADC,
$$h^2 = b^2 - x^2.$$

In the triangle ADB,
$$h^2 = c^2 - (a - x)^2.$$
$$\therefore b^2 - x^2 = c^2 - (a - x)^2$$
or $$b^2 = c^2 - a^2 + 2ax.$$

Draw AD perpendicular to BC produced.

In the triangle ADC,
$$h^2 = b^2 - x^2.$$

In the triangle ADB,
$$h^2 = c^2 - (a + x)^2.$$
$$\therefore b^2 - x^2 = c^2 - (a + x)^2$$
or $$b^2 = c^2 - a^2 - 2ax.$$

From the triangle ACD,

$$\frac{x}{b} = \cos C;$$

From the triangle ACD,

$$\frac{x}{b} = \cos (180° - C)$$

$$= - \cos C;$$

substituting,

$$c^2 = a^2 + b^2 - 2ab \cos C.$$

substituting,

$$c^2 = a^2 + b^2 - 2ab \cos C.$$

The formula $c^2 = a^2 + b^2 - 2ab \cos C$ is therefore true for every triangle ABC whether C is acute or obtuse; but remember that, if C is obtuse, the value of $\cos C$ is negative.

If $C = 90°$, $\cos C = 0$ and the formula becomes $c^2 = a^2 + b^2$ (Pythagoras).

If C is acute, $c^2 < a^2 + b^2$; if C is obtuse, $c^2 > a^2 + b^2$ and we have a simple method of finding whether an angle of a triangle is acute or obtuse.

The cosine formula may be put as $\cos C = \dfrac{a^2 + b^2 - c^2}{2ab}$, which is the form needed for solving a triangle given its sides.

The following results are, of course, also true.

$$b^2 = a^2 + c^2 - 2ac \cos B \quad \text{or} \quad \cos B = \frac{a^2 + c^2 - b^2}{2ac};$$

$$a^2 = b^2 + c^2 - 2bc \cos A \quad \text{or} \quad \cos A = \frac{b^2 + c^2 - a^2}{2bc}.$$

N.B. Do not use the cosine formula to solve an isosceles triangle.

Example 3. *Given $b = 3$, $c = 4.8$ and $A = 120°$, solve the triangle ABC.*

$$a^2 = b^2 + c^2 - 2bc \cos A = 9 + 23.04 - 28.8 \cos 120°.$$
$$\cos 120° = - \cos 60° = -\tfrac{1}{2};$$
$$\therefore \ a^2 = 32.04 + 14.4 = 46.44 \quad \text{and} \quad a = 6.815.$$

Using

$$\frac{a}{\sin A} = \frac{b}{\sin B},$$

$$\frac{6.815}{\sin 120°} = \frac{3}{\sin B}$$

$$\therefore \ \sin B = \frac{3 \sin 120°}{6.815} = \frac{3 \sin 60°}{6.815}$$

and

$$B = 22° \ 25'.$$

N.B. The angle B cannot be obtuse because a triangle cannot have two obtuse angles.

$$\therefore \ C = 180° - 120° - 22° \ 25' = 37° \ 35'.$$

N.B. As a check, notice that since c is greater than b, C should be greater than B.

Example 4. Given a = 4.12, b = 6.82, C = 42° 30', solve the triangle ABC.

$$c^2 = a^2 + b^2 - 2ab \cos C = (4.12)^2 + (6.82)^2$$
$$- 2(4.12)(6.82) \cos 42° 30'$$
$$= 16.97 + 46.51 - 41.43$$
$$= 22.05.$$
$$\therefore \ c = 4.695.$$

From the sine formula,

$$\frac{\sin A}{4.12} = \frac{\sin 42° 30'}{4.695}.$$

N.B. Calculate the angle opposite the smaller of the given sides. This cannot be obtuse.

$$\therefore \ \sin A = \frac{4.12 \sin 42° 30'}{4.695}$$

and $\qquad\qquad A = 36° 21'.$

Finally $\qquad\quad B = 180° - 42° 30' - 36° 21'$
$$= 101° 9'.$$

Example 5. Given a = 7.12, b = 6.43, c = 8.27, find the angles of the triangle ABC.

N.B. Start by finding the smallest angle which cannot be obtuse; continue with the sine formula and find the smaller of the two remaining angles.

$$\cos B = \frac{7.12^2 + 8.27^2 - 6.43^2}{2(7.12)(8.27)}$$
$$= \frac{50.69 + 68.39 - 41.34}{2(7.12)(8.27)}$$
$$= \frac{77.74}{2(7.12)(8.27)}.$$
$$B = 48° 42'.$$

From $\qquad\quad \dfrac{\sin A}{a} = \dfrac{\sin B}{b}.$

$$\sin A = \frac{7.12 \sin 48° 42'}{6.43}.$$
$$A = 56° 18'.$$

Finally $\qquad\quad C = 180° - 48° 42' - 56° 18'$
$$= 75° 0'.$$

Exercise 43.2a

Solve the following triangles:

1. $a = 3.7, b = 4.2, C = 110°.$
2. $a = 8.4, b = 2.6, C = 24°.$
3. $b = 21.2, c = 14.3, A = 60° 22'.$
4. $c = 2.81, a = 4.32, B = 95° 28'.$

5. $c = 4.73$, $a = 14.2$, $B = 80°$.

6. $a = 2.43$, $b = 7.12$, $c = 6.43$.

7. $a = 3.12$, $b = 4.82$, $c = 4.82$.

8. $a = 7.64$, $b = 6.82$, $c = 12.3$.

9. $a = 4.13$, $b = 18.1$, $c = 17.6$.

10. $a = 5.42$, $b = 11.2$, $c = 10.9$.

Exercise 43.2b

Solve the following triangles:

1. $a = 4.2$, $b = 4.8$, $C = 130°$.

2. $a = 11.2$, $b = 13.4$, $C = 32°$.

3. $b = 20.8$, $c = 12.6$, $A = 48° 15'$.

4. $c = 3.21$, $a = 5.26$, $B = 96° 43'$.

5. $c = 5.62$, $a = 15.3$, $B = 70°$.

6. $a = 3.71$, $b = 8.42$, $c = 7.15$.

7. $a = 4.62$, $b = 7.31$, $c = 7.31$.

8. $a = 11.2$, $b = 12.4$, $c = 21.6$.

9. $a = 18.4$, $b = 22.2$, $c = 12.7$.

10. $a = 26.7$, $b = 33.3$, $c = 19.4$.

An example illustrating the solution of a problem by means of the sine and cosine formulae is now given.

Example 6. The angles of elevation of the top Y of a vertical mast from two points A, B on the same level as its foot X are 32° 14' and 22° 28' respectively. If the height of the mast is 52 m and the bearings of A and B from X are 270° and 220° respectively, find the distance AB and the bearing of B from A.

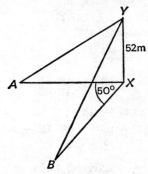

Fig. 43.7

From the triangle AXY,

$$\frac{AX}{52} = \cot 32° 14'.$$

$$\therefore \ AX = 82.47.$$

From the triangle BXY,

$$\frac{BX}{52} = \cot 22° 28'.$$

$$\therefore BX = 125.7.$$

But

$$\begin{aligned}
AB^2 &= AX^2 + XB^2 - 2AX.XB \cos A\widehat{X}B \\
&= (82.47)^2 + (125.7)^2 - 2(82.47)(125.7) \cos 50° \\
&= 6802 + 15\ 800 - 13\ 330 \\
&= 9272.
\end{aligned}$$

$$\therefore AB = 96.29 \quad \text{or} \quad 96.3 \text{ m.}$$

From

$$\frac{\sin X\widehat{B}A}{AX} = \frac{\sin 50°}{96.29},$$

$$\sin X\widehat{B}A = \frac{82.47 \sin 50°}{96.29}.$$

$$\therefore X\widehat{B}A = 41° 1'.$$
$$X\widehat{A}B = 180° - 50° - 41° 1'$$
$$= 88° 59'.$$

The bearing of B from A is $90° + 88° 59' = 178° 59'$.

The area of a triangle

The area of a triangle is equal to half the product of base and altitude.

In Fig. 43.8, the area, $\Delta = \frac{1}{2}BN.AC$.

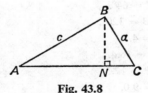

Fig. 43.8

From the triangle ABN, $\dfrac{BN}{c} = \sin A$.

$$\therefore BN = c \sin A \quad \text{and} \quad \Delta = \tfrac{1}{2}bc \sin A.$$

By symmetry, $\frac{1}{2}ac \sin B$ and $\frac{1}{2}ab \sin C$ are also equal to Δ.

To find the area of a triangle, therefore, multiply half the product of any two sides by the sine of the included angle. Another formula, not proved here, for the area of a triangle is $\sqrt{s(s-a)(s-b)(s-c)}$, where s is the semi-perimeter and equals $\frac{1}{2}(a+b+c)$.

Example 7. *The sides of a triangle are 8.1 m, 6.3 m and 4.8 m. Find its area,*
and the smallest angle.

$$s = \tfrac{1}{2}(8.1 + 6.3 + 4.8) = 9.6.$$
$$\therefore \triangle = \sqrt{(9.6)(1.5)(3.3)(4.8)}$$
$$= 15.10 \text{ m}^2 = 15.1 \text{ m}^2 \text{ (to 3 s.f.)}.$$

The smallest angle, x, is opposite the smallest side.
$$\therefore \tfrac{1}{2}(6.3)(8.1) \sin x = 15.1.$$
From which $x = 36° \, 18'.$

Exercise 43.3a

Find the areas of the following triangles:

1. $a = 2.4$, $b = 3.2$, $C = 60°$.
2. $a = 4.6$, $b = 7.8$, $C = 82°$.
3. $a = 6.2$, $c = 4.8$, $B = 71°$.
4. $a = 4.2$, $c = 7.1$, $B = 110°$.
5. $a = 2.4$, $b = 3.2$, $c = 4.1$.
6. $a = 7.8$, $b = 9.2$, $c = 11.3$.
7. $a = 8.23$, $b = 9.16$, $c = 10.24$.
8. $a = 7.12$, $b = 11.23$, $c = 16.41$.

Exercise 43.3b

Find the areas of the following triangles:

1. $a = 2.8$, $b = 7.1$, $C = 30°$.
2. $a = 4.7$, $b = 8.2$, $C = 74°$.
3. $a = 6.18$, $b = 8.2$, $C = 64°$.
4. $a = 8.16$, $c = 7.24$, $B = 120°$.
5. $a = 2.8$, $b = 3.1$, $c = 4.7$.
6. $a = 7.4$, $b = 12.2$, $c = 16.4$.
7. $a = 8.12$, $b = 11.4$, $c = 12.3$.
8. $a = 7.0$, $b = 8.0$, $c = 9.0$.

Exercise 43.4. Miscellaneous

1. Y is due north of X. The bearings of Z from X and Y are $26° \, 30'$ and $42° \, 40'$ respectively. Given that XYZ are all at the same level and that $XY = 1$ km, find the distance YZ.

2. A ship steams 4 km due north from a point and then 3 km on a bearing of $040°$. How far is the ship from the point?

3. The angles of elevation of the top of a tower from two points A, B in line with its foot and on the same level are $27°$ and $32°$ respectively. If the distance AB is 21 m, find the height of the tower.

4. The distance between two houses as the crow flies is 500 m. To walk from one to the other it is necessary to go 350 m due north and then 300 m on another road. What is the bearing of the second road?

5. First slip is standing 10 m from the batsman's wicket at an angle of 160° with the pitch. How far is he from the bowler's wicket?

6. The distance from a tee T to the hole H on a golf course is 400 m. A golfer drives 230 m but his shot is 10° off the direct line. How far is his ball from the hole?

7. In the triangle ABC, $b = 3$, $c = 5$ and $A = 120°$. Find a.

8. In the triangle ABC, $b = 3.1$, $c = 1.4$ and $A = 46°$. Find B.

9. A and B are two observation stations, B being 10 km due west of A. At 09.00h a ship is observed to be due north of A and at a bearing of 064° from B. At 10.00h it is observed to be at a bearing 032° from A and 071° from B. Find the course and speed of the ship.

10. In the triangle ABC, $a = 14.6$, $B = 52°\ 10'$ and $C = 77°$. Calculate the perimeter of the triangle.

11. Find the least angle of a triangle whose sides are 6.2 m, 7.3 m and 8.4 m· Find also its area.

12. In the triangle ABC, $A = 50°$, $b = 6$ m and $c = 8$ m. Calculate the angles B and C and the area of the triangle.

13. A road runs east and west. X and Y are two points of the road 150 m apart. The bearings of a point Z from X and Y are 215° and 202° respectively. Calculate the distance YZ.

14. The sides of the triangle ABC are given in metres by $a = 4.8$, $b = 7.2$ and $c = 9.3$. Calculate the angle B and the area of the triangle.

15. A boat travelling in a direction 222° reaches a point P from which two buoys X and Y are seen both due south of P. Y is known to be 200 m due south of X. The boat continues on the same course and reaches a point Q from which the bearing of X is 140° and that of Y is 165°. Calculate the distance PX.

16. Calculate the angle between the diagonals of a parallelogram whose sides are 5.1 and 2.5 cm and contain an angle of 70°.

17. The hands of a clock are 4 cm and 5 cm long. Find the distance between the tips of the hands at 16.00h.

18. Two circles of radii 5 cm and 6 cm have their centres 8 cm apart. Calculate the acute angle between the tangents at one of the two points of intersection of the circles.

19. The sides of a rhombus are each 4.2 cm. One angle of the rhombus is 58°. Calculate the lengths of the diagonals.

20. $ABCD$ is a trapezium with AB, DC parallel. $AB = 2$ cm, $BC = 3$ cm, $CD = 6$ cm and the angle BCD is 120°. Calculate the length of AD.

21. An observer M is 30 m from the base X of a tower XY and on the same level as X. The tower has a vertical mast YZ. If the angles of elevation of Y and Z from X are 40° and 50° respectively, find the height of the mast.

22. The angles of elevation of a balloon from two points A, B which are 0.3 km apart, are 62° and 48° respectively. If the balloon is vertically above the line AB, find its height.

23. A boat steaming due north is 2 km away in a direction 070°. 5 minutes later the bearing of the boat is 040°. Find the speed of the boat.

24. In the triangle ABC, $a = 6$ cm, $b = 7$ cm and $c = 8$ cm. Find the length of the line joining A to the point on BC distant 3 cm from B.

25. In Fig. 43.9, CT is the tangent at C to the circle and $AB = 4$ cm. Calculate the length of CT.

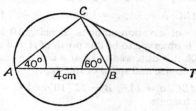

Fig. 43.9

26. The sines of the angles of a triangle are in the ratios $2 : 3 : 4$. Find the ratios of the cosines.

27. A yacht sails 2 km due north and then 3 km on a bearing of 030°. How far is it from the starting point?

28. Observations are taken from two points P and Q which are on the same horizontal ground as the foot F of a vertical flagstaff. If the angle $PFQ = 48°$ and the angles of elevation of the top of the flagstaff from P and Q are 32° and 40° respectively, find PQ given that the flagstaff is 24 m high.

29. The area of an acute-angled triangle is 1.6 cm². If two of the sides are 3.2 and 1.8 cm long respectively, find the angle included between these sides.

30. From a point P the bearings of two landmarks L and M are 308° and 222° respectively. L is 500 m from M on a bearing of 008°. Calculate the distance PL.

31. A vertical tower AB is 40 m high. X and Y are two points on the same level as the foot A of the tower, X being west and Y north-west of the tower. The angles of elevation of the top of the tower from X and Y are 25° and 32° respectively. Find the distance XY.

32. A rhombus has sides of length 6 cm and the length of one of the diagonals is 7 cm. Find the angles of the rhombus.

CALCULUS

CALCULUS

Differentiation

$$\frac{d}{dx} \text{(constant)} = 0$$

$$\frac{d}{dx}(Kx) = K$$

$$\frac{d}{dx}(x^2) = 2x$$

$$\frac{d}{dx}(x^3) = 3x^2$$

$$\frac{d}{dx}(x^n) = nx^{n-1}$$

$$\text{velocity} = \frac{ds}{dt}$$

$$\text{acceleration} = \frac{dv}{dt}$$

Integration

$$\int k \, dx = kx + C$$

$$\int kx \, dx = \tfrac{1}{2}kx^2 + C$$

$$\int x^n \, dx = \frac{x^{n+1}}{n+1} + C, n \neq -1$$

The area under a curve $= \int y \, dx$

The volume of rotation about the x-axis $= \pi \int y^2 \, dx.$

44 *The Derived Function*

The gradient of a straight line

The gradient of a line referred to axes of x and y is the ratio of the increase of y between any two points of the line to the increase of x between the same two points. If y decreases as x increases, the gradient is negative. Written as a fraction, the gradient equals

$$\frac{\text{increase of } y}{\text{increase of } x}.$$

Consider the line $y = 2x + 3$.

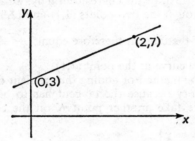

Fig. 44.1

When $x = 0$, $y = 3$. So the point where $x = 0$ and $y = 3$ lies on the line. This point is written (0,3).

N.B. The x-coordinate is always written first.

When $x = 2$, $y = 7$; so another point on the line is (2,7). The increase of y between these points is $7 - 3 = 4$. The increase of x between these points is $2 - 0 = 2$. The gradient of the line is $\frac{4}{2}$ or 2.

N.B. The reader should prove by similar triangles, that the gradient is the same whatever points are taken on the line.

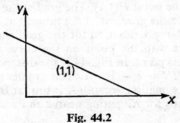

Fig. 44.2

Now consider the line $2y + x = 3$.

Two points on this line are (1,1) and (3,0). The increase of x between these points is $3 - 1 = 2$. The *decrease* of y between these points is $1 - 0 = 1$. The gradient of the line is therefore $\dfrac{-1}{2} = -\frac{1}{2}$.

In general, a line which makes an acute angle with the positive x direction has a positive gradient; a line which makes an obtuse angle with the positive x direction has a negative gradient.

The gradient of a curve

The gradient of a curve at a point is equal to the gradient of the tangent at that point. Consider the point (1,1) on the curve $y = x^2$.

Draw the tangent at this point and choose any two convenient points on this tangent. If we draw the curve accurately, we should find that the tangent passes through the two points (1,1) and (2,3).

The gradient of the tangent therefore equals $\dfrac{3 - 1}{2 - 1}$ or 2. This is also the gradient of the curve at the point (1,1).

This is a drawing method of finding the gradient of a curve, but it is not very satisfactory because the tangent has to be drawn by guess-work. Suppose we take another point K on the curve and move it

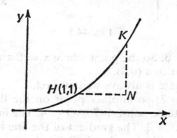

Fig. 44.3

nearer and nearer the point $H(1,1)$. The gradient of the line HK will get nearer and nearer to the gradient of the tangent at H and we should be able to find a good approximation for the gradient of the tangent.

Suppose we start with the point on the curve where $x = 2$. Since $y = x^2$, $y = 4$ at this point. In Fig. 44.3, the distance $KN = 4 - 1 = 3$ and the distance $HN = 2 - 1 = 1$. The gradient of HK is 3 which is not a good approximation because K is too far from H. We will now take several positions for K, getting nearer and nearer H, and tabulate the results.

The gradient of the line HK becomes very nearly equal to 2 but will never become exactly 2 as K cannot be made to coincide with H. Both

The x coordinate of K	1.5	1.4	1.3	1.2	1.1	1.05	1.01	1.001
The y coordinate of K	2.25	1.96	1.69	1.44	1.21	1.1025	1.0201	1.002001
KN	1.25	0.96	0.69	0.44	0.21	0.1025	0.0201	0.002001
HN	0.5	0.4	0.3	0.2	0.1	0.05	0.01	0.001
The gradient of HK	2.5	2.4	2.3	2.2	2.1	2.05	2.01	2.001

quantities KN and HN would then become zero and the fraction $\frac{0}{0}$ cannot be evaluated.

Suppose now we have a further addition to the table and consider the point whose x coordinate is $(1 + h)$.

The x coordinate of $K = 1 + h$.
The y coordinate of $K = (1 + h)^2 = 1 + 2h + h^2$.
$$KN = 2h + h^2.$$
$$HN = h.$$
The gradient of HK $= \dfrac{2h + h^2}{h}$.

The point K cannot be made to coincide with H and therefore h is not zero. The fraction $\dfrac{2h + h^2}{h}$ equals $(2 + h)$, unless $h = 0$.

The gradient of the line HK is therefore $(2 + h)$, which is 1 greater than the x coordinate of the point $K(= 1 + h)$. Notice that the bottom line of the table is always one greater than the corresponding number of the top line.

The gradient $(2 + h)$ becomes very nearly equal to 2 as h becomes small. We say that the gradient of the line tends to 2 as h tends to 0. (Or: gradient $\longrightarrow 2$ as $h \longrightarrow 0$.)

We may also say that the limit of the gradient is 2 as h tends to 0. So the gradient of the curve $y = x^2$ at the point $(1,1)$ is 2.

The gradient of $y = x^2$ at any point

Suppose that H is the point on $y = x^2$ whose x coordinate is x, and that K is another point on $y = x^2$ whose x coordinate is $(x + h)$.

The y coordinate of H is x^2.
The y coordinate of K is $(x + h)^2$.
$$KN = (x + h)^2 - x^2 = 2hx + h^2.$$
$$HN = (x + h) - x = h.$$

The gradient of $HK = \dfrac{2hx + h^2}{h}$
$$= 2x + h \text{ (if } h \text{ is not zero).}$$

As h tends to 0, the gradient tends to $2x$.

The gradient of the tangent is $2x$, or the gradient of the curve $y = x^2$ at the point (x,y) is $2x$.

It follows that the gradient of the curve at the point $(1,1)$ is 2; that the gradient at $(2,4)$ is 4; that the gradient at $(3,9)$ is 6 and so on.

Example 1. *Find the gradient of the curve* $y = 2x^2 - 3x - 4$ *at the point* (x,y).

Suppose that (x,y) and $(x + h, y + k)$ are two points on the curve.

Then $\quad\quad\quad\quad y + k = 2(x + h)^2 - 3(x + h) - 4 \quad\quad\quad$ (i)

and $\quad\quad\quad\quad\quad y = 2x^2 - 3x - 4 \quad\quad\quad\quad\quad\quad\quad$ (ii)

Subtract (ii) from (i) $\quad k = 4xh + 2h^2 - 3h$

Therefore $\quad\quad\quad\quad \dfrac{k}{h} = 4x + 2h - 3 \quad$ (if h is not zero).

$\dfrac{k}{h}$ is the gradient of the chord and tends to $(4x - 3)$ as h tends to 0.

The gradient of the curve at the point (x,y) is therefore $(4x - 3)$. For example, $(2,-2)$ is a point on the curve. The gradient of the curve at this point is $4(2) - 3 = 5$.

Exercise 44.1a

Find the gradients of the following curves at the points specified.

1. $y = 3x^2$ at $(1,3)$. $\quad\quad\quad\quad\quad$ **2.** $y = x^2 + x$ at $(1,2)$.

3. $y = x^2 + x + 1$ at $(1,3)$. $\quad\quad$ **4.** $y = 2x^2 + x$ at $(1,3)$.

5. $y = 2x^2$ at $(2,8)$. $\quad\quad\quad\quad\quad$ **6.** $y = x^2 - x$ at $(2,2)$.

7. $y = 7x$ at (x,y). $\quad\quad\quad\quad\quad$ **8.** $y = 2x^2$ at (x,y).

9. $y = 3x^2 + x + 1$ at (x,y). $\quad\quad$ **10.** $y = 4x^2 - x - 2$ at (x,y).

Exercise 44.1b

Find the gradients of the following curves at the points specified.

1. $y = 4x^2$ at $(1,4)$. $\quad\quad\quad\quad\quad$ **2.** $y = 3x^2 + x$ at $(1,4)$.

3. $y = x^2 - x + 1$ at $(1,1)$. $\quad\quad$ **4.** $y = 3x^2 - 1$ at $(1,2)$.

5. $y = 3x^2$ at $(2,12)$. $\quad\quad\quad\quad$ **6.** $y = 2x^2 - x$ at $(2,6)$.

7. $y = 4x$ at (x,y). $\quad\quad\quad\quad\quad$ **8.** $y = 3x^2$ at (x,y).

9. $y = 2x^2 + x + 1$ at (x,y). $\quad\quad$ **10.** $y = 6x^2 - 2x$ at (x,y).

The derived function

The gradient of a curve at the point (x,y) on it is called its derived function. The derived function of x^2 is $2x$. The derived function of $(2x^2 - 3x - 4)$, as proved in the worked example, is $(4x - 3)$. The derived function of x^2 is written $D(x^2)$ and the derived function of $(2x^2 - 3x - 4)$ is written $D(2x^2 - 3x - 4)$.

$$\therefore D(x^2) = 2x,$$

and
$$D(2x^2 - 3x - 4) = 4x - 3.$$

The derived function of a constant

The graph of $y = c$ is a straight line parallel to the axis of x and its gradient is zero.

$$\therefore D(\text{any constant}) = 0.$$

The derived function of *ax*, where *a* is constant

Suppose (x,y) and $(x + h, y + k)$ are two points on the line $y = ax$.

Then
$$y + k = a(x + h) \tag{i}$$

and
$$y = ax \tag{ii}$$

Subtract (ii) from (i),
$$k = ah.$$

$$\therefore \frac{k}{h} = a.$$

$$\therefore D(ax) = a.$$

The derived function of *ax²*, where *a* is constant

Suppose (x,y) and $(x + h, y + k)$ are two points on the curve $y = ax^2$.

Then
$$y + k = a(x + h)^2 \tag{i}$$

and
$$y = ax^2. \tag{ii}$$

Subtract (ii) from (i),
$$k = 2axh + ah^2.$$

$$\therefore \frac{k}{h} = 2ax + ah \text{ (if } h \text{ is not zero).}$$

The limit of the gradient as h tends to 0 is $2ax$.

$$\therefore D(ax^2) = 2ax.$$

Notice that
$$D(ax^2) = aD(x^2).$$

The derived function of *ax² + bx + c*, where *a, b, c,* are constants

Suppose that (x,y) and $(x + h, y + k)$ are two points on the curve $y = ax^2 + bx + c$.

Then
$$y + k = a(x + h)^2 + b(x + h) + c$$
$$= ax^2 + 2axh + ah^2 + bx + bh + c \tag{i}$$

and
$$y = ax^2 + bx + c \tag{ii}$$

Subtract (ii) from (i), $k = 2axh + ah^2 + bh$.

$$\therefore \frac{k}{h} = 2ax + ah + b \text{ (if } h \text{ is not zero).}$$

The limit of the gradient as h tends to 0 is $(2ax + b)$.

$$\therefore D(ax^2 + bx + c) = 2ax + b.$$

Notice that $\qquad D(ax^2 + bx + c) = aD(x^2) + bD(x) + D(c)$.

Example 2. *Find the derived function of $3x^2 - 7x - 1$.*

$$\begin{aligned} D(3x^2 - 7x - 1) &= 3(Dx^2) - 7D(x) - D(1) \\ &= 3(2x) - 7(1) - 0 \\ &= 6x - 7. \end{aligned}$$

Example 3. *Find the derived function of $(3x + 2)(2x - 1)$.*

$$\begin{aligned} (3x + 2)(2x - 1) &= 6x^2 + x - 2. \\ D(6x^2 + x - 2) &= 6D(x^2) + D(x) - D(2) \\ &= 6(2x) + 1 - 0 \\ &= 12x + 1. \end{aligned}$$

Exercise 44.2a

Find the derived functions of the following.

1. $4x^2 - 6x$.   2. $\frac{3}{2}x^2 - 1$.   3. $3x - x^2$.

4. $x^2 + \frac{3}{4}x + \frac{7}{8}$.   5. $1 - 6x - 2x^2$.   6. $4x^2 - 4x + 1$.

7. $(x - 1)(x - 2)$.   8. $(x + 6)(2x - 1)$.   9. $(2x - 1)(3x + 1)$.

10. $x(4x + 3)$.

Exercise 44.2b

Find the derived functions of the following.

1. $2x^2 - 8x$.   2. $\frac{1}{2}x^2 - \frac{1}{4}$.   3. $2x - x^2$.

4. $x^2 + \frac{2}{3}x + \frac{1}{6}$.   5. $1 - 4x - 3x^2$.   6. $3x^2 + 3x + 1$.

7. $(x + 1)(x - 1)$.   8. $(2x - 3)(x - 4)$.   9. $(2x + 1)(3x - 1)$.

10. $x\left(x + \dfrac{1}{x}\right)$.

The derived function of x^3

Suppose that (x,y) and $(x + h, y + k)$ are two points on the curve $y = x^3$.

Then $\qquad y + k = (x + h)^3 = (x + h)(x^2 + 2xh + h^2)$

$$= x^3 + 3x^2h + 3xh^2 + h^3 \qquad\qquad \text{(i)}$$

and $\qquad\qquad y = x^3 \qquad\qquad\qquad\qquad\qquad\qquad\qquad \text{(ii)}$

Subtract (ii) from (i), $k = 3x^2h + 3xh^2 + h^3$.

$$\therefore \frac{k}{h} = 3x^2 + 3xh + h^2 \text{ (if } h \text{ is not zero).}$$

The limit of the gradient as h tends to 0 is $3x^2$.

$$\therefore D(x^3) = 3x^2.$$

The derived function of $\frac{1}{x}$

Suppose that (x,y) and $(x + h, y + k)$ are two points on the curve $y = \frac{1}{x}$.

Then $\qquad y + k = \dfrac{1}{x + h}$ $\qquad\qquad$ (i)

and $\qquad\qquad y = \dfrac{1}{x}$ $\qquad\qquad\qquad$ (ii)

Subtract (ii) from (i), $k = \dfrac{1}{x + h} - \dfrac{1}{x} = \dfrac{x - (x + h)}{x(x + h)}$

$$= - \frac{h}{x(x + h)}.$$

As h tends to 0, the gradient $\left(\dfrac{k}{h}\right)$ tends to $-\dfrac{1}{x^2}$

$$\therefore D\left(\frac{1}{x}\right) = -\frac{1}{x^2}.$$

The derived function of $\frac{1}{x^2}$

Suppose that (x,y) and $(x + h, y + k)$ are two points on the curve $y = \dfrac{1}{x^2}$.

Then $\qquad y + k = \dfrac{1}{x + h^2}$ $\qquad\qquad$ (i)

and $\qquad\qquad y = \dfrac{1}{x^2}$ $\qquad\qquad\qquad$ (ii)

Subtract (ii) from (i), $k = \dfrac{1}{(x + h)^2} - \dfrac{1}{x^2} = \dfrac{x^2 - (x + h)^2}{x^2(x + h)^2}$

$$= \frac{- 2xh - h^2}{x^2(x + h)^2}$$

$$\therefore \frac{k}{h} = \frac{- 2x - h}{x^2(x + h)^2} \text{ (if } h \text{ is not zero).}$$

As h tends to 0, the gradient $\left(\dfrac{k}{h}\right)$ tends to $\dfrac{-2x}{x^4}$ or $-\dfrac{2}{x^3}$.

$$\therefore D\left(\frac{1}{x^2}\right) = -\frac{2}{x^3}.$$

General rule

Tabulating the results already proved, and remembering that $x^1 = x$ and that $x^0 = 1$;

$$D(x^3) = 3x^2.$$
$$D(x^2) = 2x.$$
$$D(x^1) = 1.$$
$$D(x^0) = 0.$$
$$D(x^{-1}) = -1x^{-2}.$$
$$D(x^{-2}) = -2x^{-3}.$$

The reader will guess from these results that $D(x^4) = 4x^3$ and that $D\left(\dfrac{1}{x^3}\right) = D(x^{-3}) = -3x^{-4}$.

In fact, it is true that $D(x^n) = nx^{n-1}$ and the reader may assume that this formula is true for all values of n, even fractional.

Notation

If $f(x)$ denotes any function of x, that is any expression containing x, we call its derived function $D\{(fx)\}$. The derived function of $f(x)$ may also be called $f'(x)$.

If $y = f(x)$, the most common way of writing its derived function is $\dfrac{dy}{dx}$. This notation will be used in future. The reader must realise that $\dfrac{dy}{dx}$ is not an ordinary fraction which can be cancelled; that the d of dy is not a multiple (compare $\sin y$) and that the dy cannot be separated from its denominator, dx. The expression $\dfrac{dy}{dx}$ actually compares the rate of change of y with that of x. Finding the derived function of an expression is called differentiating the expression. The advantage of the notation $\dfrac{dy}{dx}$ is that it tells us what quantities are being compared; in fact, that we have differentiated y with respect to x. For example, if we know that $z = t^3 - t^2$, then $\dfrac{dz}{dt} = 3t^2 - 2t$, or we may write the result in the form $\dfrac{d}{dt}(t^3 - t^2) = 3t^2 - 2t.$

Worked Examples

Example 4. *Find the derived function of* $2x^3 - \dfrac{3}{x}$.

$$D\left(2x^3 - \frac{3}{x}\right) = D(2x^3) - D\left(\frac{3}{x}\right) = 2D(x^3) - 3D(x^{-1})$$
$$= 2(3x^2) - 3(-1x^{-2})$$
$$= 6x^2 + \frac{3}{x^2}.$$

Alternative method. Let $\qquad y = 2x^3 - \dfrac{3}{x}$.

Then $\qquad\qquad\qquad\qquad \dfrac{dy}{dx} = 6x^2 + \dfrac{3}{x^2};$

or simply, $\qquad\qquad \dfrac{d}{dx}\left(2x^3 - \dfrac{3}{x}\right) = 6x^2 + \dfrac{3}{x^2}.$

Example 5. *Find the derived function of* $2\sqrt{x} - \dfrac{3}{\sqrt{x}}$.

$$D\left(2\sqrt{x} - \frac{3}{\sqrt{x}}\right) = 2D(x^{\frac{1}{2}}) - 3D(x^{-\frac{1}{2}})$$
$$= 2(\tfrac{1}{2})(x^{-\frac{1}{2}}) - 3(-\tfrac{1}{2})x^{-\frac{3}{2}} = x^{-\frac{1}{2}} + \tfrac{3}{2}x^{-\frac{3}{2}} = \frac{1}{\sqrt{x}} + \frac{3}{2\sqrt{x^3}}$$

Alternative method. Let

$$y = 2\sqrt{x} - \frac{3}{\sqrt{x}} = 2x^{\frac{1}{2}} - 3x^{-\frac{1}{2}}.$$
$$\frac{dy}{dx} = x^{-\frac{1}{2}} + \tfrac{3}{2}x^{-\frac{3}{2}}.$$
$$= \frac{1}{\sqrt{x}} + \frac{3}{2\sqrt{x^3}}.$$

Example 6. *Find the derived function of* $\dfrac{(1 + x)(1 + 2x^2)}{x}$.

Let
$$y = \frac{(1 + x)(1 + 2x^2)}{x} = \frac{1 + x + 2x^2 + 2x^3}{x} = \frac{1}{x} + 1 + 2x + 2x^2.$$

Then $\qquad\qquad\qquad \dfrac{dy}{dx} = -\dfrac{1}{x^2} + 2 + 4x.$

$$\left[\text{or,} \qquad \frac{d}{dx}\left\{\frac{1}{x} + 1 + 2x + 2x^2\right\} = -\frac{1}{x^2} + 2 + 4x.\right]$$

Exercise 44.3. Miscellaneous

Find the derived functions of:

1. $3x^3 - 1$. $\qquad\qquad$ **2.** $1 - \dfrac{1}{x^2}$. $\qquad\qquad$ **3.** $\left(x + \dfrac{1}{x}\right)^2$.

4. $x + 2x^2 + 3x^3$.   **5.** $(x + 2)(x^2 + 1)$.   **6.** $\dfrac{x + 1}{\sqrt{x}}$.

7. $\sqrt{x} + x$.   **8.** $\dfrac{x^2 + 1}{x}$.   **9.** $1 - \dfrac{1}{x}$.

10. $1 + \dfrac{1}{x} + \dfrac{1}{x^2}$.

Find the following:

11. $\dfrac{d}{dx}(x^3 + x^2)$.   **12.** $\dfrac{d}{dx}(2 + 7x + 3x^2 - 4x^3)$.

13. $\dfrac{d}{dx}\left(\dfrac{1}{x} - \dfrac{1}{x^2}\right)$.   **14.** $\dfrac{d}{dx}\left(\dfrac{1}{\sqrt{x}} + x\right)$.

15. $\dfrac{d}{dx}\left(x + 2 + \dfrac{1}{x}\right)$.   **16.** $\dfrac{d}{dy}(y^3 - y)$.

17. $\dfrac{d}{dt}(3t^2 + t^3)$.   **18.** $\dfrac{d}{dz}(z^2 + 3z + 1)$.

19. $\dfrac{d}{dv}(3v - v^2 + 7v^3 + 1)$.   **20.** $\dfrac{d}{dt}(t^3 - 3t - 1)$.

Differentiate:

21. $1 + 6x + 2x^2 + 3x^3$.   **22.** $\dfrac{1}{x^2} + \dfrac{2}{x} + 3$.

23. $x - \dfrac{1}{\sqrt{x}}$.   **24.** $x^{\frac{5}{2}}$.

25. $x + 6x^2 + 7x^3$.   **26.** $(x + 1)(x^2 + 1)$.

27. $\dfrac{x^2 + x + 1}{x}$.   **28.** x^7.

29. $\dfrac{1}{x^7}$.   **30.** $x^7 + \dfrac{1}{x^7}$.

31. Find the gradient of the curve $y = 3x^2 - 2x - 1$ at the point (1,0).

32. Find the gradient of the tangent to the curve $y = x^3 + x$ at the point (2,10).

33. Find the coordinates of the point on the graph of $y = x^2 - x$ at which the tangent is parallel to $y = x$.

34. Find the gradient of the curve $xy = 1$ at the point (1,1).

35. Given that $xy = 1$, find $\dfrac{dy}{dx}$ in terms of x.

36. Simplify $\dfrac{d}{dt}\{t^3 - 6t^2 + 4t - 1\}$.

37. Find the point on the curve $y = 1 - x^2$ at which the tangent is parallel to $y = 2x$.

38. If $x^2y = 4$, find $\dfrac{dy}{dx}$ in terms of x.

39. If $u + v = 5$, find $\dfrac{du}{dv}$.

40. If $y = x^2$, find $\dfrac{dy}{dx}$ (i) in terms of x; (ii) in terms of y.

41. Simplify $\dfrac{d}{dz}\left(z - \dfrac{1}{z}\right)$.

42. If $y = x^2$, prove that $x\dfrac{dy}{dx} = 2y$.

43. If $pv = 40$, find the value of $\dfrac{dp}{dv}$ when $p = 10$.

44. If $z = t + t^2$, find $\dfrac{dt}{dz}$.

45. Find the points on the curve $y = x^3 - x^2$ at which the tangent is parallel to the axis of x.

45 Applications of the Derived Function

Maxima and minima

At the point P on the curve shown in Fig. 45.1, the tangent is parallel to the axis of x. The gradient of this tangent is zero and therefore the

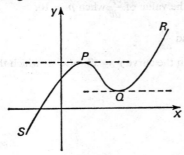

Fig. 45.1

value of the derived function at this point is zero. Such a point is called a **maximum point**. The value of y at such a point is called a maximum value of y. The value of y at a maximum point is larger than at points on either side of it; it is not necessarily the greatest value of all. For example, in the curve shown the value of y at P is less than that at R.

The gradient at the point Q is also zero. At this point, the value of y is less than at points immediately on either side of it. It is not least of all possible values of y, because, for example, the value of y at S is smaller still. Such a point is called a **minimum point** and the value of y at such a point is called a minimum value of y.

Slightly before P, the gradient of the curve is positive; at P, it is zero; slightly after P, it is negative.

So at a maximum point, the gradient changes from positive to negative through the zero value.

Slightly before Q, the gradient is negative; at Q, it is zero; slightly after Q, it is positive.

So at a minimum point, the gradient changes from negative to positive through the zero value.

Maximum and minimum points are called **turning points**; the values of y at these points are called turning values.

At a maximum point, $\dfrac{dy}{dx}$ is zero; the value of $\dfrac{dy}{dx}$ changes from positive to negative as x increases through the point.

At a minimum point, $\frac{dy}{dx}$ is zero; the value of $\frac{dy}{dx}$ changes from negative to positive as x increases through the point.

There is one other type of point at which $\frac{dy}{dx} = 0$. This is shown in Fig. 45.2, and is not a turning point because the gradient does not change sign (or turn).

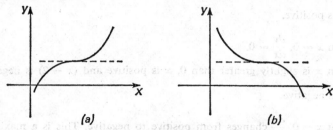

(a)   (b)

Fig. 45.2

In Fig. 45.2(a), $\frac{dy}{dx}$ is positive both before and after its zero value.

In Fig. 45.2(b), $\frac{dy}{dx}$ is negative both before and after its zero value.

A point where $\frac{dy}{dx} = 0$ but does not change sign is called a **point of inflexion.**

Example 1. Find the maximum or minimum value of $(x^2 - 4x - 6)$.

If
$$y = x^2 - 4x - 6,$$
$$\frac{dy}{dx} = 2x - 4.$$
$$\therefore \frac{dy}{dx} = 0, \text{ when } x = 2.$$

When x is slightly less than 2, $\frac{dy}{dx}$ is negative.

When $x = 2$, $\frac{dy}{dx} = 0$.

When x is slightly greater than 2, $\frac{dy}{dx}$ is positive.

Therefore the point where $x = 2$ is a minimum point.
The value of $(x^2 - 4x - 6)$ at this point is $4 - 8 - 6$ or -10.
$\therefore$ the minimum value of $(x^2 - 4x - 6)$ is -10 and occurs when $x = 2$.

Example 2. Find the maximum and minimum values of $x^2(x - 3)$.

If
$$y = x^2(x - 3) \quad \text{or} \quad x^3 - 3x^2,$$

$$\frac{dy}{dx} = 3x^2 - 6x \quad \text{or} \quad 3x(x - 2).$$

$$\therefore \frac{dy}{dx} = 0, \text{ when } x = 0 \text{ or when } x = 2.$$

(i) When x is slightly less than 0, x is negative and $(x - 2)$ is negative.

$\therefore \dfrac{dy}{dx}$ is positive.

When $x = 0$, $\dfrac{dy}{dx} = 0$.

When x is slightly greater than 0, x is positive and $(x - 2)$ is negative.

$\therefore \dfrac{dy}{dx}$ is negative.

So, at $x = 0$, $\dfrac{dy}{dx}$ changes from positive to negative. This is a maximum point and the corresponding value of y is 0.

(ii) When x is slightly less than 2, x is positive and $(x - 2)$ is negative.

$\therefore \dfrac{dy}{dx}$ is negative.

When $x = 2$, $\dfrac{dy}{dx} = 0$.

When x is slightly greater than 2, x is positive and $(x - 2)$ is positive.

$\therefore \dfrac{dy}{dx}$ is positive.

So, at $x = 2$, $\dfrac{dy}{dx}$ changes from negative to positive.

This is a minimum point and the corresponding value of y is $4(2 - 3)$ or -4.

Therefore the maximum value of $x^2(x - 3)$ is 0 and the minimum value is -4.

Example 3. Find the turning values of $x^3(x - 4)$.

If
$$y = x^3(x - 4) \quad \text{or} \quad x^4 - 4x^3,$$

$$\frac{dy}{dx} = 4x^3 - 12x^2 \quad \text{or} \quad 4x^2(x - 3).$$

$$\therefore \frac{dy}{dx} = 0, \text{ when } x = 0 \text{ or when } x = 3.$$

(i) When x is slightly less than 0, x^2 is positive and $(x - 3)$ is negative.

$\therefore \dfrac{dy}{dx}$ is negative.

When $x = 0$, $\dfrac{dy}{dx} = 0$.

When x is slightly greater than 0, x^2 is positive and $(x - 3)$ is negative.

$\therefore \dfrac{dy}{dx}$ is negative.

Since $\dfrac{dy}{dx}$ does not change sign, $x = 0$ is a point of inflexion.

(ii) When x is slightly less than 3, x^2 is positive and $(x - 3)$ is negative.

$\therefore \dfrac{dy}{dx}$ is negative.

When $x = 3$, $\dfrac{dy}{dx} = 0$.

When x is slightly greater than 3, x^2 is positive and $(x - 3)$ is positive.

$\therefore \dfrac{dy}{dx}$ is positive.

$\therefore \dfrac{dy}{dx}$ changes from negative to positive.

So $x = 3$ gives a minimum point and the minimum value of y is $3^3(3 - 4)$ or -27.

The curve has no maximum point.

Example 4. A farmer encloses sheep in a rectangular pen, using hurdles for three sides and a long wall for the fourth side. If he has 100 metres of hurdles, find the greatest area he can enclose.

Suppose the lengths of the sides of the rectangle are x metres (parallel to the wall) and y metres.

Then $$x + 2y = 100.$$
and so $$x = 100 - 2y.$$

If A m^2 is the area of the rectangle,

$$A = xy.$$
$$\therefore A = (100 - 2y)y = 100y - 2y^2.$$
$$\therefore \frac{dA}{dy} = 100 - 4y.$$
$$\therefore \frac{dA}{dy} = 0, \text{ when } y = 25.$$

When y is slightly less than 25, $\dfrac{dA}{dy}$ is positive.

When $y = 25$, $\dfrac{dA}{dy} = 0$.

When y is slightly greater than 25, $\dfrac{dA}{dy}$ is negative.

So $\dfrac{dA}{dy}$ changes from positive to negative and the point where $y = 25$ is a maximum point.

When $y = 25$, $x = 100 - 2y = 50$.
The value of A is 25×50, i.e. 1250 m².
The greatest area the farmer can enclose is 1250 m².
N.B. that when a curve has one turning point only, a maximum value is the greatest value; a minimum value is the least value.

Example 5. A right circular cylinder is to be made so that the sum of its radius and its height is 6 m. Find the maximum volume of the cylinder.
Let r metres be the radius and h metres the height.

Then
$$h + r = 6 \quad \text{or} \quad h = 6 - r.$$

Let V be the volume in m³.

$$\therefore \ V = \pi r^2 h = \pi r^2 (6 - r) = \pi(6r^2 - r^3).$$

We need to find the maximum value of V.

$$\frac{dV}{dr} = \pi(12r - 3r^2) = 3\pi r(4 - r).$$

$$\therefore \ \frac{dV}{dr} = 0, \text{ when } r = 0 \text{ or when } r = 4.$$

(i) When r is 0, the value of V is 0 and this is obviously not the greatest value.

(ii) When r is just less than 4, r is positive and $(4 - r)$ is positive.

$\therefore \ \dfrac{dV}{dr}$ is positive.

When $r = 4$, $\dfrac{dV}{dr} = 0$.

When r is just greater than 4, r is positive and $(4 - r)$ is negative.

$\therefore \ \dfrac{dV}{dr}$ is negative.

V is therefore a maximum when $r = 4$ and $h = 2$.
The maximum volume is $\pi r^2 h$, i.e. 32π m³.

Exercise 45.1a

Find maximum and minimum values of the following (questions **1–6**).

1. $x^2 - 6x$.   **2.** $1 - 2x - x^2$.   **3.** $x^2 - 2x - 3$.

4. $3x - x^3$.   **5.** $x - \sqrt{x}$.   **6.** $x + \dfrac{1}{x}$.

7. The sum of two numbers is 24. What is their maximum product?

8. A rectangle has a perimeter of 12 m. Find its maximum area.

9. The volume of a right circular cylinder, open at one end, is to be 8π cm³. Find its minimum surface area.

10. The sum of two numbers is 20. Find the numbers so that the sum of their squares is a minimum.

Exercise 45.1b

Find maximum and minimum values of the following (questions 1–6).

1. $x^2 - 8x$.   **2.** $1 - 4x - x^2$.   **3.** $x^2 - 4x - 1$.

4. $12x - x^3$.   **5.** $x + \sqrt{x}$.   **6.** $4x + \dfrac{1}{x}$.

7. The sum of two numbers is $2k$. What is their maximum product?

8. A rectangle has a perimeter of $4k$ metres. What is its maximum area?

9. The volume of a right circular cylinder, closed at both ends, is to be 16π cm^3. Find its minimum surface area.

10. The sum of two numbers is $2k$. Find the numbers so that the sum of their squares may be a minimum.

Velocity

Figure 45.3 shows a graph plotting the distance gone by a body, s metres

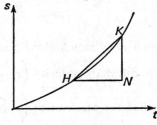

Fig. 45.3

(vertical axis) against the time taken, t seconds (horizontal axis). Such a graph is called a distance-time graph.

Suppose H, K are any two points on the graph and lines through H parallel to the t axis and through K parallel to the s axis meet at N. The length KN represents the distance gone between the two points and the length HN represents the time taken to cover this distance.

The gradient of the line $\left(= \dfrac{KN}{HN} \right)$ therefore represents the average velocity of the body during this interval.

This is true however close K is to H. As the distance between H and K gets smaller, the average velocity becomes closer to the velocity at H. The actual velocity at H is, in fact, the limit of the average velocity as K moves along the graph towards H. The limit of the line HK as K moves towards H is the tangent at H.

Therefore the gradient of the tangent at H is equal to the velocity at that instant.

The gradient may be found by drawing and by measurement from the graph. It may also be found by differentiation if the relation between s and t is given.

The velocity v at time t is given by $v = \dfrac{ds}{dt}$.

Example 6. *The distance moved by a point along a line in t seconds is given in metres by $s = t^2 + 4t$. Find (i) the initial velocity, (ii) the velocity after 2 s, (iii) the average velocity for the first 2 s.*

$$s = t^2 + 4t.$$

$$v = \frac{ds}{dt} = 2t + 4.$$

The velocity at time t is $(2t + 4)$ m/s.
(i) The initial velocity is the velocity when $t = 0$ and is 4 m/s.
(ii) The velocity after 2 s is found by putting t equal to 2, and is 8 m/s.
(iii) The distance gone in 2 s is found by putting $t = 2$ in $s = t^2 + 4t$.
The distance gone in 2 s is therefore $4 + 8$, i.e. 12 m. The average velocity is 6 m/s.

Acceleration

We have seen that when a body moves along a straight line, the rate of change of distance is called the velocity $\left(v = \dfrac{ds}{dt}\right)$. The rate of change of velocity is called the **acceleration** $\left(a = \dfrac{dv}{dt}\right)$. If the rate of change of velocity is constant, then the acceleration is constant. Suppose, for example, that the velocity of a body after 1 s is 4m/s, after 2 s is 6 m/s, after 3 s is 8 m/s. Then there is a steady increase in velocity of 2 m/s in each second. The acceleration is said to be 2 m/s per second or 2 m/s^2.

If a body is slowing down, it is said to have a negative acceleration or a **retardation**.

In the same way that the velocity is equal to the gradient of the distance-time graph, the acceleration is equal to the gradient of the velocity-time graph.

$$\therefore \text{ the acceleration, } a = \frac{dv}{dt}.$$

Example 7. *The distance s metres gone by a body in t seconds is given by $s = t^3 - 6t^2 + 4t$. Find (i) its initial velocity, (ii) its velocity after 2 seconds, (iii) its acceleration after 2 seconds.*

$$s = t^3 - 6t^2 + 4t.$$

$$v = \frac{ds}{dt} = 3t^2 - 12t + 4.$$

(i) When $t = 0$, $v = 4$. The initial velocity is 4 m/s.

(ii) When $t = 2$, $v = 12 - 24 + 4 = -8$.

The velocity after 2 s is -8 m/s (i.e. 8 m/s reversed in direction).

(iii) $a = \dfrac{dv}{dt} = 6t - 12$.

When $t = 2$, $a = 0$.

The acceleration after 2 s is zero.

Example 8. *The velocity v m/s of a point moving along a straight line, is given by $v = 4 - t^2$, where t is the time in seconds. Find the acceleration of the point when it is momentarily at rest.*

The point is momentarily at rest when $v = 0$, i.e. when $4 - t^2 = 0$. The values of t are 2 and -2 and so the point is momentarily at rest after 2 s. (The value -2 means that the point was at rest 2 seconds ago.)

$$v = 4 - t^2.$$

$$a = \frac{dv}{dt} = -2t.$$

Putting $t = 2$, the acceleration after 2 s is -4.

So when the body is at rest, it has a retardation of 4 m/s².

Exercise 45.2a

1. A point moves in a straight line so that its distance from a fixed point O of that line is $27t - t^3$ after t seconds. Show that it moves away from O for 3 s.

2. The distance s metres moved by a point in t seconds is given by
$$s = t^3 + 3t^2 + 4.$$
Find the velocity and acceleration after 3 s.

3. The velocity v m/s of a point moving in a straight line is given after t seconds by $v = 3t^2 + 4t$. Find the acceleration after 2 s.

4. The distance s metres moved by a particle along a line in t seconds is given by $s = 3t + t^2$. Find its velocity and acceleration after n seconds.

5. The velocity v m/s of a point moving in a straight line after t seconds is given by $v = t^2 - t$. Find the acceleration after 3 s and also find when the acceleration is zero.

Exercise 45.2b

1. A point moves in a straight line so that its distance from a fixed point O in the line is $32t - 2t^2$ cm after t seconds. Show that it moves away from O for 8 s.

2. The distance s metres moved by a point in t seconds is given by
$$s = t^3 + 3t^2 + 2t.$$
Find the velocity and acceleration after 2 s.

3. After t seconds, the velocity, v m/s, of a point moving in a straight line is given by $v = 4t^2 + 5t$. Find the acceleration after 3 s.

4. The distance s metres moved by a particle in t seconds is given by $s = 4t + 2t^2$. Find the velocity and acceleration after n seconds.

5. The velocity v m/s of a point moving in a straight line is given in terms of the time, t seconds, by $v = t^2 - 4t$. Find the acceleration after 3 s and also find when the acceleration is zero.

Rate of change

$\dfrac{dy}{dx}$ compares the rate of change of y with that of x; if $\dfrac{dy}{dx} = 6$, y is increasing 6 times as fast as x, numerically; if x and y are distances and x is increasing at 4 m/s, then y is increasing at 24 m/s.

If $\dfrac{dy}{dx} = -6$, then y decreases 6 times as fast as x increases.

Example 9. *The volume of a sphere of radius r is $\frac{4}{3}\pi r^3$. If the radius of a soap bubble is increasing at 0.1 cm/s, find the rate of increase of its volume when the radius is 2 cm.*

$$V = \tfrac{4}{3}\pi r^3$$

$$\frac{dV}{dr} = 4\pi r^2.$$

When $r = 2$, $\dfrac{dV}{dr} = 16\pi.$

The volume increases 16π times as fast as the radius.
The radius increases at 0.1 cm/s.
Therefore the volume increases at 1.6π cm³/s.

Example 10. *Water is poured into an inverted cone of semi-vertical angle $45°$, at a constant rate of 2 m³ per minute. At what rate is the depth of water increasing when it is 2 m deep?*

When the depth is x metres, the radius of cross section is $x \tan 45°$ or x metres. The volume of water is $\frac{1}{3}\pi x^2(x)$ or $\frac{1}{3}\pi x^3$ m³.

$$V = \tfrac{1}{3}\pi x^3.$$

$$\frac{dV}{dx} = \pi x^2.$$

When $x = 2$, $\dfrac{dV}{dx} = 4\pi.$

The volume increases 4π times as fast as the depth.
The volume increases at 2 m³ per minute.

Therefore the depth increases at $\dfrac{2}{4\pi}$ metres per minute $= \dfrac{1}{2\pi}$ metres per minute.

Exercise 45.3a

1. Find the rate of increase of the area of a square when the length of side is 2 cm and is increasing at 0.3 cm/s.
2. Find the rate of increase of the volume of a cube when the length of side is 3 cm and is increasing at 0.1 cm/s.
3. Find the rate of increase of the volume of a cube when the length of side is 2 cm and the area of a face is increasing at 1 cm²/s.
4. Find the rate of increase of the surface area of a sphere when the radius is 4 cm and is increasing at 0.1 cm/s.
5. Find the rate of increase of the volume of a sphere when the radius is 3 cm and is increasing at 0.2 cm/s.

Exercise 45.3b

1. Find the rate of increase of the area of a square when the length of side is 3 cm and is increasing at 0.2 cm/s.
2. Find the rate of increase of the volume of a cube when the length of side is 2 cm and is increasing at 0.2 cm/s.
3. Find the rate of increase of the volume of a cube when the length of side is 3 cm and the area of a face is increasing at 0.5 cm²/s.
4. Find the rate of increase of the surface area of a sphere when the radius is 2 cm and is increasing at 0.4 cm/s.
5. Find the rate of increase of the volume of a sphere when the radius is 2 cm and is increasing at 0.4 cm/s.

Exercise 45.4. Miscellaneous

1. Find the gradient of the curve $y = x(x - 2)^2 + 1$, at the point $(3,4)$ and determine the points at which the tangent is parallel to the axis of x.
2. A cylindrical vessel with one end open is to have a volume of 100 m³. Find the least area of sheet metal needed.
3. A piece of wire of length 1 m is cut into two pieces and each is bent into the form of a square. Find the least possible sum of the two areas.
4. Find any maximum or minimum values of $x(x - 1)^2 + 1$, distinguishing between them.
5. A square sheet of metal has sides of length 8 cm. Equal square pieces are removed from each corner and the remaining piece is bent into the form of an open box. Find the maximum volume of the box.
6. Show that the gradient of $y = 1 + 4x + 4x^2$ at the point where $x = 4$ is three times the gradient at the point where $x = 1$.
7. Find the maximum and minimum values of $x^2(x^2 - 2)$, distinguishing between them.
8. A point starts from rest and moves in a straight line. Its velocity after t s is $(6t - t^2)$ m/s. Find the initial acceleration and the time when the acceleration is zero.

9. The length of a rectangular block is twice the width and its volume is 72 m³. Find the minimum surface area of the block.

10. Find the maximum and minimum values of $2x^2 - x^4$, distinguishing between them.

11. A cylinder is such that the sum of its height and circumference of base is 6 m. Find the greatest possible volume of the cylinder.

12. Two concentric circles are such that the radius of the larger is always twice the radius of the smaller. If the radius of the smaller is increasing at 0.1 cm/s, find the rate at which the area between the circles is increasing when the radius of the smaller is 12 cm.

13. A sphere is expanding so that its volume is increasing at 2 cm³/s. Find the rate of increase of its radius when the radius is 10 cm.

14. If $y^2 = 4x$, find $\dfrac{dy}{dx}$ when $x = 4$.

15. Find the turning values of $24x + 3x^2 - x^3$, distinguishing between them.

16. If the pressure and volume of a gas are connected by the relation $pv^{1\cdot4} = 20$, find $\dfrac{dp}{dv}$.

17. If $s = 6 + 4t - t^2$, find the velocity and acceleration after 2 s.

18. Find the points on the graph of $y = 3x - x^2$ at which the tangent makes equal angles with the axes.

19. The graph of $y = Ax^2 + Bx + C$ passes through the origin and its gradient there is 2. The graph also passes through the point (1,1). Find A, B and C.

20. If $\dfrac{d}{dx}(x^3 - x) = \dfrac{d}{dx}(x^2 - 1)$, find x.

46 Integration

The inverse of differentiation

We know that the derived function of x is 1. Therefore, if we wish to find an expression whose derived function is 1, one answer is x. But it is not the only answer, because $x + 2$, $x + 3$, and $x + 18\frac{1}{2}$ are all expressions whose derived functions equal 1. In fact, $(x + c)$ is the general solution, where c is called an **arbitrary constant** and may be equal to any number at all. Its precise value may only be found if we are given further information.

Fig. 46.1 shows a number of straight lines, each of gradient 1. Any

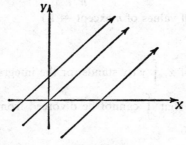

Fig. 46.1

line parallel to these lines will also have gradient 1. The equation of any such line is $y = x + c$. Therefore if $\frac{dy}{dx} = 1$, $y = x + c$.

If we are also told that $y = 2$ when $x = 0$, by substituting in $y = x + c$, we see that $c = 2$, and therefore $y = x + 2$.

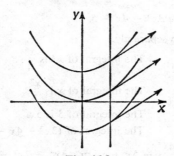

Fig. 46.2

Again, the derived function of x^2 is $2x$. The general expression whose derived function equals $2x$ is therefore $x^2 + c$, where c is an arbitrary constant. The curves $y = x^2 + c$ are 'parallel' curves and are shown in Fig. 46.2. The curves are all exactly the same shape with the vertex moved along the axis of y. All the curves have the same gradient for any given value of x.

The process of obtaining the function whose derived function is a given expression, is called integrating the expression. Integration is the inverse of differentiation.

$$\text{If } \frac{dy}{dx} = 1, \qquad\qquad\qquad y = x + c;$$

$$\text{if } \frac{dy}{dx} = x, \qquad\qquad\qquad y = \frac{x^2}{2} + c;$$

$$\text{if } \frac{dy}{dx} = x^n, \qquad\qquad\qquad y = \frac{x^{n+1}}{n+1} + c.$$

(This applies for all values of n except -1.)

Notation

If y is a function of x, $\int y\,dx$ stands for the integral of y with respect to x. The integral sign $\int$ cannot be divorced from dx if the integral is with respect to x.

For example:

$$\int x\,dx = \frac{x^2}{2} + c;$$

$$\int t^2\,dt = \frac{t^3}{3} + c;$$

$$\int (v^2 + 3v + 1)\,dv = \frac{v^3}{3} + \frac{3v^2}{2} + v + c.$$

Example 1. *Integrate $3x^3 - 4x - 5$.*

Integrate each term separately.

The integral of x^3 is $\dfrac{x^4}{4}$.

The integral of x is $\dfrac{x^2}{2}$.

The integral of 5 is $5x$.

The integral of $(3x^3 - 4x - 5)$ is

$$\frac{3x^4}{4} - \frac{4x^2}{2} - 5x + c \quad \text{or} \quad \tfrac{3}{4}x^4 - 2x^2 - 5x + c.$$

Exercise 46.1a

Integrate:

1. x^3.
2. $x^3 + 2x^2$.
3. $x^2 + 6$.
4. $\dfrac{1}{x^2}$.

5. $ax^2 + bx + c$.
6. $(x - 1)(x - 2)$.
7. $\dfrac{1}{x^3}$.
8. x^5.

9. The gradient of a curve which passes through the point (1,1) is given by $2 + 2x - x^2$. Find the equation of the curve.

10. If $\dfrac{dy}{dx} = 6x - 2$, and $y = 2$ when $x = 0$, find y in terms of x.

Exercise 46.1b

Integrate:

1. $\sqrt{x}$.
2. $x^2 + 3x + 4$.
3. $x^2 - 4x$.
4. $\dfrac{1 + x}{x^3}$.

5. $\dfrac{a}{x^2} + b$.
6. $x(x + 1)$.
7. $x^2(x + 2)$.
8. $\dfrac{x^4 + 1}{x^2}$.

9. The gradient of a curve which passes through the point (2,1) is given by $1 + 2x - 3x^2$. Find the equation of the curve.

10. If $\dfrac{dy}{dx} = x^2 + 1$, and $y = 1$ when $x = 0$, find y in terms of x.

Velocity and acceleration

If we know the distance moved along a straight line by a point in time t, the velocity is found by differentiating the distance with respect to the time. Inversely, if we know the velocity in terms of the time, the distance may be found by integration. Similarly, if we know the acceleration in terms of the time, integration gives the velocity.

Example 2. The acceleration of a point moving in a straight line is given by $a = 3t + 4$. Find formulae for the velocity and distance, given that $s = 0$ and $v = 8$ when $t = 0$.

$$a = 3t + 4.$$

$$\therefore \frac{dv}{dt} = 3t + 4.$$

Integrating,
$$v = \frac{3t^2}{2} + 4t + c.$$

When $t = 0$, $v = 8$.
$$\therefore c = 8.$$

Since
$$v = \frac{3t^2}{2} + 4t + 8,$$

$$\frac{ds}{dt} = \frac{3t^2}{2} + 4t + 8.$$

Integrating, $$s = \frac{t^3}{2} + 2t^2 + 8t + c.$$

Since $s = 0$ when $t = 0$, $c = 0$.

$$\therefore v = \frac{3t^2}{2} + 4t + 8 \quad \text{and} \quad s = \frac{t^2}{2} + 2t^2 + 8t.$$

Exercise 46.2a

(s, v, a, t, represent distance, velocity, acceleration and time in metre-second units.)

1. If $v = 3t + 1$ find s, given that $s = 0$ when $t = 0$.
2. If $a = t^2$, find v, given that $v = 2$ when $t = 0$.
3. If $a = t$, find s, given that when $t = 0$, $s = 0$ and $v = 4$.
4. If $v = t^2 + t + 1$, find s, given that $s = 0$ when $t = 0$.
5. If $a = t^2 + t$, find v, given that $v = 4$ when $t = 1$.
6. A stone falling under its own weight has a constant acceleration of 9.8 m/s². If it starts from rest, find its velocity after 4 s.
7. Find the velocity of a stone falling under gravity for 3 s, if its initial velocity is 8 m/s down.
8. Find the distance fallen by a stone in 4 s starting from rest.
9. Find the distance fallen by a stone in 2 s if its initial velocity is 8 m/s down.
10. A stone is thrown vertically upwards with a velocity of 9.8 m/s. Find how long it takes to reach its highest point.

Exercise 46.2b

1. If $v = t^2 + 1$, find s, given that $s = 0$ when $t = 0$.
2. If $a = t + 1$, find v, given that $v = 4$ when $t = 0$.
3. If $a = t^2$, find s, given that when $t = 0$, $s = 0$ and $v = 2$.
4. If $v = 2t^2 + 3t + 1$, find s, given that $s = 0$ when $t = 0$.
5. If $a = t^2 - t$, find v, given that $v = 3$ when $t = 2$.
6. A stone falls from rest. Find its velocity after 3 s.
7. A stone has a velocity of 8 m/s down. Find its velocity after 2 s.
8. Find the distance fallen by a stone in 2 s, if it starts from rest.
9. Find the distance fallen by a stone in 4 s if its initial velocity is 6 m/s down.
10. A stone is thrown vertically upwards with a velocity of 19.6 m/s. Find how high it rises.

Area of a region

Consider the region bounded by the curve $y = x^2$, the lines $x = 1$, $x = 2$ and the x-axis. Suppose that HN, KM are the ordinates at $x = 1$ and $x = 2$ and let PQ be an ordinate somewhere between HN and KM.

Suppose the distance of PQ from the axis of y is x, and let the area $NHPQ$ be A. This area A is obviously a function of x and is zero when $x = 1$. When $x = 2$, A is equal to the area required.

Suppose that PQ moves to another position $P'Q'$ whose distance from the axis of y is $(x + h)$.

Since P and P' are on the curve $y = x^2$,

$$PQ = x^2 \quad \text{and} \quad P'Q' = (x + h)^2.$$

The increase in A due to the movement from PQ to $P'Q'$ lies between the area of two rectangles, one of length x^2, the other of length $(x + h)^2$ and each of breadth h.

∴ (the increase in A) ÷ h lies between x^2 and $(x + h)^2$.

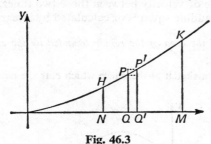

Fig. 46.3

Now suppose that h tends to zero. (The increase in A) ÷ h tends to x^2. But (the increase in A) ÷ h tends to the derived function of A as h tends to 0.

$$\frac{dA}{dx} = x^2,$$

and
$$A = \tfrac{1}{3}x^2 + c.$$

When $x = 1$, $A = 0$.

Substituting, $\qquad 0 = \tfrac{1}{3} + c \quad \text{and} \quad c = -\tfrac{1}{3}.$

$$\therefore A = \tfrac{1}{3}x^3 - \tfrac{1}{3}.$$

The area required is the value of A when $x = 2$, which is $(\tfrac{8}{3} - \tfrac{1}{3})$ or $\tfrac{7}{3}$. Therefore the area of the region is $2\tfrac{1}{3}$ square units.

In general the problem of finding the area of the region under a curve $y = f(x)$ between two given ordinates $x = h$ and $x = k$ is equivalent to the following problem.

Find the value of A when $x = k$, given that $\dfrac{dA}{dx} = f(x)$ and that $A = 0$ when $x = h$.

Area of the region under a velocity–time graph

The area of the region under a velocity–time graph represents the integral of the velocity with respect to the time. We have already seen that this is equal to the distance travelled. If we know a number of corresponding readings of velocity and time, the area of the region may be found by addition of the small squares of the region. A small square is included in the addition if more than half of it is in the required region, otherwise it is excluded. If we know the relation between velocity and time, the distance can be calculated exactly by integration.

Area of the region under an acceleration–time graph

Similarly the area of the region under an acceleration–time graph is equal to the integral of the acceleration with respect to the time. Therefore this area gives the velocity. The area between two times or ordinates equals the change of velocity between those two times. This may either be computed by adding squares or calculated by integration.

Example 3. *Find the area of the region bounded by the curve $y = 2x - x^2$ and the x-axis.*

Fig. 46.4 shows a sketch of the curve which cuts the axis of x at $(0,0)$ and $(2,0)$.

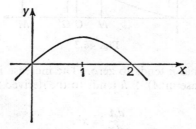

Fig. 46.4

We wish to find A when $x = 2$, given that $\dfrac{dA}{dx} = 2x - x^2$ and that $A = 0$ when $x = 0$.

$$\frac{dA}{dx} = 2x - x^2.$$

Integrating, $A = x^2 - \frac{1}{3}x^3 + c.$

When $x = 0$, $A = 0$. $\therefore c = 0.$

$$A = x^2 - \frac{1}{3}x^3.$$

When $x = 2$, $A = 4 - \dfrac{8}{3} = \dfrac{4}{3}.$

Example 4.   *Find the area of the region bounded by the curve $y = x^2$, the line $y = x$ and the x-axis.*

Fig. 46.5 shows a sketch of the curve and line. They meet where $x = x^2$, i.e. at the points $(0,0)$ and $(1,1)$.

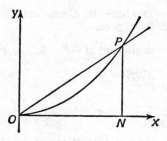

Fig. 46.5

To find the area of the region under the curve between $x = 0$ and $x = 1$,

we must find the value of A when $x = 1$, given that $\dfrac{dA}{dx} = x^2$ and that $A = 0$

when $x = 0$.

Integrating, $\qquad A = \tfrac{1}{3}x^3 + c$.
Since $A = 0$, when $x = 0$, $c = 0$. $\therefore A = \tfrac{1}{3}x^3$.
When $x = 1$, $A = \tfrac{1}{3}$.
The area of the triangle $OPN = \tfrac{1}{2}$

(The area of the region bounded by the curve $y = x^2$, the line $y = x$ and the x-axis is

$$(\tfrac{1}{2} - \tfrac{1}{3}) = \tfrac{1}{6} \text{ square units.}$$

(This may be verified by counting squares. Be careful to draw as large a figure as possible on your graph paper.)

Example 5. A body moving in a straight line starts from rest and accelerates at 2 m/s² for 4 s. It then decelerates uniformly and comes to rest in another 6 s. Find the distance gone.

The gradient of the velocity–time graph is the acceleration. When the acceleration is constant, the graph is a straight line.

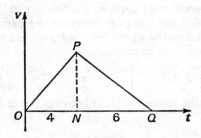

Fig. 46.6

The graph consists of two lines OP, of gradient 2, and PQ. The area under the graph equals the distance gone.

$$\frac{PN}{ON} = \text{the gradient of } OP = 2. \quad \therefore \ PN = 8.$$

The area of $OPQ = \frac{1}{2}(10)8 = 40$.
The distance gone is 40 m.

The definite integral

In Example 3, we found the area of the region bounded by the curve $y = 2x - x^2$ and the x-axis.

$$\frac{dA}{dx} = 2x - x^2 \quad \text{and therefore} \quad A = \int (2x - x^2) \, dx.$$

From this equation, $A = x^2 - \dfrac{x^3}{3} + c.$

We know that $A = 0$ when $x = 0$ and we wish to find the value of A when $x = 2$.

Substituting, $\qquad\qquad\qquad A = 2^2 - \dfrac{2^3}{3} + c$

and $\qquad\qquad\qquad\qquad\quad 0 = 0 + c.$

Subtracting: $\qquad\qquad\quad A = \left(2^2 - \dfrac{2^3}{3}\right) - 0.$

A therefore equals the value of $\left(x^2 - \dfrac{x^3}{3}\right)$ when $x = 0$ subtracted from the value of $\left(x^2 - \dfrac{x^3}{3}\right)$ when $x = 2$.

$A = \displaystyle\int (2x - x^2) \, dx$ is called an **indefinite integral** and must contain an arbitrary constant.

$A = \displaystyle\int_0^2 (2x - x^2) \, dx$ is called a **definite integral** and means that the integration is performed between the limiting values 0 and 2 for x.

To evaluate a definite integral, integrate the expression; from the value of the integral at the upper limit subtract the value at the lower limit. The work is set out as shown.

$$A = \int_0^2 (2x - x^2) \, dx = \left[x^2 - \frac{x^3}{3}\right]_0^2 = \left(4 - \frac{8}{3}\right) - 0 = \frac{4}{3}.$$

The area under the curve $y = f(x)$ between the ordinates $x = a$ and $x = b$ is given by

$$A = \int_a^b y \, dx \quad \text{or} \quad \int_a^b f(x) \, dx.$$

Example 6. Find the area of the region bounded by the curve $y = x^2 + 1$, the ordinates $x = 1$, $x = 2$ and the x-axis.

$$\frac{dA}{dx} = x^2 + 1.$$

$$\therefore A = \int_1^2 (x^2 + 1)\, dx = \left[\frac{x^3}{3} + x\right]_1^2$$

$$= \left(\frac{8}{3} + 2\right) - \left(\frac{1}{3} + 1\right) = 3\tfrac{1}{3}.$$

Exercise 46.3a

1. Find the area under the curve $y = x + x^2$ between $x = 1$ and $x = 3$.
2. Find the area between the curve $y = x - x^2$ and the axis of x.
3. Find the area under the curve $y = \sqrt{x}$ between $x = 0$ and $x = 4$.
4. Find the area under the curve $y = \dfrac{1}{x^2}$ between $x = 1$ and $x = 4$.
5. Find the area under the curve $y = x^2 + 4$ between $x = 1$ and $x = 2$.
6. Find the area between the line $y = 2x$ and the curve $y = x^2$.
7. A body moving in a straight line starts with a velocity of 4 m/s and moves with constant acceleration of 2 m/s^2 for 8 s. Find the distance gone.
8. A body starts from rest and accelerates at 4 m/s for 5 s. It then decelerates uniformly and stops in a further 10 s. Find the distance gone.

Exercise 46.3b

1. Find the area under the curve $y = 2x + x^2$ between $x = 1$ and $x = 3$.
2. Find the area between the curve $y = (x - 1)(3 - x)$ and the axis of x.
3. Find the area under the curve $y = \sqrt{x}$ between $x = 1$ and $x = 9$.
4. Find the area under the curve $y = 1 + \dfrac{1}{x^2}$ between $x = 1$ and $x = 4$.
5. Find the area under the curve $y = 3x^2 + 1$ between $x = 1$ and $x = 3$.
6. Find the area between the line $y = 4x$ and the curve $y = 4x^2$.
7. A body moving in a straight line starts with a velocity of 10 m/s and moves with constant acceleration of 3 m/s^2 for 6 s. Find the distance gone.
8. A body starts from rest and accelerates at 2 m/s^2 for 6 s. It moves with constant velocity for 2 s and then decelerates uniformly to stop in another 4 s. Find the distance gone.

Solid of revolution

If the region under a curve and included by two ordinates is rotated about the axis of x, the resulting solid is called a **solid of revolution**. A section of this solid by a plane perpendicular to the axis of x is a circle.

The region under a line parallel to the axis of x gives a right circular cylinder.

The region under a line through the origin gives a right circular cone. We shall now find the volume of a solid of revolution, given the equa-

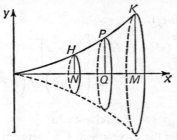

Fig. 46.7

tion of the curve and the bounding ordinates.

As an example, consider the curve $y = x^2$ between the ordinates $x = 1$ and $x = 2$. Suppose the bounding ordinates are HN and KM and that PQ is any ordinate between them.

If we rotate PQ, we shall get a circle. If this circle moves from the position in which HN is a radius to the position when KM is a radius, it will trace out the solid required. Let V be equal to the volume between the starting position and the position when the distance of PQ from the y-axis is equal to x. Then V is a function of x which is zero when $x = 1$; we wish to find V when $x = 2$. If PQ moves to a position $P'Q'$ whose distance from the axis of y is $(x + h)$, V is increased by a solid whose plane ends are circles of radii x^2 and $(x + h)^2$. This solid lies between that of a cylinder of radius x^2 and length h and that of a cylinder of radius $(x + h)^2$ and length h. The increase in V therefore lies between $\pi h x^4$ and $\pi h(x + h)^4$.

$\therefore$ (the increase in V) $\div h$ lies between πx^4 and $\pi(x + h)^4$.

$\therefore$ (the increase in V) $\div h$ tends to πx^4 as h tends to 0.

But (the increase in V) $\div h$ tends to the derived function of V as h tends to 0.

$$\therefore \frac{dV}{dx} = \pi x^4.$$

Integrating,

$$V = \frac{\pi x^5}{5} + c.$$

But $V = 0$ when $x = 1$.

$$\therefore 0 = \frac{\pi}{5} + c.$$

$$\therefore V = \pi\left(\frac{x^5 - 1}{5}\right).$$

When $x = 2$,

$$V = \pi\left(\frac{32 - 1}{5}\right) = \frac{31\pi}{5}.$$

The problem of finding the volume of the solid obtained by rotating about the axis of x the area under the curve $y = f(x)$ between $x = h$ and $x = k$ is equivalent to the following problem.

Find V when $x = k$, given that $\dfrac{dV}{dx} = \pi\{f(x)\}^2$ and that $V = 0$ when $x = h$.

This volume may also be expressed as a definite integral. The volume of a solid formed by rotation about the x-axis of the area under the curve $y = f(x)$ between $x = h$ and $x = k$ is given by

$$V = \pi \int_h^k y^2 \, dx \quad \text{or} \quad \pi \int_h^k \{f(x)\}^2 \, dx.$$

Example 7. *Find the volume of a cone of height h and base radius r.*

The cone is formed by the revolution about the axis of x of the region under a line which passes through the origin. The gradient of the line is $\dfrac{r}{h}$ and so its equation is $y = \dfrac{r}{h}x$.

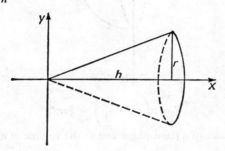

Fig. 46.8

We are asked to find V when $x = h$, given that $\dfrac{dV}{dx} = \pi\dfrac{r^2}{h^2}x^2$ and that $V = 0$ when $x = 0$.

Integrating, $\qquad\qquad V = \pi\dfrac{r^2}{h^2}\dfrac{x^3}{3} + c.$

Since $V = 0$ when $x = 0$, $c = 0$.

$$\therefore V = \frac{\pi r^2}{h^2}\frac{x^3}{3}.$$

When $x = h$ $\qquad\qquad V = \dfrac{\pi r^2}{h^2}\dfrac{h^3}{3} = \tfrac{1}{3}\pi r^2 h.$

Example 8. *By rotating a quadrant of the circle $x^2 + y^2 = r^2$ about the axis of x, find a formula for the volume of a sphere.*

By rotating a quadrant of a circle, we shall get a hemisphere.

We wish to find V when $x = r$, given that $\dfrac{dV}{dx} = \pi y^2$ and that $V = 0$ when $x = 0$.

$$\frac{dV}{dx} = \pi y^2 = \pi(r^2 - x^2).$$

Integrating, $\qquad V = \pi\left(r^2 x - \dfrac{x^2}{3}\right) + c.$

N.B. r^2 is a constant.

When $x = 0$, $V = 0$ and therefore $c = 0$.

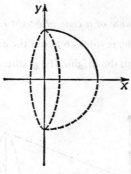

Fig. 46.9

When $x = r$, $\qquad V = \pi\left(r^3 - \dfrac{r^3}{3}\right) = \tfrac{2}{3}\pi r^3.$

This is the volume of a hemisphere and so the volume of a sphere of radius r is $\tfrac{4}{3}\pi r^3$.

Alternative method. Since

$$\frac{dV}{dx} = \pi y^2.$$

$$V = \pi \int_0^r y^2\, dx.$$

But $y^2 = r^2 - x^2$, and therefore

$$V = \pi \int_0^r (r^2 - x^2)\, dx$$

$$= \pi\left[r^2 x - \frac{x^3}{3}\right]_0^r$$

$$= \pi\left[\left(r^3 - \frac{r^3}{3}\right) - 0\right] = \tfrac{2}{3}\pi r^3.$$

The volume of the sphere is $\tfrac{4}{3}\pi r^3$.

Exercise 46.4a

Find the volume of rotation about the axis of x of the region bounded by the x-axis and the given curves and ordinates.

1. $y = \sqrt{x}$, $x = 1$, $x = 4$.
2. $y = 4x^2$, $x = 0$, $x = 1$.
3. $y = \dfrac{1}{x}$, $x = 1$, $x = 2$.
4. $y = 4 - x^2$, $x = 1$, $x = 2$.
5. $y = 2 + x$, $x = 2$, $x = 6$.

Exercise 46.4b

Find the volume of rotation about the axis of x of the region bounded by the x-axis and the given curves and ordinates.

1. $y = \sqrt{x}$, $x = 1$ and $x = 9$.
2. $y = 2x^2$, $x = 0$, $x = 2$.
3. $y = \dfrac{1}{x}$, $x = 1$, $x = 3$.
4. $y^2 = r^2 - x^2$, $x = \frac{1}{2}r$, $x = r$.
5. $y = a + x$, $x = 0$, $x = a$.

Exercises 46.5. Miscellaneous Integration

1. Find the area of the region enclosed by the axis of x, the curve $y = x^2 + x + 1$ and the ordinates $x = 1$ and $x = 2$.
2. A point moves in a straight line so that after t seconds its velocity is $(3t^2 + 4t)$ m/s. Find the distance gone in the third second.
3. The region under the curve $y^2 = x^3$ between $x = 0$ and $x = 1$ is rotated about the axis of x. Find the volume of the solid of revolution so formed.
4. A point moves in a straight line and its acceleration in m/s² is given by $a = t + 2$ where t is in seconds. If the point starts from rest, find the distance gone in 4 s.
5. Show that the curves $y^2 = 4x$ and $x^2 = 4y$ meet at $(4,4)$. Draw a rough sketch and find the area of the region bounded by the curves.
6. Find the volume formed by the revolution about the x-axis of that part of the ellipse $4y^2 = 9 - x^2$ between $x = 0$ and $x = 3$.
7. Integrate $x^2(1 + x)^2$.
8. A point moves in a straight line so that its acceleration in m/s² after t seconds is given by $a = 3t + 2$. If its initial velocity is 4 m/s, find its velocity after 2 seconds.
9. Find the area of the region bounded by the curve $y = 2x - 3x^2$ and the line $y = x$.
10. Find the area of the region bounded by the curve $y = 4 - x^2$ and the axis of x.

11. If $\dfrac{dy}{dx} = 2 - x^2$ and $y = 3$ when $x = 2$, find y in terms of x.

12. Find the area between the curve $y = x^2(5 - x)$ and the axis of x.

13. Calculate the volume of the solid formed by rotating the region bounded by $y = x(2 - x)$ and the axis of x about that axis.

14. The following table gives the velocity at various times of a decelerating body.

t (seconds)	0	1	2	3	4	5	6
v (m/s)	30	20	14	9	5	1.5	0

Find the distance gone in the six seconds.

15. The following table gives corresponding values of velocity and time:

t (seconds)	0	1	2	3	4	5	6
v (m/s)	10	13	18	24	32	41	54

Find the distance gone in the six seconds.

16. The velocity of a point moving in a straight line is given by $v = 2t^2 + 3t$. Find the acceleration after 2 s and the distance gone in the third second.

17. Find the volume formed by rotating the region bounded by $y = 3x$, the axis of x and the line $x = 2$ about the axis of x.

18. Find the area of the region bounded by the curve $y = 4x^2$ and the straight lines $x = 1$, $x = 2$, $y = 0$.

19. Find the area of the region bounded by the curve $y = 3x^2 - x^3$ and the axis of x.

20. Integrate $(3x + 1)^2$.

21. The acceleration of a particle moving in a straight line is given in m/s² by $a = t + 3$ where t is in seconds. By drawing a graph and calculating the area under the graph, find the velocity of the particle after 3 s given that the initial velocity is 2 m/s.

22. The gradient of a certain curve is $x^2(3 - x)$ and the curve passes through the point $(1,1)$. Find the equation of the curve.

23. By rotating a quadrant of the ellipse $\dfrac{x^2}{a^2} + \dfrac{y^2}{b^2} = 1$ about the axis of x,

find a formula for the volume of an ellipsoid.

24. The area under a certain curve cut off by the ordinates at 0 and x is equal to $4x - x^3$. Find the equation of the curve.

25. If $f'(x) = 3x - x^2$ and $f(2) = 4$, find $f(x)$.

26. Evaluate $\displaystyle\int_{2}^{3} (x^2 + x)\, dx$.

27. Evaluate $\displaystyle\int_{-1}^{1} (1 - t^2)\, dt.$

28. Evaluate $\displaystyle\int_{-1}^{2} (3u + u^2)\, du.$

Revision Exercise

Some of the questions in this exercise are taken from GCE and other examinations, reprinted by kind permission of the following Boards:

Associated Examining Board	(AEB)
London University Schools Examination Council	(L)
Oxford University Delegacy for Local Examinations	(O)
Oxford and Cambridge Board	(O&C)
MEI Project	(MEI)
SMP Project	(SMP)
London Chamber of Commerce, Commercial Education Scheme	(LCC)

1. A transistor radio can be bought for £25.50 cash, or by a deposit of £5.10 and twelve monthly payments of £2.07. By what percentage does the hire-purchase cost exceed the cash cost?

2. Light travels at approximately 3×10^5 km/s. How far does light travel in a day? How far away, in kilometres, is a star if light from that star takes one year to reach the Earth?

3. A certain examination paper has 5 questions, all compulsory, and is marked out of 120. The first question carries 30 marks, and the marks for the others are in the ratio 3:4:5:6. Find the marks awarded for each question. (LCC)

4. A traveller changes £300 into francs at 8.40 francs to the £, spends 2200 francs, then changes the remaining francs back into sterling at 8.80 francs to the £. How much, correct to the nearest penny, does he receive?

5. If $p + q = 5(2p - q)$, find the value of $\dfrac{p}{q}$.

6. The expenses of a tea-shop are £P per day plus x pence for each tea served. The teas are sold for y pence each. Find a formula to give the profit for a day in which n teas are served. (O&C)

7. Solve the equation $(y - 3)(y + 5) = 20$.

8. A hollow cylindrical tube has length $5x$ cm, and the internal and external radii of the circular section are x and $3x$ cm respectively. Find an expression for the volume of the tube.

9. A newsagent buys papers at £x per dozen and sells them at y pence each. Find an expression for his profit in a day when he buys a newspapers and sells b of them. (Those unsold he gives away.)

10. In a certain year the rate of a dialled telephone call was x seconds for 1p. Later in the year the rate was changed to $(x + 6)$ seconds for $1\frac{1}{2}$p. Write down expressions for the cost in pence of a 3-minute call at each of the two rates.

As a result of this change, the cost of a 3-minute call is increased by $1\frac{1}{2}$p. Form an equation and solve it to find the two possible values of x. (SMP)

11. (i) Find the value of x for which $(x + 2)^2 = x^2 + a^2x + 2a$ when $a = 1$.
 (ii) Find the values of a for which the above equation is satisfied when $x = -1$.
 (iii) Show that when $a = 2$, the equation is satisfied by all values of x.
 (iv) Find a value of a for which the equation is satisfied by no value of x. (L)

12. Given that the volume of a frustum of a cone is $\dfrac{\pi h}{3}(R^2 + Rr + r^2)$, calculate the value of r required to give a volume of 3322 units when $h = 21$ and $R = 9$. (Take π to be $3\frac{1}{7}$.) (AEB)

13. The operation $*$ is defined by

$x * y$ is the last digit of the product xy,

e.g. $3 * 2 = 6$, $3 * 4 = 2$, $3 * 6 = 8$. Draw up a table showing the value of $x * y$ for all values of x, y in the set S, where $S = \{1,3,5,7,9\}$.

Show that S is not a group under $*$, and, assuming the associative property, find a subset of S which is a group under $*$.

14. The operation $*$ is defined on the set $\mathbb{Z}_+$ of positive integers and zero by

$x * y$ is the positive (or zero) difference of x and y,

e.g. $2 * 5 = 3$, $6 * 2 = 4$, $3 * 3 = 0$.
 (i) Find any three integers x, y, z to show that the operation $*$ is not associative over $\mathbb{Z}_+$.
 (ii) Find the identity element, and the inverse of every element.
 (iii) Solve the equation $7 * (x * 6) = 2$.

15. If $f:x \mapsto x^2 - 2x$, find

 (i) the range, if the domain is $\{-2, 0, 2\}$;
 (ii) the domain, if the range is $\{-1, 0, 3\}$.
 Illustrate each mapping by a Papy-graph.

16. A consumers' association tests two brands of margarine X and Y and assigns points as below. Two housewives Mrs A and Mrs B give weightings as shown to the qualities assessed.

	Brand X	Brand Y	Mrs A	Mrs B
			(weighting)	
Taste	100	105	1	3
Appearance	100	115	2	2
Wrapping	120	100	3	1

Calculate the weighted mean for each brand of margarine for Mrs A and Mrs B, to determine which brand each would prefer.   (LCC)

17. A certain country classified its population in 5 groups by income, and recorded how much the railways were used by each income group. It was found that:

> Group I (the richest) travelled 45% of the passenger-kilometres,
> Group II travelled 25% of the passenger-kilometres,
> Group III travelled 15% of the passenger-kilometres,
> Group IV travelled 10% of the passenger-kilometres

and Group V (the poorest) travelled 5% of the passenger-kilometres.

Draw a suitable statistical diagram to display what proportion of the passenger-kilometres was travelled by each of these economic groups.   (LCC)

18. The ages of the cars in a certain village were recorded and classified as below.

Age in years	Number of cars	Age in years	Number of cars
Under 1	60	4 and under 5	45
1 and under 2	60	5 and under 6	35
2 and under 3	55	6 and under 7	25
3 and under 4	50	Over 7	30

(i) Draw a cumulative frequency curve, and use this to estimate the median and quartiles of the ages of the cars in this village.

(ii) Why is it not possible to calculate an estimate of the mean age of these cars?

19. Two cards are drawn at random from nine cards bearing the numbers 1,2,3,4, . . ., 9. A two-digit number is formed by placing these two cards side-by-side. Find the probability that this two-digit number is a prime.

20. In a simplified form of Bingo for use at a children's party there are thirty similar balls numbered 1 to 30 in a box. These are drawn out at random one by one and are not replaced. Each child has a card with five numbers between 1 and 30 on it, and whenever a ball with one of those numbers on it is drawn out the child crosses off that number. The first child to cross off all five numbers on his card wins. On this occasion there are only two children playing, A and B; A's card has 1, 8, 18, 19, 27 on it and B's has 1, 8, 10, 13, 29.

(i) Write down the probabilities that on the first draw (a) A will cross off a number; (b) neither A nor B will cross off a number.

(ii) Find the probability that A will cross off a number on both the first and second draws.

(iii) Find the probability that on the first draw A will cross off a number but B won't. Show also that the probability that this happens and that then, on the second draw, B will cross off a number but A won't is $\frac{3}{290}$.   (MEI)

21.

$$\begin{array}{cc} & \text{Tomorrow} \\ & \text{Fine Wet} \\ \text{Today } \begin{array}{c} \text{Fine} \\ \text{Wet} \end{array} & \begin{pmatrix} \frac{2}{3} & \frac{1}{3} \\ \frac{1}{2} & \frac{1}{2} \end{pmatrix} \end{array}$$

The matrix above, denoted by $\mathbf{M}$, is one whose entries are all probabilities relating to changes in the weather. For example the entry $\frac{2}{3}$ in the top left-hand corner indicates that if today's weather is fine then the probability of tomorrow's being fine is $\frac{2}{3}$.

(i) Write down the probability that if today is fine, tomorrow will be wet.

(ii) Assume that today is Sunday and that it is fine. Calculate the probabilities

(a) that Monday and Tuesday will both be fine;

(b) that Monday will be wet and Tuesday will be fine.

(iii) Calculate $\mathbf{M}^2$. Interpret the entry in the bottom right-hand corner of this matrix as a probability.

(iv) It is desired to find numbers x and y such that $x + y = 1$ and such that

$$(x \quad y)\begin{pmatrix} \frac{2}{3} & \frac{1}{3} \\ \frac{1}{2} & \frac{1}{2} \end{pmatrix} = (x \quad y).$$

Verify that $x = \frac{3}{5}$ and $y = \frac{2}{5}$.   (O)

22. Two vessels, P and Q, contain different liquids. One-sixth of the liquid in P is run into a jar R, and one-tenth of the liquid in Q is run into a jar S. The contents of R and S are then poured into Q and P respectively. This process is repeated every minute. (It can be assumed that no liquid is spilt.)

If p_0, q_0 are the volumes of liquid in P and Q at time T minutes and p_1, q_1 are the volumes of liquid in P and Q at time $(T + 1)$ minutes, write down expressions for p_1 and q_1 in terms of p_0 and q_0.

Show that $\begin{pmatrix} p_1 \\ q_1 \end{pmatrix} = \mathbf{M}\begin{pmatrix} p_0 \\ q_0 \end{pmatrix}$, where $\mathbf{M}$ is the matrix $\begin{pmatrix} 5/6 & 1/10 \\ 1/6 & 9/10 \end{pmatrix}$.

If $\begin{pmatrix} p_2 \\ q_2 \end{pmatrix} = \mathbf{N}\begin{pmatrix} p_0 \\ q_0 \end{pmatrix}$, where p_2, q_2 are the volumes at time $(T + 2)$

minutes, find the elements of the matrix $\mathbf{N}$.

Find also the ratio of p to q, if $\mathbf{M}\begin{pmatrix} p \\ q \end{pmatrix} = \begin{pmatrix} p \\ q \end{pmatrix}$, and state the meaning

of your results.   (L)

23. The Trans-Can Airline has eight Viscounts, six Tridents and two Caravelles. East Atlantic Airways have nine Viscounts, one Trident and seven Caravelles, and the Il-Oil Company has two Viscounts, eleven Tridents and no Caravelles.

   (i) Express this information as a 3×3 matrix **A**.
   (ii) The Viscount carries 50 passengers, the Trident carries 140 and the Caravelle 80. Write down a suitable product of two matrices which when evaluated will determine the number of passengers that each airline is able to carry when all its aircraft are full.

   (iii) Evaluate this product.
   (iv) Describe what information would be given by premultiplying your matrix **A** by the matrix (1 1 1).   (SMP)

24. An aircraft is to carry a load of relief food supplies from a depot which holds two kinds of food package, X and Y, which are as shown in the table below.

	Mass in kg	Vol. in litres	No. of ration units
Package X	15	40	250
Package Y	25	20	350

   The load can have at most a mass of 1500 kg and a volume of 3000 litres. On a day when the depot holds 70 of each kind of package it is required to make up a load with a maximum number of ration units. Assuming x packages of type X and y of type Y are included, write down and simplify (i) the inequalities (other than $x \geqslant 0, y \geqslant 0$) which govern the values of x and y; and (ii) an expression in x and y which has to be maximized.

   Show on graph paper, by means of shading, the region in which permissible pairs (x, y) lie. Show that there are two solutions involving 85 packages in all, and find which of these is the better solution of the problem.   (O)

25. In planning the layout of the cabin of a new airliner, it is proposed to instal x complete rows of first-class seats and y complete rows of economy-class seats. The conditions governing the installation are:

   (a) Each row of first-class seats occupies 1.6 m of the length of the cabin, and each row of economy-class seats occupies 1 m; the total length of cabin available is 40 m.

   (b) Each row of first-class seats holds 6 passengers, whereas each row of economy-class seats holds 8 passengers; the airline requires that the cabin can seat at least 192 passengers altogether.

   (c) Each first-class passenger is entitled to 20 kg of baggage, and each economy-class passenger is entitled to 15 kg of baggage; the total baggage carried must not exceed 3840 kg.

   Write down each of these conditions as an inequality involving x and y.

Illustrate all three inequalities graphically, labelling each boundary line with its equation, and using a scale of 1 cm to represent 2 rows of seats on each axis. Outline on your diagram the region containing points representing numbers of rows of seats which satisfy all three inequalities simultaneously.

From your graph find the minimum permissible number of economy-class passengers under these conditions   (MEI)

26. An amateur dramatic society is putting on a small entertainment. The audience can be seated partly in easy chairs and partly in upright chairs for which tickets will be cheaper. The number of easy chairs is x and the number of upright chairs is y, and previous box-office experience allows us to assume that $y \leqslant 2x$.

The hall has a floor area of 90 m²; easy chairs require 1 m² and upright chairs require 0.6 m² of space. The organisers must allow for an audience of at least 100, but there are only 50 easy chairs available. Write down these relations in the form of inequalities involving x and y.

Illustrate all four relations on a graph. Assuming that easy chairs must be put in rows of ten and that upright chairs must be put in rows of twelve, mark clearly on your diagram the points which represent possible numbers of each chair which will satisfy all the conditions.

If tickets for easy chairs cost 40p and those for upright chairs cost 35p, which layout will be more profitable to the society, assuming that all seats are sold?   (MEI)

27. Given that $(x - p)(x - q) > 0$, which of the following statements *must* be true?
    (i) $x = p$ or $x = q$.
    (ii) $x > p$ and $x > q$.
    (iii) $x > p$ or $x > q$.
    (iv) $x < p$ and $x < q$.

28. Write down the converse of each of the following theorems.
    (i) The square on the hypotenuse of a right-angled triangle is equal to the sum of the squares on the other two sides.
    (ii) The opposite angles of a cyclic quadrilateral are supplementary.
    (iii) The angle in a semicircle is a right angle.

29. O, A, and B are three points not in a straight line. Given that $OA = \mathbf{a}$ and $OB = \mathbf{b}$, draw a sketch showing O, A, and B, and mark on your sketch points C and D, where $OC = \frac{1}{2}\mathbf{a} + \mathbf{b}$, $OD = \mathbf{a} + \frac{1}{2}\mathbf{b}$. Use vectors to prove that AB is parallel to DC.

30. Two vectors $\mathbf{a}$ and $\mathbf{b}$ are not parallel. Given that
$$x\mathbf{a} + y\mathbf{b} = \mathbf{b} + y(\mathbf{a} - \mathbf{b}),$$
find the scalars x and y (i) by algebra; (ii) by geometry, using properties of a parallelogram.

31. If $x = 3a + 5b$ and $y = 4a - 3b$, find **a** and **b** in terms of **x** and **y**.

32. $OABCDE$ is a regular hexagon. If $\overrightarrow{OA} = $ **a**, $\overrightarrow{OB} = $ **b**, find $\overrightarrow{OC}$, $\overrightarrow{OD}$, and $\overrightarrow{OE}$ in terms of **a** and **b**.

33. $OABCWXYZ$ is a cuboid (rectangular box) with W above O, X above A, Y above B and Z above C. If $OA = $ **a**, $OC = $ **c**, and $OW = $ **w**, find the position vectors of B, X, Y and Z in terms of **a**, **c** and **w**.

34. Show that $(x + 4)^2 - (x + 5)(x + 3) \equiv 1$. Using this identity, find the inverse of the matrix

$$\begin{pmatrix} 984 & 985 \\ 983 & 984 \end{pmatrix}$$

35. Show that the matrix

$$\begin{pmatrix} x - 1 & x^2 - 1 \\ 1 & x + 1 \end{pmatrix}$$

is a singular matrix for all values of x.

36. Show that the matrix

$$\begin{pmatrix} x + 5 & 4 \\ x & x + 1 \end{pmatrix}$$

is not a singular matrix for any real value of x.

37. The vertices of the triangle T_1 are given by the matrix $\begin{pmatrix} 1 & 2 & 4 \\ 1 & 3 & 3 \end{pmatrix}$.

The matrix $\mathbf{R} = \begin{pmatrix} 0 & 1 \\ -1 & 0 \end{pmatrix}$ maps T_1 into T_2, and the matrix $\mathbf{S} = \begin{pmatrix} 1 & 2 \\ 0 & 1 \end{pmatrix}$

maps T_1 into T_3. Mark two coordinate axes and show the three triangles T_1, T_2 and T_3.

What is the image of T_3 under the transformation described by $(\mathbf{R} . \mathbf{S}^{-1})$, and what is the image of T_2 under the transformation $(\mathbf{S} . \mathbf{R}^{-1})$?

38. For each of the following equations, find an expression for x in terms of the other letters, simplifying your answer where possible.

(i) $ax = b$;   (ii) $ax = b(x + c)$;

(iii) $\dfrac{a}{x + b} = \dfrac{b}{x + a}$;   (iv) $\sqrt{(x^2 + 9)} = x + b$.

For each of these equations to have a unique solution, it is necessary in one case only that $a \neq 0$, in another case only that $b \neq 0$, and in a third case only that $a \neq b$. State to which equation each of these conditions applies, and explain why all three conditions are necessary for the remaining equation to be solved.   (MEI)

39. A and B are two fixed points, and P is a variable point lying in a fixed plane contaniing A and B. Draw six *separate* sketches showing the locus of P for each of the following cases:

  (i) $PA = PB$;

 (ii) $PA < PB$;

(iii) angle $ABP = 40°$;

(iv) the area of the triangle APB is constant;

 (v) $AP^2 + BP^2 = AB^2$;

(vi) $AP^2 = AB^2 + BP^2$.   (MEI)

40. Two parallel lines l_1, l_2 are 8 cm apart. The image of a point P when reflected in l_1 is denoted by $l_1(P)$; when reflected in l_2 by $l_2(P)$. Show that for every point P in the plane of l_1 and l_2, $l_1(l_2(P))$ and $l_2(l_1(P))$ are the same distance d cm apart, and find d.

Index

Answers

Exercise 1.1a (p. 9).

1. (i) 101, 111, 1000, 1001, 11 001, 11 011, 100 000, 111 100.
   (ii) 12, 21, 22, 100, 221, 1000, 1012, 2020.
   (iii) 10, 12, 13, 14, 100, 102, 112, 220.
   (iv) 5, 7, 10, 11, 31, 33, 40, 74.
   (v) 5, 7, 8, 9, 21, 23, 28, 50.
2. 3, 5, 9, 13, 16, 24.   3. 7, 9, 15, 64, 127, 512.
4. 4, 10, 17, 26, 27, 54.   5. 3, 5, 17, 77, 155, 1000.
6. 1001, 1011, 1111, 10 000, 1 000 000, 10 000 000.
7. The last digits are 000.   8. Two, seven, fourteen.   9. Nine.   10. Eight, 3.
11. 100 011; 101 101; 10 000 111; 1 100 100; 1 011 010; 11; 101; 1011; 11 001; 1 100 000; 110; 11 010.
12. 330; 314; 264; 4230; 3124; 33; 41; 55; 12; 417.

Exercise 1.1b (p. 10)

1. (i) 10, 1011, 10 000, 10 010, 10 100, 100 000, 101 000, 1 100 100.
   (ii) 2, 102, 121, 200, 202, 1012, 1111, 10 201.
   (iii) 2, 23, 100, 102, 110, 200, 220, 1210.
   (iv) 2, 13, 20, 22, 24, 40, 50, 114.
   (v) 2, 11, 13, 15, 17, 26, 31, 79.
2. 6, 13, 15, 16.   3. 5, 10, 20, 100, 1000.
4. 101, 11 101, 1 000 101, 1 101 000, 1 111 111.
5. 11.   6. 111 111, 1.   7. Six.   8. Seven
9. 122.   10. Seven, 2.   11. 127.   12. 6.

Exercise 1.2a (p. 11)

1. 0.01, 0.11, 0.0001, 0.0101.   2. 0.11, 0.001, 0.011, 0.0101.
3. 100.1, 1000.001, 10 000.001, 10 001.0001.   4. 0.011   5. 1.01.

Exercise 1.2b (p. 12)

1. 0.011, 0.111, 0.00001, 0.11111.   2. 0.01, 0.0001, 0.0İ00İ, 0.0İ.
3. 0.0̇0̇1̇. 0.0̇1̇1̇, 0.0̇1̇1̇, 0.000111̇.   4. 1.11.   5. 0.001, 0.125

Exercise 1.3a (p. 15)

1.

+	0	1	2	3	4	5
0	0	1	2	3	4	5
1	1	2	3	4	5	0
2	2	3	4	5	0	1
3	3	4	5	0	1	2
4	4	5	0	1	2	3
5	5	0	1	2	3	4

×	0	1	2	3	4	5
0	0	0	0	0	0	0
1	0	1	2	3	4	5
2	0	2	4	0	2	4
3	0	3	0	3	0	3
4	0	4	2	0	4	2
5	0	5	4	3	2	1

2.

+	2	4	6	8	10
2	4	6	8	10	0
4	6	8	10	0	2
5	7	9	11	1	3
7	9	11	1	3	5
9	11	11	3	5	7

×	1	3	5	7	9	11
2	2	6	10	2	6	10
4	4	0	2	4	0	2
5	5	3	1	11	9	7
7	7	9	11	1	3	5
9	9	3	9	3	9	3

3. 2, 3, 4, 1, 4, 4.   **4.** 1, 4, 4, 1.
5. 1; 2, 5 or 8; 4; 1, 3, 5 or 7; no solution; 1 or 4; 1 or 5; 3 or 4.

Exercise 1.3b (p. 16)

4. 2, 4, 2, 4, 2; 2, 4, 0, 0, 0.   **5.** 3; no solution; no solution.
                                     0 or 4; no solution; no solution.
                                     1; 1, 2, 3, 4 or 5; 1.

Exercise 1.4 (p. 16)

1. 686.   **2.** $0.0\dot{1}00\dot{1}$.   **3.** Yes.   **4.** Yes.   **5.** No.
6. The last two digits are zeros.
7. The sum of the first, third, fifth, etc., digits differs from the second of the second, fourth, etc., digits by a multiple of ten (which may be zero, of course).   **8.** 257_{eight}.
9. 2.   **10.** 11.   **11.** 1 or 10.   **12.** 11, 12, 13, (not 14), etc.
13. Yes.   **14.** Yes.   **15.** 6.   **16.** 15, 15.   **17.** 31, 31.
18. 0.11235 . . . in each case. Fibonacci sequence.
19. (i) $x = 5, y = 4$; $x = 7, y = 2$; $x = 9, y = 7$. (ii) $x = 6$ or 8.
20. (i) $x = 5, z = 3$; $x = 9, z = 5$, (ii) $x = 7$ or 8. (iii) $x = 6, z = 2$ or 5.

Exercise 2.1a (p. 19)

1. (i) 900. (ii) 3000. (iii) 40. (iv) 1400. (v) 1500. (vi) 0.32. (vii) 10. (viii) 15. (ix) 0.16. (x) 167. (xi) 375. (xii) 1 00 000.
2. (i) $35 \times 74 = 2590$. (ii) $36 \times 75 = 2700$. (iii) $685 \div 19 \simeq 36.1$. (iv) $586 \div 0.26 \simeq 2250$.
3. (i) 500; 7; 0.7; 0.08; 0.008. (ii) 470; 67; 0.75; 0.080; 0.0078. (iii) 4770; 66.6; 0.749; 0.0802; 0.00775.   **4.** 10; 13; 13.5; 13.50; 13.495.
5. (i) 16.7; 16.1; 16.0; 0.1; 0.0. (ii) 16.75; 16.07; 16.05; 0.07; 0.05. (iii) 16.705; 16.075; 16.074; 16.005; 0.048.

Exercise 2.1b (p. 20)

1. (i) 2400. (ii) 25 000. (iii) 5.4. (iv) 4200. (v) 4.8. (vi) 18. (vii) 10. (viii) 87.5. (ix) 129 000. (x) 62.5. (xi) 1300. (xii) 0.1.
2. $59^2 \simeq 3480$; $670^2 \simeq 449\,000$; $3.9^3 \simeq 59.3$; $\sqrt{230} \simeq 15.3$; $\sqrt{0.047} \simeq 0.217$.
3. (i) 600; 6; 0.06; 0.05; 6000. (ii) 570; 56; 0.57; 0.0051; 0.0050. (iii) 5670; 5610; 56 000; 0.0561; 0.00560.
4. 3.14; 3.142; 3.1416; 3.14159; 3.141593.
5. (i) 0.7; 0.7; 0.7; 0.7. (ii) 0.07; 0.01; 0.00; 0.70; 0.70. (iii) 0.007; 0.007; 0.700; 0.700; 0.701.

Exercise 2.2a (p. 22)

1. (i) 7.56×10^3. (ii) 7.56×10^5. (iii) 7.56×10^9. (iv) 7.56×10. (v) 7.56×10^{-2}. (vi) 7.56×10^{-3}. (vii) 7.56×10^{-7}. (viii) 7.56.

2. (i) 6×10^7. (ii) 1.2×10^8. (iii) 1.3×10^2. (iv) 1.04×10^6. (v) 1.04.
(vi) 1.04×10^2. (vii) 1.04×10^{-6}. (viii) 8.24.
3. (i) 1.5×10^3. (ii) 7.5×10^2. (iii) 8×10^4. (iv) 8×10^{14}. (v) 3×10^{13}.
(vi) 5×10^{13}. (vii) 8×10^6. (viii) 8.
4. 6×10^8, 6, 6×10^{-6}, 6×10^{-8}.

Exercise 2.2b (p. 23)

1. (i) 4.38×10^3. (ii) 4.5×10^4. (iii) 4.5×10^8. (iv) 4.38×10. (v)
4.5×10^{-3}. (vi) 4.005×10^{-5}. (vii) 4.005×10^4. (viii) 4.005×10^{10}.
2. (i) 1.2×10^8. (ii) 2.1×10^6. (iii) 1×10^7. (iv) 1×10^3. (v) 1×10^{-5}.
(vi) 2.4×10^{-1}. (vii) 1.08×10. (viii) 1.08×10^{-5}.
3. (i) 2.5×10. (ii) 2.5×10^6. (iii) 4×10. (iv) 4×10^5. (v) 3.5×10^{-6}.
(vi) 5×10^2. (vii) 2.5×10^6. (viii) 2×10^6.
4. 1×10^6, 1×10^3, 1.6×10^{-5}, 1.25×10^{-7}.

Exercise 2.3 (p. 23)

1. (i) 22, 26. (ii) 30.25, 42.25. (iii) 7.8, 9.2.　　**2.** 1125, 1925.
3. 5, 7; 6.25, 12.25; 15.7, 22.0; 19.6, 38.5.
4. 422, 614; 338, 434; 13.0, 14.7.　　　　**5.** 5.9, 9.3.
6. 34, 38; 17, 19; 578, 722.　**7.** 300, 500; 6.4, 10.　**8.** 211, 414; 121, 184.
9. 0.30095, 0.30105; 0.6019, 0.6021; 0.90285, 0.90315; 0.301032, 0.301035.
10. 354 kg, 358 kg; 88.875 kg, 89.125 kg.　　**11.** £4355, £4485.
12. 631, 634; 622, 643.

Exercise 3.1a (p. 27)

1. (i) 2.420. (ii) 7.48. (iii) 9.68. (iv) 11.88. (v) 16.28.
2. (i) 4.32. (ii) 6.12. (iii) 8.28. (iv) 9.72. (v) 13.14.
3. (i) 4.2. (ii) 6.475. (iii) 10.675. (iv) 12.95 (v) 15.4.
4. (i) 2.713. (ii) 5.363. (iii) 10.617. (iv) 12.167. (v) 13.547.
5. (i) 2.538. (ii) 9.139. (iii) 14.364. (iv) 15.535. (v) 21.157.

Exercise 3.1b (p. 27)

1. (i) 28.8. (ii) 62.4. (iii) 124.8. (iv) 172.8. (v) 235.2.
2. (i) 67.2. (ii) 952. (iii) 1.288. (iv) 1512. (v) 2.044.
3. (i) 3.0. (ii) 4.625. (iii) 7.625. (iv) 9.250. (v) 11.00.
4. (i) 253.75. (ii) 501.7. (iii) 99.325. (iv) 113.82. (v) 0.1267.
5. (i) 301.04. (ii) 92.288. (iii) 1442.7. (iv) 1.560. (v) 43.788.

Exercise 3.2a (p. 28)

1. (i) 1.2. (ii) 1.05. (iii) 3.1. (iv) 1.5. (v) 2.2.
2. (i) 1.125. (ii) 1.077. (iii) 3.207. (iv) 1.530. (v) 2.256.
3. (i) 5.76. (ii) 6.72. (iii) 7.44. (iv) 6. (v) 7.04.
4. (i) 5.847. (ii) 5.868. (iii) 6.978. (iv) 4.339. (v) 5.331.
5. (i) 58.468. (ii) 0.0587. (iii) 0.0698. (iv) 4.339. (v) 533.13.

Exercise 3.2b (p. 29)

1. (i) 0.8. (ii) 0.6. (iii) 0.8. (iv) 1.3. (v) 0.9.
2. (i) 0.827. (ii) 0.585. (iii) 0.812. (iv) 1.368. (v) 0.8654.
3. (i) 4.966. (ii) 3.310. (iii) 3.862. (iv) 5.379. (v) 3.103.
4. (i) 5.035. (ii) 3.315. (iii) 3.908. (iv) 4.787. (v) 3.403.
5. (i) 50.347. (ii) 3315.1. (iii) 3.908. (iv) 0.0479. (v) 340.3.

Exercise 3.3a (p. 30)

1. (i) 49. (ii) 1.69. (iii) 5.76. (iv) 19.36. (v) 40.96.
2. (i) 2500. (ii) 196. (iii) 67 600. (iv) 0.1936. (v) 0.4096.
3. (i) 2.236. (ii) 7.071. (iii) 1.1401. (iv) 3.6055. (v) 3.0983.
4. (i) 22.36. (ii) 70.71. (iii) 0.3605. (iv) 0.114. (v) 97.97.
5. (i) 0.25. (ii) 0.4167. (iii) 0.294. (iv) 2.2727. (v) 0.119.

Exercise 3.3b (p. 30)

1. (i) 36. (ii) 2.25. (iii) 6.25. (iv) 53.29. (v) 77.44.
2. (i) 6400. (ii) 324. (iii) 72 900. (iv) 0.505. (v) 0.384.
3. (i) 2.8284. (ii) 8.944. (iii) 1.449. (iv) 4.583. (v) 2.846.
4. (i) 28.284. (ii) 89.442. (iii) 0.458. (iv) 0.1449. (v) 0.9.
5. (i) 4.5454. (ii) 0.03125. (iii) 0.233. (iv) 1.7543. (v) 0.0109.

Exercise 3.4a (p. 31)

1. (i) 1.389. (ii) 2.052. (iii) 2.571. (iv) 2.544. (v) 2.757.
2. (i) 7.8784. (ii) 5.6382. (iii) 3.064. (iv) 1.7. (v) 0.48608.
3. (i) 0.14. (ii) 0.2094. (iii) 0.2792. (iv) 3.3796. (v) 2.7720.
4. (i) 1.408. (ii) 2.912. (iii) 3.356. (iv) 5.8891. (v) 15.896.
5. (i) 46.083. (ii) 17.544. (iii) 5.2219. (iv) 1.9629. (v) 0.49374.

Exercise 3.4b (p. 31)

1. (i) 1.8116. (ii) 2.113. (iii) 2.121. (iv) 3.8958. (v) 8.1672.
2. (i) 6.7613. (ii) 4.532. (iii) 2.121. (iv) 1.8172. (v) 0.71504.
3. (i) 0.1834. (ii) 0.2180. (iii) 2.335. (iv) 4.2725. (v) 0.71504.
4. (i) 1.8753. (ii) 2.3315. (iii) 3. (iv) 9.2213. (v) 93.48.
5. (i) 27.048. (ii) 11.831. (iii) 4.2426. (iv) 2.0051. (v) 71.741.

Exercise 3.5 (p. 31)

1. (i) 41.485. (ii) 136.86. (iii) 547.44. (iv) 1204.4.
2. (i) 1.4453. (ii) 0.16621. (iii) 0.66484. (iv) 0.05097.
3. (i) 1.5909. (ii) 19.091. (iii) 0.02068. (iv) 11.932.
4. (i) 1.7837. (ii) 6.1791. (iii) 0.1954. (iv) 15.447.
5. (i) 0.74933. (ii) 2.3695. (iii) 0.050452. (iv) 14.956.
6. 0.32961.   7. 0.55137.   8. 0.45212.   9. 0.37620.   10. 1.7333.
11. 3.2219.   12. 82.664.   13. 110.63.   14. 1.4142.   15. 0.7953.
16. 4.452.   17. 78.156.   18. 8.9517.   19. 19.435.   20. 14.19.

Exercise 4.1a (p. 35)

1. 23.2.   2. 0.465.   3. 27.6.   4. 260.   5. 445.
6. 4430.   7. 0.845.   8. 42.1.   9. 0.0222.   10. 0.0627.

Exercise 4.1b (p. 35)

1. 9.14.
2. 175.
3. 0.223.
4. 0.00168.
5. 1.41.
6. 0.000115.
7. 0.296.
8. 2.30.
9. 3.07
10. 318.

Exercise 4.2a (p. 37)

1. 1.3806.
2. $\bar{3}$.3436.
3. $\bar{4}$.2863.
4. $\bar{1}$.4403.
5. $\bar{3}$.7037.
6. 5.35.
7. 0.424.
8. 0.375.
9. 0.299.
10. 0.0168.

Exercise 4.2b (p. 37)

1. 0.3885.
2. $\bar{4}$.1537.
3. $\bar{1}$.1333.
4. $\bar{3}$.8111.
5. $\bar{1}$.4722.
6. 3.18.
7. 0.918.
8. 0.832.
9. 0.574.
10. 2.09.

Exercise 4.3. (p. 39)

1. 0.0780.
2. 2.58.
3. 48.7.
4. 4990 cm³.
5. 0.223.
6. 0.945.
7. 1.70.
8. 153.
9. 1.21.
10. 7.90.
11. 0.748.
12. 52.5.
13. 0.00835.
14. 123.
15. 1.11.
16. 0.878.
17. 0.656.
18. 21.8.
19. 0.919.
20. 1190.
21. 0.876.
22. 29.7.
23. 13.7.
24. 14.2.
25. 0.859.
26. 1.22.
27. 0.703.
28. 0.0518.
29. 0.165.
30. 0.00102.

Exercise 5.1a (p. 44)

1. 11.6 cm².
2. $4\frac{8}{13}$cm.
3. 7.2 cm.
4. 180 cm².
5. 4.5 cm.
6. 62 cm³.
7. 20 m³.
8. 0.9.
9. 4000 kg.
10. 40 500 cm².

Exercise 5.1b (p. 45)

1. 42.4 cm².
2. 6 cm.
3. 13 cm.
4. 49.1 cm³.
5. 4 m.
6. 100 m³.
7. 156 cm³.
8. 108 m.
9. 600 m³.
10. 1000 kg.

Exercise 5.2a (p. 49)

1. 4.46 cm.
2. 3.66 cm.
3. 1.99 cm.
4. 106 g.
5. 4330 cm³.
6. 377 cm².
7. 5.82 cm.
8. 283 cm².
9. 11.3.
10. 96.8 cm³.
11. 12 cm.
12. 110 cm³.

Exercise 5.2b (p. 50)

1. 154 cm.
2. 405 m².
3. 3140.
4. 9.08 cm³.
5. 302 cm².
6. 2.93 cm.
7. 4.67 cm.
8. 9.65.
9. 264 cm³.
10. 2980 cm³.
11. 28 cm.
12. 7 cm.

Exercise 5.3 (p. 52)

1. $\frac{2}{3}$ cm.
2. 2730 m².
3. 900.
4. 0.0015 cm.
5. $\frac{4}{27}$ cm.
6. 11.3 cm².
8. 2.59.
9. 100.

10. 1920 g.   **11.** 16.7 m/s, 2 min.   **12.** 1040 m³.
13. 99.3 cm².   **14.** 1670.   **15.** 720 m³, 384 m².
16. 16 cm.   **17.** 2.19 m.   **18.** 16.7 cm.   **19.** 16.4 cm².
20. £1230.   **21.** 0.87 cm.   **22.** 6360.   **23.** 4½ cm³.
24. 2.02 cm.   **25.** 2.88 cm.   **26.** 19.5 m³, 52.8 m².

27. 7.68.   **28.** $\sqrt{\dfrac{9}{4\sqrt3}}$.   **29.** 5.54 m, 12.8 m³.
30. 7.14 m, 57.1 m³.

Exercise 6.1a (p. 55)

1. $\frac{4}{150}$.   **2.** $\frac{3}{10}$.   **3.** $\frac{1}{3}$.   **4.** $\frac{1}{3}$.   **5.** $\frac{1}{24}$.
6. $\frac{1}{8}$.   **7.** 40.   **8.** $\frac{1}{10}$.   **9.** $\frac{7}{1000}$.   **10.** $\frac{1}{100}$.

Exercise 6.1b (p. 55)

1. $\frac{3}{8}$.   **2.** $\frac{3}{7}$.   **3.** $\frac{5}{48}$.   **4.** $\frac{10}{9}$.   **5.** $\frac{250}{21}$.
6. $\frac{5}{7}$.   **7.** $\frac{500}{41}$.   **8.** $\frac{1}{50}$.   **9.** $\frac{7}{50}$.   **10.** $\frac{75}{2}$.

Exercise 6.2a (p. 56)

1. 8 cm, 14 cm, 16 cm.   **2.** 9 cm, 21 cm, 30 cm.
3. £2, £3.   **4.** £70, £105, £175.   **5.** 54, 36, 24.
6. 20.   **7.** £80, £40, £20.   **8.** £4, £3, £1.50.
9. 8 : 12 : 21.   **10.** 50, 25, 50.

Exercise 6.2b (p. 57)

1. 54, 72, 120.   **2.** £750, £500, £250.   **3.** £3.75, £4, £4.25.
4. 9 cm, 12 cm, 15 cm.   **5.** 21.   **6.** 21.   **7.** £9, £12.
8. £5, £2.50.   **9.** £15, £10.   **10.** 21 : 14 : 24.

Exercise 6.3a (p. 58)

1. 35.4.   **2.** 3.47.   **3.** 47½.   **4.** 74.   **5.** 4⅜.
6. 4⅖.   **7.** 38.   **8.** 36.   **9.** 37½.   **10.** £4.27½.

Exercise 6.3b (p. 58)

1. 1.19.   **2.** 45.6.   **3.** 5.21.   **4.** 17½.   **5.** 91½.
6. 70.4.   **7.** 21.   **8.** 4.2.   **9.** £2.12½.   **10.** 8.82.

Exercise 6.4a (p. 60)

1. 20.   **2.** 25.   **3.** £0.60.   **4.** £5.62½.   **5.** £54.
6. ½.   **7.** £60.   **8.** 12.   **9.** £20.   **10.** 9$\frac{1}{11}$.

Exercise 6.4b (p. 60)

1. 20.   **2.** £320.   **3.** 37½.   **4.** 38⅞.   **5.** £25.
6. 16⅔.   **7.** £3.   **8.** £22.50.   **9.** 6¼% fall.   **10.** 476.

Exercise 6.5 (p. 65)

1. £48.	**2.** £750, £1500, £3000.	**3.** 264.	**4.** 16.
5. $9\frac{1}{3}$.	**6.** 8.16.	**7.** 12.5.	**8.** 800.	**9.** £5.75.
10. £75.	**11.** £5.50.	**12.** $66\frac{2}{3}$.	**13.** £80.	**14.** $9\frac{1}{11}$.
15. £8000.	**16.** 78.5.	**17.** 100.	**18.** $8\frac{1}{3}$, £374.	**19.** 200.
20. 50.	**21.** 6.	**22.** 51.2.	**23.** £11, 229.	**24.** £1100, $28\frac{4}{7}$.
25. £1550.	**26.** £2.70.	**27.** £50.	**28.** 92.7.	**29.** 46.1.
30. 80%, 94p.	**31.** 16.	**32.** 57.9.	**33.** 25; 25.	**34.** $43\frac{3}{4}$.
35. $33\frac{1}{3}$.			

Exercise 7.1a (p. 67)

1. £136.   **2.** £18.15.   **3.** $3\frac{1}{2}$.   **4.** $2\frac{1}{2}$.   **5.** 5 years.
6. $2\frac{1}{2}$ years.   **7.** £1640.   **8.** £200.   **9.** £500.   **10.** £505.

Exercise 7.1b (p. 68)

1. £80.80.   **2.** £22.   **3.** $2\frac{1}{4}$.   **4.** $3\frac{1}{4}$.   **5.** 4 years.
6. $2\frac{1}{2}$ years.   **7.** £848.   **8.** £750.   **9.** £97.60.   **10.** £500.

Exercise 7.2a (p. 69)

1. £24.97.   **2.** £45.18.   **3.** £9.45.   **4.** £46.26.   **5.** £27.82.

Exercise 7.2b (p. 70)

1. £43.10.   **2.** £17.05.   **3.** £24.05.   **4.** £22.41.   **5.** £37.87.

Exercise 7.3 (p. 71)

1. £27 595.	**2.** £545.	**3.** £219.	**4.** £1169.86.
5. 3.1%.	**6.** 14.2 years.	**7.** 4.7.	**8.** 4th.
9. £5065.	**10.** 120 days.	**11.** £215.04.	**12.** £170.30.
13. £941.90.	**14.** £1650, £2220.	**15.** 8 years.	**16.** £45.98.
17. £34.30.	**18.** 15 years.	**19.** 49 719.	**20.** £65.
21. £19 350.	**22.** £700.	**23.** £34.20.	**24.** £3193.
25. 16.4.	**26.** £98.70.	**27.** 78%, £3120.	**28.** $\dfrac{£y^4}{x^3}$.
29. £0.07$\frac{1}{2}$.	**30.** £108.	**31.** £318.36.	**32.** £726.
33. £36.	**34.** 10 years.	**35.** £268.	

Exercise 8.1a (p. 73)

1. £185.76.   **2.** £287.04.   **3.** 75p.   **4.** 3.5p.   **5.** £260.

Exercise 8.1b (p. 74)

1. £353.44.   **2.** £236.28.   **3.** 77p.   **4.** £4 800 000.   **5.** £360.

Exercise 8.2a (p. 75)

1. £924.   **2.** £924, 25%.   **3.** £3300.   **4.** £2162.60.   **5.** £2441.

Exercise 8.2b (p. 76)

1. £653.40.   **2.** £1200.   **3.** £4000.   **4.** £1804.   **5.** £6200.

Exercise 8.3a (p. 76)

1. £5.12.   **2.** £6, 48p.   **3.** £4, £4.32.   **4.** £11   **5.** £4500, £25 500.

Exercise 8.3b (p. 77)

1. £20.   **2.** £20, £1.60.   **3.** £3.78.   **4.** £1800, £180.   **5.** £4.35.

Exercise 8.4a (p. 78)

1. £97.50.   **2.** £40.   **3.** £350.   **4.** £145, £11.
5. £1300.   **6.** £532.40, £14.96   **7.** £52.50.
8. £20.   **9.** £1050.   **10.** 5000.

Exercise 8.4b (p. 79)

1. £90.   **2.** £24.   **3.** £555.   **4.** 107.   **5.** £270.
6. 160.   **7.** £40.   **8.** £18.   **9.** £1.50.   **10.** £4.

Exercise 8.5a (p. 80)

1. $4\frac{4}{9}$.   **2.** $6\frac{2}{13}$.   **3.** $7\frac{1}{2}$.   **4.** $5\frac{5}{11}$.   **5.** $5\frac{5}{17}$.

Exercise 8.5b (p. 80)

1. $4\frac{3}{4}$.   **2.** $6\frac{2}{3}$.   **3.** 5.   **4.** $4\frac{16}{21}$.   **5.** 8.

Exercise 8.6 (p. 81)

1. Flat, £65.   **2.** £667.70.   **3.** £72.   **4.** £129.60, £158.40, 3.3%.
5. £287.   **6.** £2620.   **7.** £81.37 or £81.38.
8. £10 520.88 to £10 521.12.   **9.** £800.   **10.** £800.
11. $77\frac{7}{9}$.   **12.** £22.   **13.** £245, £9.45.   **14.** £1.   **15.** £79.

Exercise 9.1a (p. 87)

1. $2a(2 + 3b - 4b^2)$.   **2.** $5h(h + 2g - 4g^2)$.   **3.** $2cd(c - 2d)$.
4. $6a^2b^2(a^2 - 3b)$.   **5.** $8xy(x - 3y)$.   **6.** $6uv(u + 2v^2)$.

7. $\dfrac{a + 3b}{3a - 4b}$.   **8.** $\dfrac{2(a - 2b)}{3(a + 3b)}$.   **9.** $\frac{3}{2}$.

10. $\dfrac{x - b}{a - x}$.   **11.** $\dfrac{x}{a - x}$.   **12.** $\dfrac{x + b}{a + b}$.

Exercise 9.1b (p. 87)

1. $a(x + y + z)$.   **2.** $a(x - 2y + 3z)$.   **3.** $ax(a - 4 + 3x)$.
4. $bc(ab - ac - bc)$.   **5.** $4a^2x^2(1 + 2ax)$.   **6.** $3ac(2b - 3a)$.

7. $\dfrac{3(x+3y)}{2(x-3y)}$. **8.** $\dfrac{x+2y}{2x-y}$. **9.** $\dfrac{2x}{x-3y}$.

10. $\dfrac{y-z}{x-z}$. **11.** $\dfrac{y}{x-y}$. **12.** $\dfrac{x+y}{x-y}$.

Exercise 9.2a (p. 88)

1. $(x+4)(x-4)$. **2.** $(p+3q)(p-3q)$. **3.** $(5x-y)(5x+y)$.
4. $(5a+4b)(5a-4b)$. **5.** $(a+2b-2c)(a-2b+2c)$.
6. $(3a+2b-2c)(3a-2b+2c)$. **7.** $(5a-5b+c)(5a-5b-c)$.
8. $(a-b+c-d)(a-b-c+d)$.
9. $(2a-2b+c-d)(2a-2b-c+d)$. **10.** $(1+3x)(1-3x)$.
11. $(2+5a)(2-5a)$. **12.** $(1+a-b)(1-a+b)$.
13. $(1+3a-3b)(1-3a+3b)$. **14.** $(4+a-b)(4-a+b)$.
15. $(5+4a-4b)(5-4a+4b)$. **16.** $(6+z)(6-z)$.
17. $3(6+z)(6-z)$. **18.** $(z^2+1)(z+1)(z-1)$.
19. $(x^2+4)(x+2)(x-2)$.
20. $(2a-2b+5c-5d)(2a-2b-5c+5d)$.

Exercise 9.2b (p. 88)

1. $(x+7)(x-7)$. **2.** $(a+8b)(a-8b)$. **3.** $(5p+8q)(5p-8q)$.
4. $(p+3q^2)(p-3q^2)$. **5.** $(1+7x)(1-7x)$.
6. $(3a-3b+c)(3a-3b-c)$.
7. $(a-b+3c-3d)(a-b-3c+3d)$. **8.** $(7+2x)(7-2x)$.
9. $(p+7r-7s)(p-7r+7s)$. **10.** $(3+a-b)(3-a+b)$.
11. $(1+4x^2)(1+2x)(1-2x)$. **12.** $7(1+3x)(1-3x)$.
13. $(4p+7q)(4p-7q)$. **14.** $(6+5z)(6-5z)$.
15. $(4+7a-7b)(4-7a+7b)$. **16.** $(x^4+4)(x^2+2)(x^2-2)$.
17. $(5a-5b+7c-7d)(5a-5b-7c+7d)$. **18.** $(a+9b^2)(a-9b^2)$.
19. $(2a-b+9c-3d)(2a-b-9c+3d)$.
20. $(12a+8b+5p+10q)(12a+8b-5p-10q)$.

Exercise 9.3a (p. 90)

1. $(x+2)(x+3)$. **2.** $(x-10)(x+2)$. **3.** $(x-3)(x+2)$.
4. $(x-2)(x-3)$. **5.** $(x-5)(x-3)$. **6.** $(x-6)(x+2)$.
7. $(x+9)(x+2)$. **8.** $(x^2+6)(x^2+4)$. **9.** $(x+y)(x+3y)$.
10. $(x^2-5y)(x^2+y)$. **11.** $(x+1)(2x-3)$. **12.** $(2x+3y)(x-2y)$.
13. $(3x+2)(x-3)$. **14.** $(5x+2y)(x+3y)$. **15.** $(3x-2)(2x-5)$.
16. $(3+x)(2-x)$. **17.** $(3-2x)(4+3x)$. **18.** $(2x+5)(x+3)$.
19. $(2x+5)(x-3)$. **20.** $(4x+5)(3x-2)$.

Exercise 9.3b (p. 90)

1. $(x+4)(x+2)$. **2.** $(t-9)(t+2)$. **3.** $(u-6)(u+1)$.
4. $(v+5)(v+2)$. **5.** $(y-6)(y+4)$. **6.** $(x-11y)(x+y)$.
7. $(x+8y)(x+2y)$. **8.** $(x^2+4)(x^2+1)$. **9.** $(x+5y)(x+y)$.
10. $(z^2-4x)(z^2+2x)$. **11.** $(2p-7)(p+1)$. **12.** $(3a+2b)(2a+5b)$.

13. $(3l - 4m)(2l - 3m)$.   **14.** $(7x + 2)(x - 3)$.   **15.** $(4x^2 + 1)(3x^2 + 2)$.
16. $(4 + x)(3 - x)$.   **17.** $(3 - 2x)(1 + x)$.   **18.** $(4y + 1)(3y + 2)$.
19. $(5 + 2x)(3 - x)$.   **20.** $(5 + 4x)(2 - 3x)$.

Exercise 9.4a (p. 91)

1. $(x + y)(h + m + n)$.   **2.** $(x + q)(p - 6)$.   **3.** $(x + y)(1 - a)$.
4. $(x - q)(c - d)$.   **5.** $(a - x)(b - y)$.   **6.** $(a - 1)^2(a + 1)$.
7. $(a + b + c)(h + p)$.   **8.** $(2a - b)(x + 1)$.   **9.** $(2a - b)(x - 1)$.
10. $(a + b)(x + y)(x - y)$.   **11.** $(b + c)(a - b - c)$.
12. $(a - 3)(b - 2c)$.   **13.** $(x + y)(x - y - 6)$.
14. $(x - y)(x + y - 6)$.   **15.** $(x + y)(x - y - 1)$.
16. $(x - y)(x + y - 1)$.   **17.** $(x - 4)(x - y)$.   **18.** $(x + 5)(x - y)$.
19. $(a + b)(c + d + 1)$.   **20.** $(a + b)(c + d - 1)$.

Exercise 9.4b (p. 91)

1. $(a + b)(p - 2q)$.   **2.** $(a - b)(p - 2q)$.
3. $(x + y)(a + b - 2)$.   **4.** $(x + y)(a + b - x - y)$.
5. $(a + 1)(a^2 + 1)$.   **6.** $(1 + c)(3ab + 2c)$.
7. $(p - q)(p + q + 4)$.   **8.** $(p - q)(p + q - 5)$.
9. $(h + k)(h - k - 6)$.   **10.** $(a + 2b)(a - 2b - c)$.
11. $(a - 3b)(a + 3b - x)$.   **12.** $(a + b)(a^2 + a + 1)$.
13. $(x - 1)(2 - x)$.   **14.** $(h - k)(a - b)$.
15. $(h - k)(ah + ak + b)$.   **16.** $(y - 1)(x - 1)$.
17. $(a - b)(a + c)$.   **18.** $(a - b)(a - c)$.
19. $(x - 2b)(x - c)$.   **20.** $(z - 2a)(z + b)$.

Exercise 9.5a (p. 92)

1. $(2a + b)(4a^2 - 2ab + b^2)$.   **2.** $(z + 1)(z^2 - z + 1)$.
3. $(z^2 + 1)(z^4 - z^2 + 1)$.   **4.** $(z + 1)(z - 1)(z^4 + z^2 + 1)$.
5. $(a - bc)(a^2 + abc + b^2c^2)$.   **6.** $(a - 3bc)(a^2 + 3abc + 9b^2c^2)$.
7. $(4h - k)(16h^2 + 4hk + k^2)$.   **8.** $(4h + 3)(16h^2 - 12h + 9)$.
9. $(a - 2)(a^2 + 2a + 4)$.   **10.** $(2 - 3b)(4 + 6b + 9b^2)$.

Exercise 9.5b (p. 92)

1. $(3a + b)(9a^2 - 3ab + b^2)$.   **2.** $(z + 2)(z^2 - 2z + 4)$.
3. $(z^2 + 2)(z^4 - 2z^2 + 4)$.   **4.** $(z^2 - 2)(z^4 + 2z^2 + 4)$.
5. $(h - 2mn)(h^2 + 2mnh + 4m^2n^2)$.
6. $(1 - 3bc)(1 + 3bc + 9b^2c^2)$.   **7.** $(4 - h)(16 + 4h + h^2)$.
8. $(4h - 3)(16h^2 + 12h + 9)$.   **9.** $(5a - 1)(25a^2 + 5a + 1)$.
10. $(2 - 5x)(4 + 10x + 25x^2)$.

Exercise 9.6a (p. 92)

1. $(a + 1)(a + 1 - x)$.   **2.** $(x + 3)(x + 1 + a)$.
3. $(x + y + z)(x + y - z)$.   **4.** $(2z + 2x + 1)(2z - 2x - 1)$.

5. $(z + 2)(z + 1 + a)$. **6.** $(x + 1)(a - x - 3)$.
7. $(x + p + y)(x + p - y)$. **8.** $(4x + 1)(x + 1 - y)$.
9. $(1 + a - b)(1 - a + b)$. **10.** $(x + 5)(x + 2 + y)$.

Exercise 9.6b (p. 93)

1. $(a + 2)(a + 3 + x)$. **2.** $(x + 1)(x + 2 + a)$.
3. $(x + 2 + z)(x + 2 - z)$. **4.** $(z + x + 3)(z - x - 3)$.
5. $(p - 6)(p + 1 + q)$. **6.** $(y + 1)(a - y - 1)$.
7. $(x + 3p + 3y)(x + 3p - 3y)$. **8.** $(2x + 1)(x + 1 + y)$.
9. $(4 + a + b)(4 - a - b)$. **10.** $(x + 4)(x + 3 + y)$.

Exercise 9.7 (p. 93)

1. $(y - 7)(y + 6)$. **2.** 199. **3.** 9180.
4. $(2a - 3b)(a - 6b)$. **5.** 9. **6.** 81.
7. 9. **8.** 4. **9.** $(x + 3)(a - y)$.
10. $(x + 1)(ax + a + 1)$. **11.** $(1 + 3p + q)(1 - 3p - q)$.
12. $(x + 1)(x - 3)$. **13.** $(x + 1)(y + 1)(x - 1)(1 - y)$.
14. $(x - y)(x + y + 1)(x^2 + y^2 + x + y + 2)$. **15.** $(y - 8)(x + 5)$.

16. $x + 2$. **17.** $\dfrac{1}{x + 1}$. **18.** $\left(3a - \dfrac{1}{a}\right)^2$.

19. $(R - r)(R - 3r)$. **20.** $a(b - 2)(b^2 + 2b + 4)$.
21. $x(y + 4)(y^2 - 4y + 16)$.
22. $(1 - 5ab)(1 + 5ab + 25a^2b^2)$.
23. $(3 + 2abc)(9 - 6abc + 4a^2b^2c^2)$.
24. $(a - b + x - y - z)(a - b - x + y + z)$.
25. $(4c + a - b - d)(4c - a + b + d)$.
26. $(3a - 3b + 7c - 7d)(3a - 3b - 7c + 7d)$.
27. $(a + 7)(a + 2 - x)$. **28.** $x(x + 2)(x + 3)(x + 7)$.
29. $3(a + 1)(a + 3)(a^2 + 2a + 3)$. **30.** $a(b^2 + b + 1)$.
31. $x - 3$. **32.** $(a - x)(a + x - 1)$.
33. $(a^2 + b^2)(x^2 + y^2)$. **34.** $(lx + my)(lx + my - 3)$.
35. $4\pi r(3R^2 + 3Rr + r^2)$. **36.** $(x + 5)(x^2 + 4)$.
37. $\frac{1}{2}mu(2v + 3u)$. **38.** 2.
39. $(x - 1)(x - 2)(x - 3)$. **40.** $(x + 1)(x - 1)(x + 4)$.

Exercise 9.8a (p. 95)

1. 10. **2.** -20. **3.** $(x + 1)(x - 1)(x - 4)$. **4.** $-6, 11$.
5. 3. **6.** $-1, 1$. **7.** $(x - 3y)(x + y)(x + 2y)$.
8. $(x - 1)(x + 3)(x - 2)$. **9.** 1, 2. **10.** $k = 7, (x + 1)(x + 3)$.

Exercise 9.8b (p. 96)

1. 28. **2.** 16. **3.** $(x - 1)(x + 2)(x + 3.)$. **4.** 0.
5. 2, -1. **6.** 7, 14. **7.** $(x + y)(x + 2y)(x + 3y)$.
8. $(x - 1)(x + 3)(x + 4)$. **9.** $-1, -2$. **10.** $k = 1, (x + 2)(x + 3)$.

Exercise 10.1a (p. 97)

1. $(500 - pP - qQ)$. **2.** £$49x$. **3.** $\pi r^2 h$ cm^3.

4. $\pi y(y + 2x)$. **5.** $2\pi r(r + h)$. **6.** $\dfrac{xy}{100}$.

7. $2z(x + y)$ m^2. **8.** $4x$ minutes past noon; $4y$ minutes to noon.

9. $\dfrac{xm + 12y}{x + 1}$. **10.** $\left(\dfrac{P - Q}{P}\right)100$.

11. $\dfrac{(pP + qQ)(100 + z)}{1000\,(p + q)}$ pence. **12.** $\dfrac{hk}{x^2}$.

13. $\left(\dfrac{x}{u} + \dfrac{y}{v}\right)$ hours; $\dfrac{(x + y)uv}{xv + yu}$ km/h. **14.** $(2pr + \pi r^2)$ m^2. **15.** $\dfrac{xs}{100g}$.

Exercise 10.1b (p. 98)

1. $z(x - 2z)(y - 2z)$ cm^3. **2.** $\dfrac{xyz}{n}$. **3.** $(xg + yh)$ g.

4. $\dfrac{1000c}{1000x + y}$. **5.** $\dfrac{3x}{2v}$ hours; $\tfrac{4}{3}v$ km/h.

6. $2r(p + q) - xy - 2cd$. **7.** $x(50 + w)$ pence.

8. $(10x + 8y)$ pence. **9.** $\dfrac{xp + yq}{x + y}$. **10.** $\dfrac{px - qy + rz}{p - q + r}$.

11. $\dfrac{19x}{12V}$ hours. **12.** £$\left(Aa + 7Ab + \dfrac{Ca}{2} + \dfrac{7Cb}{2}\right)$.

13. $4x + 6y - 4p - 6q$. **14.** $\dfrac{pqs}{xyz}$. **15.** $\sqrt{\dfrac{xy - A}{\pi}}$ m.

Exercise 10.2a (p. 100)

1. $(x - 1)(x + 1)(x - 2)$. **2.** $6(x - 1)(x + 1)$.
3. $(x - a)(x + a)(x - b)$. **4.** $(x - a)^2(x + a)$.
5. $(x^2 + a^2)(x + a)(x - a)$. **6.** $y(y - 1)(y - 2)$.
7. $z(z - 1)^2$. **8.** $z(z - 1)(z + 1)$.
9. $3(t - 1)(t + 1)$. **10.** $(p - 1)(p - 2)(p - 3)$.

Exercise 10.2b (p. 100)

1. $(x - 1)(x - 2)(x + 2)$. **2.** $6(x + 1)^2$.
3. $(x + a)(x - a)(x - 1)$. **4.** $x(x - a)^2$.
5. $(x^2 + 1)(x + 1)^2$. **6.** $y(y - 1)(y + 1)$.
7. $3y(y - 1)(y + 1)$. **8.** $z(z - 1)(z + 1)$.
9. $6(a - b)(a + b)$. **10.** $(x - 1)(x + 1)(x - 2)(x + 2)$.

Exercise 10.3a (p. 101)

1. $\dfrac{17}{12x}$. **2.** $\dfrac{2x}{(x+1)(x-1)}$. **3.** $\dfrac{2x}{(2-x)(4-x)}$. **4.** $\dfrac{x}{12}$.

5. $\dfrac{(x+1)^2}{x(x+2)}$. **6.** $\dfrac{x+3}{12x}$. **7.** $\dfrac{3-x}{2(x+1)(x-1)}$.

8. $\dfrac{1}{(a-c)(c-b)}$. **9.** $\dfrac{x+4}{x(x+1)(x+2)}$. **10.** $\dfrac{3}{x(x+2)}$.

Exercise 10.3b (p. 102)

1. $\dfrac{1}{6x}$. **2.** $\dfrac{x+6}{4(x+2)}$. **3.** $\dfrac{x+3}{(x-3)(1-x)}$. **4.** $\dfrac{11x-19}{12}$.

5. $\dfrac{2(x+1)}{x(x+2)(x+3)}$. **6.** $\dfrac{2(x+1)}{3x}$. **7.** $\dfrac{11-4x}{(x+2)(x-2)}$.

8. 0. **9.** $\dfrac{3-x}{x(x-1)(x+1)}$. **10.** $\dfrac{3}{x(x-2)}$.

Exercise 10.4a (p. 103)

1. $\dfrac{d-b}{a-c}$. **2.** $\dfrac{ab}{b-a}$. **3.** $a+b$. **4.** $\dfrac{bc}{b+c-a}$.

5. $\dfrac{ab}{2b-a}$. **6.** $\dfrac{bl-am}{m-l}$. **7.** $-a-b$. **8.** $\dfrac{ap}{q-p}$.

9. $\dfrac{ab}{c(a+b)}$. **10.** $\sqrt{b^2+a^2}$. **11.** $\dfrac{A-2\pi r^2}{2\pi r}$. **12.** $\dfrac{A-\pi t^2}{2\pi t}$.

13. $\dfrac{fv}{v-f}$. **14.** $\dfrac{T^2g}{4\pi^2}$. **15.** $\dfrac{A-\pi r^2}{\pi r}$. **16.** $\dfrac{2(s-ut)}{t^2}$.

17. $\sqrt{v^2-2fs}$. **18.** $\sqrt{r^2-(y-b)^2}+a$.

19. $\dfrac{bx^2-a}{x^2-1}$. **20.** $\dfrac{2(S-an)}{n(n-1)}$.

Exercise 10.4b (p. 104)

1. $\dfrac{2b-a}{a-b}$. **2.** $\dfrac{a}{a-1}$. **3.** $\dfrac{a}{1-b}$. **4.** $\dfrac{cd}{c+d}$.

5. $\dfrac{2ab-bc-ac}{a+b-2c}$. **6.** $\dfrac{c-d}{ad-cb}$. **7.** $a+b$.

8. $\dfrac{bp-aq}{q-p}$. **9.** $\dfrac{bc+ca+b^2}{b+c}$. **10.** $\sqrt{b^2-y^2}$.

11. $\dfrac{S}{n}-\dfrac{1}{2}(n-1)d$. **12.** $-\dfrac{ax^2+c}{x}$. **13.** $\dfrac{100I}{PT}$.

14. $100\left(\dfrac{A}{P}-1\right)$. **15.** $P-\dfrac{550H}{v}$. **16.** $\dfrac{v^2-u^2}{2s}$.

17. $\dfrac{uv}{u+v}$.   **18.** $\dfrac{a}{b}\sqrt{^2b-y^2}$. **19.** $\dfrac{s}{t}-\tfrac{1}{2}ft$. **20.** $\dfrac{v^2-u^2}{2f}$.

Exercise 10.5 (p. 105)

1. $\tfrac{1}{1000}\pi x^2 y$.   **2.** $\dfrac{2y^2-x^2}{xy}$.   **3.** $\dfrac{101n}{x+y}$.   **4.** $21\tfrac{1}{4}$.

5. $\dfrac{mx}{m+n},\dfrac{nx}{m+n}$. **6.** $\dfrac{pq(x+y)}{qx+py}$ km/h.   **7.** $\dfrac{3\sqrt{3A}}{\pi}$.

8. $\dfrac{25Ex+3My}{300(E+M)}$. **9.** $\dfrac{xp-yq}{x+y}$. **10.** $\dfrac{2x^2-2x+1}{x-2},\dfrac{y^2+2y+2}{y-2}$.

11. $\dfrac{1}{2(x^2-1)}$. **12.** 0.   **13.** $\dfrac{x^2+2x+4}{2x}$.

14. $\dfrac{1}{abc}$.   **15.** $x(a^2+x^2)(a-x)(a+x)$.

Exercise 11.1a (p. 107)

1. $\tfrac{6}{5}$. **2.** 1. **3.** 5. **4.** 1. **5.** 4.
6. 12. **7.** 2. **8.** $\tfrac{12}{13}$. **9.** 5. **10.** -2.
11. 4. **12.** -15. **13.** $-\tfrac{7}{9}$. **14.** $-\tfrac{61}{62}$. **15.** -7.
16. 2. **17.** $-\tfrac{1}{3}$. **18.** 3. **19.** 0. **20.** $\tfrac{1}{4}$.

Exercise 11.1b (p. 107)

1. $\tfrac{8}{3}$. **2.** 2. **3.** $\tfrac{17}{4}$. **4.** 1. **5.** -3.
6. 40. **7.** 3. **8.** $\tfrac{60}{47}$. **9.** -13. **10.** $\tfrac{17}{5}$.
11. $-\tfrac{6}{7}$. **12.** -6. **13.** $\tfrac{8}{7}$. **14.** $-\tfrac{97}{87}$. **15.** $-\tfrac{5}{3}$.
16. 7. **17.** $-\tfrac{1}{3}$. **18.** -2. **19.** No solution. **20.** $-\tfrac{6}{25}$.

Exercise 11.2a (p. 109)

1. $x<\tfrac{6}{5}$. **2.** $x>1$. **3.** $x<4$. **4.** $x>1$. **5.** $x<4$.
6. $x<4$. **7.** $x<\tfrac{1}{2}$. **8.** $x>12$. **9.** $x<6$. **10.** $x>\tfrac{12}{13}$.

Exercise 11.2b (p. 109)

1. $x<\tfrac{8}{3}$. **2.** $x>2$. **3.** $x<\tfrac{17}{4}$. **4.** $x>1$. **5.** $x>-3$.
6. $x>-3$. **7.** $x>3$. **8.** $x>40$. **9.** $x<6$. **10.** $x>\tfrac{60}{47}$.

Exercise 11.3a (p. 111)

1. 16. **2.** 24.   **3.** 7 km. **4.** 40 and 80. **5.** $\tfrac{3}{4}$.
6. 36. **7.** $10y+x$. **8.** 48 km. **9.** 220. **10.** £1600.

Exercise 11.3b (p. 111)

1. 60 km. **2.** 40. **3.** £1400. **4.** 37. **5.** 2 km.
6. 400. **7.** 300. **8.** 336 km. **9.** 8. **10.** 26 and 27.

Exercise 11.4a (p. 112)

1. $0 < x \leqslant 4$.   **2.** $0 < x < 45$.   **3.** $0 < x < \frac{58}{7}$.
4. $0 < x < 8$.   **5.** $0 < x \leqslant 4$.

Exercise 11.4b (p. 113)

1. $x \nless 2\frac{1}{2}$.   **2.** $0 \leqslant x < 18$.   **3.** $0 \leqslant x \leqslant 8$.
4. $0 \leqslant x \leqslant 15$.   **5.** $x \ngtr 340$.

Exercise 11.5a (p. 114)

1. 6, 2.   **2.** $2\frac{1}{2}$, 2.   **3.** $2\frac{1}{2}$, 1.   **4.** 2, 1.   **5.** 1, 1.
6. 2, 1.   **7.** 2, 3.   **8.** $\frac{3}{5}$, $-\frac{1}{5}$.   **9.** 1, 1.   **10.** 2, 1.
11. $\frac{16}{7}$, $-\frac{2}{7}$.   **12.** 2, 3.   **13.** 2, 3.   **14.** 13.   **15.** $\frac{8}{3}$, $-\frac{17}{3}$.
16. 36.   **17.** $\frac{1}{2}$, 1.   **18.** 5, 18.   **19.** ± 1, ± 4.   **20.** 12.

Exercise 11.5b (p. 114)

1. 4, 2.   **2.** $\frac{5}{3}$, 3.   **3.** 1, 1.   **4.** 2, 4.   **5.** 1, 1.
6. 2, 1.   **7.** 3, 2.   **8.** $-\frac{3}{5}$, $-\frac{4}{5}$.   **9.** 1, 1.   **10.** 3, 2.
11. $\frac{1}{3}$, $-\frac{2}{9}$.   **12.** 3, 4.   **13.** 4, 2.   **14.** $-3\frac{1}{2}$.   **15.** 2, 5.
16. 32.   **17.** $\frac{19}{14}$, -19.   **18.** 4, -5.   **19.** ± 2, ± 3.   **20.** 11.

Exercise 11.6a (p. 116)

1. 100, 2.   **2.** 400.   **3.** 2000, 3000.   **4.** 40p.   **5.** 39.
6. 1 penny.   **7.** £7.   **8.** £1.20.   **9.** 3 km.   **10.** 16, 30.

Exercise 11.6b (p. 117)

1. 8p.   **2.** 300.   **3.** 3, 4.   **4.** 64p.   **5.** 78.
6. 40.   **7.** 600.   **8.** 50p.   **9.** 4 km.   **10.** 45, 60.

Exercise 11.7a (p. 119)

1. $A = 3$, $B = -5$, $C = 1$.   **2.** $A = 2$, $B = 6$, $C = 3$.
3. Yes.   **4.** Neither.   **6.** $A = 2$, $B = -1$.   **7.** $A = B = 1$.
8. $A = \frac{9}{2}$, $B = \frac{1}{2}$, $C = -4$.   **9.** $A = \frac{9}{2}$, $B = \frac{1}{2}$, $C = -4$.

Exercise 11.7b (p. 119)

1. $A = 4$, $B = -1$, $C = 5$.   **2.** $A = 1$, $B = 0$, $C = 1$.
3. Yes.   **4.** Neither.   **5.** $A = 3$, $B = -2$.   **6.** $A = B = \frac{3}{4}$.
8. $A = \frac{4}{3}$, $B = \frac{1}{6}$, $C = -\frac{1}{2}$.   **9.** $A = \frac{4}{3}$, $B = -1$. $C = \frac{1}{3}$.
10. $A = \frac{1}{2}$, $B = \frac{1}{4}$, $C = \frac{1}{4}$, $\frac{1}{2}x + \frac{1}{4}$, $\frac{1}{4}$.

Exercise 11.8 (p. 119)

1. $\frac{5}{2}$.   **2.** 3.   **3.** 4 or -4.   **4.** 5 or -7.   **5.** $x < 1$.
6. $0 < x < \frac{4}{3}$.   **7.** $x > 0$ or $x < -2$.   **8.** $x > 1$ or $x < -3$.

9. 1.2, -3.5.   **10.** 2, $-\frac{2}{7}$.   **11.** 2 km.   **12.** 2.5 km.
13. 12.   **14.** 20.   **15.** 3.   **16.** $\frac{13}{17}$.
17. 25, 77.   **18.** 973.   **19.** £3.

Exercise 12.1a (p. 122)

1. 1, 3.   **2.** 3, 2.   **3.** 6, -1.   **4.** -5, -2.   **5.** 5, -2.
6. 6, -2.   **7.** 4, -2.   **8.** 9, 1.   **9.** 9, -2.   **10.** 2, $\frac{1}{2}$.
11. $\frac{1}{2}$, $\frac{1}{3}$.   **12.** $\frac{1}{2}$, $\frac{2}{3}$.   **13.** $-\frac{2}{3}$, -4.   **14.** 2, $-2\frac{1}{2}$.   **15.** 3, -6.
16. 2, $-\frac{3}{2}$.   **17.** 4, $\frac{5}{9}$.   **18.** 5, -1.   **19.** 5, $-\frac{1}{2}$.   **20.** 5, $-\frac{1}{2}$.

Exercise 12.1b (p. 123)

1. 7, -1.   **2.** 8, -2.   **3.** 2, 12.   **4.** 7, -3.   **5.** 5, -4.
6. -9, -2.   **7.** -3, -6.   **8.** -4.   **9.** 3, $\frac{1}{2}$.   **10.** $\frac{2}{3}$, $-\frac{1}{2}$.
11. $\frac{4}{3}$, $\frac{3}{2}$.   **12.** 5, $-\frac{2}{3}$.   **13.** 7, $\frac{5}{2}$.   **14.** 2, $-\frac{7}{2}$.   **15.** 2, -3.
16. 1, $-\frac{18}{7}$.   **17.** 3, $-\frac{1}{2}$.   **18.** $-2\frac{1}{2}$.   **19.** 2, $-\frac{2}{3}$.   **20.** 2, $-\frac{2}{3}$.

Exercise 12.2a (p. 124)

1. $-2 < x < 2$.   **2.** $x < -5$ or $x > 5$.   **3.** $-\frac{3}{2} < x < \frac{3}{2}$.
4. $x < 2$ or $x > 3$.   **5.** $-2 < x < 1$.   **6.** $x < -2$ or $x > \frac{1}{2}$.
7. $-3 < x < -\frac{1}{2}$.   **8.** $x < 1$ or $x > 2$.   **9.** $-1 < x < 4$.
10. True for all values of x, except $x = 1$.

Exercise 12.2b (p. 124)

1. $x < -4$ or $x > 4$.   **2.** $-6 < x < 6$.   **3.** $x < -\frac{2}{3}$ or $x > \frac{2}{3}$.
4. $3 < x < 4$.   **5.** $x < -1$ or $x > 2$.   **6.** $-2 < x < -\frac{3}{2}$.
7. $x > -\frac{5}{2}$ or $x < -3$. **8.** $-1 < x < 3$.   **9.** $x < 1$ or $x > 3$.
10. Not true for any value of x.

Exercise 12.3a (p. 125)

1. 5, 12.   **2.** 4, 9.   **3.** 12 m.   **4.** 6.
5. 17, 19.   **6.** 7, 8.   **7.** 1 km/h.   **8.** 20.
9. 7 cm $<$ length $<$ 9 cm.   **10.** $x > 5$.

Exercise 12.3b (p. 126)

1. 3, 13.   **2.** 4, 8.   **3.** 4 cm.   **4.** 8.
5. 2, 7.   **6.** 5.   **7.** 2 km/h.   **8.** 4.
9. Less than 6 cm. **10.** $0 < x < 4$.

Exercise 12.4a (p. 128)

1. 2.77, -1.27.   **2.** 0.77, -0.43.   **3.** 4.70, -1.70.
4. 1.29, 0.31.   **5.** 0.81, -0.53.   **6.** -1.43, -0.23.
7. $-1, 0.75$.   **8.** -1.27, -0.39.   **9.** -3.37, 2.37.
10. 2.62, 0.38.   **11.** 3.87, 0.13.   **12.** ± 1.58.

13. $-3.09, 2.09.$ **14.** $0.77, -7.77.$ **15.** $-5.83, -0.17.$
16. $-2.30, 1.30.$ **17.** $1.62, -0.62.$ **18.** $-0.38, -2.62.$
19. $0.86, -0.19.$ **20.** $0, -2.$

Exercise 12.4b (p. 129)

1. $3.83, -1.83.$ **2.** $1.85, -0.18.$ **3.** $2.30, -1.30.$
4. $-0.69, 0.29.$ **5.** $0.77, -0.43.$ **6.** $1.30, -0.88.$
7. $-2.28, -0.22.$ **8.** $-1.40, 0.90.$ **9.** $1.79, -2.79.$
10. $4.79, 0.21.$ **11.** $5.92, 0.85.$ **12.** $0.58, -2.58.$
13. $1.41, -3.91.$ **14.** $6.16, -0.16.$ **15.** $2.39, -1.89.$
16. $-3.62, 2.62,$ **17.** $0.41, -4.07.$ **18.** $3.79, -0.79.$
19. $-3.62, -1.38.$ **20.** $-3.55, 1.88.$

Exercise 12.5a (p. 130)

1. 2, 3 or 3, 2.   **2.** $3, 1$ or $-\frac{2}{3}, \frac{25}{3}.$ **3.** $\frac{1}{2}, \frac{1}{3}.$
4. 1, 1.   **5.** $2, 1$ or $-\frac{1}{2}, -\frac{1}{2}.$

Exercise 12.5b (p. 131)

1. 3, 4 or 4, 3.   **2.** $1, 2$ or $\frac{3}{2}, \frac{1}{2}.$ **3.** $1, \frac{1}{2}.$
4. $\pm 2, \pm 3.$ **5.** $\frac{4}{3}, \frac{5}{3}.$

Exercise 12.6a (p. 132)

1. 32 km/h.   **2.** 11.   **3.** 40.   **4.** 18.   **5.** $3\frac{1}{2}$p.

Exercise 12.6b (p. 132)

1. 50 km/h.   **2.** 21.   **3.** 20p.   **4.** 20 m.   **5.** 440.

Exercise 12.7 (p. 133)

1. $1, -1.$ **2.** $-1\frac{1}{2}.$ **3.** $1.17, -0.95.$ **4.** $a, 2-a.$ **5.** $\frac{1}{3}, \frac{2}{3}.$
6. $-2b \pm 3c.$ **7.** $0.58, -0.41.$ **8.** 12.   **9.** 12.   **10.** 18.
11. 48 km.   **12.** 300.   **13.** 5.   **14.** 20 days.   **15.** 32.

Exercise 13.5 (p. 142)

1. 13.40 h.   **2.** 12 km.   **3.** 10.58 h.   **4.** 69 cm, 3400 m.
5. 6.5 km.   **6.** 122.5 m, 10 s.   **7.** 0.4; 1.75; $x < -1.7$ or
$-0.3 < x < 2.$ **8.** $2.3, 2, -1.2, 0.7, 2.4$
9. $-0.8 < x < 1.3.$ **10.** $-1.2, 0.6, 2.6.$
11. 0.6; $-0.55, 0.8.$ **12.** $-0.9 < x < 4.25.$ **13.** $0.35, -2.85,$
$2x^2 + 5x - 2 = 0$ **14.** $-1.7, 2.2.$ **15.** $-1.5, 4.7.$ **16.** 0.3 or $1, 1.$
17. 1.41.   **18.** $1.25, 2.8.$ **19.** $0.3 < x < 0.7$ or $x > 4.1.$
20. $-1, 8.$

Exercise 14.1a (p. 148)

7. (a) $x > 0, y > 0, 4x + 5y < 20$.   (b) $x < 2, y > 0, y - x < 1$.
8. (1, 1), (1, 2), (2, 1).   **9.** (2, 2), (2, 3), (2, 4), (2, 5), (4, 3).
10. (4, 8), (0, 12).

Exercise 14.1b (p. 149)

7. (a) $y > 0, y < x, x + y < 2$.   (b) $x > 0, y > x - 1, 2y + x < 2$.
8. (1, 3), (1, 4).   **9.** (2, 2), (2, 4), (4, 2).   **10.** ($1\frac{1}{2}$, $1\frac{1}{2}$).

Exercise 14.2a (p. 152)

1. 15, from (10, 5), (9, 6) or (8, 7).   **2.** 5 ha.   **3.** £2880.
4. 1 kg of A, 2 kg of B.   **5.** 22 m, 46 m; 24 m, 42 m; 22 m × 46 m.

Exercise 14.2b (p. 153)

1. (200, 300).   **2.** 7 kg.   **3.** F. T. first (6,0, 6,1, 6,2, 6,3, 7,0, 7,1, 7,2).
4. £90.   **5.** 10, 28, 4.

Exercise 14.3 (p. 156)

1. 350; 200, 100.   **2.** (3, 4), (3, 5).   **3.** (i) (15, 10) or (16, 8) or (17, 6)...
or (20, 0).   (ii) (15, 10).   (iii) (0, 25).
4. (0, 4), (1, 4), (2, 4), (3, 4); (1, 1), (1, 2), (1, 3), (2, 1) (2, 2), (3, 0).
5. 250 kg A, 240 kg B.   **6.** 150 tonnes of x, 98 tonnes of y.
7. 150, 100, 200 tonnes; £6500. No.   **8.**

	B_1	B_2	B_3
Q_1	500	1500	1000
Q_2	1500	0	0.

Exercise 15.1a (p. 159)

1. $\frac{1}{27}$.   **2.** 4.   **3.** $\frac{1}{4}$.   **4.** 8.   **5.** $\frac{1}{8}$.
6. 8.   **7.** $\frac{1}{3}$.   **8.** $\frac{1}{1000}$.   **9.** 4.   **10.** $\frac{13}{12}$.

Exercise 15.1b (p. 159)

1. $\frac{1}{8}$.   **2.** 2.   **3.** $\frac{1}{2}$.   **4.** $\frac{1}{5}$.   **5.** $\frac{1}{25}$.
6. $\frac{1}{27}$.   **7.** $\frac{1}{16}$.   **8.** $\frac{1}{64}$.   **9.** $\frac{8}{3}$.   **10.** $\frac{5}{4}$.

Exercise 15.2a (p. 160)

1. 4.   **2.** 4.   **3.** $\frac{3}{2}$.   **4.** 3.   **5.** $\frac{1}{2}$.
6. $\frac{3}{2}$.   **7.** $\frac{1}{2}$.   **8.** $\frac{1}{3}$.   **9.** $\frac{5}{2}$.   **10.** $\frac{5}{6}$.
11. 2.   **12.** 3.   **13.** 1.   **14.** 2.   **15.** 0.
16. 1   **17.** $\frac{1}{100}$.   **18.** 27.   **19.** 2.   **20.** 2.

Exercise 15.2b (p. 160)

1. $\frac{1}{2}$.   **2.** 3.   **3.** -1.   **4.** -2.   **5.** $\frac{5}{2}$.
6. $\frac{3}{2}$.   **7.** $\frac{3}{2}$.   **8.** $\frac{1}{3}$.   **9.** $\frac{2}{3}$.   **10.** $\frac{1}{2}$.

11. 0.   **12.** 5.   **13.** 4.   **14.** -3.   **15.** $-\frac{1}{2}$.
16. 10 000.   **17.** $\frac{1}{10}$.   **18.** 16.   **19.** 3.   **20.** 64.

Exercise 15.3a (p. 161)

1. 2.   **2.** 1.   **3.** 1.   **4.** 4.   **5.** 4.
6. 2.   **7.** 3.   **8.** 3.   **9.** $\frac{1}{2}$.   **10.** $2\frac{1}{2}$.

Exercise 15.3b (p. 162)

1. 2.   **2.** 3.   **3.** 1.   **4.** 3.   **5.** 2.
6. 1.   **7.** 3.   **8.** $\frac{1}{2}$.   **9.** $\frac{2}{3}$.   **10.** 2.

Exercise 15.4 (p. 163)

1. 2.302.   **2.** $\dfrac{100}{x^2}$.   **3.** $\dfrac{6^3}{x^2}$.   **4.** $\dfrac{x^2}{4}$.   **5.** $\dfrac{10}{x}$.

6. (i) $\dfrac{x^3}{3}$,   (ii) $\frac{1}{4}$,   (iii) $8x^{-12}$.   **7.** $100x^{\frac{3}{4}}$.

8. (i) $\frac{1}{9}$.   (ii) 4.   (iii) $\frac{3}{2}$.   **9.** 1.892.   **10.** 1.531.   **11.** 1.062.
12. -3.   **13.** 0.179.   **14.** 32.   **15.** 1.   **16.** 0.3010, 0.7782.
17. $a^{-\frac{1}{2}}$.   **18.** 0 or 1.585.   **19.** 2.   **20.** 2.   **21.** 1.
22. 2.52.   **23.** -3.   **24.** 8.63.   **25.** 0, 0.631.   **26.** 0, 1.

27. 2^{6x+1}.   **28.** 243.   **29.** $\dfrac{10}{y^2}$.   **30.** $\dfrac{z^4}{y^2}$.   **31.** (i) 3.   (ii) $\frac{2}{3}$.

    (iii) 6.   **32.** $\frac{3}{2}p^4$.   **33.** $24z^4y^2$.   **34.** 1.255.   **35.** $2\frac{1}{3}$.

Exercise 16.1a (p. 166)

5. 48.   **6.** 18.   **8.** 64 kg.   **9.** 1.73 : 1.   **10.** 30 km.

Exercise 16.1b (p. 167)

3. $\frac{1}{3}$.   **4.** $\frac{1}{2}$.   **5.** 84.   **6.** 20.   **8.** 21.
9. 54 kg.   **10.** 8 : 1.

Exercise 16.3 (p. 172)

1. $1\frac{1}{4}$.   **2.** 225 kg.   **3.** 2.   **4.** £65.   **5.** 40 000.
6. 280 hours.   **7.** $7\frac{1}{2}$ pence.   **9.** 35 km/h.   **10.** directly.   **11.** $4\sqrt{6}$ m/s.
12. $7\frac{1}{9}$.   **13.** doubled.   **14.** 1152.   **15.** 1 : 3.   **16.** £3$(y - x)$.
17. 19 kg.   **18.** 126 kg.   **19.** 100.   **21.** 40 km/h.
22. 12 newtons.   **23.** 8 : 1.   **24.** 400 : 1.

Exercise 17.1a (p. 175)

1. 81.   **2.** -109.   **3.** 22.   **4.** 21.   **5.** 8, 13.
6. 2, 9.   **7.** 348.   **8.** $236\frac{1}{4}$.   **9.** $5n - 1$.   **10.** $13 - 3m$.

Exercise 17.1b (p. 176)

1. -47.   **2.** 99.   **3.** 100.   **4.** No.   **5.** 15.

6. 6.   **7.** 2500.   **8.** $\dfrac{3n(n+1)}{2}$.   **9.** $\dfrac{61-n}{10}$.   **10.** $2n$.

Exercise 17.2a (p. 177)

1. $2(7)^{39}$.   **2.** $\dfrac{1}{2^{12}}$.   **3.** 9.   **4.** 18.   **5.** 36.

6. 10, 20.   **7.** $2^{22}-4$.   **8.** $\dfrac{3^{17}+1}{4(3^{14})}$.   **9.** $6(3)^{n-1}$.   **10.** $5(-\tfrac{1}{2})^{n-1}$.

Exercise 17.2b (p. 178)

1. $5(3)^{23}$.   **2.** $-\dfrac{1}{3^{16}}$.   **3.** 22.   **4.** No.   **5.** 70.

6. 21, 63.   **7.** $3(2^{18}-1)$.   **8.** $\dfrac{1-3^{2n}}{2}$.   **9.** $\dfrac{9}{2^{31}}$.

10. $2^{n-1}xy^{n-1}$.

Exercise 17.3 (p. 179)

1. $\dfrac{n(n+1)}{2}$; 32.   **2.** 780.   **3.** £276.   **4.** 19.

5. £7970.   **6.** 19 years.   **7.** 32, 44 etc.   **8.** 45, 135, 405.   **9.** 9.

10. $-\tfrac{1}{2}$, 7; $\tfrac{8}{9}$, $\tfrac{81}{8}$; 465.   **11.** 54, $\dfrac{3^5}{2}$.   **12.** 4, 6, 8.   **13.** 2, 3^n-1.   **14.** 3.

17. $\dfrac{x(r-q)+y(p-r)}{p-q}$.   **18.** 12th; $82\tfrac{1}{2}$.   **19.** 1683.

20. 1368.   **21.** $a\dfrac{\left(\dfrac{b}{a}\right)^{\frac{n}{n-1}}-1}{\left(\dfrac{b}{a}\right)^{\frac{1}{n-1}}-1}$.

22. 3675.   **23.** 30.   **24.** $\dfrac{n^2-n+2}{2}$; $\dfrac{n}{2}(n^2+1)$.

25. $x-\dfrac{y}{3}$, $x+\dfrac{y}{3}$.   **26.** 210 log 2.   **27.** 66 660.   **28.** $1\tfrac{1}{2}$.

29. 3, 27.   **30.** 13.

Exercise 18.1a (p. 188)

1. {1, 4, 6}, {1, 2, 4, 5, 6}.   **2.** {1, 2, 3,}, {1, 2, 3, 4, 5,}.
3. {1, 2, 3}, {1, 2, 3, 5}.   **4.** Ø, {1, 2, 3, 4, 5, 6, 7, 8}.
5. {3}, {4}, {5}, {3, 4}, {4, 5}, {3, 5}.   **6.** 14
7. e.g. all multiples of seven less than 34.
8. e.g. all perfect squares between 64 and 144 inclusive.
9. 5, 6.   **10.** 1.   **12.** {2, 3, 5, 7}, {1, 2, 3, 4, 5, 6, 7}.
13. {6, 12, 18}, {3, 6, 9, 12, 15, 18}.
15. (iii) {3, 5}, {2, 3, 5, 7}.   (v) $A \cap B = C$, $A \cup B = D$.

Exercise 18.1b (p. 189)

1. {2, 4, 8, 10}, {2, 3, 4, 6, 8, 10}.   2. {1, 2, 3, 4}, {1, 2, 3, 4, 5}.
3. 2, 3, No.   4. {11, 13, 17, 19, 23, 29}, {10, 12, 14, 16, 18, 20, 22, 24, 26, 28}, {12, 15, 18, 21, 24, 27}.   5. Ø, {1, 2, 4, 5, 6, 7, 8, 12, 18}.
6. {10, 20}, {5, 10, 15, 20, 25}.   8. $A \cap B = C$, $A \cup B = D$.
10. Repeated factor in 24 and 36.

Exercise 18.2 (p. 194)

1. {3, 5, 6, 7}, {2, 4, 6, 8}, {1, 3, 5, 7}.   2. {1, 2, 3, 4, 5, 7, 8, 9, 10}.
3. {James, Jack}.   4. {ash, elm}.   5. {red, green, blue, black, white}.
6. (i) {1, 2, 3, 4, 5, 6}.   (ii) {3, 4}.   (iii) {5, 6, 7}.   (iv) {1, 2, 7}.
   (v) 6.   7. (i) {1, 2, 4, 8, 16}.   (ii) ϕ.   (iii) 6.   (iv) {0, 8, 16}.
   (v) {0}.   8. (ii) is correct.   (i) {4, 5, 6, 7}.   (iii) {4}.   (iv) 0.   (v) 8.
16. 15.   17. 32.   18. 5.   19. 38.   20. 75.   21. 2.   22. 25.   23. 29.
24. 21.   25. 0.   26. 7.   27. 12, 4.   28. $15 \leqslant n \leqslant 45$ 29. $1 \leqslant n \leqslant 5$
30. 144.   31. > 137.   32. > 218.

Exercise 19.1a (p. 199)

1. 11, 7, 28, 786, 36; closed, but not associative nor commutative.

2. 6, 6, 6, $\sqrt{27} \sqrt{24}$, commutative but not associative. Not closed.

3. 3, 0, 3.   4. (i), (ii), (iv), (vi), (viii) closed.   5. 23.   6. 4, 5, -8 or $+6$.

Exercise 19.1b (p. 200)

1. (i) 1, 1, 4, 16, closed and commutative.   2. 2, 2, associative and commutative.   3. 2, 3, 1. No.   5. 8, 7, 14, 12, 11; 11.

Exercise 19.2 (p. 202)

1. (i) 0, -2.   (ii) 1, $\frac{1}{2}$.   (iii) 0, $-\frac{2}{3}$.   (iv) 0, No inverse.

2. No identities.   3. $\begin{pmatrix} 0 \\ 0 \end{pmatrix}$, $\begin{pmatrix} -2 \\ -1 \end{pmatrix}$.   4. Rotate through 0°; rotate through $-90°$.

Exercise 19.3 (p. 203)

3. Yes.   5. Yes.

Exercise 20.1 (p. 206)

1. {2, 4, 6}, {$+1$, -1, $+\sqrt{2}$, $-\sqrt{2}$, $+\sqrt{3}$, $-\sqrt{3}$}, {$y : y < 3$}, {$y : \leqslant 1$}.
2. onto, into, onto, onto.   3. one-many, one-many.
4. {0, 1, 4, 5, 6, 9}, 28177, 281918.   5. {George III}.

Exercise 20.2 (p. 208)

r ≡ reflexive,　s ≡ symmetric,　t ≡ transitive

1. t.　　　**2.** s, t.　　**3.** t.　　**4.** r, s, t.　　**5.** t.　　　**6.** r, s, t.
7. s.　　　**8.** r, s, t.　**9.** t.　　**10.** t.　　　**11.** none.　**12.** r, s, t.

Exercise 20.3a (p. 212)

1. $\{-3, 0, 3, 6\}$, $\{3, 2, 1, 0, -1\}$, $\{\frac{1}{4}, \frac{1}{2}, 1, 2, 4\}$, {all positive rationals of the form $\frac{1}{K}$, where K is an integer}, $\{y : \frac{1}{25} \leqslant y \leqslant 1\}$.

2. $x \mapsto \frac{1}{3}x$, $x \mapsto 1 - x$, $x \mapsto 4x$, $x \mapsto \frac{1}{x}$, $x \mapsto \sqrt[3]{x}$.

3. $D = \{0, 1, 2\}$, $D = \{x : 0 \leqslant x \leqslant 1\}$.

Exercise 20.3b (p. 212)

1. $\{0, 1, 2, 3\}$, $\{-2, -1, 0, 1, 2\}$, $\{y : 1 \leqslant y < 8\}$, $\mathbb{Z}_-$; $\{y : 0 < y \leqslant 1\}$.

2. $x \mapsto \frac{1}{4}x$, $x \mapsto -x$, $x \mapsto 2x$, $x \mapsto \frac{1}{x} - 1$, $x \mapsto \frac{1}{2}\sqrt[3]{x}$.

3. $\mathbb{R}_+$, $\{x : 0 < x < 180\}$.

Exercise 20.4a (p. 213)

1. $8, 12, \frac{2}{3}, 2, 2, \frac{10}{3}$.　　　　**2.** 49, 14, 2.　　　　**4.** $x \mapsto x^2 + 4x + 5$.

Exercise 20.4b (p. 213)

1. 0,1; 2,4; 4,6.　　　**2.** $\frac{1}{5}, \frac{9}{4}$; $-\frac{5}{18}, 2$.　　　**4.** $x \mapsto x^2 + 8x + 16$.
5. $-3 \pm \sqrt{7}$.

Exercise 20.5 (p. 214)

1. (i), (iii) and (v).　　　　**2.** (i), (ii) and (iv).
3. The first relates two elements in $\mathbb{Z}$; the second is a statement which is either true or false.　　　**4.** Yes; one $\mapsto$ un or une.　　　**5.** r, s and t.
6. (i) Many-one; domain $\{x : 0 < x \leqslant 10\}$, where x kg is the mass of a parcel; range $\{y : y = 55, 70, 85 \ldots 160\}$, where y pence is the cost of postage.
7. (i) Many-one; domain $\{x : 0 < x < 25\}$, where x km is the length of a journey; range $\{y : y = 5, 9, 13 \ldots 101\}$, where y pence is the cost of a journey.　　　**8.** $y : 0 < y \leqslant \frac{1}{2}$.
9. $\{x : 10 < x < 1000\}$. **10.** $\mathbb{Z}$.　　　**11.** $\{y : y > 1\}$.
12. $-1, 2$.　　　　**13.** $-1, 4$.　　　**14.** $-\frac{1}{4}, 1$.　**16.** 1, 2.
17. $-\frac{1}{2}, 1$. Only 1.　　**18.** $-4, 5$.　　　**19.** $x \mapsto 9x^2 + 1$.

20. $f : x \mapsto x^2$, $g : x \mapsto \frac{1}{x}$, $h : x \mapsto x + 2$.

Exercise 21a (p. 220)

1. All except (ii) and (iii), no inverses.
2. (i) $\frac{3}{4}$.   (ii) No solutions   (iii) 1.   (iv) 3.
3. (i) T_6.   (ii) Inverses are T_1, T_2, T_3, T_5, T_4, T_6.
4. (i) S_0 is identity element; S_1, S_2, S_3 do not have inverses.   (ii) No identity element.

Exercise 21b (p. 221)

1. (i), (iv) and (v).   2. (i) No solution.   (ii) 3.   (iii) 2, 5.   (iv) 2, 5.
3. T_0 leaves the cloth unchanged.

4. $X^{-1} = \dfrac{1}{a+b-1}\begin{pmatrix} b & b-1 \\ a-1 & a \end{pmatrix}, \begin{pmatrix} 1 & 0 \\ 0 & 1 \end{pmatrix}$

5.

	e	a	b	c		e	a	b	c
e	e	a	b	c	e	e	a	b	c
a	a	e	c	b	a	a	e	c	b
b	b	c	e	a	b	b	c	a	e
c	c	b	a	e	c	c	b	e	a

Exercise 22.1a (p. 224)

1. (i) 2.2, 2, 1.   (ii) 3.4, 4, 5.   (iii) $1\frac{5}{6}$, $1\frac{1}{2}$, 1.   (iv) 3, 2, 0
2. (i) 4.   (ii) 14.   (iii) 104.   (iv) 80. 3. (i) 4.   (ii) 4.1.   (iii) 3.9. (iv) 20.
4. (i) 4.4.   (ii) 1.9.   (iii) 2.9.   (iv) 13.4.   5. 12.

Exercise 22.1b (p. 224)

1. (i) 3.2, 3, 3.   (ii) 3, 2, 0.   (iii) $4\frac{2}{3}$, $4\frac{1}{2}$, 9.   (iv) $2\frac{1}{2}$, $1\frac{1}{2}$, 0.
2. (i) 2.   (ii) 12.   (iii) 52.   (iv) 8.   3. (i) 1.   (ii) 1.1.   (iii) 20.   (iv) 5.
4. (i) 1.8.   (ii) 2.7.   (iii) 4.1.   (iv) 24.4.   5. 2.

Exercise 22.2a (p. 226)

1. 20.7.   2. (i) 5.25.   (ii) 21.1.   3. (i) 72, 45.   (ii) 102, 6.
4. (i) 240.   (ii) 0.048.   5. \$4992.5.

Exercise 22.2b (p. 227)

1. 15.4   2. (i) 2.6.   (ii) 5.   3. (i) 82.55.   (ii) 98.89.
4. (i) 2700.   (ii) 0.0134.   5. 171.9 cm.

Exercise 22.3a (p. 229)

1. (i) 30.   (ii) 31.   (iii) 24.5.   2. 2 : 3.   3. $5\frac{2}{3}$.
4. 24.3, 23.7, 23.2, 27.3, 28.3, 25.7, 21.3, 19, 20.7. First.
5. 3.83, 1.88, 2.46, 4.96, 5.49, 5.16, 3.96, 2.52, 2.29, 2.71.

Exercise 22.3b (p. 230)

1. (i) 6.4. (ii) 6.1. (iii) 4.8. **2.** 114. **3.** Z, Y.
4. 1903, 2384, 2462, 3787, 4042, 4605, 4448.
5. 68p, 76p, 78p, 86p, 83p, 82p, 76p, 71p, 69p, 66p.

Exercise 22.4 (p. 231)

1. 23. **2.** 5. **4.** 1, 0. **5.** 0, 1. **6.** $3\frac{1}{2}$, 1. **7.** 2.04.
8. 32.49. **9.** 3.725. **10.** 23.06.

Exercise 23.3 (p. 239)

1. \$4920, (ii) 34%. **2.** 169.6 cm, 39, 2.75 cm. **3.** 48, 32.9, 2.4.
4. 30%, 26. **5.** 2.5 km.

Exercise 23.4 (p. 240)

1. 3.55. **2.** 76.7p. **3.** £3.99; affected by a few large donations.
4. 12, £190, £240 **5.** (i) £50.10, grouping. (ii) £51, 310.

Exercise 24.1a (p. 244)

1. (i) $\frac{1}{13}$. (ii) $\frac{1}{52}$. (iii) $\frac{1}{4}$. (iv) $\frac{1}{26}$. (v) $\frac{2}{13}$. (vi) $\frac{4}{13}$. (vii) $\frac{51}{52}$. (viii) $\frac{3}{4}$.
(ix) $\frac{3}{4}$. (x) $\frac{3}{13}$.
2. (i) $\frac{4}{9}$. (ii) $\frac{5}{9}$. (iii) $\frac{4}{9}$. (iv) $\frac{4}{9}$. (v) $\frac{2}{9}$. (vi) $\frac{1}{9}$. (vii) $\frac{1}{9}$. (viii) $\frac{2}{3}$.
3. (i) $\frac{1}{4}$. (ii) $\frac{4}{19}$. (iii) $\frac{1}{6}$. (iv) 1. **4.** (i) $\frac{3}{5}$. (ii) $\frac{1}{15}$. (iii) 0. (iv) $\frac{1}{15}$.
(v) $\frac{4}{15}$. (vi) $\frac{11}{15}$.
5. $\frac{1}{4}$. **6.** $\frac{5}{12}$, $\frac{4}{11}$. **7.** $\frac{5}{26}$. **8.** (i) $\frac{1}{36}$. (iii) 7.

Exercise 24.1b (p. 245)

1. (i) $\frac{5}{9}$. (ii) $\frac{4}{9}$. (iii) $\frac{9}{17}$. (iv) $\frac{10}{17}$. (v) $\frac{1}{2}$. (vi) $\frac{5}{8}$. (vii) $\frac{9}{16}$.
2. (i) $\frac{2}{5}$. (ii) $\frac{2}{5}$. (iii) $\frac{9}{10}$. (iv) $\frac{1}{10}$. (v) $\frac{1}{3}$. **3.** (i) $\frac{2}{5}$. (ii) $\frac{2}{5}$. (iii) $\frac{5}{5}$. (iv) $\frac{1}{5}$.
4. $\frac{2}{3}$. **5.** (i) $\frac{1}{6}$. (ii) $\frac{1}{3}$. (iii) $\frac{1}{3}$. (iv) $\frac{5}{12}$. **6.** $\frac{1}{25}$. **7.** $\frac{7}{26}$. (i) $\frac{6}{25}$.
(ii) $\frac{2}{5}$. (iii) $\frac{9}{25}$. **8.** (i) $\frac{1}{36}$. (ii) $\frac{4}{36}, \frac{4}{36}, \frac{10}{36}, \frac{12}{36}, \frac{9}{36}$. (iii) 5.

Exercise 24.2 (p. 249)

1. $\frac{27}{91}$. **2.** $\frac{2}{11}$. **3.** (i) $\frac{1}{5}$. (ii) $\frac{9}{20}$. (iii) $\frac{7}{10}$. **4.** (i) $\frac{1}{20}$. (ii) $\frac{1}{10}$. (iii) $\frac{1}{10}$.
(iv) 0. **5.** $\frac{3}{7}$. **6.** (i) $\frac{1}{2}$. (ii) $\frac{1}{10}$. (iii) $\frac{1}{2}$. **7.** (i) $\frac{27}{40}$. (ii) $\frac{1}{40}$. (iii) $\frac{9}{40}$.
8. (i) $\frac{3}{20}$. (ii) $\frac{1}{5}$. (iii) $\frac{2}{5}$.

Exercise 25a (p. 254)

1. (i) $\frac{4}{25}$. (ii) $\frac{9}{25}$. (iii) $\frac{12}{25}$. **2.** (i) $\frac{4}{9}$. (ii) $\frac{1}{9}$. (iii) $\frac{4}{9}$.
3. (i) $\frac{1}{16}$. (ii) $\frac{3}{8}$. (iii) $\frac{1}{64}$. (iv) $\frac{27}{64}$. (v) $\frac{27}{64}$. **4.** (i) $\frac{1}{8}$. (ii) $\frac{1}{62}$. (iii) $\frac{3}{248}$.
(iv) $\frac{1}{465}$.
5. (i) $\frac{5}{9}$. (ii) $\frac{4}{9}$. (iii) $\frac{23}{54}$. (iv) $\frac{31}{54}$. **6.** (i) $\frac{9}{40}$. (ii) $\frac{9}{20}$. (iii) $\frac{3}{20}$. (iv) $\frac{1}{60}$.
7. $\frac{13}{20}$. **8.** (i) $\frac{13}{100}$. (ii) $\frac{13}{25}$. (iii) $\frac{21}{100}$. (iv) $\frac{7}{50}, \frac{13}{34}$.

Exercise 25b (p. 256)

1. (i) $\frac{4}{25}$. (ii) $\frac{9}{25}$. (iii) $\frac{9}{100}$. (iv) $\frac{6}{25}$. (v) $\frac{2}{5}$.
2. (i) 0.49. (ii) 0.343. (iii) 0.441. (iv) 0.2401.
3. (i) 0.0001. (ii) 0.000001. (iii) 0.000297.
4. (i) 0.81. (ii) 0.2916. (iii) 0.00001. (iv) $(0.9)^9$.
5. (i) 0.43. (ii) 0.57. (iii) 0.571. (iv) 0.429.   6. 0.67.
7. $\frac{51}{64}$.   8. (i) 0.036 (ii) 0.036 (iii) 0.064, 0.2368

Exercise 26.1a (p. 265)

1. $11\frac{1}{4}°$.   2. $6x + y = 360°$.   3. 140°, 140°, 40°.   4. $y = x + z$.
6. $p + q + r = 360°$.   8. 150°.   9. $y + x = 180°$.

Exercise 26.1b (p. 266)

1. 10°. 2. $y + 5x = 360°$. 3. 130°, 50°, 130°.   5. $x + y + z = 360°$.
6. 120°.   7. 60°.   8. $x + y + z = 180°$.   10. $p + q + r = 720°$.

Exercise 26.2a (p. 271)

1. 30.   2. 100°.   3. Yes, No, Yes.   4. Yes, Yes, No.   5. 80.
6. 156°. 7. 120°, 135°.   8. 84.   9. 95.   10. 130°.

Exercise 26.2b (p. 271)

1. 20.   2. 95°.   3. Yes, No, Yes.   4. Yes, Yes, No.
5. 48.   6. $157\frac{1}{2}°$.   7. 125°, 145°.   8. 36.
9. 120°, 140°.   10. 60°.

Exercise 26.3 (p. 271)

1. 70°.   2. $(360 - x - y - z)°$.   3. $(190 - 5x)°$. 4. 15.
5. 66.   6. 140°.   7. $p + q + r$.   9. 20°.
10. 100°, 30°.   11. 80.   12. 40.

Exercise 27.2a (p. 276)

1. 65°, 100°, 80°.   2. 36°.   4. 50°.   6. 15°.   8. 30°.   9. $3x$.

Exercise 27.2b (p. 277)

1. 55°, 80°, 40°.   2. 30°.   5. 30°.   6. 40°.   7. 44°.

Exercise 27.3 (p. 280)

1. 62°, 118°, 118°.   2. 40°, 140°.

Exercise 27.4 (p. 281)

1. No.   2. $BC < CA < CD$.   3. $BC > CA > CD$.   10. CX.

Exercise 27.5 (p. 282)

1. 150°.　　**3.** 99°.　　**5.** 9°.　　**8.** 22½°.　　**9.** 18.

Exercise 28.1a (p. 284)

1. (i) 2 : 3.　(ii) 3 : 4.　(iii) 5 : 4.　(iv) 1 : 2.　　**2.** (i) 3 : 2.　(ii) 1 : 2.
(iii) 5 : 4.
3. (i) 6⅔ cm, 3⅓ cm, 20 cm, 10 cm.　(ii) 7½ cm, 2½ cm, 15 cm, 5 cm.
(iii) 6 cm, 4 cm, 30 cm, 20 cm.　(iv) 8 cm, 2 cm, 13⅓ cm, 3⅓ cm.
4. (i) 5 cm, 15 cm, 10 cm, 30 cm.　(ii) 15 cm, 5 cm, 30 cm, 10 cm.　(iii) 8 cm,
12 cm, 40 cm, 60 cm.　(iv) 12 cm, 8 cm, 60 cm, 40 cm.
5. (i) 25 : 16.　(ii) 25 : 9.　(iii) 25 : 4.　(iv) 25 : 1.

Exercise 28.1b (p. 284)

1. (i) 1 : 2.　(ii) 2 : 3.　(iii) 1 : 1.　(iv) 1 : 3.
2. (i) 7 : 8.　(ii) 4 : 5.　(iii) 3 : 2.　(iv) 7 : 10.
3. (i) 22½ cm, 7½ cm, 45 cm, 15 cm.　(ii) 18 cm, 12 cm, 90 cm, 60 cm.
(iii) 24 cm, 6 cm, 40 cm, 10 cm.　(iv) 25 cm, 5 cm, 37½ cm, 7½ cm.
4. (i) 36 : 25.　(ii) 9 : 4.　(iii) 9 : 1.　(iv) 36 : 1.

Exercise 28.2a (p. 287)

1. (i) 2 : 3.　(ii) 2 : 5.　　**2.** (i) 3 : 5.　(ii) 3 : 8.　　**3.** 2 : 1.
4. (i) 3 : 5.　(ii) 3 : 5.

Exercise 28.2b (p. 287)

1. (i) 4 : 5.　(ii) 4 : 9.　**2.** 4 : 1.　**3.** (i) 3 : 1.　(ii) 1 : 4.　(iii) 15 : 11.
(iv) 11 : 16.

Exercise 28.3 (p. 289)

1. 4.2 cm.　　**2.** 10 cm.　　**4.** 16.8 cm.　　**5.** 3 : 7.

Exercise 28.4a (p. 290)

1. 2⅓ cm, 2⅔ cm, 9 : 1.　**2.** 15 cm, 18 cm, 1 : 9.　**3.** (i) 3 : 4.　(ii) 3 : 4.
(iii) 9 : 16.
4. 324 : 625.　　　**5.** 4 : 25.　　　**9.** (i) 9 : 4.　(ii) 1 : 1.

Exercise 28.4b (p. 290)

1. 3.75 cm, 5 cm, 16 : 25.　　**3.** 5 cm, 7.5 cm, 10 cm, 4 : 25.
4. 1⅞ cm, 9 : 64, 25 : 64.　　**5.** 3 m.　**6.** 4 : 45.　**9.** 5 : 6.

Exercise 28.5a (p. 293)

1. 13 cm.　　**2.** 10 cm.　　**3.** 5 cm.　　**4.** 12 cm.　　**5.** Yes.
10. 2⅐ cm.

Exercise 28.5b (p. 293)

1. 12 cm.　**3.** 1 12/13, 11 1/13 cm.　**4.** No.　**7.** 10.3 cm.　**10.** √98 cm.

Exercise 29.1a (p. 296)

1. 8 cm.   **2.** 5 cm.   **3.** $7\frac{1}{24}$ cm.   **4.** 1 cm.   **5.** $2\sqrt{x^2 - y^2}$ cm.
8. $\dfrac{3\sqrt{3}r^2}{4}$.   **9.** $\dfrac{4\sqrt{3}}{3}$ cm.

Exercise 29.1b (p. 296)

1. 24 cm.   **2.** 3 cm.   **3.** $3\frac{1}{8}$ cm.   **4.** 17 cm.   **5.** $\sqrt{y^2 + \frac{1}{4}x^2}$ cm.
8. $\dfrac{x}{\sqrt{2}}$ cm.   **9.** $\dfrac{x}{\sqrt{3}}$ cm.

Exercise 29.2a (p. 299)

1. 15°.   **3.** 45°, 60°, 75°.   **5.** 55°, 75°, 105°, 125°.

Exercise 29.2b (p. 300)

1. 18°.   **2.** $x + 2y = 180°$.   **3.** 45°, 60°, 75°.
5. 45°, 85°, 95°, 135°.   **6.** $x + y + z = 180°$.   **7.** 70°.

Exercise 29.3 (p. 300)

1. $8\frac{1}{2}$ cm.   **7.** $2\sqrt{62}$ cm.   **8.** $2\sqrt{3}$ cm.   **13.** 2 cm.

Exercise 29.4a (p. 303)

3. 55°, 60°, 65°.   **5.** 60°.   **6.** 11°.   **7.** 24°.   **8.** 42°.   **9.** 22°.
10. 96°.

Exercise 29.4b (p. 304)

3. 66°, 63°, 51°.   **5.** $\dfrac{x + y}{2}$.   **6.** $x + \frac{1}{2}y - 90°$.   **7.** $y - \dfrac{x}{2}$.
8. $90° - 2x$.   **9.** $x - 90°$.   **10.** $360° - 2x - 4y$ or $(2x + 4y)$.

Exercise 29.5a (p. 306)

1. 12 cm.   **2.** $9\frac{3}{13}$ cm.   **4.** 12 cm.   **5.** 12 cm.   **6.** 5 cm.

Exercise 29.5b (p. 306)

1. 3 cm.   **2.** 4.8 cm.   **4.** $\sqrt{Rr}$.   **5.** $\sqrt{d_2 - (R - r)^2}$.
6. $\sqrt{d^2 - (R + r)^2}$.   **9.** 5 cm, 4 cm, 1 cm.

Exercise 29.6a (p. 308)

1. 14 cm.   **2.** 6 cm.   **3.** 2 cm.   **4.** 12.5 cm   **5.** 16 cm.   **6.** 4 cm.

Exercise 29.6b (p. 309)

1. 8 cm.   **2.** 2 cm.   **3.** 12 cm.   **4.** 8 cm.   **5.** 4 cm.   **6.** 4 cm.

Exercise 29.7 (p. 309)

14. 8 cm, 12 cm.   **15.** $6\frac{1}{2}$ cm.

Exercise 30 (p. 315)

8. Four.   **20.** 520 m.

Exercise 32.1b (p. 340)

1. Two, two, (one), two.   **3.** Four, six, eight.
4. One. Two.
5. (i), (ii), (v) and (vi), one axis of symmetry.   (iii) Four axes of symmetry.
(iv) One axis of symmetry; rotational symmetry of order five.

Exercise 33.1a (p. 347)

1. C and D.   **2.** (i) $\begin{pmatrix} 3 & -1 \\ 4 & -1 \end{pmatrix}$.   (ii) $\begin{pmatrix} 4 & 1 \\ 7 & 3 \end{pmatrix}$.

(iii) $\begin{pmatrix} -1 & 5 \\ 2 & -1 \end{pmatrix}$.   (iv) $\begin{pmatrix} 1 & -5 \\ -2 & -1 \end{pmatrix}$.   (v) $\begin{pmatrix} -2 & 3 \\ -1 & -3 \end{pmatrix}$.

(vi) $\begin{pmatrix} 3 & 6 \\ 9 & 0 \end{pmatrix}$.   (vii) $\begin{pmatrix} 4 & 1 \\ 7 & -1 \end{pmatrix}$.   (viii) $\begin{pmatrix} 12 & -11 \\ 11 & 3 \end{pmatrix}$.

(ix) $\begin{pmatrix} -1 & -9 \\ -8 & -13 \end{pmatrix}$.   **3.** $\begin{pmatrix} 3 & 3 & 2 & 0 & 3 & 2 & 3 \\ 1 & 0 & 1 & 2 & 0 & 3 & 0 \end{pmatrix}$.

4. $\begin{pmatrix} 0 & 1 & 1 & 0 & 0 \\ 1 & 0 & 1 & 0 & 1 \\ 1 & 1 & 0 & 1 & 1 \\ 0 & 0 & 1 & 0 & 1 \\ 0 & 1 & 1 & 1 & 0 \end{pmatrix}$.   **5.** $\begin{pmatrix} 0 & 1 & 1 & 0 & 0 \\ 1 & 0 & 1 & 0 & 1 \\ 1 & 1 & 0 & 1 & 1 \\ 0 & 0 & 0 & 0 & 1 \\ 0 & 0 & 1 & 1 & 0 \end{pmatrix}$.   **6.** $\begin{pmatrix} 1 & 2 & 0 \\ 1 & 2 & 2 \\ 1 & 0 & 1 \\ 0 & 1 & 3 \end{pmatrix}$.

7. $\begin{pmatrix} 3 & 4 & 2 \\ 0.3 & 0.5 & 1 \\ 0.2 & 0.1 & 0.1 \end{pmatrix}$.   **8.** $\begin{pmatrix} 50 & 100 & 50 \\ 150 & 50 & 50 \\ 100 & 10 & 70 \\ 10 & 10 & 10 \end{pmatrix}$.

Exercise 33.1b (p. 348)

1. B, C and E, D and F.   **2.** (i) $\begin{pmatrix} 1 & 1 & 2 \\ 3 & 3 & 3 \end{pmatrix}$.   (ii) $\begin{pmatrix} -1 & -1 & 2 \\ -2 & -1 & 3 \end{pmatrix}$.

(iii) $\begin{pmatrix} 1 & 2 & 1 & 2 \\ 2 & 3 & 2 & 3 \end{pmatrix}$.   (iv) $\begin{pmatrix} 4 & 8 \\ 8 & 12 \end{pmatrix}$.   (vi) $\begin{pmatrix} -2 & -4 & 1 & 2 \\ -4 & -6 & 2 & 3 \end{pmatrix}$.

(vii) $\begin{pmatrix} \frac{1}{2} & 1 \\ 1 & 1\frac{1}{2} \end{pmatrix}$.   (viii) $\begin{pmatrix} \frac{1}{2} & -1 & 0 \\ 2 & -\frac{1}{2} & 0 \end{pmatrix}$.   (ix) $\begin{pmatrix} 2 & 3 & 2 \\ 5 & 6 & 3 \end{pmatrix}$.

3.
$$\begin{pmatrix} 0 & 30 & 0 & 0 & 30 & 0 \\ 20 & 20 & 0 & 20 & 40 & 0 \\ 0 & 10 & 100 & 80 & 60 & 100 \\ 10 & 25 & 0 & 60 & 80 & 0 \end{pmatrix}.$$

4.
$$\begin{pmatrix} 1 & 1 & 0 & 0 \\ 1 & 1 & 0 & 0 \\ 1 & 1 & 0 & 0 \\ 1 & 0 & 1 & 0 \\ 1 & 1 & 0 & 0 \end{pmatrix}.$$

$$\begin{pmatrix} 12 & 6 & 5 & 1 \\ 12 & 7 & 3 & 2 \\ 11 & 7 & 3 & 1 \\ 12 & 6 & 3 & 3 \\ 11 & 6 & 3 & 2 \end{pmatrix}.$$

5. 2, 0, −4, 6.

6.
$$\begin{pmatrix} 450 & 0 & 900 & 900 \\ 0 & 480 & 960 & 640 \\ 600 & 0 & 1000 & 1000 \\ 360 & 0 & 1350 & 1080 \\ 0 & 560 & 800 & 800 \end{pmatrix}.$$

Exercise 33.2a (p. 349)

1. A.B, A.E, B.A, B.C, B.D, C.A, C.D.

2. $\begin{pmatrix} 1 & 2 \\ 5 & 8 \end{pmatrix}$, $\begin{pmatrix} 4 & 3 \\ 12 & 5 \end{pmatrix}$, $\begin{pmatrix} 4 & -1 \\ 8 & 0 \end{pmatrix}$, $\begin{pmatrix} 2 & -1 \\ 4 & 2 \end{pmatrix}$.

3. $\begin{pmatrix} 1 & 0 \\ 4 & 1 \end{pmatrix}$, $\begin{pmatrix} 7 & 10 \\ 15 & 22 \end{pmatrix}$, $\begin{pmatrix} 16 & -6 \\ 0 & 4 \end{pmatrix}$, $\begin{pmatrix} 3 & 1 \\ 6 & 1 \end{pmatrix}$.

4. (-2), (18), $\begin{pmatrix} 3 \\ 20 \end{pmatrix}$, $(14 \quad 6)$.

5. $(8 \quad 15)$, $(39 \quad 24 \quad 31 \quad 30 \quad 24 \quad 61 \quad 24)$, $\begin{pmatrix} 1 \\ 1 \\ 1 \\ 1 \\ 1 \\ 1 \\ 1 \end{pmatrix}$, $(2 \quad 2 \quad 3)$.

6. £4.50, £10, £15.50, £149.

7. $(0.12 \quad 0.8 \quad 0.7 \quad 0.2)$.

8. $(6.5 \quad 18.15 \quad 22 \quad 35.6 \quad 46 \quad 22)$.

9. $\begin{pmatrix} 15 \\ 15 \\ 15 \\ 14 \\ 13 \end{pmatrix}$, $\begin{pmatrix} 17 \\ 17 \\ 17 \\ 15 \\ 15 \end{pmatrix}$.

10. $\begin{pmatrix} 2 \\ 3 \\ \frac{1}{2} \\ 1 \end{pmatrix}$.

Exercise 33.3 (p. 357)

1. $\begin{pmatrix} 1 & -1 \\ -2 & 3 \end{pmatrix}$, $\begin{pmatrix} \frac{5}{2} & -\frac{3}{2} \\ -\frac{1}{2} & \frac{1}{2} \end{pmatrix}$, $\begin{pmatrix} \frac{3}{14} & \frac{1}{7} \\ -\frac{1}{14} & \frac{2}{7} \end{pmatrix}$, $\begin{pmatrix} 1 & 0 \\ 0 & 1 \end{pmatrix}$.

2. Only (i) has an inverse. **3.** $(3 \quad 1)$, $\begin{pmatrix} 3 & 1 \\ 2 & 4 \end{pmatrix}$, $\begin{pmatrix} 3 \\ 2 \end{pmatrix}$, $\begin{pmatrix} 3 & 4 \\ 2 & 5 \\ 1 & 6 \end{pmatrix}$.

4. (iii), (iv). **5.** (i) 0, 1, (ii) −0.2, 0.6. (iii) −0.2, −0.4.

Exercise 33.4 (p. 357)

1. (i) 1, 3. (ii) 5, 4. (iii) 1, 2. (iv) 7, 4. (v) 2, 4. (vi) 1, -2. (vii) $-\frac{1}{2}$, 2. (viii) 3, -8.

2. $\begin{pmatrix} 9 & 0 \\ 0 & 4 \end{pmatrix}$, $\begin{pmatrix} 27 & 0 \\ 0 & 8 \end{pmatrix}$, $\begin{pmatrix} \frac{1}{3} & 0 \\ 0 & \frac{1}{2} \end{pmatrix}$, $\begin{pmatrix} 3^{10} & 0 \\ 0 & 2^{10} \end{pmatrix}$.

3. $\mathbf{B}^{-1}\mathbf{AB} = \begin{pmatrix} 12 & 0 \\ 0 & 4 \end{pmatrix}$, $\mathbf{A}^4 = \begin{pmatrix} 5376 & 10240 \\ 7680 & 15616 \end{pmatrix}$.

4. $f_5 = 5$.　　　　**5.** (iv) $\mathbf{A}^4\mathbf{X} = \begin{pmatrix} 392 \\ 184 \\ 28 \end{pmatrix}$, $\mathbf{B}^4\mathbf{X} = \begin{pmatrix} 176 \\ 50 \\ 20 \end{pmatrix}$.

7. C, SF, W, M, SH; SF, W, C, SH, M.

8. (i) $V_1 = V_2 + 2I_2$, $I_1 = I_2$　(ii) $V_1 = \frac{5}{3}V_2 + 2I_2$, $I_1 = \frac{1}{3}V_2 + I_2$.
(iii) $V_1 = \frac{5}{3}V_2 + \frac{16}{3}I_2$, $I_1 = \frac{1}{3}V_2 + \frac{5}{3}I_2$.　　(iv) $V_1 = \frac{31}{9}V_2 + \frac{16}{3}I_2$,
$I_1 = \frac{8}{9}V_2 + \frac{5}{3}I_2$.　　(v) $V_1 = \frac{203}{27}V_2 + \frac{110}{9}I_2$, $I_1 = \frac{55}{27}V_2 + \frac{31}{9}I_2$.
(vi) $V_2 = V_1 - 2I_1$, $I_2 = I_1$. (vii) $V_2 = V_1 - 2I_1$, $I_2 = -\frac{1}{3}V_1 + \frac{5}{3}I_1$.
(viii) $V_2 = \frac{1}{3}V_1 - \frac{5}{3}I_1$, $I_2 = \frac{5}{3}V_1 - \frac{16}{3}I_1$.　　　(ix) $V_2 = \frac{5}{3}V_1 - \frac{16}{3}I_1$,
$I_2 = -\frac{8}{9}V_1 + \frac{31}{9}I_1$. (x) $V_2 = \frac{31}{9}V_1 - \frac{110}{9}I_1$, $I_2 = -\frac{55}{27}V_1 + \frac{203}{27}I_1$.

Exercise 34 (p. 367)

1. (i) Reflection in x-axis. (ii) Enlargement factor 2. (iii) Reduction factor $\frac{1}{2}$.
(iv) Reflection in x-axis and enlargement factor 2. (v) Reflection in
$x + y = 0$. (vi) Rotation through $+90°$. (vii) Reflection in $y = x$ and
enlargement factor 3. (viii), (ix), (xi) Shear parallel to x-axis. (x), (xii)
Shear parallel to y-axis.

2. (i) $\begin{pmatrix} 2 & 0 \\ 0 & 2 \end{pmatrix}$.　　(ii) $\begin{pmatrix} 2 & 0 \\ 0 & 1 \end{pmatrix}$.　　(iii) $\begin{pmatrix} 1 & 0 \\ 0 & -1 \end{pmatrix}$.　　(iv) $\begin{pmatrix} 0 & -1 \\ 1 & 0 \end{pmatrix}$.

(v) $\begin{pmatrix} 3 & 0 \\ 0 & -3 \end{pmatrix}$　(vi) $\begin{pmatrix} 4 & 1 \\ 1 & 1 \end{pmatrix}$.　(vii) $\begin{pmatrix} 1 & -2 \\ 0 & 1 \end{pmatrix}$.　(viii) $\begin{pmatrix} 0 & 1 \\ \frac{1}{2} & 1 \end{pmatrix}$.

3. (i) $\begin{pmatrix} 0 & 4 & 4 & 0 \\ 0 & 0 & 4 & 4 \end{pmatrix}$, $\begin{pmatrix} \frac{1}{4} & 0 \\ 0 & \frac{1}{4} \end{pmatrix}$.　　(ii) $\begin{pmatrix} 0 & \frac{1}{8} & \frac{1}{8} & 0 \\ 0 & 0 & \frac{1}{8} & \frac{1}{8} \end{pmatrix}$, $\begin{pmatrix} 8 & 0 \\ 0 & 8 \end{pmatrix}$.

(iii) $\begin{pmatrix} 0 & 0 & 1 & 1 \\ 0 & -1 & -1 & 0 \end{pmatrix}$, $\begin{pmatrix} 0 & -1 \\ 1 & 0 \end{pmatrix}$.

(iv) $\begin{pmatrix} 0 & -1 & -1 & 0 \\ 0 & 0 & -1 & -1 \end{pmatrix}$, $\begin{pmatrix} -1 & 0 \\ 0 & -1 \end{pmatrix}$.

(v) $\begin{pmatrix} 0 & 1 & 1 & 0 \\ 0 & 0 & -1 & -1 \end{pmatrix}$, $\begin{pmatrix} -1 & 0 \\ 0 & 1 \end{pmatrix}$ (vi) $\begin{pmatrix} 0 & 0 & -1 & -1 \\ 0 & -1 & -1 & 0 \end{pmatrix}$, $\begin{pmatrix} 0 & 1 \\ 1 & 0 \end{pmatrix}$.

(vii) $\begin{pmatrix} 0 & 2 & 3 & 1 \\ 0 & -1 & -4 & -3 \end{pmatrix}$, $\begin{pmatrix} \frac{3}{5} & \frac{1}{5} \\ -\frac{1}{5} & -\frac{2}{5} \end{pmatrix}$. (viii) $\begin{pmatrix} 0 & 3 & 3 & 0 \\ 0 & 0 & 2 & 2 \end{pmatrix}$, $\begin{pmatrix} \frac{1}{3} & 0 \\ 0 & \frac{1}{2} \end{pmatrix}$.

(ix) $\begin{pmatrix} 0 & 4 & 6 & 2 \\ 0 & 1 & 3 & 2 \end{pmatrix}$, $\begin{pmatrix} \frac{1}{3} & -\frac{1}{3} \\ -\frac{1}{6} & \frac{2}{3} \end{pmatrix}$. (x) $\begin{pmatrix} 0 & 3 & 1 & -2 \\ 0 & 1 & 2 & 1 \end{pmatrix}$, $\begin{pmatrix} \frac{1}{5} & \frac{2}{5} \\ -\frac{1}{5} & \frac{3}{5} \end{pmatrix}$.

(xi) $\begin{pmatrix} 0 & 2 & 2\frac{1}{2} & \frac{1}{2} \\ 0 & 4 & 3 & -1 \end{pmatrix} \begin{pmatrix} \frac{1}{4} & \frac{1}{8} \\ 1 & -\frac{1}{2} \end{pmatrix}.$

(xii) $\begin{pmatrix} 0 & -1 & -2 & -1 \\ 0 & 1 & 0 & -1 \end{pmatrix} \begin{pmatrix} -\frac{1}{2} & \frac{1}{2} \\ -\frac{1}{2} & -\frac{1}{2} \end{pmatrix}.$

4. (i) $\begin{pmatrix} 0 & 1 & 3 & 2 \\ 0 & 2 & 6 & 4 \end{pmatrix}.$ (ii) $\begin{pmatrix} 0 & -1 & 1 & 2 \\ 0 & -2 & 2 & 4 \end{pmatrix}.$

(iii) $\begin{pmatrix} 0 & 1 & -1 & -2 \\ 0 & 2 & -2 & -4 \end{pmatrix}.$ (iv) $\begin{pmatrix} 0 & 1 & 1\frac{1}{2} & \frac{1}{2} \\ 0 & 2 & 3 & 1 \end{pmatrix}.$

(v) $\begin{pmatrix} 0 & 1 & 3 & 2 \\ 0 & -1 & -3 & -2 \end{pmatrix}.$ (vi) $\begin{pmatrix} 0 & 1 & 2 & 1 \\ 0 & 0 & 0 & 0 \end{pmatrix}.$

5. (i) 2. (ii) 0. (iii) 1. (iv) 8. (v) 1. (vi) 2.

6. (i) $\begin{pmatrix} 3 \\ -1 \end{pmatrix}.$ (ii) $\begin{pmatrix} -1 \\ 2 \end{pmatrix}.$ (iii) $\begin{pmatrix} 3 \\ 0 \end{pmatrix}.$ (iv) $\begin{pmatrix} -2 \\ -1 \end{pmatrix}.$

7. $\begin{pmatrix} 0 & 2 & 2 & 0 \\ 0 & 0 & 1 & 1 \end{pmatrix}.$ 8. $\begin{pmatrix} 3 & 1 \\ 0 & 1 \end{pmatrix}, \begin{pmatrix} \frac{1}{3} & -\frac{1}{3} \\ 0 & 1 \end{pmatrix}.$

9. $\begin{pmatrix} 0 & 3 & 4 & 1 \\ 0 & 0 & 3 & 3 \end{pmatrix}, \begin{pmatrix} 0 & \frac{1}{3} & \frac{2}{9} & -\frac{1}{9} \\ 0 & 0 & \frac{1}{3} & \frac{1}{3} \end{pmatrix}.$ Enlargement by factors 3, $(\frac{1}{3})$ and shears parallel to x-axis.

Exercise 35.1a (p. 374)

1. (i) $\begin{pmatrix} 2 \\ 3 \end{pmatrix}, \begin{pmatrix} 4 \\ -2 \end{pmatrix}, \begin{pmatrix} 4 \\ 3 \end{pmatrix}.$ (ii) $\begin{pmatrix} -2 \\ -3 \end{pmatrix}, \begin{pmatrix} 2 \\ -5 \end{pmatrix}, \begin{pmatrix} 2 \\ 0 \end{pmatrix}.$

2. $\begin{pmatrix} 2 \\ -3 \end{pmatrix}, \begin{pmatrix} -1 \\ -2 \end{pmatrix}, \begin{pmatrix} -2 \\ -4 \end{pmatrix}, \begin{pmatrix} 3 \\ -1 \end{pmatrix}, \begin{pmatrix} 1 \\ 2 \end{pmatrix}; \overrightarrow{BA} = -\overrightarrow{AB}; \overrightarrow{BC} = -\overrightarrow{CB}.$

3. $\begin{pmatrix} 1 \\ 1 \end{pmatrix}, \begin{pmatrix} -1 \\ -1 \end{pmatrix}, \begin{pmatrix} 1 \\ 3 \end{pmatrix}, \begin{pmatrix} 1 \\ 3 \end{pmatrix}.$

4. (i) $\begin{pmatrix} 5 \\ 5 \end{pmatrix}.$ (ii) $\begin{pmatrix} 1 \\ 7 \end{pmatrix}.$ (iii) $\begin{pmatrix} 4 \\ 9 \end{pmatrix}.$ (iv) $\begin{pmatrix} 4 \\ 9 \end{pmatrix}.$

(v) $\begin{pmatrix} 1 \\ -1 \end{pmatrix}.$ (vi) $\begin{pmatrix} 6 \\ 4 \end{pmatrix}.$ (vii) $\begin{pmatrix} 3\frac{1}{2} \\ 4 \end{pmatrix}.$ (viii) $\begin{pmatrix} 4 \\ 17 \end{pmatrix}.$

(ix) $\begin{pmatrix} 9 \\ 8 \end{pmatrix}.$ (x) $\begin{pmatrix} 3 \\ -9 \end{pmatrix}.$ (xi) $\begin{pmatrix} 6 \\ -4 \end{pmatrix}.$ (xii) $\begin{pmatrix} 2\frac{1}{2} \\ -3 \end{pmatrix}$

5. (i) $\sqrt{2}$, 45°. (ii) 2, 180°. (iii) 2, 90°. (iv) $\sqrt{2}$, 135°. (v) $\sqrt{5}$, −153°, (vi) $\sqrt{5}$, 117°. (vii) $\sqrt{5}$, 27°. (viii) $\sqrt{10}$, −18°. (ix) $\sqrt{10}$, 72°.

6. 3.   7. 3, 2.   8. 2, 3.   10. $\begin{pmatrix} 2 \\ 1 \end{pmatrix}.$

Exercise 35.1b (p. 375)

1. $\begin{pmatrix} 0 \\ -3 \end{pmatrix}, \begin{pmatrix} -1 \\ -5 \end{pmatrix}, \begin{pmatrix} -3 \\ -5 \end{pmatrix}.$

2. $\begin{pmatrix} -3 \\ -3 \end{pmatrix}, \begin{pmatrix} 3 \\ 3 \end{pmatrix}, \begin{pmatrix} -1 \\ -1 \end{pmatrix}, \begin{pmatrix} 1 \\ -1 \end{pmatrix}.$ $\overrightarrow{AB} = -\overrightarrow{BA}, \overrightarrow{AB} = 3\overrightarrow{CD}.$

3. $\begin{pmatrix} -4 \\ 1 \end{pmatrix}, \begin{pmatrix} -4 \\ 3 \end{pmatrix}, \begin{pmatrix} -6 \\ 1 \end{pmatrix}, \begin{pmatrix} 2 \\ -3 \end{pmatrix}.$

4. (i) $\begin{pmatrix} 3 \\ 5 \end{pmatrix}.$ (ii) $\begin{pmatrix} 0 \\ -4 \end{pmatrix}.$ (iii) $\begin{pmatrix} 0 \\ 1\frac{1}{2} \end{pmatrix}.$ (iv) $\begin{pmatrix} -6 \\ 0 \end{pmatrix}.$

(v) $\begin{pmatrix} 2 \\ 4 \end{pmatrix}.$ (vi) $\begin{pmatrix} -\frac{2}{3} \\ 0 \end{pmatrix}.$ (vii) $\begin{pmatrix} -3 \\ -5 \end{pmatrix}.$ (viii) $\begin{pmatrix} 0 \\ 4 \end{pmatrix}.$

(ix) $\begin{pmatrix} \frac{1}{2} \\ -1 \end{pmatrix}.$ (x) $\begin{pmatrix} 0 \\ -\frac{1}{3} \end{pmatrix}.$ (xi) $\begin{pmatrix} 5 \\ 5 \end{pmatrix}.$ (xii) $\begin{pmatrix} -3 \\ -1 \end{pmatrix}.$

5. (i) $\sqrt{2}, -45°.$ (ii) $\sqrt{8}, -135°.$ (iii) $1, 180°.$ (iv) $1, -90°.$ (v) $\sqrt{5}, -27°.$ (vi) $\sqrt{20}, -27°.$ (vii) $\sqrt{8}, -45°.$ (viii) $\frac{1}{2}\sqrt{2}, 45°.$ (ix) $\frac{1}{2}\sqrt{5}, 27°.$ (v) and (vi), (i) and (vii), (ii) and (viii).

6. 2.   7. 2, −1.   9. 90°.

Exercise 35.2 (p. 378)

1. $\begin{pmatrix} 1 \\ 3 \end{pmatrix}, \begin{pmatrix} 5 \\ 1 \end{pmatrix}.$ 2. $\begin{pmatrix} 5 \\ 5 \end{pmatrix}, \begin{pmatrix} 0 \\ 10 \end{pmatrix}.$

3. $\begin{pmatrix} 1\frac{1}{2} \\ -\frac{1}{2} \end{pmatrix}, \begin{pmatrix} \frac{1}{2} \\ -2\frac{1}{2} \end{pmatrix}.$ 4. $\begin{pmatrix} 6.2 \\ -3.4 \end{pmatrix}, \begin{pmatrix} 9.6 \\ 2.8 \end{pmatrix}.$ 5. $\begin{pmatrix} 1 \\ 7 \end{pmatrix}, \begin{pmatrix} -6 \\ 8 \end{pmatrix}.$

Exercise 35.3 (p. 378)

1. 1.85, 0.77, 2.8, 3.8.   2. 1.9, 1.2, 1.2, 4.8.   3. 60°.   4. 120°.

5. 1.88, 3.76, 0.68.   6. (iii) $\begin{pmatrix} 4 \\ -6 \end{pmatrix}.$ 2 : 3.   7. 25 s.   8. 22 s.

9. $\begin{pmatrix} 200 \\ 170 \end{pmatrix}$ 10. $\begin{pmatrix} 25 \\ -6 \end{pmatrix}.$ 11. 66°.   12. At 78° to bank, 102 s.

Exercise 36 (p. 382)

1. (i) $\begin{pmatrix} 0 \\ 4 \end{pmatrix}.$ (ii) $\begin{pmatrix} 4 \\ 4 \end{pmatrix}.$ 5. (i) $a - b = \lambda(c - d).$ (ii) $a - b = c - d.$

(ii) $a - d = \lambda(b - c).$ (iv) $c - b = \frac{1}{2}(c + d) - a.$

7. $a, 2a + b$ and $5a + 4b.$ 8. $\overrightarrow{AG} = 2(a + b), \overrightarrow{OG} = 5a + 2b.$

9. 3 : 8.

Exercise 37.1a (p. 391)

1. $\frac{3}{5}, \frac{3}{5}, \frac{3}{4}, \frac{3}{4}, \frac{5}{4}, \frac{5}{4}$. **2.** $\frac{5}{12}, \frac{12}{13}, \frac{12}{5}, \frac{13}{12}$. **3.** $\frac{4}{5}, \frac{7}{4}$.

4. $\frac{3}{5}, \frac{3}{5}$. **5.** $\frac{1}{2}, \sqrt{3}$.

Exercise 37.1b (p. 392)

1. $\frac{8}{17}, \frac{8}{17}, \frac{8}{15}, \frac{8}{15}$. **2.** $\frac{3}{5}, \frac{4}{5}$. **3.** $\frac{5}{13}, \frac{12}{7}$. **4.** $\frac{8}{5}, \frac{8}{5}$. **5.** $1, \dfrac{1}{\sqrt{2}}$.

Exercise 37.2a (p. 394)

1. 0.3420, 0.3420, 1.1918, 0.2345, 2.002, 4.0121, 3.0031, 3.9969, 0.4107, 0.8456.

2. 30°, 60°, 39° 50′, 38°, 60°, 26° 34′, 19° 28′, 48° 11′, 45°, 63° 26′.

Exercise 37.2b (p. 394)

1. 0.4226, 1.4945, 0.3249, 0.9286, 0.306, 4.975, 4.0029, 4.0031, 2.6640, 1.1547.

2. 60°, 30°, 71° 34′, 30°, 60°, 24° 29′, 11° 32′, 53° 8′, 18° 26′, 45°.

Exercise 37.3a (p. 395)

1. 0.8. **2.** $1\frac{3}{5}$. **3.** $\frac{8}{17}$. **4.** $\frac{20}{21}$. **5.** $\dfrac{\sqrt{m^2 + n^2}}{m}$.

Exercise 37.3b (p. 395)

1. 0.6. **2.** $1\frac{2}{5}$. **3.** $\frac{8}{15}$. **4.** $\frac{29}{21}$. **5.** $\dfrac{q}{\sqrt{p^2 - q^2}}$.

Exercise 37.4a (p. 399)

1. 1. **2.** 1. **3.** 1. **4.** 4. **5.** 1. **6.** 1. **7.** 1. **8.** $\frac{4}{3}$. **9.** $\frac{1}{2}$. **10.** $\frac{1}{3}$.

Exercise 37.4b (p. 399)

1. 1. **2.** 1. **3.** 1. **4.** $1\frac{3}{4}$. **5.** 1. **6.** 1. **7.** 1. **8.** 4. **9.** $\frac{3}{4}$. **10.** $\frac{3}{4}$.

Exercise 37.5 (p. 399)

3. 2. **4.** $\frac{3}{5}$. **5.** $\frac{4}{5}, \frac{4}{5}, \frac{5}{3}, \frac{5}{3}$. **6.** 1, 1, 1.

7. $\dfrac{p}{\sqrt{q^2 - p^2}}$. **8.** $\frac{8}{17}, \frac{15}{17}, 1$. **9.** 66, 72, 39.

10. $\dfrac{\sqrt{b^2 - \frac{1}{4}a^2}}{b}, \dfrac{\sqrt{4b^2 - a^2}}{a}$. **11.** $\dfrac{h}{c}, \dfrac{\sqrt{c^2 - h^2}}{h}, \dfrac{a}{\sqrt{a^2 - h^2}}$.

12. (i), (iv), (v).

Exercise 38.1a (p. 403)

1. 14.3 cm, 4.42 cm.
2. 4.15 cm, 2.21 cm.
3. 23.5 cm, 22.0 cm.
4. 10.5 m, 5.73 m.
5. 13.0 cm, 26.1 cm.

Exercise 38.1b (p. 403)

1. 44.6 cm, 41.3 cm.
2. 75.7 cm, 72.0 cm.
3. 6.01 cm, 15.3 cm.
4. 4.73 m, 4.01 m.
5. 6.88 cm, 10.2 cm.

Exercise 38.2a (p. 404)

1. 44° 25′, 45° 35′.
2. 36° 52′, 53° 8′.
3. 70° 17′, 19° 43′.
4. 64° 18′, 25° 42′.
5. 29° 38′, 60° 22′.

Exercise 38.2b (p. 404)

1. 41° 49′, 48° 11′.
2. 62° 15′, 27° 45′.
3. 68° 4′, 21° 56′.
4. 28° 26′, 61° 34′.
5. 80° 26′, 9° 34′.

Exercise 38.3a (p. 411)

1. 69° 52′.
2. 8.32 cm.
3. 6.12 cm.
4. 24° 14′.
5. 9.3 m.
6. 376 m.
7. 8°.
8. $n = 4.13$.
9. 36.7 m.
10. 75 m.
11. 153° 26′.
12. 4.36 km, 036° 35′.

Exercise 38.3b (p. 412)

1. 73° 58′.
2. 10.5 cm.
3. 4.06 cm.
4. 56° 19′.
5. 22.4 m.
6. 308 m.
7. 9°.
8. 3.4.
9. 69.1 m.
10. 30.8 m.
11. 207°.
12. 319° 8′.

Exercise 38.4 (p. 412)

1. 12.3 m.
2. 2.56 cm.
3. 53° 8′.
4. 41° 11′.
5. 11 cm.
6. 11 cm.
7. 8° 8′.
8. 8.66 cm.
9. 555 m.
10. 73° 44′.
11. 10.2 m.
12. 14° 53′.
13. 0.37 km.
14. 41° 24′.
15. 38° 56′.
16. 77° 53′.
17. 34.6 m.
18. 20° 22′.
19. 073° 19′.
20. 5.35 m.
21. 51.9 m.
22. 119° 29′.
23. 18° 26′.
24. 4.1 cm, 5.6 cm.
25. 52.6 cm².

Exercise 39.1a (p. 416)

1. −ve.
2. +ve.
3. −ve.
4. −ve.
5. −ve.
6. −ve.
7. −ve.
8. −ve.
9. −ve.
10. −ve.

Exercise 39.1b (p. 416)

1. +ve.
2. −ve.
3. −ve.
4. −ve.
5. +ve.
6. −ve.
7. +ve.
8. +ve.
9. −ve.
10. −ve.

Exercise 39.2a (p. 418)

1. −0.1736.
2. +0.9397.
3. −1.732.
4. −0.9848.
5. −0.3420.
6. 0.5774.
7. 5.671.
8. −5.671.
9. −0.9397.
10. +0.5.
11. −2.0.
12. 1.5557.
13. −1.732.
14. −2.9238.
15. −1.1547.
16. 1.1918.
17. −1.0154.
18. 3.8637.
19. −0.4663.
20. −0.3420.
21. −0.6428.
22. 2.0.
23. −1.0154.
24. −11.43.

Exercise 39.2b (p. 418)

1. −0.3420.
2. 0.866.
3. −1.1918.
4. −0.9397.
5. −0.6428.
6. 0.8391.
7. −2.7475.
8. 0.
9. −0.866.
10. 0.6428.
11. −1.5557.
12. 2.0.
13. −2.7475.
14. −2.0.
15. −1.3054.
16. 0.8391.
17. −1.0642.
18. 2.0.
19. −0.8391.
20. −5.671.
21. −0.9848.
22. −1.0154.
23. −1.0154.
24. 2.7475.

Exercise 39.3a (p. 419)

1. $-\sin x$.
2. $-\sin x$.
3. $\cot x$.
4. $-\cot x$.
5. $\sec x$.
6. $\operatorname{cosec} x$.
7. $-\operatorname{cosec} x$.
8. $-\cot x$.
9. $-\tan x$.
10. $-\sec 2x$.

Exercise 39.3b (p. 419)

1. $-\cos x$.
2. $\cos x$.
3. $-\sin x$.
4. $-\tan x$.
5. $-\sin x$.
6. $-\sec x$.
7. $-\sec x$.
8. $-\cot x$.
9. $\sec x$.
10. $\cot 2x$.

Exercise 39.4 (p. 421)

3. 20.
4. 20.
5. 113° 35′, 246° 25′.
6. 203° 35′, 336 25′.
7. $30 < x < 150$.
8. $60 < x < 300$.
9. 44° 25′, 135° 35′.
10. 126° 52′, 233° 8′.
11. 63° 26′, 243° 26′.
12. 154° 32′, 334° 32′.
13. 158° 12′, 338° 12′.
14. 120°, 240°.
15. $\sec x$.
16. $-\operatorname{cosec} x$.
17. $\sec x$.
18. $-\cos x$.
19. $-\tan x$.
20. $-\tan x$.
21. 126° 52′.
22. 233° 8′.
23. 306° 52′.
24. 126° 52′.
25. 306° 52′.
26. 233° 8′.

27. $\sin A$, $-\cos A$, $-\tan A$, $\cos \dfrac{A}{2}$, $\cos 2A$.

28. $180° < \theta < 225°$ or $315° < \theta < 360°$.
29. 197° 28′, 342° 32′; 101° 32′, 258° 28′; 50° 12′, 230° 12′.
30. $-\frac{1}{4}\sqrt{7}$.

Exercise 40 (p. 426)

9. (i) 3, 2.85, 2.65, 2.4. (ii) 26. (iii) 18.
10. 9.5
11. (i) 11. (ii) 33.
12. 32.
13. (i) 3.25, 3.564, 2.83, 2.63. (ii) 86.
14. 63.
15. 13.

Exercise 41.1a (p. 434)

1. 42° 4′.   **2.** 21° 48′.   **3.** 43° 36′.   **4.** 84° 4′.   **5.** 21° 48′.
6. 33° 42′.   **7.** 26° 34′.   **8.** 33° 42′.   **9.** 26° 34′.   **10.** 39° 48′.

Exercise 41.1b (p. 435)

1. 90°.   **2.** 45°.   **3.** 90°.   **4.** 43° 18′.   **5.** 45°.
6. 30° 58′.   **7.** 36° 52′.   **8.** 36° 52′.   **9.** 45°.   **10.** 51° 20′.

Exercise 41.2 (p. 435)

1. 46° 12′, 38° 13′, 18°.
2. 20° 42′.   **3.** 10.6 m.   **4.** 61° 53′.
5. 69° 18′.   **6.** 41° 24′.   **7.** 5.54 m.
8. 67° 22′.   **9.** 78° 13′.   **10.** 25° 40′.
11. 1 in 8.5.   **12.** 28° 24′.   **13.** 56° 26′.
14. 3.57 m, 63° 12′, 67° 13′.
15. 10.58 cm, 61° 53′, 69° 18′, 56° 14′, 41° 24′.
16. 5.75 m; 73° 13′, 81° 12′, 62° 10′.   **17.** 16 m.
18. $\sin^{-1}(\sin \alpha \cos \beta)$.   **19.** 46° 9′, 48° 11′.   **20.** 20° 42′.

Exercise 42.1a (p. 440)

1. 3.49 cm.   **2.** 0.075 cm.   **3.** 11.1 cm².   **4.** 15.3 cm².
5. 48 cm.   **6.** 4.9 cm.   **7.** 10.6 : 1.
8. 22° 1′. 157° 59′.   **9.** 2.04.   **10.** 25 cm².

Exercise 42.1b (p. 441)

1. 3.49 cm.   **2.** 0.11 cm.   **3.** 17.5 cm².   **4.** 25.6 cm².
5. 44.3 cm.   **6.** 4.27 cm.   **7.** 14.7 : 1.
8. 18° 12′, 161° 48′.   **9.** 81° 22′.   **10.** 2.5.

Exercise 42.2a (p. 443)

1. 2670 km.   **2.** 3190 km.   **3.** 2570 km.   **4.** 1440 km/h.
5. 9⅓ min.   **6.** 77°.   **7.** 6680 km.   **8.** 5570 km.
9. 1.732 : 1.   **10.** 2560 km.

Exercise 42.2b (p. 444)

1. 3560 km.   **2.** 5520 km.   **3.** 2040 km.   **4.** 639 km.
5. 1 hr 44 min.   **6.** 54′.   **7.** 4440 km.   **8.** 10 010 km.
9. 1.085 : 1.   **10.** 1580 km.

Exercise 42.3 (p. 444)

1. 9.16 cm.   **2.** 3.89 cm.   **3.** 1.4 cm.   **4.** 15.4 cm².
5. 14.3 cm.   **6.** 65 cm².   **7.** 283 cm².   **8.** 45.5 m.

9. 6° 22′.   **10.** 180°.   **11.** 41° 24′.   **12.** $2R \sin \theta$.
13. 3330 km.   **14.** 63° 17′.   **15.** 6300 km.
16. 12.9 cm, 69.8 cm².   **17.** 1278 km, 1274 km.
18. 6.1 m.   **19.** 45° 50′.   **20.** 1120 km.   **21.** 222 cm².
22. 3 cm².   **23.** 31.4 cm².   **24.** 19° 12′.   **25.** 3330 km.
26. 41.2 km, 18.5 km.   **27.** 5470 km.   **28.** 4230 km.
29. 112 cm².   **30.** 12 000 km.

Exercise 43.1a (p. 450)

1. $b = 4.00$, c $= 7.66$, $A = 88°$.   **2.** $A = 46°$, $a = 5.22$, $c = 5.39$.
3. $b = 11.6$, $A = 26°$, $a = 5.5$.   **4.** $b = 4.13$, $c = 3.79$, $A = 64°$.
5. $C = 56° \ 30'$, $a = 41.1$, $c = 34.7$.
6. Impossible.   **7.** $c = 9.92$, $C = 120° \ 39'$, $B = 21° \ 21'$.   **8.** Impossible.
9. $B = 32°$, $C = 122°$, $c = 8.32$ or $B = 148°$, $C = 6°$, $c = 1.03$.
10. $B = 63° \ 9'$, $C = 65° \ 51'$, $c = 3.17$ or $B = 116° \ 51'$, $C = 12° \ 9'$, $c = 0.73$.

Exercise 43.1b (p. 451)

1. $b = 14.5$, $c = 16.9$, $A = 87°$.   **2.** $A = 77°$, $c = 3.46$, $a = 8.28$.
3. $A = 47°$, $a = 19.2$, $b = 25.0$.
4. $A = 66° \ 30'$, $b = 14.3$, $c = 12.5$.
5. $a = 32.2$, $c = 20.2$, $C = 38° \ 9'$.
6. $B = 55° \ 56'$, $C = 82° \ 34'$, $c = 7.18$ or $B = 124° \ 4'$, $C = 14° \ 26'$, $c = 1.81$.
7. $B = 33° \ 15'$, $C = 34° \ 45'$, $c = 4.37$.   **8.** Impossible.
9. $B = 26° \ 39'$, $C = 131° \ 6'$, $c = 7.39$ or $B = 153° \ 21'$, $C = 4° \ 23'$, $c = 0.78$.
10. Impossible.

Exercise 43.2a (p. 453)

1. $c = 6.48$, $A = 32° \ 28'$, $B = 37° \ 32'$.
2. $c = 6.12$, $B = 9° \ 57'$, $A = 146° \ 3'$.
3. $a = 18.8$, $C = 41° \ 21'$, $B = 78° \ 17'$.
4. $b = 5.37$, $C = 31° \ 22'$, $A = 53° \ 10'$.
5. $b = 14.16$, $C = 19° \ 13'$, $A = 80° \ 47'$.
6. $A = 19° \ 50'$, $B = 96° \ 20'$, $C = 63° \ 50'$.
7. $A = 37° \ 46'$, $B = 71° \ 7'$, $C = 71° \ 7'$.
8. $A = 33° \ 48'$, $B = 29° \ 45'$, $C = 116° \ 27'$.
9. $A = 13° \ 10'$, $B = 90° \ 20'$, $C = 76° \ 30'$.
10. $A = 28° \ 23'$, $B = 78° \ 52'$, $C = 72° \ 45'$.

Exercise 43.2b (p. 454)

1. $c = 8.16$, $A = 23° \ 13'$, $B = 26° \ 47'$.
2. $c = 7.11$, $A = 56° \ 38'$, $B = 91° \ 22'$.
3. $C = 37° \ 8'$, $B = 94° \ 37'$, $a = 15.6$.
4. $b = 6.47$, $A = 53° \ 50'$, $C = 29° \ 27'$.

5. $C = 21° 32'$, $A = 88° 28'$, $b = 14.4$.
6. $B = 96° 30'$, $A = 26°$, $C = 57° 30'$.
7. $B = C = 71° 35'$, $A = 36° 50'$.
8. $A = 22° 24'$, $C = 132° 26'$, $B = 25° 10'$.
9. $A = 56°$, $B = 89° 6'$, $C = 34° 54'$.
10. $A = 53° 18'$, $B = 91° 4'$, $C = 35° 38'$.

Exercise 43.3a (p. 456)

1. 3.32.	2. 17.8.	3. 14.1.	4. 14.0.
5. 3.84.	6. 35.6.	7. 35.8.	8. 32.6.

Exercise 43.3b (p. 456)

1. 4.97.	2. 18.5.	3. 22.8.	4. 25.6.
5. 4.18.	6. 42.1.	7. 44.9.	8. 26.8.

Exercise 43.4 (p. 456)

1. 1.61 km.   2. 6.59 km.   3. 57.9 m.   4. 079° 43′ (or 280° 17′).
5. 31.6 m.   6. 178 m.   7. 7.   8. 108° 40′.
9. 2.78 km/h; 100° 9′.
10. 47.8.   11. 45° 40′; 21.9 m².
12. 47° 58′, 82° 2′, 18.4 m².   13. 546 m.
14. 49° 45′, 17 m².   15. 181 m.   16. 50° 33′.
17. 7.8 cm.   18. 87° 8′.   19. 4.07, 7.34.   20. 6.08.
21. 10.6 m.   22. 0.81 km.   23. 18.7 km/h.   24. 6.89 cm.
25. 6.61 cm.   26. 14 : 11 : −4.   27. 4.84 km.   28. 28.7 m.
29. 33° 45′.   30. 280 m.   31. 60.8 m.
32. 71° 22′, 108° 38′.

Exercise 44.1a (p. 466)

1. 6.	2. 3.	3. 3.	4. 5.
5. 8.	6. 3.	7. 7.	8. $4x$.
9. $6x + 1$.	10. $8x - 1$.		

Exercise 44.1b (p. 466)

1. 8.	2. 7.	3. 1.	4. 6.
5. 12.	6. 7.	7. 4.	8. $6x$.
9. $4x + 1$.	10. $12x - 2$.		

Exercise 44.2a (p. 468)

1. $8x - 6$.	2. $3x$.	3. $3 - 2x$.	4. $2x + \frac{3}{4}$.
5. $-6 - 4x$.	6. $8x - 4$.	7. $2x - 3$.	8. $4x + 11$.
9. $12x - 1$.	10. $8x + 3$.		

Exercise 44.2b (p. 468)

1. $4x - 8$.   **2.** x.   **3.** $2 - 2x$.   **4.** $2x + \frac{2}{3}$.
5. $-4 - 6x$.   **6.** $6x + 3$.   **7.** $2x$.   **8.** $4x - 11$.
9. $12x + 1$.   **10.** $2x$.

Exercise 44.3 (p. 471)

1. $9x^2$.

2. $\dfrac{2}{x^3}$.

3. $2x - \dfrac{2}{x^3}$.

4. $1 + 4x + 9x^2$.

5. $3x^2 + 4x + 1$.

6. $\dfrac{1}{2\sqrt{x}} - \dfrac{1}{2x^{\frac{3}{2}}}$.

7. $\dfrac{1}{2\sqrt{x}} + 1$.

8. $1 - \dfrac{1}{x^2}$.

9. $\dfrac{1}{x^2}$.

10. $-\dfrac{1}{x^2} - \dfrac{2}{x^3}$.

11. $3x^2 + 2x$.

12. $7 + 6x - 12x^2$.

13. $-\dfrac{1}{x^2} + \dfrac{2}{x^3}$.

14. $-\dfrac{1}{2x^{\frac{3}{2}}} + 1$.

15. $1 - \dfrac{1}{x^2}$.

16. $3y^2 - 1$.   **17.** $6t + 3t^2$.   **18.** $2z + 3$.
19. $3 - 2v + 21v^2$.   **20.** $3t^2 - 3$.   **21.** $6 + 4x + 9x^2$.

22. $-\dfrac{2}{x^3} - \dfrac{2}{x^2}$.

23. $1 + \dfrac{1}{2x^{\frac{3}{2}}}$.

24. $\dfrac{5}{2}x^{\frac{3}{2}}$.

25. $1 + 12x + 21x^2$.

26. $3x^2 + 2x + 1$.

27. $1 - \dfrac{1}{x^2}$.

28. $7x^6$.

29. $-\dfrac{7}{x^8}$.

30. $7x^6 - \dfrac{7}{x^8}$.

31. 4.   **32.** 13.   **33.** $1, 0$.   **34.** -1.

35. $-\dfrac{1}{x^2}$.

36. $3t^2 - 12t + 4$.

37. $(-1, 0)$.

38. $-\dfrac{8}{x^3}$.

39. -1.

40. $2x,\ 2\sqrt{y}$.

41. $1 + \dfrac{1}{z^2}$.

43. $-2\frac{1}{2}$.

44. $\dfrac{1}{1 + 2t}$.

45. $(0, 0)$ and $(\frac{2}{3}, -\frac{4}{27})$.

Exercise 45.1a (p. 478)

1. -9.   **2.** 2.   **3.** -4.   **4.** ± 2.
5. $-\frac{1}{4}$.   **6.** ± 2.   **7.** 144.   **8.** 9 m^2.
9. 12π cm^2.   **10.** $10, 10$.

Exercise 45.1b (p. 479)

1. -16.   **2.** 5.   **3.** -5.   **4.** ± 16.
5. $-\frac{1}{4}$.   **6.** ± 4.   **7.** k^2.   **8.** k^2 m^2.
9. 24π cm^2.   **10.** k, k.

Exercise 45.2a (p. 481)

2. 45 m/s, 24 m/s^2.

3. 16 m/s^2.

4. $3 + 2n$, 2.

5. 5 m/s^2, $\frac{1}{2}$.

Exercise 45.2b (p. 482)

2. 26 m/s, 18 m/s^2.

3. 29 m/s^2.

4. $4 + 4n$, 4.

5. 2 m/s^2, 2.

Exercise 45.3a (p. 483)

1. 1.2 cm^2/s.

2. 2.7 cm^3/s.

3. 3 cm^3/s.

4. 3.2π cm^2/s.

5. 7.2π cm^3/s.

Exercise 45.3b (p. 483)

1. 1.2 cm^2/s.

2. 2.4 cm^3/s.

3. $2\frac{1}{4}$ cm^3/s.

4. 6.4π cm^2/s.

5. 6.4π cm^3/s.

Exercise 45.4 (p. 483)

1. 7, (2, 1), $(\frac{2}{3}, \frac{59}{27})$.

2. 94.7 m^2.

3. 312.5 cm^2.

4. 1, min, $1\frac{4}{27}$, max.

5. $\frac{1024}{27}$ cm^3.

7. 0, max, -1, min.

8. 6 m/s^2; 3 s.

9. 108 m^2.

10. 1, max; 0, min.

11. $\frac{8}{\pi}$.

12. 7.2π cm^2/s.

13. $\dfrac{1}{200\pi}$ cm/s.

14. $\frac{1}{2}$.

15. 80, max, -28 min.

16. $-\dfrac{28}{v^{2.4}}$.

17. 0, -2 m/s^2.

18. (1, 2) and (2, 2).

19. -1, 2, 0.

20. 1 or $-\frac{1}{3}$.

Exercise 46.1a (p. 487)

(arbitrary constants omitted)

1. $\dfrac{x^4}{4}$.

2. $\dfrac{x^4}{4} + \dfrac{2x^3}{3}$.

3. $\dfrac{x^3}{3} + 6x$.

4. $-\dfrac{1}{x}$.

5. $a\dfrac{x^3}{3} + b\dfrac{x^2}{2} + cx$.

6. $\dfrac{x^3}{3} - \dfrac{3x^2}{2} + 2x$.

7. $-\dfrac{1}{2x^2}$.

8. $\dfrac{x^6}{6}$.

9. $y = 2x + x^2 - \dfrac{x^3}{3} - \dfrac{5}{3}$.

10. $y = 3x^2 - 2x + 2$.

Exercise 46.1b (p. 487)

1. $\frac{2}{3}x^{\frac{3}{2}}$.

2. $\frac{x^3}{3} + \frac{3x^2}{2} + 4x$.

3. $\frac{x^3}{3} - 2x^2$.

4. $-\frac{1}{2x^2} - \frac{1}{x}$.

5. $-\frac{a}{x} + bx$.

6. $\frac{x^3}{3} + \frac{x^2}{2}$.

7. $\frac{x^4}{4} + \frac{2x^3}{3}$.

8. $\frac{x^3}{3} - \frac{1}{x}$.

9. $y = 3 + x + x^2 - x^3$.

10. $y = \frac{x^3}{3} + x + 1$.

Exercise 46.2a (p. 488)

1. $s = \frac{3t^2}{2} + t$.

2. $\frac{t^3}{3} + 2$.

3. $\frac{t^3}{6} + 4t$.

4. $\frac{t^3}{3} + \frac{t^2}{2} + t$.

5. $\frac{t^3}{3} + \frac{t^2}{2} + 3\frac{1}{6}$.

6. 39.2 m/s.

7. 37.4 m/s.

8. 78.4 m.

9. 35.6 m.

10. 1 s.

Exercise 46.2b (p. 488)

1. $\frac{t^3}{3} + t$.

2. $\frac{t^2}{2} + t + 4$.

3. $\frac{t^4}{12} + 2t$.

4. $\frac{2t^3}{3} + \frac{3t^2}{2} + t$.

5. $\frac{t^3}{3} - \frac{t^2}{2} + \frac{7}{3}$.

6. 29.4 m/s.

7. 27.6 m/s.

8. 19.6 m.

9. 102.4 m.

10. 19.6 m.

Exercise 46.3a (p. 493)

1. $12\frac{2}{3}$.

2. $\frac{1}{6}$.

3. $5\frac{1}{3}$.

4. $\frac{3}{4}$.

5. $6\frac{1}{3}$.

6. $\frac{4}{3}$.

7. 96 m.

8. 150 m.

Exercise 46.3b (p. 493)

1. $16\frac{2}{3}$.

2. $1\frac{1}{3}$.

3. $17\frac{1}{3}$.

4. $3\frac{3}{4}$.

5. 28.

6. $\frac{2}{3}$.

7. 114 m.

8. 84 m.

Exercise 46.4a (p. 497)

1. $\frac{15\pi}{2}$.

2. $\frac{16\pi}{5}$.

3. $\frac{\pi}{2}$.

4. $\frac{53\pi}{15}$.

5. $\frac{448\pi}{3}$.

Exercise 46.4b (p. 497)

1. 40π.

2. $\frac{128\pi}{5}$.

3. $\frac{2\pi}{3}$.

4. $\frac{5}{24}\pi r^3$.

5. $\frac{7}{3}\pi a^3$.

Exercise 46.5 (p. 497)

1. $4\frac{5}{6}$. **2.** 29 m. **3.** $\frac{\pi}{4}$. **4.** $26\frac{2}{3}$ m. **5.** $5\frac{1}{3}$.

6. $\frac{9\pi}{2}$. **7.** $\frac{x^3}{3} + \frac{x^4}{2} + \frac{x^5}{5}$. **8.** 14 m/s.

9. $\frac{1}{54}$. **10.** $\frac{32}{3}$. **11.** $\frac{5}{3} + 2x - \frac{x^3}{3}$. **12.** $\frac{625}{12}$.

13. $\frac{16\pi}{15}$. **16.** 11 m/s², $20\frac{1}{6}$ m. **17.** 24π.

18. $\frac{28}{3}$. **19.** $\frac{27}{4}$. **20.** $3x^3 + 3x^2 + x$.

21. $15\frac{1}{2}$ m/s. **22.** $y = \frac{1}{4} + x^3 - \frac{x^4}{4}$.

23. $\frac{4}{3}\pi ab^2$. **24.** $y = 4 - 3x^2$.

25. $\frac{2}{3} + \frac{3x^2}{2} - \frac{x^3}{3}$. **26.** $8\frac{5}{6}$. **27.** $\frac{4}{3}$. **28.** $7\frac{1}{2}$.

Revision Exercise (p. 501)

1. 17.4%. **2.** 2.592×10^{10} km; 9.46×10^{12} km. **3.** 30, 15, 20, 25, 30.
4. £36.36. **5.** $\frac{2}{3}$. **6.** £$\{\frac{1}{100}n(y - x) - P\}$. **7.** −7 or 5.
8. $40\pi x^3$ cm³. **9.** £$(\frac{1}{100}by - \frac{1}{12}ax)$. **10.** 24, 30.
11. $-\frac{2}{3}$, 0 or 2; e.g., −2. **12.** 5. **13.** {1, 3, 7, 9}.
14. (ii) 0, itself. (iii) 1, 11, 15. **15.** {0, 8}; {−1, 0, 1, 2, 3}.
16. Mrs A prefers Brand X, Mrs B prefers Brand Y. **18.** 3.1; 1.5; 5.

19. $\frac{5}{18}$. **20.** $\frac{1}{6}, \frac{11}{15}, \frac{2}{87}, \frac{1}{10}$. **21.** $\frac{1}{3}; \frac{4}{9}; \frac{1}{6}; \begin{pmatrix} \frac{11}{18} & \frac{7}{18} \\ \frac{7}{12} & \frac{5}{12} \end{pmatrix}$;

Pr (day after tomorrow is wet, given that today is wet).

22. $\begin{pmatrix} \frac{32}{45} & \frac{13}{75} \\ \frac{13}{45} & \frac{62}{75} \end{pmatrix}$. 3:5. **23.** (iii) $\begin{pmatrix} 1400 \\ 1150 \\ 1640 \end{pmatrix}$ **24.** (63, 22) is the better.

25. 88. **26.** (40, 72) best. **27.** None. **30.** $\frac{1}{2}, \frac{1}{2}$.
31. $\frac{1}{29}(3x + 5y), \frac{1}{29}(4x - 3y)$. **32.** $2b - 2a, 2b - 3a, b - 2a$.

33. $a + c, a + w, a + c + w, c + w$. **34.** $\begin{pmatrix} 984 & -983 \\ -985 & 984 \end{pmatrix}$.

37. T_2, T_3. **38.** (i) $\frac{b}{a}$; (ii) $\frac{bc}{a - b}$, (iii) $-(a + b)$, (iv) $\frac{9 - b^2}{2b}$. **40.** 32 cm.